# The Military Revolution and Revolutions in Military Affairs

# De Gruyter Studies in Military History

Edited by
Jörg Echternkamp and Adam Seipp

# Volume 3

# The Military Revolution and Revolutions in Military Affairs

—

Edited by
Mark Charles Fissel

**DE GRUYTER**
OLDENBOURG

ISBN 978-3-11-151888-6
e-ISBN (PDF) 978-3-11-066141-5
e-ISBN (EPUB) 978-3-11-065759-3
ISSN 2701-5629

**Library of Congress Control Number: 2022936647**

**Bibliographic information published by the Deutsche Nationalbibliothek**
The Deutsche Nationalbibliothek lists this publication in the Deutsche Nationalbibliografie; detailed bibliographic data are available on the internet at http://dnb.dnb.de.

© 2024 Walter de Gruyter GmbH, Berlin/Boston
This volume is text- and page-identical with the hardback published in 2023.
Cover image: *Theuerdank* (Nuremberg 1517), Woodcut engraving from the editor's collection
Typesetting: Integra Software Services Pvt.

www.degruyter.com

*In Memoriam*

Thomas Garden Barnes 1930–2010
Conrad Sebastian Robert Russell (5th Earl Russell) 1937–2004
Buchanan Sharp 1942–2020

# Preface and Acknowledgments

While addressing "military revolutions" and "revolutions in military affairs" substantively, this collection contextualizes itself within a vibrant historiography. Conversations with scholars who have pondered the emerging cross-disciplinary literature confirm that an historiographical approach is imperative, particularly due to the field's confluence with the RMA from the 1980s onwards. We showcase five historians focusing primarily on the early-modern world and then feature two "participants" in the RMA. The introduction classifies our contributors within the military revolution debate. Chapters one through five constitute original research representative of the current state of the field. Chapter six bridges the military revolution essays with offerings (chapters seven and eight) by practitioners of the revolution in military affairs. A salient point of the introduction is that historiography is as important in coming to terms with the military revolution as is adding to specialized studies.

Roughly five hundred published secondary works detail the military revolution. A structured, thematic map to the historiography is available electronically (and as a supplement to the concise bibliographies presented at the end of each chapter in this anthology) [Fissel, Mark Charles. "Military Revolutions." In *Oxford Bibliographies in Military History*, edited by Kaushik Roy. New York: Oxford University Press, https://www.oxfordbibliographies.com/view/document/obo-9780199791279/obo-9780199791279-0212.xml]. The crisp and tightly organized Oxford Bibliographies enable graduate students, for example, to construct coherent schemata of complex academic fields. An entire historical topic is fit (sometimes uncomfortably) into uniform compartments while observing strict word count limitations. Arranging sources in this fashion was exasperating because military history (like the reality it interprets) is rarely tidy or straightforward. The constraints of format frequently stymied a full inclusion of a theme's historiography. Inevitably excellent scholarship went without mention due to editorial impediments restricting the number of sources allocated to each topic. An historiographical approach involves hitting a moving target, which is why the editor's introduction attempts to present a framework of historical interpretation at the outset. Nevertheless, it is advantageous that an electronic reference tool dovetails with the essays and bibliographies of this De Gruyter volume. The bibliography will be updated periodically to recognize new publications and ensuing realignments of how military revolutions are conceptualized.

The above-mentioned challenge of regimenting a rapidly evolving and considerable historiography made more impressive the writings of those who had already interfaced the military revolution with the revolution in military affairs, notably Clifford J. Rogers, Carlo Alberto Cuoco and Michael Thompson (individuals partially obscured by cosmic dueling between Geoffrey Parker and Jeremy Black). Friends assisted the publication of this volume by proposing original ideas, critiquing drafts, and offering recommendations. They deserve more than a mention but understand the necessity of brevity in acknowledgments: D.J.B. Trim, Steve Walton, Wayne E. Lee, David J. Ulbrich, Jim O'Neill, Mark H. Danley, Vladimir Shirogorov, Hyeok Hweon Kang, the late Dennis Showalter, Kenneth Swope, Hubert Van Tuyll, Jeremy Black, Idan Sherer, Eduardo de Mesa, and Master Chief Bill Wells. De Gruyter has been gracious in publishing these studies. Elise Wintz, Rabea Rittgerodt, Jana Fritsch, Ian Copestake, Ulla Schmidt, Helene Remiszewska, and a host of others receive accolades for getting us through the production process during difficult times.

This anthology appears under the pall of a pandemic. The casualties included Buchanan Sharp, the editor's undergraduate mentor (and second reader on his doctoral dissertation). Buchanan's passing compounded the losses incurred by the deaths of Conrad Russell in 2004 and Thomas Garden Barnes in 2010. Professors gone on to emeritus status reach a point where they are obliged to render homage to those who made possible their career. The present writer was undeservedly sponsored by gifted (and compassionate) historians. It is only in retrospect that the magnitude of their tutelage has been comprehended. Vignettes spring to mind. Huddling over quarter sessions records in Trowbridge with Buchanan; the unabashedly paternal consultations in Tom's living room when life interfered with history-writing; Conrad appearing unannounced on the doorstep one evening in Belsize Park to inquire about my illness. It was he who suggested I research the Bishops' Wars as my dissertation topic. The common denominator that bound us was Berkeley. Buchanan had matriculated there as undergraduate and as graduate. Tom, having provided guidance to the cusp of the terminal degree, good-naturedly allowed me (and my books) to use his office in Dwinelle Hall as a teaching venue. Conrad disseminated at Cal his then-gestating revisionist views on the Stuarts a year before he was recruited by Yale. Also at the University of California for a semester in the late 1970s was John Alexander Guy. He is still among the living, happily. Like Conrad he prepared our cadre for research in Old Blighty. John's seminars drew up on his exhaustive knowledge of the Chancery Lane Public Record Office, which he coupled with tutorials on paleography and diplomatic. John Guy's guidance proved invaluable from the first day I entered the archives. He frequently put a roof over my head and fed me at his family's table. Of the four superb teachers with whom I was blessed, only John remains. This acknowledgment aims to proffer thanks to him.

# Contents

**Preface and Acknowledgments —— VII**

Mark Charles Fissel
**Introduction —— 1**

Hyeok Hweon Kang
**Difference in an Age of Parity: Technology and Global Military History —— 29**

Aliaksandr Kazakou
**Gunpowder Revolution in the East of Europe and the Battle of Orsha, 1514 —— 65**

Wayne E. Lee
**To Stop a Cannonball: Ottoman Fortress Design and Comparing Military Revolutions, 1350–1730 —— 113**

James O'Neill
**Firearms and Fieldworks: Military Transformation and the End of Gaelic Ireland —— 177**

Vladimir Shirogorov
**A True Beast of Land and Water: The Gunpowder Mutation of Amphibious Warfare —— 207**

Mark Charles Fissel
**From the Gunpowder Age Military Revolution to a Revolution in Military Affairs —— 313**

Mark D. Mandeles
**To Dream the Impossible Dream: Feasibility of Deliberate Government Guidance of Revolutions in Military Affairs —— 369**

João Vicente
**The Dilemma of Human Interference in War: The Coming Revolution of Autonomous Air Warfare —— 405**

**Appendix: On the Cover Illustration —— 447**

**List of Contributors —— 457**

**Index —— 459**

Mark Charles Fissel
# Introduction

Are wars won by clever application of new technologies? Or, does tactical insight and leadership make possible the best use of weaponry? Perhaps less spectacular forces, behind the curtain so to speak, are at work, such as the material weight of organizations, institutions, and logistical capabilities? Naturally, soldiers inquire as to whether formulas that lead to victory are there to be discovered. Scholars look for patterns of causation and to interpret the contexts and consequences of war. The literature on "military revolutions" was composed to elucidate the above-mentioned questions and gravitated to musings about what appears to be the central development in human warfare, namely the exploitation of gunpowder in the service of violence. What was primarily a specialized discussion by professional historians was then appropriated by practitioners and think-tankers to explain (and forecast) warfare at the end of the twentieth century. That body of work constructed a phenomenon labeled as the revolution in military affairs (RMA). What follows below in the present work recognizes the primacy of the "gunpowder revolution" by proffering five original studies on military revolutions in the early modern world. These emphasize regional adaptation and cross-cultural interaction across geographical peripheries. Treatment of the early modern gunpowder revolution is then bridged with the twentieth century by what are considered a tripartite of crucial linkages. The consequent revolution(s) in military affairs that extended from the 1980s forward that was qualitatively and quantitatively different from the military revolution of the late 1400s to mid-1600s. That transformation was largely the result of a trio of long-term developments: naval military revolutions, nationalism, and industrialization. Once arrived in the twentieth century, the volume concludes with a pair of essays. An RMA theorist looks at a theme that has bedeviled historians of military revolution, specifically the role played by government in the dynamic of military revolution. Finally, an RMA practitioner, an aviator with experience in NATO, offers the view from the cockpit and comments on autonomous lethal airborne systems and our future.

Military revolutions may or may not be actual pivotal transformations in how war is waged, but as a minimum they are powerful conceptual constructs. So potent, they may validate the postmodernist conviction that language shapes corporeality as well as human perception of reality. Terminology and abstractions more than illumine history. The academic polemics of the military revolution debate inspire, shape, and legitimize violence itself in the near-contemporary incarnation of revolutions in military affairs. The concepts themselves demand scrutiny, in addition to their use as tools to demystify war's dynamics. The military

https://doi.org/10.1515/9783110661415-001

revolution's confluence of historiography and reality have in turn animated diverse (and expanding) scholarly communities. The imperative of comprehending the military revolution's contradictory literature is further spurred on by the profound consequences of avowed contemporary RMAs.[1] Wayne E. Lee puts it succinctly: "Ideas matter in warfare."[2]

The study of military revolutions in general, and RMAs particularly, seeks strings of causation that reveal the social and technological processes that improved the efficiency of organized human conflict. Scholars' focus has been, first, intellectual (in achieving a historicist rendering of the art of war), and secondly the utilitarian application of the latter to inform contemporary military science. In delineating the historical record, practitioners have gathered insights into warfare's next incarnation and apply that knowledge to advantage. However, as Jeremy Black cautions, the "usefulness of a concept . . . does not demonstrate its accuracy."[3] The flurry of military revolution case studies published over the last several decades inclined scholars to overlook a "fundamental issue, one seen with all processes of conceptualisation, categorisation and analysis, namely what is the purpose of this approach, and why it was framed and developed."[4] The contradictions stem from academicians' use of the concept to develop (usually, national) histories, whilst prognosticators angle for the next super-weapon. The consequential elusiveness of consensus regarding an adequate definition of military revolution has been its curse; adherents admit the deficiencies of the concept and numerous "exceptions to the rule" but ultimately take refuge in the concept's "usefulness" somewhat like the "fertile error." This leads Black to judge early modern military revolution "a weak concept poorly-applied" but still to encourage experimentation with the general concept.[5] Geoffrey Parker's original *magnum opus* on the subject is now greatly expanded and forthcoming from Cambridge University Press. Parker retains his influential following which believe that the longevity and ubiquity of the concept guarantees that continuing scholarship will refine the idea of "military revolution" and that

---

[1] The editor owes much to Clifford J. Rogers' "'Military Revolutions' and 'Revolutions in Military Affairs,'" 21–35. See also Carlo Alberto Cuoco, *The Revolution in Military Affairs*, and Michael J. Thompson, "Military Revolutions and Revolutions in Military Affairs", 82–107. Fissel, "Military Revolutions."

[2] Lee, *Warfare and Culture in World History*, 1; in at least two publications, Professor Rogers pioneered a systematic comparative methodology for juxtaposing military revolution schools of thought upon the multifarious historical examples of military revolutions in action. For example: Rogers, "The Military Revolution in History and Historiography," 1–35, 77.

[3] Black, "Rethinking Military Revolutions," unpaginated.

[4] Black, "Rethinking Military Revolutions," unpaginated.

[5] Black, "Rethinking Military Revolutions," unpaginated.

reliable variations on the theme will emerge as scholarship moves forward. Tonio Andrade, a Sinologist, proposed a more balanced global framework in *The Gunpowder Age*. While questioning the validity of the military revolution concept, Andrade concludes that advent of firearms and artillery was sufficiently momentous that the debate still generates light despite the distracting heat.[6]

Parker's refinement and amplification of Roberts' original thesis revolved around gunpowder, particularly as implemented by predominantly Western powers. Gunpowder remains a common denominator in the debate, and the term "gunpowder revolution" appears in the pages that follow. Our opening essays document applications of gunpowder technology in regions regarded as "peripheral" to mainstream military revolution studies. These reveal hybrid types of martial innovation, flexible and adaptive best practices, prompting close consideration of the ebb and flow of techniques along frontiers bordering expansionary states. A recurrent theme in our anthology is engagement between cultures in conflict with neighbors, especially Northern and Eastern European regions once considered "peripheral": Russia, Ireland, Lithuania, Poland, and Belarus. "Peripheries" are set within a global context by comparison with mighty East Asia, and then commentary upon the Ottoman Empire. Initially, military revolutions were regarded more as products of technological innovation than of cultural consequence. Increasingly it has been recognized that strategic culture largely shapes technology, though the interplay is reciprocal. What appeared linear and tangible (such as bulletproofing) is a series of intricate socio-economic interactions that yielded temporary, sometimes illusory senses of military superiority.[7]

There exist various species of military revolutions, sometimes identifiable by region and culture, as well as by physical and environmental (battle space) characteristics such as centuries-old seagoing naval revolutions, and new RMAs such as cyberwarfare. Our volume is particularly intrigued with physical features where land and water meet, a less compelling venue than the realms of air combat, warfare in outer space or within cyber-reality. The volume's fifth chapter calls attention to gunpowder revolutions in amphibious warfare. Clearly, the gunpowder revolution of the late medieval and early modern era engulfed the globe, artillery as well as of firearms, on land, on sea, and at the littoral. How gunpowder weapons were integrated in these physical environments and used was determined by the best practices enshrined within a given strategic culture. Finally, evolutionary *moyen duree* is the "dark matter" backdrop to military revolutions. Continuity

---

[6] See Parker's "Is the 'Military Revolution' Dead Yet?" Andrade, *The Gunpowder Age*. See however, remarks touching on Parker and Roberts by Henninger, "La guerre de cent ans, les clés d'une révolution militaire."

[7] Sharman, *Empires of the Weak*.

sustains. "Revolutions" are exceptional and do not exist except in contrast to streams of tradition and stability. Therefore, though largely focused upon the gunpower age, our essays do not confine themselves to the sixteenth century. Indeed, as this anthology developed collaborators exchanged with the editor their ideas about connections between military revolutions and the omnipresent RMA. Resonances fascinate us, such as in early military revolutions, e.g., archetypal ordnance, the symbiosis between technology and tactical innovations, comparative insights via collective representations of the geopolitical nature of military revolutions, all of which are referenced when the contributors deem the topics relevant.

Those initial five essays that frame the military revolution regionally, emulating the geographical approach advocated by Parker and Black (and applied efficaciously by Kaushik Roy and others), are acutely attuned to frontiers and the above-mentioned "peripheries." We believe that environments inculcated certain skills (e.g., marksmanship, mastery of solving technical challenges associated with sailing, cohesive and choreographed tactical behavioral capabilities, metallurgical proficiencies, etc.) in populations residing therein. External conditions such as land, sea, and littoral dictated circumstances, obviously. But human initiative factored as well, as strategic cultures developed the capabilities of individuals and societies, collectively. Strategic predicaments (recognized by cultures) suggested approaches for assimilating extant military revolutions. Frequently those perceptions were flawed, a fact that undercuts the mechanistic precision of technology-based military revolutions.[8]

Vulnerable and occasionally indecisive peripheral military powers (as described by Wayne E. Lee, James O'Neill, Aliaksandr Kazakou, Hyeok Hweon Kang, and Vladimir Shirogorov) could not rest secure in mastering a single type of military revolution-era defensive scheme. "Eastern" strategic cultures thrived upon borderlands. Warfare fostered state development and vice versa, but did so well before any orthodox Western European model of military revolution (1500–1650) influenced them (consider the discussions herein regarding the Baltic States and Russia). Case studies are of greater value when conjoined with thorough analysis of the polities with which they shared spheres of influence. A literature has developed around perimetric military revolutions, early works by Kenneth Chase and Jeremy Black springing to mind among others, particularly in regions of south Asia and the Indian subcontinent. Once, historians of military revolutions were inclined to be "national" historians; today, the field demands a comprehensive perspective on global history (if not to mention also economics and anthropology).

---

**8** Sharman, *Empires of the Weak*, 146–149.

Michael Roberts' classic military revolution template showcased a regionally aggressive Scandinavian nation-state, Sweden. Aliaksandr Kazakou demonstrates, below, that regional proximity did not foster isomorphic military revolutions. However, terrain features that promoted imitation (of both an organizational nature as well as technical science) via technological transfer are coastlines and ports. Seagoing interfaces clearly empowered Roberts' Sweden as well as Geoffrey Parker's Low Countries. Most strikingly such characteristics are found in other early military revolutions, in Asia, noting that Korea, Japan, China, and Southeast Asia also possess lengthy littorals as well. The "aquatic connection" is neither prerequisite nor a major mover in military revolution, but the amphibious dimension (riverine as well as seagoing as is evident in Shirogorov's and Kazakou's essays) invites closer consideration, for example the Muscovite strategic calculation of placing artillery at river crossings to impede Tatar incursions. The connection between amphibious operations and military revolutions is a particularly pressing field of research since orthodox models of military revolution focused on land warfare, and then naval warfare only comparatively later in the historiography.[9]

Rival polities that abutted aquatic and terrestrial frontiers adapted firearms and fortifications to fit their respective strategic cultures. When these ways of war were fought on peripheries the resultant hybrid military revolutions exhibited both affinities and striking contrasts. Individual strategic cultures reacted to the gunpowder revolution in their own ways, adapting and adopting, not necessarily copying. Case studies of peripheries necessitate more macrocosmic regional comparisons, some on a global scale. Much historiography concentrates on great nation states and empires, and nationalism (addressed in our "bridge" section) remains an undercurrent. Our essays fix upon regions; the characteristic military revolution problem of identification with nations is quite understandably nearly unavoidable in the conceptualization of all seven authors. Equally traditionally, chronology gets its due as well. The historical analyses herein lay bare detailed interactions between adversaries as they occurred in chronological sequence, as well as generalize about competing cultures and polities in ways that might captivate social scientists.

Hyeok Hweon Kang leads the way, and ranges across Asia. He delves primarily into firepower, particularly on the visceral level of armies inflicting violence and what actually occurred in combat. How were pragmatic applications of gunnery

---

[9] David Trim's trailblazing "Medieval and Early-Modern Inshore, Estuarine, Riverine and Lacustrine Warfare," 357–419, established common denominators that set the stage for Vladimir Shirogorov's essay.

channeled in the midst of fierce clashes when a profusion of diverse adversaries made flexibility (and innovation) a matter of survival? Similarly, Aliaksandr Kazakou showcases three strikingly different Eastern Europe powers, each coming to terms with its respective gunpowder revolution. In his narrative Muscovy figures prominently, both within the East-West periphery as well as on the frontiers of empire-building. In this way Kazakou highlights imperial integration as a counterbalance to the volume's broader emphasis on hybridity and the particularism of regional variations. Wayne E. Lee's geographical and chronological panoramas of Ottoman frontiers (1350–1730) illuminate precisely such syngergisms. Lee's comparative approach (which characterizes his publications in general) presumes no binary configurations of East and West when analyzing Ottoman military architecture, a topic Geoffrey Parker's 2019 lecture identified as demanding additional comparative exploration. Chronicling one of the darker stories of colonial exploitation, James O'Neill's study provides a variation on the ubiquitous East-West configurations of a supposedly more advanced continental Western European art of war imposing itself upon an "indigenous" defense. O'Neill draws attention to the complexities of seaborne imperialism as an aquatically empowered monarchy attempted to colonize a proximate island homeland of a different culture. O'Neill's notable conclusion is not that (like continental Europe) Ireland experienced an orthodox military revolution, but rather the course of that transformation was shaped by unique practical necessity and a native strategic culture.

Mark D. Mandeles and Colonel João Vicente close out our anthology, offering insights (and perspectives) useful to historians of gunpowder revolutions. Modes of inquiry, such as assessment of bureaucratic limitations and institutional isomorphism, reconcile the conflicting mentalities and agendas of military revolutions and RMAs. Mandeles' involvement in the study of seagoing vessels and airpower provides him an understanding of the practical processes of implementing military innovation, especially with reference to state power. He showcases the balance between the realities of strategic culture and best practices as well as the parameters in which state-sponsored efforts and technological capabilities interact. We conclude with the remarks of a practitioner of the RMA, Colonel Vicente, who suggests that the cycle of military revolutions and RMAs is reaching a conclusion of historic (and fateful) significance. Immersed in the application of airpower and the newest airborne technology, Vicente reveals a frightening future for humankind as a direct result of military revolutions and the RMAs. The conclusions of these latter authors are treated in greater detail in the editor's "bridge" chapter linking our early modern studies with the RMA.

## Hyeok Hweon Kang

Historiographical momentum has passed from historians of Europe to globalists, particularly to a new generation of Asianists, personified by Hyeok Hweon Kang. European tactical use of firearms sometimes paralleled contemporaneous wars raging amongst Japanese, Mongolian, Korean and Chinese armies. Kang confronts Asia's and Europe's resonances and divergences. He examines Asian dimensions expounded upon by Sun Laichen, Peter Lorge, Tonio Andrade, Andrew de la Garza, and others. Against backdrops of innovative technology emerging in regional contact zones, Kang takes his place among an historiographical cadre skilled in researching ascendant Asian achievements within their own context, thus demystifying Asian military revolutions for a wide audience. Recent scholarship, much of which is available only in Korean and Chinese, proves that efficacious adaptation of gunpowder weaponry (which of course had been invented by Asians) proliferated, particularly during protracted struggles in East Asia. While students of the European art of war have frequently fixed on sheer firepower (e.g., volley fire), what has been less explored is marksmanship, certainly a legitimate measure of a culture's assimilation of gunpowder weapons.[10] Taking infantry gunnery as a key measure, Kang documents how East Asian drillmasters emphasized accuracy as the ideal in handling a gun properly. East Asian drill was rigorous and its treatise literature oftentimes more practical than that of Europeans. Training – for example the loading of weapons and emphasis placed on accuracy – was not "less" but rather "different," in method and emphasis, in contrast with European praxis. Rapidity of fire and the sheer volume of lead unleashed may have been the objectives for some European armies, but East Asians valued marksmanship. Sharpshooters and snipers were to be found among European ranks, but these were specialists with custom-made weapons incorporating rifled barrels and precision gunlocks. Kang's description of the adaptive use of firearms reflects the predicament of the Ottomans who (like the Koreans) faced diverse foes on varying terrain: "[T]he Ottoman technique of volley firing was effective against the tightly packed Habsburg infantry . . . [A]gainst the fast-moving, scattered cavalry of the Mongols/Jurchens, Rajputs, and Uzbeks, the Ming, Mughal and Safavid musketeers did not have the incentive to practice volley firing."[11] Like Andrew de la Garza and Aliaksandr Kazakou, Kang focuses upon battle itself. Kang's theme of "parity" echoes in the rough chronological coincidence of Kazakou's battle of

---

[10] We know that artillerymen devoted detailed study to aiming their guns. See Bennett and Johnston, *The Geometry of War 1500–1750*.
[11] Roy, *Military Transitions in Early Modern Asia, 1400–1750*, 200.

Orsha (1514) with the battle of Chaldiran (1514), where in this case (but not in others) Ottoman firepower made the difference, reversing the course of a contest initially being won by the Shah.[12]

Kang's application of the term parity invites comparison of military revolutions without attaching value judgments or "ranking." He demonstrates that ascribing military developments to cultural influences is not as productive as focusing closely upon how armies fought. Furthermore, Kang's research reveals that state centralization, though common across the globe, was neither uniform nor inevitable. Historians of Asia have proven definitively that gunpowder revolution did not necessarily force reformation of the state, thus Roberts' model cannot be paradigmatic globally. What was universal, it would appear, is the state's preoccupation with gunpowder weaponry. Kang cautions against inchoate assumptions about cultural superiority and inferiority in technical matters. Instead of generalizations about state-empowered military bureaucratization and aptitude, variances proceeded from how the respective military cultures conceived how best to exploit gunpowder weaponry amongst infantry.

## Wayne E. Lee

Hyeok Hweon Kang's comparative approach in assessing Asian handheld gunpowder weapons complements Wayne E. Lee's comparative treatment of the Ottomans and their enemies.[13] The Sultan's widely varying imperial interfaces with European and non-European adversaries exemplifies pragmatically driven technical solutions. Military revolution's propulsion, or dynamism, even when qualified by comparing Europe with Asia, is primarily molded by administrative, political and cultural factors, though these interact with evolving technological advances.[14] Lee's approach to poliorcetics proves Kang and Shirogorov's resonant cultural adaptations of military technology, uncovering most importantly timeline patterns rooted in specific conflicts. Cultures make military choices (frequently haphazardly). Process requires calculation based upon understandings of a fluid situation as well as a culture's orientation or collective awareness of its

---

12 Oman, *A History of the Art of War in the Sixteenth Century*, 612–615; Khan, *Gunpowder and Firearms*, 6, 143–144.
13 Perhaps four major powers (if we now add Korea to the Ming, Ottomans, and Japanese) had produce volley-firing infantry before Nassau's experiment. See Roy, *Military Transition in Early Modern Asia*, 88; Subrahmanyam and Parker, "Arms and the Asian," 12–42; see also this volume, 113–171.
14 Morillo, "Guns and Government," 100.

strategic situation. Lee simultaneously embellishes Parker's emphasis on defensive warfare by underscoring how fortifications interacted in non-linear ways with early modern gunpowder weapons. Lee highlights Clifford Rogers' striking observation that from 1325 to 1500 (in other words, before the *trace italienne* and traditionally accepted parameters of the military revolution were in place), profound changes in design occurred so frequently and in such variation that a quarter century in this period saw more substantive change in artillery technology than in the entirety of the three centuries from 1500 to 1800![15] Lee's photographs, drawings, and diagrams depict "individualized" structures crafted via Turkish experience and modified according to terrain and strategic purpose, but also perhaps occasionally hamstrung by bureaucratic inertia. Lee aligns with Kaushik Roy's regional contextualization of Ottoman siege warfare, in the past sometimes misrepresented. The Ottomans retrofitted edifices according to their empire-building configurations, accommodating topography while anticipating diverse foes. The present volume reveals much about the intensity of Ottoman frontier preparations (and Muscovy's imaginative responses).[16]

## Aliaksandr Kazakou

Aliaksandr Kazakou surveys Baltic Europe and Muscovy, inquiring how firearms and cannon were employed upon that landscape by differing polities. Kazakou's comparison of a trio of sixteenth-century states identifies varied patterns of strategic culture in Eastern Europe and how their interactions conditioned their adaptations of military revolutionary tools (both hardware and "software"). In the Grand Duchy of Muscovy, firearms and cannon are first mentioned circa 1382. Moscow emerged as a hub of ordnance manufacture, the recruitment of Italian and German technicians stimulated a Muscovite arms industry. Kazakou notes that the governors assembled comparatively sophisticated artillery parks because ordnance had become omnipresent in Muscovy's armies. Contrastingly, the Grand Dukes of Lithuania were slow to embrace cannon (and allowed their nobles to acquire guns privately). In the 1500s, the Grand Duchy of Lithuania imported Polish ordnance rather than building foundries. Domestic politics blunted Lithuanian preparedness in artillery fortification. According to Kazakou, Lithuanian efforts at integrating cannon into their military establishment were undermined by administrative

---

15 Rogers, "Gunpowder Artillery in Europe, 1326–1500," 37–71; see also Smith and DeVries, *The Artillery of the Dukes of Burgundy 1363–1477*.
16 Roy, *Military Transitions in Early Modern Asia*, 29–30, 37, 90–99.

weaknesses stemming from corruption, poor financing, and bureaucratic inefficiency. Regarding expanding and contracting states, political organization varied considerably in even in confined contact zones. Kazakou also illustrates how strategic calculation and fiscal priorities shaped the degree to which a state embraced a Robertsian military revolution. Strategic cultures collectively first chose, and then undertook the uncharted challenges of implementation. The societal contexts described by Kazakou, specifically political limitations (e.g., fiscal capabilities) applied to Eastern Europe, Ireland, and to Asian polities as well as Western Europe. The wealth of the state mattered little if its administrative capabilities were hamstrung by political division. Political expediency, then as now, precipitated a present mindedness that could cripple strategic anticipation of future threats. Undue caution and complacency undermine strategic calculation. Again, the field of action is within human agency, subject to the dominion of (flawed) human cognizance. Kazakou reaffirms other essays herein: armed forces were willing to experiment with technological innovations that showed promise. However, socio-cultural context determined implementation. Military science in Eastern Europe, as elsewhere, was not subject to technological determinism. Thus an inverse of gunpowder usage existed wherein the Muscovites did not find caliver and musket so efficacious. Aside from the battle of the Ugra River (1480) Kazakou finds little evidence of extensive battlefield use of firearms. The Poles also remained conservative in that instead of using pikemen to protect musketeers, Poland's vaunted cavalry guarded infantry's tactical role (a luxury that James O'Neill would point out was not afforded Irish armies). This traditionalism had tangible ramifications on the battlefield, for example when Polish troops faced Moldavian forces possessing a four-fold superiority in artillery. Persistent conservatism and inertia in embracing military revolutionary innovation has been explained in Vladimir Shirogorov's recent monograph.

# Vladimir Shirogorov

Vladimir Shirogorov has proffered conclusions about military revolution:

> The true Western-style military transformation did not start to penetrate the subcontinent before the second half of the seventeenth century. It flourished only after the Petrine reforms in Russia in the first third of the eighteenth century. But it is evident that the main work of the state- and nation-building had been completed in Eastern Europe long before that time, within the last third of the fifteenth century and the second third of the sixteenth. The political and social transformation of Eastern Europe did not run simultaneously with the Military Revolution; the latter lagged far behind the former. The causation link of the

Military Revolution between technological development, expressed in the Gunpowder Revolution – the emergence of firearms, and the political and social upheaval of the nation-building is evident in Western Europe. In Eastern Europe, it is seemingly absent.[17]

While "naval revolution" will be a focus of a chapter below, suffice it to say that Europe's global military impact came from oceanic expansion (again, making Roberts' initial avoidance of the subject a bit of a riddle). Shirogorov shows how an amphibious military revolution developed among Eastern European strategic cultures as reflected in the timelines of their adaptation of gunpowder revolution. He explains how firearms for amphibious operations transitioned from tactical and logistical tools into a means to achieve operational and strategic goals. Military revolution was practiced within varied aquatic systems (a field of study championed by Jeremy Black decades ago, subsequently taken up by D.J.B. Trim). Deltas, riverine systems, littorals and other amphibious battle spaces are addressed as strategically important contact zones. Shoreline peripheries can be barricades or gateways.[18]

It is within the littoral, in tandem with the ports that punctuate it, that the world economies interact. Shirogorov's "transit warfare" discloses the mercantile nature of early modern warfare, paving the way for consideration of the Commercial Revolution and the post-1650 art of war emanating from Europe, especially via the Atlantic powers. His essay describes the riverine strategies (and tactics) of expansionary states, revealing the intimate relationship between economics (in the form of commerce) and force projection. Violence and trade are interwoven manifestations of state expansion. Shirogorov's concept of transit warfare fought amongst polities at different stages of economic development occurred in an era of unprecedented international competitiveness. The build-up of new types of "amphibian" forces required intensive recruitment and training of human resources by nations committed to overseas (and overland) expansion.

Shirogorov formulates an operational taxonomy of early-modern amphibious applications of firearms and artillery. The incorporation of guns in seagoing, riverine, and shoreline operations was technology-enabled, but dictated by strategic concerns and geography as is evident in the Western European, Eastern European, and Ottoman adaptations of gunpowder weaponry. In short, Shirogorov's interpretation of offensive operations complements nicely Wayne E. Lee's treatment of defensive adaptation in that corner of the globe. Competition among Eastern European states permeates the essays of Lee, Kazakou, and Shirogorov.

---

17 Shirogorov, *War on the Eve of Nations*, 364.
18 Trim and Fissel, eds, *Amphibious Warfare 1000–1700*, 19–20, 52–53, 112, 118, 357–413, 429–432.

Michael Roberts' 1966 embellishment of his 1956 thesis, concentrated even more intently upon Swedish expansionism, drew attention to this state-of-affairs in the Baltic regions.[19]

More recently, Marshall Poe has documented military revolutions originating in the Baltic and Eastern Europe, where peripheral polities hazarded the Scylla of the Ottomans on one hand and Charybdis of Scandinavia (e.g., Swedish adventurism) on the other.[20] The entire region, including Greater Hungary, Muscovy, Lithuania, and Poland, comprised bewildering intersections of politics, strategy, geography, and geopolitics. Unsurprisingly, each combatant state forged its own strategic culture. Greater Hungary witnessed the traditional military revolution of firearms proliferation and assimilation of the *trace italienne* fortification. Survival in the midst of conflict disposed Hungarian forces to exploit the potential of the military revolution, not contemplative imitation of Western Europe for emulation's sake.[21] The centrality of the Hungarian-Ottoman rivalry compels us to again consider Lee's characterization of the Ottoman military revolution. Furthermore, Kazakou's (and Shirogorov's) essays herein embellish Marshall Poe's configurations of Russian military revolution, which attempt to "resolve the current conundrum in military revolution scholarship concerning which came first – enhanced state power (as a precondition for the rapid growth of military power) or the military revolution (as the source of enhanced state power)".[22] Poe locates "three halting stages" emblematic of "punctuated equilibrium."[23] These involved a "diffusionary process" that resulted in more than one version of a Muscovy-centered military revolution.[24] Poe's trinity should be considered in tandem with David C. Jones's dual Muscovite military revolutions.[25] Russia's military revolutions, certainly within the confluence of Western European, Eastern European, Ottoman and far eastern Asian military revolutions, followed their uniquely Russian path primarily for reasons of material (as well as strategic) culture. Natural resources and topography created a Russian agricultural system that produced social classes different from those of Western Europe. Like the Ottomans, Russia confronted strategic problems stemming directly from peripheral geography. Both

---

[19] Roberts, "The Military Revolution," 195–225.
[20] Poe, "The Consequences of the Military Revolution in Muscovy," 603–618.
[21] Kelenik, "The Military Revolution in Hungary," 117–159.
[22] Dunning and Smith, "Moving Beyond Absolutism," 19–43; Dunning and Smith, 43, evaluating Marshall Poe's contributions.
[23] On punctuated equilibrium, see below, 313–315, 321, 328–335, 356.
[24] Poe, "The Consequences of the Military Revolution in Muscovy," 603, 607.
[25] Jones, "Muscovite-Nomad Relations on the Steppe Frontier before 1800 and the Development of Russia's 'Inclusive' Imperialism," 109–140.

empires faced a "Tatar problem" posed by adversaries skilled in comparatively "unorthodox" arts of war.²⁶ Muscovy especially, Janus-like, arguably pursued multiple military revolutions, military sciences for combatting central Europeans, complemented by adaptations for diverse enemies to the east (e.g., the aforementioned Tatars).²⁷ As Kazakou, Shirogorov and others prove, Muscovy's unique peripheral location obliged its military to adapt different weapons and flexible tactics against widely varied adversaries. These tools were gunpowder weapons and their associated tactics. Further resonances between Shirogorov's essay and the Robertsian Military Revolution are the tactical consequences of firepower. The latter theme permeates the essay of James O'Neill. While topography in the East was not at all like that of O'Neill's Ireland (discussed below), still the principle of flexible deployment of gunpowder weapons in varieties of terrain holds. And beyond "terrain" is the aquatic application of gunpowder that Shirogorov finds crucial. Maritime commercial outreach, throughout the globe, tied gunpowder to colonialism, which returns us to the work of James O'Neill.

## James O'Neill

James O'Neill's substantial corpus of work on the Gaelic art of war illumines Ireland's "peripheral" strategic culture. Irish soldiers selected contemporary practices best suited to that island's terrain and strategic situation. His compelling revisionist portrait of the tactical and organizational transformation of Hugh O'Neill, Earl of Tyrone's, army would have impressed the late Michael Roberts. Though unique, the Irish military revolution resonated with continental European developments, Tyrone's reforms occurring roughly simultaneously with those of the Nassaus in the United Netherlands and predating those of Gustavus Adolphus in Sweden. Tyrone's sweeping military revolution incorporated hardware, particularly infantry-borne firearms but was complemented by training that empowered innovative Irish tactics. In combination, the unexpected skills and equipment of these Irish troops brilliantly confounded the English invaders. The "primacy of skill," echoing H. H. Kang's observations, must be acknowledged.²⁸ Individual human proficiencies had to be coordinated, in other words a "collective

---

**26** Roy, *Military Transition in Early Modern Asia, 1400–1750*, 54.
**27** Jones, "Muscovite-Nomad Relations on the Steppe Frontier before 1800 and the Development of Russia's 'Inclusive' Imperialism," 109–140.
**28** For a contemporary equivalent, see Ágoston, "The Ottoman Empire and the Technological Dialogue Between Europe and Asia," 27–40.

synchronized discipline" was imposed, oftentimes through a leader who understood the strategic culture in which his men could be trained in innovative ways.[29] Frequently the impetus for military revolution derived from an inspired and charismatic commander, a component of the political culture that fostered institutional change.[30] The Irish military revolution thus underscores affinities among sixteenth-century military revolutions. Like the Swedish and Dutch military revolutions, a visionary military leader served as a catalyst (largely through organizational skills) in transforming the armed forces. "Tyrone's reforms revolutionized the means and manner in which the Irish under his command fought." The Irish way of war in the late 1500s was "open warfare," tactically an advanced "hybrid that emphasized the power of modern firearms while maintaining operational and tactical mobility . . . The revolutionary changes in infantry combat seen in continental Europe were modified to accept the realities of fighting within the Irish landscape."[31] The author cites the incorporation of terrain and culture in any reckoning of the effects of military revolution. The resultant marriage of firearms and drill, with its tactical and strategic ramifications, equaled military revolutions occurring not so far away, in the Low Countries and in the Baltic. Sixteenth-century military revolutions throughout the globe, especially on frontiers and peripheries, followed the path of unique strategic cultures in their embrace of firepower, resonant tactical innovation, and strategic empowerment.

Consistent with Shirogorov's assertion that (despite the contradictions and inchoate definitions) military revolutions are "knowable," O'Neill's essay offers an explicable chronology for the development of an Irish way of war. His rendering reaffirms Roberts and Parker regarding (a) the assimilation of firearms accompanied by drill, (b) the ensuing expansion of tactical dimensions, (c) the resultant transformation of strategy, (d) the adaptation of new fieldworks and fortifications, and (e) confirmation of Parker's assertion that political factors exercised an overriding effect on outcomes. And, like Kazakou, O'Neill explains military revolution via warfare's *histoire événementielle*. In the dueling strategies of the Nine Years War, Tyrone chose weapons and tactics accordingly. In 1595 the Earl dismantled his own castle and those of several adherents, "thus denying the crown any target on which to concentrate their forces." On the English side, "Mountjoy's grueling chess-like fort building and garrisoning in Ulster" precipitated the showdown at Kinsale.[32] Ireland's theater of conflict thus

---

29 See Lee, *Waging War*, 238–240.
30 Di Cosmo, "State Formation and Periodization in Inner Asian History," 1–40.
31 O'Neill, *The Nine Years War*, 195–196.
32 Correspondence with Eric Klingelhofer, May 2018; see also his *First Forts* chapter which covers more broadly English sixteenth-century defenses.

created (not merely staged) a unique military revolution, demonstrating that Irish infantry as well as Irish military architecture did indeed come to terms with military revolution but within an Irish context (primarily geographic and cultural). Like Lee's Ottoman examples, strategic culture contoured fortification as well as firepower. Navies, cavalry, artillery both mobile and fixed, and masonry defenses, did not affect the exercise of force in Ireland as they did elsewhere, due to geography and demography. O'Neill's conclusions mesh with what Kang and Lee suggest regarding strategic culture in general and strategic calculation specifically. The Irish predicament differed radically from the challenges facing the Ottomans. But in both cases, the strategic culture and context shaped the reaction to the *trace italienne*. Ireland's clannish and confederated political framework constrained full exploitation of an Irish way of war that blossomed in the 1590s under Tyrone's guidance. In this sense, strategic culture shaped by political circumstances (rather than purely military considerations) explains Irish defeat in the seventeenth century. Similarly, state formation patterns in Europe's easternmost borderlands predominated the practice of warfare before any orthodox Western European military revolution formula had a significant impact, according to Kazakou and Shirogorov. "Long before the Gunpowder Revolution started to influence the military transformation, social and political changes in Eastern Europe directly, . . . nation-building was in a full run in the subcontinent," as Shirogorov reminds us.[33] In Ireland, England's dogged imposition of imperialist rule confounded Gaelic state formation. Indeed, assertion of political and economic independence motivated Tyrone's pursuit of the benefits of the military revolution. Within the cauldron of conflict among expansionary states, recipes differed. Theories describing linear unfolding of a protypical military revolution rarely if ever represent accurately historical realities.[34]

Assessment of Irish resistance prompted the claim that of the two early European-wide military revolutions which were posited (late fifteenth century and eighteenth century), Ireland only experienced the latter.[35] O'Neill's research corrects assertions that the Irish military system only partially assimilated a standardized military revolution based upon the model amalgamated from the combined scholarship of Michael Roberts and Geoffrey Parker. In fact, Ireland's strategic culture and political situation prompted a unique (and "peripheral")

---

**33** Shirogorov, *War on the Eve of Nations*, 364.
**34** Loeber and Parker, "The 'Military Revolution' in Seventeenth-century Ireland," 169–191.
**35** Lenihan, "Conclusion: Ireland's Military Revolution(s)," 345–369, reinforcing Murtagh, "The Seventeenth-Century Military Revolution and Ireland," 75–81, which underscore Stradling's assertion that military revolutions benefited the largest European states at the expense of smaller polities, in Stradling, "'A Military Revolution,'" 271–278.

military revolution akin to but not copied from the heralded Western European military revolutions.

In concordance with his fellow contributors, O'Neill weighs the gunpowder revolution defensively as well as offensively. Irish fortifications, like Lee's Ottoman poliorcetics, were eminently hybrid, responding to fiscal and topographical realities, not strictly geometric ideals touted to be formulas for unassailable defenses.[36] O'Neill points out that English commanders left to us unreliable and loose terminology categorizing Irish defenses. Of the visual evidence, a pair of drawings (Inisloughlan and the Blackwater Fort) indicate that "neither were constructed along modern designs." When substantial defenses were advantageous to Irish resistance, they were built, as when the *trace italienne* was incorporated into Dunboy Castle. Mountjoy's assessment of the Moyry Pass in 1601 could be applied to Irish defensive strategy in general, namely that they took the "time therein to do whatsoever industry could add to the natural strength of the place."[37] The latter observation resonates Lee's Ottoman commentary. The particularities of terrain, extant edifices, availability of resources and strategic culture merged in shaping fortifications, not according to a paradigmatic design capable of deflecting all manner of projectiles. Lee's analyses, herein and elsewhere, of clever Ottoman adaptation illustrate this point. Kazakou and Shirogorov demonstrate that even within generally defined regions, pragmatic improvisation coupled with informed adaptation precipitated variant military solutions.

## Collective Insights into the Gunpowder Revolution

Gunpowder revolution, in the guise of smoking arquebuses, or roaring cannon cutting swathes through infantry as at Ravenna (1512), delivered victories throughout the globe. Of course, one needed to know how to deploy gunpowder weapons to enjoy the fruits of victory; the King of Scots had siegecraft sorted out but not

---

[36] Extending cultural comparisons in a "peripheral" imperial context is Segovia, *The Fortifications of Cartagena de Indias*.
[37] Quoted in the essay below penned by Jim O'Neill, 196. This example makes clear why archaeology is vital to explaining Irish military history.

artillery configurations on a battlefield, as at Flodden (1513).[38] Global narratives of warfare reveal ebbs and flows between technological and tactical parity manipulated by human capabilities. Cross-cultural tactical applications are exemplified by Günhan Börekçi's motley janissaries that mastered volley fire skills equivalent to the cream of other Eurasian armies.[39] The unusual social origins of janissary recruitment sustained "an ideal force to equip with firearms."[40] Slave servitor status produced cohesive and integrated (psychologically as well as ambidextrously) teams of gunmen capable of repelling the empire's myriad enemies. Kang finds evidence of volley fire tailored for Asian warfare just as the Ottomans crafted technology and tactics to meet their mutable and multifarious needs. Western European military revolution orthodoxies of mathematically calculated countermarches and volley fire cannot be configured comparatively in terms of chronological development or critiqued by technique.[41] Kazakou, like Shirogorov, chronicles how Eastern European states' soldiers' adoption of the gunpowder revolution remained within the parameters of entrenched strategic cultures and individual dexterity. Tactical discharge of firearms involved more than mechanics. "Gunpowder did not *require* synchronization of effort. Rather a cultural prejudice about those who would wield it suggested that disciplined synchronization was the only way to make it effective."[42]

How did primacy of skill evolve in the progression of the abilities of "empires of the weak" or among peoples situated on perilous peripheries surrounded by hostiles bearing variegated arms? Military practice has ever been fluid, constantly mutating, unpredictable, and matter of survival subject to the catastrophic whims of fortune and unseen calamities of warfare. Comparative exercises, wherein implicitly exclusionary definitions lurk, imply single best practices in war. Unilateral paradigmatic realities, such as the first deliverable thermonuclear device, probably exist, if only temporarily. Yet, variations will inevitably and inescapably be rooted in terrain, resources, and strategic culture in terms of space (geographically) and time (via developmental pace). For example, a model "Eurasian way of war" constructed from comparison of practices in China and Byzantium in the 600s suggests that military revolutions need not be based on mutually exclusive

---

**38** Fissel, *English Warfare 1511–1642*, 21–23, 32; Phillips, *The Anglo-Scots Wars 1513–1550*, 86, 117–133, 256, 258; Phillips, "Scotland in the Age of the Military Revolution 1488–1560," 182–208.
**39** Börekçi, "A Contribution to the Military Revolution Debate," 407–438.
**40** Lee, *Waging War*, 241.
**41** Parker, "The Limits to Revolutions in Military Affairs," 331–372; Subrahmanyam and Parker, "Arms and the Asian," 14–42; Chase, *Firearms*, 75, 147, 165, 189; Lee, *Waging War*, 243.
**42** Lee, *Waging War*, 240.

factors such as technological determinism or cultural dynamics.[43] An historiography has developed around battlefield adaptions to topography and opponent. This is particularly true of the early gunpowder age, where the application of firearms technology differed widely among cultures. Imposition of chronological patterns of uniformity glosses over the complexities and accidents of operational history and distorts historical reality. Shirogorov's conclusions bear relevance to both the classic military revolution and the revolution in military affairs because military history is teleologically useful as its outcomes (such as victory and defeat) are largely "knowable." Furthermore, he reveals how twists and turns in campaigning and combat wrought state formation. In the process, Shirogorov argues that broad chronologies are always qualified by *histoire événementielle*, particularly that combatants should be compared painstakingly, especially at crucial moments in the historical narrative. This approach is likewise championed by Kang, Kazakou, O'Neill, and Lee. O'Neill's essay demonstrates that chronology, specifically strategic timelines, shaped military revolutions. Fundamental to Shirogorov's thesis is the conviction that time is not homogeneous. For Shirogorov and his fellow contributors, the devil is in the details. Understanding how some nation-states emerged from the period 1450 to 1500 whilst other polities crumbled is found in the oft-overlooked intricacies of warfare (not from monolithic paradigms such as the military revolution or sweeping studies that postulate centuries of developmental determinism). Warfare birthed nations but did not necessarily do so always along the lines of more familiar western "great nation-states." Our contributors encourage historians to bestow upon Eastern Europe a revisionist realignment akin to that enjoyed currently by Asian military revolutions. In the borderlands of the East warfare fostered state development, but it did so before the orthodox model of military revolution (1500–1650) overtook the Baltic States and Russia. Qualitative variations of causation are evident in the chaos of battle (as Kang and Kazakou also argue). Shirogorov illumines how the catalyst of warfare sculpted Eastern Europe. Kazakou's essay about complicated East-West interfaces reveals what peripheral states countenanced. The reader benefits from contemplating Lee, Kazakou, and Shirogorov as an integrated interpretation.

A theme uniting the essays of Hyeok Hweon Kang, Wayne E. Lee, Aliaksandr Kazakou, Vladimir Shirogorov, and James O'Neill is how military revolutions occurred (or did not occur) in conflicts involving expanding states. Lee's Ottomans, along with Kazakou's and Shirogorov's Muscovites, overtly pursued imperialist policies. The early sixteenth century witnessed tremendous triumphs for the

---

**43** Graff, *The Eurasian Way of War*, 4–13, 176–181.

Ottoman Empire and Muscovy; for the former, Damascus (1516), Cairo (1517), Belgrade (1521), Rhodes (1522), Baghdad (1534), Buda (1541), and the latter, Pskov (1510), Smolensk (1514), Kazan (1552), and Astrakhan (1556).[44] Sustained strategies of conquest produced conditions in which both the offensive and defensive facets of military revolution were burnished.[45] With conquest came strategic challenges, such as occupation, as Lee evinces.[46] Mounting resistance to imperialist aggression could occasion indigenous military revolution. For example, both O'Neills (author and historical protagonist) immerse themselves in the ingenuity and assimilation of an Irish military revolution created as a direct response to Elizabethan colonization. The defensive features of military revolutions have not received attention equal to that lavished on offensive warfare (just as the literature on revolutions overshadows scholarship on counter-revolution). Tangentially, the Western European orthodoxy fixated upon centralizing state formation. States, such as the Ottoman Empire, which used decentralization to their strategic advantage, appeared aberrational. Geoffrey Parker's linkage of army size and public expenditure on the *trace italienne* made military revolution a Moloch that could only be satisfied by centralized nation-states. Azar Gat reverses the chain of causation, showing that state formation facilitated the innovations of military revolution, not the other way round.[47] A newer "reversal of fortune" has been proffered by J.C. Sharman. Did the more ancient and formidable empires of the East intimidate interlopers such as the Portuguese, Spanish, English, and Dutch to inflict frightening and idiosyncratic levels of violence (as outnumbered insurgents sometimes do) in compensation for their perceived comparative "weakness"?[48]

Perhaps military revolutions' quest for formulaic models overlooks the qualitative distinctions in the application of violence when exchanged between parties unequal in human and material resources. Consider the centuries of aggression associated with English incursions into Ireland. Jim O'Neill's analysis of Tyrone's "hybrid" successes casts doubt upon the efficacy of England's pair of military revolutions (one initiated under the Tudors, namely by Henry VIII; and a second, under the Stuarts, specifically parliament's building of a war machine for the waging of civil war against the Crown). "Tyrone created a hybridised

---

44 On this "world of war" see most specifically Gunn, *The English People at War in the Age of Henry VIII*, 136, and more generally: Oman, *A History of the Art of War in the Sixteenth Century*.
45 Day, *Conquest. How Societies Overwhelm Others*, 112–131.
46 On the more general issues of conquest and occupation, see Lee, "Conquer, Extract, and Perhaps Govern," 236–260.
47 Discussed by Lee, *Waging War*, 245–246.
48 This is, unavoidably, an oversimplification of the complex reasoning found in Sharman's *Empires of the Weak*.

army, by retaining Irish advantages in tactical and operational mobility, while grafting the benefits of the gunpowder revolution in Europe . . . [T]he Irish could not always depend on terrain and fortifications to protect their shot."[49] The disaster at Kinsale in 1600/1 did not result from any deficiencies in Ireland's assimilation of the weapons of the military revolution. The explanation lies in the political and strategical dimensions of each country's strategic culture. The way in which the Nine Years War unfolded (1593–1603) must be understood before one judges Oliver Cromwell's devastating incursion into Ireland in 1649 as an inevitable success based upon the adoption of the military revolution during Britain's civil wars, referenced above.[50]

One also ought to reassess what have been characterized as tradition-bound and idiosyncratic military revolutions. Conflicts between Ireland and England mirror not only what transpired in those islands but also how Russian military revolutions (including those of Muscovy and her neighbors) played out among geographically proximate but distinct strategic cultures. Similar characterizations apply to the Ottoman military revolution, as well. The conservatism of respective strategic cultures (and, perhaps, skepticism of ideas of progress regarding technology transfer?) inhibited those societies from choosing among the full range of options available in that particular historical context. Jim O'Neill's 2017 monograph offers a comparative perspective that illumines explicitly "tradition-bound" states. The "peripheral" features of warfare to the northwest of the Channel parallels Eastern European developments. Russian conservatism, it is argued, inhibited Muscovy's adoption of contemporary pike and shot tactics. The Ottomans, too, even if technologically current and clever in their adaption of fortifications, were likewise conservative in assimilating and converting to the pike and shot infantry tactics that proved so effective in the Gaelic context.[51] The Irish were not confronted by equine nomadic people or mobile insurgency as were Ottoman and Muscovite armies.[52] Although physically on continental Europe's periphery, Ireland had the advantage of focusing on mastery of the Western European techniques of military revolution (with a little help from Spanish allies). O'Neill's exposition, herein and elsewhere, explains how the challenge-response mechanism performed in a Gaelic strategic culture. Contemporary parallels exist between the

---

49 O'Neill, 189, 191, 201.
50 The present writer has somewhat addressed this in *English Warfare 1511–1642*, though its conclusions must be revised in light of James O'Neill's publications.
51 O'Neill, *The Nine Years War*, 233.
52 Roy, *Military Transition in Early Modern Asia*, 44, 52–54, 199–200.

Irish art of war and Asian variations on the military revolution, e.g., mobile infantry utilizing the firepower of calivers, light cavalry, and earthwork defenses.[53]

The bulk of military revolution historiography focuses upon expansionary states and how they achieved dominion (which makes J.C. Sharman's revisionism all the more deserving of attention). How did polities resisting imperialist strategies use military revolutionary tactics and technology to defend themselves? Therein lies another significant conclusion found in O'Neill's Ireland. Irish strategic culture capitalized upon the advantages and disadvantages of their topography and population distribution. These factors were then exploited in exercising disciplined use of firearms to thwart the invaders. The dynamics of tailored military revolutions are evident in contact zone contests between powers of disproportionate strength. These interactions presented opportunities for technology transfer as well as mimetic administrative improvements (such as logistics and resource management), for example the phenomenon of institutional isomorphism wherein enemies become sources of inspiration.[54] Case studies are useful but especially so when adversaries of a different culture are integrated into the analysis. Historians such as Kaushik Roy and Jeremy Black (both daring to range from Eurasian medieval/early modern empires to contemporary insurgencies) detail hybrid warfare between (and among) conflicting societies over time, the latter work embracing also Black's placement of the military revolution debate into the larger methodological problem of periodization.[55] Roy's and Black's respective publications show how technology transfer (and organizational reform) are shaped profoundly by regional variation in strategic cultures.[56] Synthetic approaches, emphasizing developmental process, apply chronologically more broadly to RMAs as well as to preindustrial military revolutions. Indeed, David Zimmerman proposes an industrial watershed that distinguishes qualitatively (and to a degree, quantitatively) between pre-modern military revolution and modern warfare.

---

53 Kang, "Big Heads and Buddhist Demons," 127–189; Cheng-Heng, "Military Revolution Marches Steppe: Gunpowder Weapons and Walled-Cities in the Inner Asian Steppe, 1600–1760"; also, de la Garza, *The Mughal Empire at War. Babur, Akbar and the Indian military revolution*; see also the comparative study of "agrarian bureaucratic Asian empires" in Roy, *Military Transition in Early Modern Asia*.
54 Lee, *Waging War*, 243.
55 See also further argument in favor of chronological continuity in Black, "From Alexander the Great at Issus to Hannibal at Cannae 333–216 BCE," 41–48.
56 See for example the format as well as the conclusions in Sondhaus, *Strategic Culture and Ways of War*.

# From Gunpowder's Military Revolution to Revolutions in Military Affairs

The "Geoffrey Roberts" (using Laurent Henninger's conflation of the pair of sequential fundamental theories of the classic military revolution) early modern military revolution and the 1980s to 1990s RMAs ascribe implicitly (and sometimes explicitly) a causal role to technological determinism. While the latter interpretation has waned in influence in fields such as ancient history, it maintains its dominance elsewhere, for example RMA scholars' analyses of the refinement of drones and cyberwarfare. Not all advocates of techno-centric views of RMAs warn against mono-causal, or determinist, explanations of military revolutions.[57] We should differentiate too between technological causation by the interaction of interworked technologies, and the primarily monocausal theories that stem from splendid examples (the stirrup, the Argive shield, etc.). A general consensus has emerged, however, that societal context and strategic culture predominantly shape warfare. That qualification extends to military revolutions as well as RMAs.

As we explore in the present writer's chapter six, historians and archaeologists understandably can fetishize nautical material culture. Shuttered gunports, astrolabes, bronze rams, sleek hull designs, aircraft carriers, each possesses aesthetic presence fitting of technology that transformed the *status quo* of naval warfare. However, the magnificent seagoing vessels left unmentioned by Michael Roberts, on a more macrocosmic scale, provide linkage from the sixteenth-century military revolution to the RMA because they also carry along with their armaments and goods embodiments of their strategic cultures. Aquatic warfare manifests strategic culture as well as mobile weaponry. Vladimir Shirogorov cites commercial dimensions impelling Western European States to innovate, sometimes spontaneously. So, logically, the topics of commerce, nationalism, industrialization, and imperialism emerge, leading the reader to revolutions in military affairs and future war. Imperial ambitions owed much to the increased potency of amphibious warfare, and that efficacy is examined by Shirogorov. Because the original military revolution model failed to address combat upon the waters until Geoffrey Parker and others rewrote the equation, Shirogorov's bridgehead establishing the relationship between the gunpowder revolution and amphibious warfare links what the editor and others see as a second Western European military revolution. Overseas expansion ultimately financed revolutionary industrialization to

---

57 Hacker, "Military Institutions, Weapons, and Social Change," 768–834.

create the circumstances that Mark Mandeles and Colonel Vicente describe in their essays.[58]

## Mark D. Mandeles

Those engaged in procurement debates or researching future war study history because military revolutions, Mark Mandeles observes, "are identified most clearly with hindsight;" current practitioners are generally oblivious to faded and failed innovations in the past.[59] The present volume, aiming to decipher how the traditional views of military revolution dovetail with contemporaneous RMAs, strives to avoid the sin of ahistorical scholarship. The RMA literature has, understandably, been critiqued as "a self- conscious attempt to use history alternatively to 'prove' the RMA thesis and to illuminate the transhistorical dynamics of radical change in the nature of war."[60] Mandeles renders a verdict on the perennial problem of gauging how state bureaucracy contends with the hard physical reality of technology. As John Guilmartin put it (to soldiers as much as to historians), "a downside to these impressive technological capabilities [is that] improved tactical capabilities breed heightened strategic expectations."[61] Past victories can induce a torpor-like faith in current martial practices. As Kaushik Roy, Andrew de la Garza, Tonio Andrade, and scholars of Asian military systems remind us, success can be, in the strategic long-run, fatal.

## João Vicente

Colonel João Vicente, a practitioner of "future war," suggests how the storied saga of military revolution will likely end and its implication for humankind. From his perspective as an airman as well as a theorist, Vicente heralds a juncture where history may have become irrelevant to the reality of warfare. The rise of independent robotic weapons systems empowered by artificial intelligence and unencumbered by morality may at first glance appear to take us back to technological determinism. But our organizational structures (a feature of strategic culture)

---

[58] See Headrick, *The Tools of Empire*, on the role of industrialization in powering late nineteenth-century European imperialism.
[59] Mandeles, *The Future of War*, 34.
[60] Latham, "Warfare Transformed," 231.
[61] Guilmartin, "Technology and Strategy," 11–46, 23.

have in fact brought humankind to the precipice. The conduct of warfare is largely determined organizationally, a lesson clearly embedded in Jim O'Neill's chapter, and brought up to date in Mark Mandeles' essay on the bureaucratic common denominators shared among military revolutions and RMAs. Carlo Alberto Cuoco, whose pioneering work connecting the military revolution with RMAs inspired this volume, quotes Andrew Marshal: "The term revolution is not meant to insist that change will be rapid . . . but only that the change will be profound, that the new methods of warfare will be far more powerful than the old. Innovations in technology make a military revolution possible, but the revolution itself takes place only when new concepts of operations develop, and, in many cases, new military organizations are created."[62] The latter point is a pillar of Mark Mandeles' work on RMA organizational culture.[63] Whether discussing Tyrone's organizational reforms as analyzed by Jim O'Neill, or the forging of new asymmetric tactics by irregular forces operating in present-day Yemen, the definition of hybrid warfare encompasses both that which is regular and irregular. A receptive socio-cultural environment permits catalytic reactions (frequently taking the form of hybridization) of technological innovation to occur and be guided. Certainly Wayne E. Lee shows us that it was Ottoman social organization that created fortuitous conditions for military revolution. The diverse eastern political entities expounded upon by Kazakou and Shirogorov likewise show how socio-economic realities translated to battlefield performance. Hyeok Hweon Kang's achievement in comparing and contrasting Asian states demonstrates Cuoco's assertion applies globally. However, as Kang makes clear, in those regional contact zone interactions technology transfers occurred and did so dynamically. The role of technological innovation remains crucial even if not the primary mover. To give voice to Cuoco again, "technological innovation, though not being sufficient in itself, is a necessary enabler of revolutionary change in warfare: in other words, an RMA must follow from some 'efficient' adaptation of doctrine and organization to a given technology."[64] Our essays suggest that Cuoco's generalization is universal and ties the gunpowder age's innovations to contemporary military practice. How does our pair of "future war" essays by Mandeles and Vicente relate to the traditional elements of military revolutions such as transforming infantry and the decisive possibilities of tactical adaptation? The overcoming of physical constraints (especially distance) that came with industrialization led to a digital world that has rendered peripheries somewhat extinct. The cyberworld is fundamentally a solitary venue, and future

---

62 Cuoco, *The Revolution in Military Affairs*, 17, quoting Marshall directly, note 22.
63 Mandeles, *The Future of War*.
64 From Cuoco, *The Revolution in Military Affairs*, 17; compare with the five essays in volume 7 of *Vulcan* (published by Brill) vol. 7 (2019).

war is fought within interconnected battle spaces. The military revolution's connection to the world of RMAs receives further treatment below, including revisiting the scholarship of Mandeles and Vicente.

## Bibliography

Ágoston, Gábor. "The Ottoman Empire and the Technological Dialogue Between Europe and Asia: The Case of Military Technology and Know-How in the Gunpowder Age." In *Science between Europe and Asia*, edited by Feza Gunergun, and Dhruv Raina, 27–40. Dordrecht: Springer, 2011.

Andrade, Tonio. *The Gunpowder Age. China, Military Innovation, and the Rise of the West in World History*. Princeton: Princeton University Press, 2016.

Bennett, Jim, and Stephen Johnston. *The Geometry of War 1500–1750*. Oxford: Museum of the History of Science, 1996.

Black, Jeremy. "From Alexander the Great at Issus to Hannibal at Cannae 333–216 BCE: Redating the Military Revolution, a Preliminary Note." In *Military History. Some Introductions Designed to Begin a Debate*, Società Italiana di Storia Militare, Jeremy Black, 41–48. Rome 2020.

Black, Jeremy. "Rethinking Military Revolutions" *The Critic* (20 January 2020), https://thecritic.co.uk/rethinking-military-revolutions.

Black, Jeremy. *War and Technology*. Bloomington: Indiana University Press, 2013.

Börekçi, Günhan. "A Contribution to the Military Revolution Debate: The Janissaries' Use of Volley Fire during the Long Ottoman-Habsburg War of 1593–1606 and the Problem of Origins." *Acta Orientalia* (Academiae Scientiarum Hungaricae) 59.4 (December 2006): 407–438.

Chase, Kenneth. *Firearms. A Global History to 1700*. Cambridge: Cambridge University Press, 2003.

Cheng-Heng Lu. "Military Revolution Marches Steppe: Gunpowder Weapons and Walled-Cities in the Inner Asian Steppe, 1600–1760." Unpublished paper, Annual Society for Military History Conference, Louisville, Kentucky, 7 April 2018.

Di Cosmo, Nicola. "State Formation and Periodization in Inner Asian History." *Journal of World History* 10.1 (1999): 1–40.

Cuoco, Carlo Alberto. *The Revolution in Military Affairs: Theoretical Utility and Historical Evidence*. Research Institute for European and American Studies. Athens, 2010.

Day, David. *Conquest. How Societies Overwhelm Others*. Oxford: Oxford University Press, 2008.

De la Garza, Andrew. *The Mughal Empire at War. Babur, Akbar and the Indian Military Revolution, 1500–1605*. London: Routledge, 2016.

Dunning, Chester, and Norman S. Smith. "Moving Beyond Absolutism: Was Early Modern Russia a 'Fiscal-Military' State?" *Russian History/Histoire Russe* 33.1 (Spring 2006): 19–43.

Fissel, Mark Charles. *English Warfare 1511–1642*. London: Routledge, 2001.

Fissel, Mark Charles. "Military Revolutions." In *Oxford Bibliographies in Military History*, edited by Kaushik Roy. New York: Oxford University Press, https://www.oxfordbibliographies.com/view/document/obo-9780199791279/obo-9780199791279-0212.xml.

Graff, David A. *The Eurasian Way of War. Military practice in seventh century China and Byzantium*. London: Routledge, 2016.

Guilmartin, John F, Jr. "Technology and Strategy: What Are the Limits?" In *Two Historians in Technology and War*, edited by Michael Howard and John F. Guilmartin, Jr., 11–46. Carlisle Barracks, 20 July 1994, US Army War College, Strategic Studies Institute.

Gunn, Steven. *The English People at War in the Age of Henry VIII*. Oxford: Oxford University Press, 2018.

Hacker, Barton C. "Military Institutions, Weapons, and Social Change: Toward a New History of Military Technology" *Technology and Culture* 35.4 (October 1994): 768–834.

Headrick, Daniel R. *The Tools of Empire: Technology and European Imperialism in the Nineteenth Century*. New York: Oxford University Press, 1981.

Henninger, Laurent. "La guerre de cent ans, les clés d'une révolution militaire", https://www.youtube.com/watch?v=hN8kdhSlsxU.

Jones, David R. "Muscovite-Nomad Relations on the Steppe Frontier before 1800 and the Development of Russia's 'Inclusive' Imperialism." In *Empires and Indigenes: Intercultural Alliance, Imperial Expansion and Warfare in the Early Modern World*, edited by Wayne E. Lee, 109–140. New York: NYU Press, 2011.

Kang, Hyeok Hweon. "Big Heads and Buddhist Demons: The Korean Musketry Revolution and the Northern Expeditions of 1654 and 1658." *Journal of Chinese Military History* 2 (2013): 127–189.

Kelenik, Jozsef. "The Military Revolution in Hungary." In *Ottomans, Hungarians and Habsburgs in Central Europe: The Military Confines in the Era of Ottoman Conquest* edited by Pál Fodor, Géza Dávid, 117–159. Leiden: Brill, 2000.

Khan, Iqtidar Alam. *Gunpowder and Firearms. Warfare in Medieval India*. Oxford: Oxford University Press, 2004.

Klingelhofer, Eric, ed. *First Forts: Essays on the Archaeology of Proto-colonial Fortifications*. Leiden: Brill, 2010.

Latham, Andrew. "Warfare Transformed: A Braudelian Perspective on the 'Revolution in Military Affairs'" *European Journal of International Relations* 8 (2002): 231–266.

Lee, Wayne E. *Waging War. Conflict, Culture, and Innovation in World History*. Oxford: Oxford University Press, 2016.

Lee, Wayne E. *Warfare and Culture in World History*. New York: NYU Press, 2011.

Lee, Wayne E. "Conquer, Extract, and Perhaps Govern: Organic Economies, Logistics, and Violence in the Preindustrial World." In *A Global History of Early Modern Violence and its Restraint*, edited by Erica Charters, Marie Houllemare, and Peter H. Wilson, 236–260. Manchester: Manchester University Press, 2020.

Lenihan, Pádraig. "Conclusion: Ireland's Military Revolution(s)." In *Conquest and Resistance. War in Seventeenth-Century Ireland* edited by Padraig Lenihan, 345–369. Leiden: Brill, 2001.

Loeber, Ralph, and Geoffrey Parker. "The 'Military Revolution' in Seventeenth-century Ireland." In *Success is Never Final. Empire, War, and Faith in Early Modern Europe*, Geoffrey Parker, 169–191. Basic Books: New York, 2002.

Mandeles, Mark. *The Future of War. Organizations as Weapons*. Dulles: Potomac Books, 2005.

Morillo, Stephen. "Guns and Government: A Comparative Study of Europe and Japan." *Journal of World History* 6.1 (Spring 1995): 75–106.

Murtagh, Harman. "The Seventeenth-Century Military Revolution and Ireland." *An Cosantóir Review* (1995): 75–81.

Oman, Sir Charles. *A History of the Art of War in the Sixteenth Century*. New York: Dutton, 1937; reprinted 1979.

O'Neill, James. *The Nine Years War 1593–1603*. Dublin: Four Courts Press, 2017.
Parker, Geoffrey. "The Limits to Revolutions in Military Affairs: Maurice of Nassau, the Battle of Nieuwpoort (1600), and the Legacy." *Journal of Military History* 71.2 (April 2007): 331–372.
Parker, Geoffrey. "Is the 'Military Revolution' Dead Yet?' Lecture. Annual meeting of the Society for Military History, Columbus, OH (June 2019), https://www.youtube.com/watch?v=P8JonajoenM.
Phillips, Gervase. *The Anglo-Scots Wars 1513–1550*. Woodbridge: Boydell, 1999.
Phillips, Gervase. "Scotland in the Age of the Military Revolution 1488–1560." In *A Military History of Scotland*, edited by Edward M. Spiers, Jeremy A. Crang, and Matthew J. Strickland, 182–208. Edinburgh: Edinburgh University Press, 2012.
Poe, Marshall. "The Consequences of the Military Revolution in Muscovy: A Comparative Perspective." *Comparative Studies in Society and History* 38.4 (October 1996): 603–618.
Roberts, Michael. "The Military Revolution." In *Essays in Swedish History*, edited by Michael Roberts, 195–225. Minneapolis: University of Minnesota Press, 1966.
Rogers, Clifford J. "Gunpowder Artillery in Europe, 1326–1500: Innovation and Impact", in *Technology, Violence, and War*, edited by Robert S. Ehlers Jr., Sarah K. Douglas, and Daniel P.M. Curzon, 37–71. Leiden: Brill, 2019.
Rogers, Clifford J. "The Military Revolution in History and Historiography." In *The Military Revolution Debate*, edited by Clifford J. Rogers, 1–35. Boulder: Westview, 1995.
Rogers, Clifford J. "'Military Revolutions' and 'Revolutions in Military Affairs': A Historian's Perspective." In *Toward a Revolution in Military Affairs?*, edited by T. Gongora and H. von Rickhoff, 21–35. Westport: Greenwood, 2000.
Roy, Kaushik. *Military Transition in Early Modern Asia, 1400–1750*. London: Bloomsbury, 2014.
Segovia, Rodolfo. *The Fortifications of Cartagena de Indias. Strategy and History*. Bogotá El Ancora Publishers, 2009.
Sharman, J.C. *Empires of the Weak. The Real Story of the European Expansion and Creation of the New World Order*. Princeton: Princeton UP, 2019.
Shirogorov, Vladimir. *War on the Eve of Nations. Conflicts and Militaries in Eastern Europe, 1450–1500*. Lanham: Rowman and Littlefield/ Lexington, 2021.
Smith, R.D., and K. DeVries. *The Artillery of the Dukes of Burgundy, 1363–1477*. Woodbridge: Boydell, 2005.
Sondhaus, Lawrence. *Strategic Culture and Ways of War*. New York: Routledge, 2006.
Stradling, R.A. "'A Military Revolution': The Fall-Out from the Fall-In." *European History Quarterly* 24 (1994): 271–278.
Subrahmanyam, Sanjay, and Geoffrey Parker. "Arms and the Asian: Revisiting European Firearms and their place in Early Modern Asia." *Revista de Cultura* 26 (2008): 14–42.
Thompson, Michael J. "Military Revolutions and Revolutions in Military Affairs: Accurate Descriptions of Change or Intellectual Constructs?" *Strata. revue d'histoire des étudiants diplômés de l'Université d'Ottawa / Strata. University of Ottawa Graduate Student History Review* 3 (2011): 82–107.
Trim, D.J.B. "Medieval and Early-Modern Inshore, Estuarine, Riverine and Lacustrine Warfare." In *Amphibious Warfare 1000–17000. Commerce, State Formation and European Expansion*, edited by D.J.B. Trim, and M.C Fissel, 357–413. Leiden: Brill, 2006.
Zimmerman, David. "Neither Catapults nor Atomic Bombs: Technological Determinism and Military History from a Post-Industrial Revolution Perspective." *Vulcan* 7 (2019): 45–61.

Hyeok Hweon Kang
# Difference in an Age of Parity: Technology and Global Military History

## Introduction

A recondite subject of military history has mattered on an order of world-historical magnitude. The musketry volley tactic – a technique for rotating the ranks to fire in turns – figures at the crux of the military revolution. As first conceived by Michael Roberts, military revolution theory posited that in sixteenth-century Europe, tactical innovations around the use of matchlock muskets led to revolutionary changes in military organization and the state at large.[1] The musket, as argued, was a marginally useful weapon that required gargantuan military, fiscal, and administrative efforts to deploy. It did not just pressure Europeans to invent drill, adopt standing armies, and grow centrally governed nation-states, but, as Geoffrey Parker further argued, the cascading changes around the gun even enabled the rise of the West in global warfare, allowing countries in Europe to control nearly 35 percent of it by 1800.[2]

This ambitious paradigm of world history has faced many criticisms, inviting *inter alia* charges of technological determinism by other Europeanists in the field of the history of technology.[3] The most transformative critique, however, emerged from the ranks of global historians and specialists of non-Europe, especially East Asia. Namely, works by Tonio Andrade, Sun Laichen, Kenneth Swope, Peter Lorge, and others have shown that East Asians were similarly innovative in their use of gun technology.[4] With regards to muskets, for instance, which remains central to the literature, they evinced that the Japanese employed volley-firing by

---

[1] Roberts, "The Military Revolution, 1560–1660," 13–35. I thank Mark Fissel, Wayne E. Lee, Kenneth Swope, Peter Lorge, James O'Neill, Keith Roberts and Philip Tom for their generous source suggestions and insightful feedback on this paper.
[2] Parker, *The Military Revolution*; Rogers, ed., *The Military Revolution Debate*; Yerxa, ed., *Recent Themes in Military History*, 11–48. For the most comprehensive bibliography on the military revolution, see Fissel, "Military Revolutions."
[3] Hall and DeVries, "The 'Military Revolution' Revisited," 500–507. Also see Raudzens, "War-Winning Weapons," 403–434.
[4] For an overview, see Andrade, *The Gunpowder Age*, and Lorge, *The Asian Military Revolution*. Also see Laichen, "Military Technology Transfers from Ming China and the Emergence of Northern Mainland Southeast Asia (c. 1390–1527)," 495–517; Swope, *A Dragon's Head and a Serpent's Tail*; and Chase, *Firearms*. Also relevant are works on South Asia and Southeast Asia: Gommans, *Mughal Warfare*; and Charney, *Southeast Asian Warfare, 1300–1900*.

the 1590s,⁵ and the Chinese held the world's "first unequivocal" record – i.e., codified in the form of a manual – from 1560.⁶ In seventeenth century Korea, where both Chinese and Japanese influences coalesced with the local practices, a new musketry corps also formed that could not only volley fire but also aim with precision.⁷

This new scholarship has interwoven a global military history that emphasizes not so much European exceptionalism as a world of connections and parallels that generated similarities across Eurasia. Importantly, Tonio Andrade, doyen of said scholarship, employed the term "Age of Parity" (1550–1700) to capture this new understanding. As shown in his seminal *Gunpowder Age*, many innovations emerged in early modern Europe, to be sure, but China, Japan, and Korea kept up comparable levels of military developments vis-à-vis Europe well into the eighteenth century.⁸ Andrade summarizes: "Europeans did have advantages in deep-water naval warfare and fortress architecture, but East Asians fielded dynamic and effective forces, defeating European troops not just by superior numbers but also by means of excellent guns, effective logistics, strong leadership and better (or at least equivalent) drill and cohesion."⁹

The global turn in military history is commendable, but it is not with its flaws. Historians Frank Jacob and Gilmar Visoni-Alonzo criticize recent works for abusing the concept of military revolutions to "find similarities with regard to military progress" between Europe and East Asia.¹⁰ This, they argue, ends up reintroducing West-centrism through the back door. In a review of *Gunpowder Age*, economic historian Roy B. Wong also weighs in, stressing that the problem, rather, is in conflating two terms – parity and similarity. Wong argues that a state of parity does not preclude the existence of differences, and he puts it in concrete terms by comparing muskets to mechanical clocks. China adopted both from Europe – thus achieving parity; but this did not mean that its military or

---

5 Stravos, "Military Revolution in Early Modern Japan," 250. For more information on Japanese musketry and the Japanese military revolution, see Brown, "The Impact of Firearms on Japanese Warfare, 1543–98," 236–253; Varley, "Oda Nobunaga, Guns, and Early Modern Warfare," 105–125; Morillo, "Guns and Government," 95–100. There is also evidence of Ottoman volley fire by 1605. Börekçi, "A Contribution to the Military Revolution Debate," 407–438. For more on Ottoman firearms, see Ágoston, "Firearms and Military Adaptation," 85–124.
6 Andrade, *The Gunpowder Age*, 166.
7 Andrade, *The Gunpowder Age*, 185. For more on Korean musketry, see Kang, "Big Heads and Buddhist Demons," 127–189; Andrade, Kang, and Cooper, "A Korean Military Revolution?," 51–84.
8 Andrade, *The Gunpowder Age*, 5, 125, 135–236, 299.
9 Andrade, *The Gunpowder Age*, 5.
10 Jacob and Visoni-Alonzo, *The Military Revolution in Early Modern Europe*, 18.

economy was "changing to become more like Europe's in any consequential manner."[11] In comparing technologies between two regions, then, it is important to situate them in their respective contexts and reflect on broader differences in society, political economy, and institutions among others.

Better understanding difference is key to move forward, and I herewith address technological difference – i.e., in the manner of producing and employing muskets – between the two world regions. Albeit in passing, disparities in the use of muskets between Europe and Asia have been noted by historians, including Geoffrey Parker.[12] To date, however, only works in Japanese and Korean language have addressed them squarely, and with proper attention to the rich troves of data found in their respective archives. In a 2008 study, Japanese historian Kubota Masashi argued for a military divergence between Japan and Europe, based on a comparison of musket use: Japanese gunners, he showed, focused on accuracy, while their European peers trained instead to increase rate of fire.[13] At the heart of Kubota's notion of divergence was technological difference: in Japan, accurate yet slow-firing muskets – known as fowling pieces – found military use; in Europe, fast-firing yet inaccurate muskets – or shoulder arms – did. This difference, in turn, led to other wide-ranging consequences: a strong preference developed in Japan for marksmanship and individual skills training; and in Europe, for mass firing and drill (collective training).[14] Korean historian Kim Yŏngchung furthered Kubota's work with the case of Chosŏn Korea (1392–1910). Not unlike their Japanese counterparts, Korean musketeers also diverged from Europeans in their use of sporting guns as infantry weapons; as a result, they, too, emphasized accuracy at the expense of volley fire, and individual training in lieu of drill.[15]

These works divulge an unresolved point of debate in the globalist discourse on the Military Revolution. Just as the debate in Western academe was undergoing

---

11 Wong, "Tonio Andrade. The Gunpowder Age," 466.
12 Andrade, "The Arquebus Volley Technique in China, c. 1560," 122, n. 24.
13 Kubota refuted Parker's assertion that the Japanese under Oda Nobunaga had developed their own method of volley fire analogous to the Dutch invention. Kubota, *Nippon no gunji kakumei*.
14 Kubota, *Nippon no gunji kakumei*. 38–56. Kubota also mentions other differences, such as the prolonged peace in Japan and the lack of a strong cavalry as stimulus for musketry. Regarding the latter, warhorses in Japan were apparently the size of contemporary ponies, and this rendered the Japanese cavalry weak. Musketeers there thus had less incentive to mass fire against their mounted counterparts. Kubota, *Nippon no gunji kakumei*, 48, 250.
15 Kim, "17 segi Chosŏn ŭi p'osu hullyŏn pangsik kwa kunyul" 17C 조선의 포수 훈련방식과 군율, 167–97. Also see Chang, "17 segi chŏnban Chosŏn ŭi p'osu yangsŏng kwa unyŏng" 17세기 전반 朝鮮의 砲手 養成과 運用.

transformation by globalists and Asianists, a separate strain of scholarship emerged, ironically, in Japanese and Korean language, in response to the original theses by Roberts and Parker.[16] To be sure, written without the benefit of the recent Anglophone scholarship, Kubota and Kim show a certain parochiality with their understanding of the Military Revolution literature: Kubota uses a strictly Europe-based definition of concepts, concluding that the Japanese military revolution "terminated at the minimum level necessary"; Kim also misunderstands the nature of Korean musketry when he calls it a form of ("Eastern") "martial art" rather than ("Western") "drill."[17]

These conclusions notwithstanding, the possibility of a "military technological divergence" merits a close investigation: did important differences indeed exist between Europe and East Asia? If so, how do we reconcile them – and their implications about divergence – with the notion of parity, as proposed by global military historians?

The following analysis answers by focusing on musket (in)accuracy. It outlines, first, the differences in musket production and usage between Europe and East Asia, one by one, and then cross-examines both to understand the implications of said differences. By using new comparative data and a balanced evaluation of both sides, it will show that important differences – akin to those laid out by Kubota and Kim – did occur, but they stemmed not from a fundamental disparity in military or technological aptitude, as suggested, but from variances in the manner of employing effective firepower in an age of parity.

## Europe and the De-accurizing of Muskets

Let us begin with an extreme yet illustrative example: Prussian musketeers under Frederick the Great (1712–1786). The Prussian way of deploying the musket was far from efficient in the sense of killing with the least ammunition, or in achieving any sense of accuracy. In terms of rate of fire, however, it was unparalleled in Europe and probably the early modern world. The Prussians adapted every aspect of their musket use – including gun design, training, and

---

**16** See notes 13–15 above. Also see Kuba, *Higashi Ajia no heiki kakumei* 東アジアの兵器革命; No, "16–17 segi choch'ong ŭi toip kwa Chosŏn ŭi kunsajŏk pyŏnhwa" 16–17 세기 조총의 도입과 조선의 군사적 변화; No, "Kihoek nonmun chŏnjaeng ŭi sidaejŏk yangsang" 기획논문 전쟁의 시대적 양상; Chŏng, "Hwasŏng kongsimdon ŭi yurae wa kinŭng" 화성 공심돈의 유래와 기능, 1–35; Chŏng, "Hwasŏng ŭi pangŏ sisŏl kwa ch'ongpo" 화성의 방어시설과 총포, 133–161.
**17** Kubota, *Nippon no gunji kakumei*, 243, 255; Kim, "17 segi Chosŏn ŭi p'osu hullyŏn pangsik kwa kunyul," 178.

tactics – for speed and convenience. They not only drilled relentlessly, to fire and load their pieces like automatons. But various expedients accumulated over time, such as knocking their muskets on the ground instead of ramming the charge home, exploiting double-ended ramrods for quick-loading, and firing away hastily with no aim.[18] As a result, Prussians purportedly fired up to three to four rounds per minute in battle, and as many as six rounds blank in drill.[19] "This rapidity," as Frederick liked to boast, "made a Prussian equal to three adversaries," capable of firing "three shots per minute more than their enemies."[20]

Speed came at a great cost of accuracy, however. By one metric from 1742, it was reported that Frederick's men would use as many as 540 balls (or 33 pounds of lead) per enemy death at the Battle of Chotusice.[21] The statistic may be exaggerated. Still, the Prussians were such that many contemporaries ridiculed their obsession with quick-firing: French general Comte de Guibert (1743–1790), for one, likened their musketry to "burning powder at the birds."[22] Yet, as the critics themselves admitted, there was no denying that the Prussians were widely admired and imitated throughout Europe.[23] Their way, in fact, was the very culmination of a European emphasis on mass firing, which – as argued here – traded speed for accuracy.

Military historians and historians of technology have long recognized that small arms in Europe emphasized volume of fire, not accuracy. Historian Bert Hall put it clearly: the European gun was designed "not to minimize its limitations of accuracy, but to maximize its advantages as a quick-firing weapon."[24] When it came to speed, indeed, various innovations came to accelerate fire in Europe; if anything, the oft-repeated chronology of firearms inventions shows this much: the musket, first fired with a smoldering match in the seventeenth century, grew faster and more convenient with the flintlock mechanism, and eventually, paper cartridges and the percussion cap.[25] The resulting uptick in "efficiency" was, apparently, revolutionary. As Philip Hoffman explains in his *Why did Europe Conquer the World*, it translates to a "labor productivity growth of 1.5 percent per year which rivals overall labor productivity growth rates in

---

18 Quimby, *The Background of Napoleonic Warfare*, 114–115, 118–120.
19 Rothenberg, *The Art of Warfare in the Age of Napoleon*, 17–19.
20 Jomini, *The Art of War*, 279; Grant, *The Cavaliers of Fortune*, 121.
21 Hall, *Weapons and Warfare in Renaissance Europe*, 136.
22 Quimby, *The Background of Napoleonic Warfare*, 119.
23 Quimby, *The Background of Napoleonic Warfare*, 118.
24 Hall, *Weapons and Warfare in Renaissance Europe*, 140.
25 Carmen, *A History of Firearms*, 101–104; Hughes, *Firepower*, 27.

modern economies and far exceeds what one would expect even at the onset of the Industrial Revolution."[26]

So much for speed. But how did this spirit of haste impinge on accuracy? The question has gone ignored because so far, musket marksmanship has been considered impossible in Europe and by extension, the rest.[27] This section shows, however, that the former's preference for speed did foreclose possibilities of pursuing higher accuracy. The musket in Europe, as we will see, divested from accuracy in three main ways: first, they adapted the material aspects of the gun to accelerate, not accurize, fire; they also adopted distinctive ways of usage – namely, shortcuts to loading, and an emphasis on indiscriminate fire – to trade quality for quantity of shot.

Take first the case of musket manufacture. It takes no ballistic science to figure that a firearm, shot at a distance with any sense of accuracy, would meet such basic standards as having a straight bore, tightly fitting bullets, and a pair of gun sights. Surprisingly however, military muskets in Europe rarely met these standards. Regarding the bore, George Hanger (1751–1824) – an officer in the British army who was a vocal advocate of musket accuracy – lamented that military pieces in service were "bored with a most shameful and scandalous neglect." In fact, Hanger would go so far to suggest that these guns "are all crooked, and if, per chance, a few be tolerably straight bored, they are bent in folding the loops on"; and that as such, "a soldier . . . may with *just as much hopes of hitting*, fire at the moon as at the body of a man at the distance of two hundred yards."[28]

Other militaries in Europe may have commissioned better barrels. But what was surely widespread was the large windage – or gap between the bullet and the barrel – which allowed even "crooked" muskets to be loaded and fired. As early as 1590, soldier and military writer Sir John Smythe (1534–1607) explained the issue of windage in English muskets and its effect on accuracy.

> [H]ow truly soever they take their sights at point and blank, the air doth work very great effect with their bullets, which are lower by four or five bores than the height of their pieces, to carry them from the mark or marks that they are shot at. Moreover, by proof they may find that in giving their volleys of musket shot but only twelve scores at either horsemen or footmen that are in motion, they shall work no great annoyance, by reason

---

**26** Hoffman, *Why Did Europe Conquer the World?* 57–58.
**27** Hall, *Weapons and Warfare in Renaissance Europe*, 140. Also see Krenn, Kalaus and Hall, "Material Culture and Military History," 101–109.
**28** Hanger, *Reflections on the menaced invasion*, 196, 201–202.

that the bullets, beings so much lower than the height of their pieces, as is aforesaid, do naturally mount and fly uncertainly.[29]

Smythe wrote that English guns in late sixteenth century were made with large windage – to the extent that their bullets were smaller by "four or five bores than the height of their pieces." The large gap meant that even with careful aim at close range, English muskets could not give accurate fire, whose bullets would "mount and fly uncertainly" and wide from "the mark or marks that they are shot at."[30]

Designing guns with greater tolerances was a widespread practice, and an intentional one: the aforementioned Prussians also preferred it, and on average, windage on a late eighteenth-century European flintlock was as much as .05 inches.[31] To be sure, large windage conferred an advantage for rapid firing: it let the ammunition roll down the barrel with ease, which saved time in reloading. As Smythe noted, however, loose-fitting bullets harmed accuracy. This is an amply proven point in historical and modern ballistic experiments. In the early eighteenth century, mathematician Benjamin Robins (1707–1751) showed that bullets fired from a loose barrel were prone to rub against a side of the muzzle and deviate wildly from their intended path upon fire. Modern firing experiments conducted by David Miller and others also corroborate that windage diminishes accuracy, not the least when combined with bent barrels.[32]

What is perhaps most surprising about European muskets, however, is their lack of gun sights. Before the age of modern telescopes, all aiming was done with the naked eye, and a surer shot attained by matching the front bead sight with a standing back sight. This technology was commonplace and old, tracing back to as early as 1450 in Europe. But curiously, while Europeans kept furnishing sporting guns with sighting devices, they did not for military equipment.[33] In 1855, General Sir Howard Douglas reflected: "Till within the last

---

**29** Smythe, *Certain Discourses Military*, 66.
**30** Smythe, *Certain Discourses Military*, 66.
**31** Stephenson, *The Last Full Measure*, 91.
**32** Robins and Curtis, *New Principles of Gunnery*, 210–217. For a general discussion of smoothbore ballistics, see Hall, *Weapons and Warfare*, 142–143. Hanger uses Robins' experiments as "proofs" to argue in favor of smoothbores. An expert marksman and colonel, he argued that a smoothbore musket in general use – when straightened and better-bored – could "shoot as true and correct as the best rifle-gun in the kingdom" within the range of 60 yards to 200 yards. For his discussion of straightening smoothbores, see Hanger, *Reflections on the menaced invasion*, 159, 167, 197; for his use of Robins' experiments as "proofs" for his argument, see 202–203. For modern experiments, see Miller, "Ballistics of 17th century muskets."
**33** Peterson, *Encyclopedia of Firearms*, 297–299.

twenty years, no *sight* was considered necessary for common musket – the stud at the muzzle being sufficient for the purpose of taking aim."[34] Indeed, some military muskets continued to feature front sights, but these no longer provided an aiming function per se: as historian William Wells also noted, they were shaped in a rectangular form, and served merely to "indicate the direction where the muzzle was pointing rather than providing accurate aim."[35]

How could Europe, where matchlock muskets were first invented, become so resigned on accuracy? It bears mentioning here that European gunmakers could certainly make exquisite matchlocks with upright barrels, tightly fitting bullets, and accurized sights. This is more than amply evidenced by the still existing specimens made for the nobility and stored in the various museums and private collections. Among these fine guns, too, were rifled pieces that greatly enhanced accuracy by imparting a beneficial spin on the bullet.[36]

Rather than technological deficiency, then, I argue that a series of "choices" – wittingly or not – had been made that rendered European military matchlocks to be crude machines for rapid firing. The reasons behind the curious development require a separate investigation,[37] but it suffices here to understand this: the European musket had a lopsided development towards speed and convenience, and whatever accuracy had been possible or achieved – for military, strategic purposes – was eventually forgone or lost.[38]

With guns that were designed against accuracy, it certainly made sense for their users to maximize their potential for a higher rate of fire instead. Many techniques developed, for instance, in shortcutting the loading action. It is well known that after the seventeenth century, new recruits in Western nations

---

**34** Douglas, *A Treatise on Naval Gunnery*, 528.
**35** Douglas, *A Treatise on Naval Gunnery*, 528; Wells, *Shots That Hit*, 5.
**36** Wilson, "Some Important Snap Matchlock Guns," 3–10.
**37** Perhaps this was because early modern Europe was engaged in constant warfare, and the pressing demand for military firearms outstripped the state's ability to impose quality control on its burgeoning and frequently engaged armies. Or perhaps it had to do with industrial limitations, problems of logistics, and governmental funding.
**38** European muskets actually retrogressed in accuracy during the course of the seventeenth century. Fine barrels, for instance, were available yet unused in the manufacture of military muskets. Hanger thus quipped, that ill-bored barrels could easily be set "straight and fresh[ly] bored by a skillful gunsmith" – and with a meager "five or six shillings per piece." Hanger, *Reflections on the menaced invasion*, 159, 167, 197. Gun sights were allegedly furnished on early military muskets but had disappeared by the late seventeenth century. Small arms in Europe also lost their resting fork and became shorter-barreled during the seventeenth century – a transformation that dovetailed with the term "musket" overtaking the "arquebus." Firth, *Cromwell's Army*, 79–80; Hall, *Weapons and Warfare*, 176–179.

underwent intensive training in the handling of their muskets, as argued by the original proponents of the military revolution. Through this so-called 'manual exercise,' said recruits trained in as many as 42 postures for musketry, as shown in *The Exercise of Arms* by Count John de Nassau.[39] It is important to note, however, that the European manual exercise was only for initial repetitive training, and that in actual drill and battle, "much of the fancy rigmarole was dropped in favor of shortcuts."[40]

The shortcuts included forgoing the use of paper wads and ramrods. The importance of wadding was amply understood by contemporaries, but it was often overlooked in practice, to the detriment of accuracy. When properly done, wadding seals the gas that leaks around the shot, and thereby increases the powder's efficiency. But as Roger Boyle, 1st Earl of Orrery (1621–1679) wrote in 1667, "soldiers seldom put in any paper, tow or grass to ram the bullet in." This, he emphasized, led to "the little execution . . . musketeers do in time of fight, though they fired at great battalions, and those also reasonable near."[41]

Overlooking ramming was also injurious and not just for accuracy. Ramming was a crucial step in musketry loading, which by lowering the bullet through the barrel, fitted it tightly against the rear and tamped the powder; this strengthened the power of the shot and prevented misfires. This method was sure, but it was slow, and it became "custom of the day" by the late seventeenth century to simply 'seat' the bullet by slamming the butt of the musket on the ground.[42] The practice allowed musketeers to fire an extra round or two a minute. It was a dangerous shortcut, however: it not only sabotaged accuracy but reduced the effective range (by as much as 75 percent by one account), and caused frequent misfires, or the notorious "flash in the pan."[43] The latter refers to a type of misfire where the trigger goes off with its attendant flash, but leaves the main charge unignited and the bullet still in the barrel. This was not uncommon in the European battlefield, and it could spell catastrophe for soldiers who unwittingly loaded their pieces more than once (sometimes as many as

---

**39** Yerxa, ed., *Recent Themes in Military History*, 14.
**40** Stephenson, *The Last Full Measure*, 90. Nafziger also notes that by the late eighteenth century, the number of steps in the standard musketry sequence was about seventeen, which was considered more practical. Nafziger, *Imperial Bayonets*, 30.
**41** Firth, *Cromwell's Army*, 82.
**42** Firth, *Cromwell's Army*, 82.
**43** There were other shortcuts such as not reversing the ramrods while loading. For more details, see Nafziger, *Imperial Bayonets*, 30. In another example, the Prussians in the late eighteenth century "did not ram the charge home" after firing their first salvo, which reduced both accuracy and effective range – the latter by as much as 75 percent. See Quimby, 118.

seven times or more); if ever fired in earnest, these barrels would ensure terrible explosions.[44]

So far we have seen that muskets in Europe were both designed and loaded for rapid firing, but surely they were still fired with some measure of "aim"? Certainly so, but the emphasis in European militaries – unlike in East Asian armies, as we will see – was not to hit the space of a human but to set the general direction – and importantly, height – of fire. The average musketeer in early modern Europe simply leveled his gun and pointed it towards the whole façade of an enemy battalion. As shown in Table 1, musketry trials in Europe were conducted almost always with targets that represented the entire width of the enemy battalion. In 1790 Hanover, for instance, targets as wide as 45.7 m (50 yards) were used, showing an indifference to the lateral dispersion of the shot. Care was taken, however, to use a taller target for cavalry, which was 2.6 m in height as opposed to 1.8 m for an infantry target.[45]

This practice of leveling was the European way of "aiming," and in fact, the British infantry would use "level" rather than "aim" as the word of command, as late as the Battle of Waterloo in 1815. Remarkably, even when the word "aim" was explicitly used, it was not what it seemed. Consider, for instance, the instructions for firing by Prussian and American officer Friedrich Wilhelm von Steuben (1730–1794).

> Take Aim! One motion – "Step back about six inches with the right foot, bringing the left toe to the front; at the same time drop the muzzle, and bring up the butt-end of the firelock against your right shoulder; place the left hand forward on the swell of the stock, and the forefinger of the right hand before the trigger; sinking the muzzle a little below a level, and with the right eye looking along the barrel."[46]

This instruction occurs in von Steuben's famed manual that became the standard drill guide for the American military until 1812. As shown, taking aim

---

[44] Nafziger, *Imperial Bayonets*, 31; Elting, *Swords Around a Throne*, 480; Grossman, *On Killing*, 19.
[45] All measurements used in Table 1 were converted into the metric system, and the original units marked in parentheses. Data was taken from the following: von Pivka, *Armies of the Napoleonic Era*, 60; Müller, *The Elements of the Science of War*, 187; Hanger, *Reflections on the menaced invasion*, 147–148; Picard and Azan, *La campagne de 1800 en Allemagne*; von Scharnhorst, *Results of Artillery and Infantry Gun Trials*, 63–64.
[46] von Friedrich Wilhelm Ludolf Gerhard Augustin Steuben, *Regulations for the Order and Discipline of the Troops of the United States*, 9.

meant merely to look along the barrel with the right eye and sink the muzzle a "little below level."[47]

How far back might we trace this peculiar practice of aiming? More research is imperative, but we learn that as early as the late sixteenth century – when matchlock muskets were first spreading across Europe – said practice was already in place. Take for instance the prescriptions of Welsh mustermaster Maurice Kyffin, who defined what it meant to be a "trained shot" in his day.

> And as I would not have a soldier in the plain field bound to stay so long as to find his perfect mark or enemy in his fight, before he discharge; so do I hold him utterly unworthy of pay, or bear the name of trained shot, that shall not . . . be sure to bestow his bullet between head and foot although it be ten or twelve score [feet] off [horizontally] . . . and find the breast high mark before he deliver his bullet. Such a one as thus can find his mark most readily, deserves great praise.[48]

As shown, leveling of the gun – when properly done – was enough to merit the title of a "trained shot." Specifically, Kyffin clarifies that the Irish gunner was expected to level his piece to the "breast high mark" and "bestow his bullet between head and foot" of the enemy, up to as much as 240 feet was tolerated in the lateral dispersion of his shot.[49] This is characteristic of the tactical limitations of leveling without aim: while ensuring a well-directed fire (and restraining the soldiers' common habit of firing too high), it left much to be desired in terms of horizontal accuracy.

One concrete way to understand European practices of aiming is target practice. While working with a different set of standards for aiming, musketeers in Europe did practice at the target, however infrequently. The likely chance that soldiers had to practice their "aim" was during the initial training of new recruits. In 1804, Hanger described as follows the standard instruction given to a "raw countryman, or a mechanic from Birmingham" in musketry:

> [H]e is consigned to the superintendence of the drill sergeant; he is first taught to walk, next to march, and hold himself tolerably erect; then a firelock is placed in his hands, which he handles at first as awkwardly as a bear would a plumb-cake. When he is taught the manual exercise, and fit to do regimental duty, they then take him to fire powder; whilst the drill sergeant is teaching him to fire either by files or by platoons, the sergeant

---

[47] von Friedrich Wilhelm Ludolf Gerhard Augustin Steuben, *Regulations for the Order and Discipline of the Troops of the United States*, 9.
[48] Maurice Kyffin, "How shot may be well trained with least charge and waste of powder," as cited in O'Neill, *Irish Sword*, 8–14.
[49] O'Neill, *Irish Sword*, 8–14.

says to him, laying his cane along the barrels of the firelocks, "Lower the muzzles of your pieces, my lads, otherwise, when you come into action, you will fire over the enemy." After this, the recruit is taken to fire ball at a target; how is he taught? Thus he is spoken to: "Take steady aim, my lad, at the bull's eye of the target; hold your piece fast to the shoulder, that it may not hurt you in the recoil; when you get your fight, pull smartly.[50]

As shown, after learning to march and handle the musket, a new recruit first practiced firing en masse with just powder. The recruit then coordinated with his peers and took care not to fire too high, as consistent with the emphasis in Europe for leveling. After this came target practice where the soldiers used real ammunition for the first time and trained to take "steady aim" at the bull's eye.[51] Yet, as we discussed above with Table 1 – and Hanger also clarifies, the target used in these practices were "infinitely smaller" than the proverbial modern one (i.e., representing the space of a man).[52]

Target practice in Europe was also infrequent. Exact data is elusive from the early modern period, probably because they were so sparsely conducted, but we know that even during the Napoleonic times, new conscripts underwent basic instruction for just two or three weeks, and after that, were left to their own devices without any further mastery of their weapons.[53] To be sure, a "quick musketry refresher course" was held in times of war, with twenty rounds per man for target practice.[54] Annual estimates, however, were as low as "two musket shots per year in practice"[55] or "three or four rounds."[56] Perhaps the best endowed around this time were British soldiers who received each "thirty rounds ball and seventy blank" for annual practice,[57] but others had much less: Austrians in 1805 had

---

50 Hanger, *Reflections on the menaced invasion*, 152–153. Similar records of musketry training are also included in the aforementioned writing by Kyffin, where he outlined a three-step regime: 1) practice by false fires in the pan, 2) learn to fire with powder at the target, and 3) fire in groups of different configurations.
51 Hanger, *Reflections on the menaced invasion*, 152–153.
52 Hanger, *Reflections on the menaced invasion*, 148. Another problem Hanger highlights is the inadequacy of training officers. As Hanger notes, the drill sergeant was often "no marksman himself." While fit to "discipline the soldier, and form him for parade and actual service in the line," he quipped that "the sergeant is just as capable of teaching him how to solve one of Sir Isaac Newton's problems as to teach him to be a marksman" (153). It seems that the Russian army in 1810 also suffered a similar problem, as an order from the Minister of War "insisted that the officers, who would train soldiers, had to be themselves trained in aimed fire." Zhmodikov and Zhmodikov, *Tactics of the Russian Army in the Napoleonic Wars*, 12.
53 Chandler, *Dictionary of the Napoleonic Wars*, 208.
54 Elting, *Swords Around a Throne*, 483.
55 Chandler, *Dictionary of the Napoleonic Wars*, 208.
56 Nafziger, *Imperial Bayonets*, 30.
57 Haythornthwaite, *Weapons and Equipment of the Napoleonic Wars*, 21.

"six live rounds per man," and Russians practiced "at a rate of three per man per year."[58] Target practice was thus infrequent in early modern Europe, and it must be noted that even when conducted, they served not to pursue high rates of marksmanship but merely to "know which eye to use in aiming," or to not stagger at the tremendous recoil of his musket.[59]

This is not to say that all European muskets were inaccurate: as mentioned above, some were rifled and exquisitely made, like those made for the nobility to hunt. Also, as we will later revisit, these fowling pieces saw limited use by the military, and though rare, these episodes caution against sweeping generalizations. Still, for our purposes here, it seems safe to say that at least in the realm of regular military usage, the musket in Europe was made, loaded, and fired for speed, rather than accuracy. But what about East Asia?

## East Asia and Musket Marksmanship

Since its first introduction in the sixteenth century, the musket in East Asia was known for accuracy. In 1545, the year when most scholars date its debut in Japan, a Portuguese marksman supposedly "shot twenty-six ducks in quick succession" and shocked the Japanese locals into reproducing his gun.[60] The veracity of this tale is questioned. Yet, there is little doubt that in the next decades the Japanese started to produce high-quality muskets and began to unleash their potential for marksmanship. In Ming China (1368–1644) and Chosŏn Korea, where these guns were used to wreak havoc, the defenders noted their unusual ability for precise shooting and named them "bird gun" (鳥銃; Ch. *niaochong*; K. *choch'ong*).

> [The bird gun] is unlike any of the other types of fire weapons . . . .. In accuracy, it can strike the center of targets as if hitting the eye of a coin and not just for expert marksmen . . . .. [It] hits the mark eight out of ten times, and can bring down birds flying amidst the trees, hence its name.[61]

If fanciful, the passage above occurs in a military manual by Qi Jiguang (戚繼光, 1528–1588), a Ming general who was known for his hard-nosed practicality. In the 1550s, when China's southeastern coast was raided by musket-toting pirates (some hailing from Japan), Qi had many encounters with the gun and eventually incorporated it into his own army. To the skeptics of his day, the

---

58 Zhmodikov and Zhmodikov, *Tactics of the Russian Army in the Napoleonic Wars*, 12.
59 Nafziger, *Imperial Bayonets*, 30; Elting, *Swords Around a Throne*, 322.
60 Andrade, *The Gunpowder Age*, 169.
61 Qi, *Ji xiao xin shu*, 56. Also see Andrade, "The Arquebus Volley," 122.

Table 1: Musketry Trials in Western Europe.

| | Year | Target size | Target Score according to Distance | | | | | | | Source |
|---|---|---|---|---|---|---|---|---|---|---|
| | | | 76 m (100 paces) | 91 m (100 yds) | 110 m (120 yds) | 150–152 m (200 paces) | 182 m (200 yds) | 225–228 m (300 paces) | 274 m (300 yds) | |
| Hanover | 1790 | 1.8 m x 45.7 m (6 ft x 50 yds; Infantry Target) | 3/4 | 75% | | 37.5% (152 m) | | 1/3 (228 m) | 33.3% | Otto von Pivka |
| | | 2.6 m x 45.7 m (8 ft 6 in x 50 yds; Cavalry Target) | 5/6 | 83.3% | | 50% ('') | | 3/8 ('') | 37.5% | |
| Britain | 1814 | Target representing a line of cavalry | | 53% ("Well exercised men") | | | 30% ("Well . . .") | | 23% ("Well . . .") | Müller |
| | | | | 40% ("Ordinary soldiers") | | | 18% ("Ordinary . . .") | | 15% ("Ordinary . . .") | |
| Britain | 1804 | Unknown; likely frontage of infantry battalion | | 60% (3/5) | 60% | | | | | Hanger |
| | | 1.8 m x 0.5 m (6 ft x 20 in; "figure of a man") | | | 62% | | | | | |

| | | | | | |
|---|---|---|---|---|---|
| France | 1800 | 1.8 m x 3 m | 60% | 40% (") | 25% (225 m) | Picard |
| | | | | Mean error of 75 cm x 60 cm (150 m) | | Picard |
| Prussia | 1810 | 1.8 m x 30.5 m (6 ft x 100 ft) | 66.7–75% | 50% | 25% (228 m) | Scharnhorst |

general qualified that musket accuracy was often spoiled by soldiers who were improperly trained; as he liked to retort, if the musketeers neglect to "hold their guns level, hold them to the side of their cheek, or use the sights, how can one hit the enemy, to say nothing of being able to hit a bird?" Conversely, then, Qi contended that with the right equipment and training the musket could indeed hit the mark and live up to its name.

Such optimism was widespread across East Asia, and the question is how – in practical terms – accuracy was pursued. Adducing new data from all three countries, this section shows that East Asians, unlike their European peers, maximized the potential of smoothbore marksmanship: they accurized gun manufacture, trained for discriminate fire, and emphasized target practice.

Let us begin with gun manufacture. Contrary to Europe, where quality guns were reserved for sporting, the expectation in Korea, Japan, and China was that artisans would deliver similarly well-made pieces (albeit not ornately decorated) for general use in the army. Again, elements such as boring, windage, and sights were crucial. In recent years, we have come to understand gun boring of this region in detail, and what seems consistent throughout is a deep concern with quality control. The aforementioned Qi, for one, was a stickler for meticulous boring, and he commissioned local smiths to bore a mere 3 cm (1 Chinese *cun* 寸) per day for as long as a month or until the inside gives off a shine. Ensuring the supply of such quality barrels was difficult – and Qi complained about some negligent workshops. Yet, the general was adamant and he specified how to tell good barrels from bad, such as inspecting that the product is of equal thickness throughout – or in the parlance of the gunsmith, "perfectly upright."[62]

A similar attitude prevailed in Chosŏn, where both texts and practices emphasized high standards for boring. Taking after Qi's works, which were widely read in Korea, military practitioners in Chosŏn upheld this job as an important determinant of barrel quality and importantly for us, accuracy. Consider, for instance, a 1813 production manual entitled *Essentials of Military Affairs* (*Yungwŏn p'ilbi* 戎垣必備), compiled at the Special Manufactory of the Military Training Agency (*Hullyŏn togam* 訓鍊都監) – one of Korea's central armies.

> The accuracy that hits the center of targets lies in the straight muzzle [of the musket]. If the muzzle is straight, loading the powder is convenient, and with an unwavering ignition unwavering, eight or nine shots out of ten can hit the mark . . .. [D]uring manufacture, forge the iron to be refined of impurities, and bore the barrel straight. The bullet

---

[62] Qi, *Ji xiao xin shu*, 56; Pye, *The Sportsman's Dictionary*, 424. For more on historical techniques of boring, see Kang, "Crafting Knowledge," 257–262.

must twist and turn unimpeded (*wanjŏn muae* 宛轉無礙), and only then could it be called a good [musket].⁶³

This manual equates a good musket with one whose barrel is immaculately bored. As described, after forging the barrel, its inside was reamed to a high polish, such that the bullet, when fired, would "twist and turn unimpeded" and make as straight an exit as possible out of the muzzle.⁶⁴ This was, of course, easier said than done, but in Korea, strict regulations were in place for borers working in military workshops, such as the one at the Military Training Agency. In a typical musket shop from the eighteenth-century, military borers comprised the most numerous group (up to 41 percent of the whole), followed only by filers (23 percent) and blacksmiths (10 percent). By 1714, said borers were required by law to write their names on every barrel crafted, so that negligent ones can be tracked down, punished, and made to re-deliver better barrels.⁶⁵

Windage and sights, too, were important. Returning to the Ming military, Qi specified that only tightly-fitting, polished lead balls (11.25g) should be used, and that they should be pushed into the muzzle with subtle force before being rammed with strength and paper wadding down to the bottom of the barrel. This careful charging of the musket ensured greater accuracy, as well as preventing misfires, increasing the musket's effective range, and allowing it to be fired downwards from higher ground. To prevent malpractices and expedients, Qi had these loading procedures tested in drill, and exhorted against harming accuracy by "trying for speed and convenience."⁶⁶ Regarding gun sights, military muskets throughout East Asia were expected to have a pair. But the best examples are found in Japan, where thousands of historical guns have survived in great condition. As summarized by arms historian and collector Sugawa Shigeo, these objects show "five styles of front and rear sights respectively," supplemented by a middle sight in either round or square shapes.⁶⁷

---

63 *Yungwŏn p'ilbi*, 17.
64 *Yungwŏn p'ilbi*, 17.
65 Consider the full text of the said law to inscribe names: "Muskets should be bored according to the old precedent – i.e., to allow only lead bullets of 3 *chŏn* in weight. If this size bullet does not enter the muzzle, then the gun is useless. The borer shall be punished with 30 beatings if 10 barrels or fewer [reported issue] and 50 beatings if 20 barrels or more. After that, they are to make the [same number of] barrels again." Kang, "Crafting Knowledge," 261.
66 Andrade, "The Arquebus Volley," 129.
67 Sugawa Shigeo, *The Japanese Matchlock*. There were also attachable sights as the "sighting ladder" that adjusted for shooting at distances of 100 paces, 150 paces, 250 paces and allegedly 1,500 paces. Rogers, "The Development of the Military Profession in Tokugawa Japan," 189.

More needs to be said on manufacture, but this suffices to show that East Asians pursued a high standard, and higher than Europeans – at least in the seventeenth and eighteenth centuries, and when it came to military muskets. But the question still lingers: could a marginally more accurate gun – as used by East Asians – actually lead to high rates of marksmanship? In 1804, when the aforementioned George Hanger wrote in favor of training musketeers at the target – in a way that is, as we will see, reminiscent of East Asian practices – he argued that a smoothbore musket, in the right hands, did not pale against the "best rifle gun in the kingdom" within 200 yards. Could this have been true?[68]

Accuracy is subjective and we must first define distance. The laws of physics prevent a smoothbore gun, regardless of how immaculately forged and bored, to deliver a straight bullet beyond 150 m or so. This limitation was certainly noted by East Asians. In China, for instance, the musket was known to be effective under 150 m (100 Chinese paces): as explained in a 1636 encyclopedia, the shot can "shatter birds into pieces within a distance of 30 paces, kill them without tearing their bodies at more than 50 paces, and was extremely weak at 100 paces."[69] Within this range, however, accuracy could vary significantly, and practitioners in East Asia believed that the right attitudes towards aiming and target practice could make a difference.

When compared to the European notion of aiming, the East Asian counterpart strikes us as starkly different: the unit of musketry aiming in East Asia was not the whole frontage of an enemy battalion but the space of a single man. This meant that shooting was discriminate, and both vertical and lateral dispersion of the shot was controlled through human effort. First, evidence of selective aiming by the average soldier abounds in all three countries of East Asia. In China and Korea, draconian measures were taken squad-by-squad to ensure that the musketeer in the line aimed at a single enemy while volley firing.[70] "If the enemy approaches en masse," Qi warned his soldiers, "each person should designate only one out of the enemy ranks and fire."[71] Interestingly, in Korea, similar yet more detailed instructions were given, along with an emphasis on countering cavalry. A manual from 1693 has it that the Korean musketeer should "aim the muzzle at the chest of the enemy," and "if the enemy is mounted, at the head of the horse, or if the enemy comes in flocks, at one out of the

---

**68** Hanger, *Reflections on the menaced invasion*, 197, 200.
**69** Sung Ying-Hsing, *T'ien-kung K'ai-wu*, 276.
**70** Qi, *Ji xiao xin shu*, 199–200. For Korea, see *Pyŏnghak chinam chuhae*, 38–39.
**71** Qi, *Ji xiao xin shu*, 199–200.

crowd."[72] Similarly, in Japan, instructions had it that while firing in unison, "a single musketeer should aim and fire towards a single enemy."[73]

But how then did East Asians aim? Despite minor differences, soldiers in China, Korea, and Japan were expected to fire from the cheek and use the front and back sights. In Qi's army, for instance, the musketeer was instructed "to rest his cheek on the stock's end, and using one eye, match the back sight with the front sight, and the front sight with the man being shot at."[74] Korea had almost identical instructions but clarified with detail that the soldier should fire also from a kneeling position and without "shaking his head, hands or the front knee."[75] The emphasis on using the gun sights was to decrease the horizontal dispersion of the shot, but leveling the gun – as the Europeans emphasized – was also important, especially for firing at longer distances. In Japan, ample pictorial as well as textual data survive of the Japanese leveling system. For instance, when firing without adjustable sights, Japanese musketeers were advised to aim their pieces at the enemy's mid-section from 109 m (1 Japanese chō 町), the chest from 218 m (2 chō), base of the head from 327 m (3 chō), middle of the forehead from 436 m (4 chō), and the crown of the head from 545 m (5 chō).[76]

No matter how specific and rigorous the instructions were, aiming was only perfected through practice. Unlike in Europe, target practice was routinized in East Asian armies from as early as the seventeenth century. While its details varied from army to army, the convention throughout the region was to fire ball at a man-sized target, if not smaller, using precise range and lead bullets, and recording hits per individual (Table 2). Regarding ammunition, while European armies were parsimonious, preferring to train by "false fires in the pan," or with clay bullets,[77] East Asians insisted on using lead ammunition to drill as close to battle conditions as possible. Also, target practice, it seems, was quite frequent. While exact data is elusive, in one example, Japanese musketeers in the Hachisuga clan (蜂須賀氏) had practice ammunition of about 375 shots per

---

**72** This manual – *Chinpŏp Ŏnhae* 陣法諺解 – was published in 1693 by Ch'oe Suk (崔橚, 1646–1708) at Hamkyŏng Provincial Governor's Office 咸鏡監營. For an annotated and translated version, see Yi, *Chinpŏp ŏnhae chuhae*, 79.
**73** Ishioka, *Naganumaryū heihō*, 167.
**74** Qi, *Ji xiao xin shu*, 59.
**75** *Pyŏnghak chinam chuhae*, 22.
**76** Anzai, *Hōjutsuka no seikatsu*, 81. More research is needed to compare leveling systems in detail. In Western Europe, "practical knowledge veterans passed to conscripts" during the Napoleonic era was as follows: "[W]hen the enemy was 50 yards away, aim at their knees . . . at 100 yards, aim at the waist; at 140 to 200, at the head." Elting, *Swords Around a Throne*, 483.
**77** Zhmodikov and Zhmodikov, *Tactics of the Russian Army in the Napoleonic Wars*, 12.

man per year. If, according to the customs of the day, they shot ten rounds per practice, they could train approximately once every ten days.[78]

The best evidence of target practice occurs in Japan, where there was a wide variety including but not limited to the *funauchi* (船打, "shooting on board"), *chōchi* (町打, "long-range shooting"), and the *kakuchi* (角打 "short-range shooting"). Long-range accuracy was the eventual goal, and the *chōchi* was conducted on large curtain-like targets by increments of 109 m (1 *chō*), up to 1090 m.[79] Yet, the Japanese practiced most rigorously at close distances, presumably to calibrate their aim, and allow the shooter to adjust to the built-in inaccuracies of each musket. To that end, the *kakuchi* used a 24 cm (8 *sun* 寸) square target, with a bull's eye of 6 cm (5 *sun*) in diameter. This small target, or *kaku*, was clasped between bamboo poles, and raised on a firing mound 27 m (15 *ken*) from the shooter, or in some cases, 90 m or 109 m. Scoring sheets from 1795 show that shooters in the Seki-ryu Gunnery School had 87.5 percent accuracy (50 percent at the bull's eye, and 37.5 percent target hits) at the *kaku* from 27 m; there were three shooters, and a total of 24 shots, of which only three missed.[80] Comparable data – at the same target and distance – can be found in the scoring sheets submitted to the Bakufu in 1830 as "proof of the readiness of the musketeers," who achieved 95 percent accuracy (22 percent at the bull's eye, and 73 percent target hits or graze).[81]

We still need to process more data from Japanese target practices, but it is clear that the standards were remarkably high. In the Aizu domain, for instance, qualifying marks to earn the license of a "trained shot" – measured on the *kaku* at the standard 27 m, and from a kneeling position – were as follows: nine hits out of ten with a 11.5 mm caliber (2.3 *momme* 匁), eight hits out of ten with a 18.8 mm caliber (10 *momme*). Further, a "skilled shot" had to perform even better, and on a smaller 18 cm square target: 30 hits out of 50 for 18.8 mm caliber, and 80 hits out of 100 with any smaller caliber.[82]

Target practice was just as important in continental East Asia. In sixteenth-century China, Qi stipulated both archers and musketeers to practice at a 32 cm (1 *chi* 尺) by 1.6 m (5 *chi*) target. While the former shot from 80 paces, or 120 m at most, the latter took it 20 paces further and fired from 150 m. "Target strikes

---

**78** Sasama, *Kakyū bushi ashigaru no seikatsu*, 124–125, as cited in Kubota, *Nippon no gunji kakumei*, 50.
**79** Udagawa, *Edo no hōjutsu*, 22.
**80** Udagawa, *Edo no hōjutsu*, 22.
**81** *On-saki teppo kakucho* [Score Sheets for Bakufu Musketeers]. For more, see Rogers, "The Development of the Military Profession," 196.
**82** Anzai, *Hōjutsuka no seikatsu*, 23.

**Table 2:** Target Practice in East Asia.

|       | Target Size (width by height) | Distance | Source (Date) |
|-------|-------------------------------|----------|---------------|
| China | 32 cm (1 *chi*) by 1.6 m (5 *chi*) | 150 m (100 Ming paces) | *Ji xiao xin shu* (1580) |
|       | 32 cm (1 *chi*) by 1.6 m (5 *chi*); bull's eye: 16 cm (5 *cun*) in diameter. | 150 m (100 Ming paces) | *Wu Bei Zhi* (1621) |
|       | 32 cm (1 *chi*) by 1.6 m (5 *chi*) | 64 m (41 Qing paces) | *Qin ding Da Qing huidian* (1655) |
| Korea | 9 cm (3 *ch'on*) by 1.6 m (1 *bal*) | 72 m (60 Korean paces) | *Pukchŏng ilgi* (1658) |
|       | 61 cm (2 *ch'ŏk*) by 2 m (7 *ch'ŏk*) | 120 m (100 Korean paces) | *Sokdaechŏn*, f. 4 (1746) |
| Japan | 24 cm (8 *sun*) by 24 cm (8 *sun*); raised by poles; bull's eye: 6 cm (5 *sun*) in diameter | 27 m (15 Japanese *ken*) | Udagawa, 22; Anzai, 39. |
|       | 24 cm (8 *sun*) by 36 cm (1 *shaku* 2 *sun*) | 55–73 m (30–40 Japanese *ken*) | |
|       | 24 cm (8 *sun*) by 1.7 m (5 *shaku* 5 *sun*) | 109–127 m (60–70 Japanese *ken*) | |

were tallied on an abacus," as Andrade explains, "and the results recorded with the name of the soldier" on a "sample assessment forms . . . with blank spaces to be filled in with the names of soldiers and grades for their performance."[83] Often, in such a rigorous manner – with assessment forms and individual grades – musketry was pitted against archery. Speaking from experience, Qi was a clear partisan of the former: "when comparing and vying on the practice field the arquebus [musket] can hit the bull's-eye ten times better than the fast-lance and five times better than the bow and arrow."[84] In 1595, Korean King Sŏnjo (宣祖, 1567–1608) seems to have been inspired by Qi. He carried out an experiment of his own, comparing his brand-new musketry corps against the time-tested archers. At the end of the day, presumably after counting the target

---

83 Andrade, "The Arquebus Volley," 127–128. Sample assessment forms can be found in Qi, *Ji xiao xin shu*, 144–145.
84 Qi, *Ji xiao xin shu*, 57, as cited in Andrade, "The Arquebus Volley," 130.

**Table 3:** Target Practice Data from Korean Musketeers, 1658.

| Date | Results | Percentage |
|---|---|---|
| 4/6 | 51 hits / 200 shots | 25.5% |
| 4/21 | N/A | N/A |
| 5/17 | 40 hits / 200 shots | 20% |
| 5/18 | 65 hits / 200 shots | 32.5% |
| 5/21 | 200 people, 3 shots/person, 600 shots total<br>3/3 accuracy = 5 people (15 hits / 15 shots)<br>2/3 accuracy = 21 people (42 hits / 63 shots)<br>1/3 accuracy = 97 people (97 hits / 291 shots)<br>154 hits / 600 shots | 25.67% |
| Total | 310 hits / 1200 shots | 25.83% |

Source: Sin, *Pukchŏng ilgi*, 73–5.

hits like Qi, the former was declared severalfold better than the bow, and Sŏnjo exclaimed: "how believable are the words of the late [Qi] that the musket [shoots] five times better than the bow!"[85]

In Korea, there is evidence that musketeers practiced shooting on an extremely narrow target – about 10 cm wide and 1.5 m high – from as far as 72 m (60 Korean paces), and successfully at that (Table 3). These musketeers were Korean auxiliaries sent to the Russo-Qing war in the Amur region, and their commander Sin Yu kept a diary of the entire expedition, including records about musketry tests. At the aforementioned target and distance, 200 Korean musketeers fired a total of 1200 shots, and averaged 25 percent accuracy, with the highest rate being 32.5 percent and the lowest 20 percent.[86] These numbers might not seem impressive at first but they shot at an incredibly narrow target – certainly smaller than the space of a man, and comparable in width only to the bull's eye of the Japanese *kaku* (yet from nearly thrice as far as its standard range of 27 m). Using Probable Errors (PE) to extrapolate on ballistic performance, we may also conjecture that these men would have scored – at a roughly man-sized target (1.6 m tall and 30 cm wide) from the same distance –

---

85 *Sŏnjo sillok*, 1594/8/2 and 1595/10/8.
86 Sin, *Pukchŏng ilgi*, 73–75.

an average accuracy of 66.2 percent. Under their best conditions (when 32.5 percent was scored), they would have boasted 79.8 percent.[87]

As shown, musketry and marksmanship were synonymous in East Asia. Unlike in Europe, where a trained shot simply leveled his piece and fired indiscriminately into a crowd, the East Asian equivalent aimed at narrow targets, representing the space of a solitary man, if not smaller. In fact, with regard to the lateral alignment of their aim, the best marksmen in seventeenth-century East Asia outmatched European sharpshooters even during the Napoleonic era; the standard French target at the time was 53.34 cm by 1.67 m, significantly larger than most targets used in East Asia.[88]

There was indeed an appreciable difference in musketry between Europe and East Asia during the early modern period. But while a difference is discernible, its exact nature and implications are not clear. Was this difference simply a matter of choice and preference in the style of musketry? Or was it a fundamental divergence in the "culture" of training and aptitude in war-making?

## Difference in an Age of Parity

Kubota and Kim saw the difference in musket use between Europe and East Asia as evidence of a fundamental divergence in capacity and "culture" – one of the reasons why the "military revolution [in Japan and Korea] terminated at the minimum level necessary."[89] Their reasoning is straightforward: because East Asian musketry was fixated on marksmanship and individual skills training, it developed as a form of "martial art" rather than "maneuver," lacking the rigorous drill, discipline, and collective training that is characteristic of European

---

[87] The assumption in this extrapolation was that given the extremely narrow target width and the nature of smoothbores, it was due to the lateral dispersion of the shot that the musketeers had missed their targets. Thus, the height was not taken into consideration.

[88] By the early nineteenth century, the standard French target was set as 1.67 m high by 53.34 cm wide at ranges of 97 m (50 toises), 195 m, and 292 m. Another source informs that the mean error of the French musket at 150 m was 75 cm vertically x 60 cm laterally. It appears that no gun sights were used. For more details, see Elting, *Swords Around a Throne*, 482–483; Haythornethwaite, *Weapons and Equipment of the Napoleonic Wars*, 19. Interestingly, there is also data suggesting that in terms of short-range accuracy (measured at 27 m at the *kaku*), the Japanese matchlock still performed slightly better than a Dutch military musket at target shooting as late as 1863. See Anzai, *Hōjutsuka no seikatsu*, 23.

[89] See notes 13–15 in this essay.

musketry.⁹⁰ This section shows, however, that their argument lacks in subtlety: a close examination of both sides reveals that this notion of a divergence does not stand up to evidence. Differences in musketry were not always hard-and-fast, and counterexamples go both ways: just as marksmen in East Asia trained at rapid, mass firing – with plenty of drill and collective training, sharpshooting was not unknown to specialist units in European armies. Importantly, said differences also did not translate to a gap in military capability, as there were both advantages and disadvantages to each style of musketry.

The first qualification: East Asians did not simply surrender speed in pursuit of accuracy. Aimed fire took longer, to be sure, but musketeers still drilled to volley fire for alacrity and consistency. Qi, for instance, not only advocated for precise aiming and target practice, but emphasized the importance of keeping up a constant hail of death. This balance of mass firing and marksmanship was achieved in China as well as Korea, where Qi's legacy was strong. In a Korean manual titled, *Annotations on the Guide to the Military Arts* (*Pyŏnghak chinam chuhae* 兵學指南註解), Qi's instructions are glossed line-by-line with great clarity.⁹¹ Out of nearly 2,000 commentaries, an explanation of "volley fire" goes as follows:

> Those who are empty reload; those who are full fire again (which is to say, fire in five turns). While the ones who have fired are loading, those who are full then fire again (it means to complete a cycle and start all over). The firing of guns will not be lacking (or continuous without ending), and there must be no firing to the point of exhaustion (all day long, without using up [ammunition]) and with no slip-ups in shooting (one after another, firing in turns without missing a turn). Bow and arrow. . .is also fired in accordance with musketry (. . .firing overnight without ending. By nightfall, the enemies are reduced, and each squad would have fired ten volleys per musketeer).⁹²

As reassured by each annotation, to "volley fire" meant to divide into a formation five-rank deep, and to cycle through the sequence of firing and loading incessantly without missing a single volley. Further reinforced by bow and arrow, such a technique would have generated a barrage of missiles, not unlike the blanket of fire that Western musketry strove for. Exact data is elusive on how

---

90 See notes 13–15 in this essay.
91 See note 71 in this essay. Qi Jiguang's manuals were circulated widely, annotated, and readapted by Koreans. For more, see Siegmund, "Circulation of Military Knowledge and its Localization"; Siegmund, "The Adaptation of Chinese military techniques to Chosŏn Korea, their validation, and the social dynamics thereof."
92 *Pyŏnghak chinam chuhae*, 58–59. My translation of the Korean annotations is provided in parentheses. With minor edits, the translation of the original Chinese text was taken from Andrade, "The Arquebus Volley," 125.

fast the Chinese and Koreans could fire, but they had various firing systems, each presumably with differing speed and density of volleys. Andrade writes of Qi's musketry drill: "[V]arious configurations of layers could be tested, as the commanders deemed suitable: two layers of five, five layers of two, one line of ten."[93]

Second, despite that individual skill was rewarded, the average musketeer in China and Korea underwent just as much collective training as their European counterparts. In fact, volley firing with aim would have required greater discipline and fire control than volley firing without it. It is scant wonder then that Qi prescribed target practice as a team exercise, training musketeers to hear for the signal of a gong or trumpet before firing gracefully in coordination. In battle, such fire control was enforced with draconian measures. In the event of a violation, the musketeer in question would either be executed summarily or earmarked for punishment after the battle. Any squad leader who failed to enforce this rule was also subject to decapitation.

Perhaps in Japan, where Kubota first conceived his notion of divergence, was different. After all, the island was home to as many as 200 gunnery schools, where musketry was prized as form of martial art, and apprentices spent years practicing gunmanship.[94] In such an environment, ever more fanciful forms of musketry emerged, and some even trained in such virtuosic skills as shooting while crossing a river, or while supine and holding the gun between the toes.[95]

But in any case, the point is still moot. While Kubota acknowledges the existence of volley fire, a closer examination indicates that its speed and discipline deserve more credit. References to musketry drill are commonplace in the Japanese military books throughout the early modern period. In a gunnery manual titled "Thoughts on Musketry Training" (習砲私議, 1823), it was emphasized that musketry drills be "regularly practiced by a method of strict testing and comparison."[96] One such drill placed Japanese musketeers at a distance of 55 m (30 ken 間) from a small torso-shaped target of 24 cm (8 sun) by 1 (1.2 shaku 尺), and instructed them to volley-fire with both alacrity and accuracy while marching forwards. The number of hits at the target, and the time interval between each salvo, was then recorded meticulously in assessment forms.[97] A similar practice occurs in a military manual from Aizu domain, which had an elaborate system of evaluating the speed

---

93 Andrade, "The Arquebus Volley," 127.
94 Anzai, *Hōjutsuka no seikatsu*, 20–21.
95 Rogers, "The Development of the Military Profession," 189.
96 Anzai, *Hōjutsuka no seikatsu*, 38.
97 Anzai, *Hōjutsuka no seikatsu*, 38.

of volley fire by 9 different grades.⁹⁸ While exact records on speed of fire are elusive, a modern trial showed that Japanese matchlocks could be fired eight to nine aimed shots in five minutes, and that with cartridges made for rapid fire (*hayago* 早碁), up to three shots per minute was possible.⁹⁹ Rich troves of Japanese materials have yet to be engaged with the Military Revolution debate, and we will likely find better data.

Conversely, in Europe, while the average infantrymen in the line aimed at speed of fire, specialist units did strive for accuracy. Often, skirmishers and other light infantry corps employed their firepower not through mass volley but aimed fire, which they achieved with accurized pieces like their East Asian counterparts, or at times, with rifled sporting guns.

From as early as the seventeenth century, for instance, some sharpshooters were employed to pick off enemy officers. During the English Civil War (1642–1651), a marksman distinguished himself at the siege of Sherborne Castle by sniping away at enemy officers with the careful aim of his "birding-piece."¹⁰⁰ Partaking in the same war, English general George Monck later wrote in support of furnishing each company with six sharpshooters, or "[s]ouldiers that carry the Fowling-pieces," and to use them while skirmishing to "shoot not at any, but the Officers of that Division."¹⁰¹ By 1808, mutual sniping of officer corps – albeit considered a breach of military etiquette – was not uncommon in the European battlefield.¹⁰²

Until the late eighteenth century, these sharpshooters were volunteers temporarily who were recruited in times of need. Many of them had learnt to shoot by sporting with their fowling pieces, which were akin to the military muskets in East Asia. Or perhaps, they had participated in such civilian shooting contests as the Schuetzen, where expert marksmen competed for prizes with both smoothbores and rifles.¹⁰³ In any case, unlike their military peers, the volunteers were generally good marksmen. In one example, German peasants in the Thirty Years' War (1618–1648) proved lethal against "plundering parties" of the Imperialist army.¹⁰⁴ "[B]eing used to kill Crowes and vermin on their own land," these peasants were "very good marksmen (especially, the aim being better, when the

---

**98** Anzai, *Hōjutsuka no seikatsu*, 38.
**99** Sugawa Shigeo, *The Japanese Matchlock*.
**100** Firth, *Cromwell's Army*, 91.
**101** Monck, *Observations upon Military & Political Affairs*, 103.
**102** Muir, *Tactics and the Experience of Battle in the Age of Napoleon*, 66 as cited in Stephenson, *The Last Full Measure*, 112.
**103** Lugs, *A History of Shooting*, 30–34; Peterson and Elman, *The Great Guns*, 84–85.
**104** Firth, *Cromwell's Army*, 87.

mark is alive)," and their fowling pieces were "a great deal surer shooters and fitter for their handling than the warlike musket."[105]

Taken together, these examples show that war-making remained variable in both regions, and that hasty generalizations are ill-advised. This, however, is not to deny that there were advantages that one style of musketry had over the other, or that an actual divergence in gunnery developed eventually – in favor of the Europeans.

Direct comparisons of military strength are difficult because the sample size of formal battles fought between Europe and East Asian powers is small during the early modern period. We can surmise with some certainty, however, that in some situations, the emphasis on marksmanship did confer an edge. If, for instance, East Asians found themselves firing against European infantrymen from a distance, they could deliver volleys from farther away and without retaliatory fire from Europeans. In this case, the latter's skirmishers could have been obstructive – especially if armed with rifles, but all other things being equal, said advantage would favor East Asians given their general superiority in aimed fire. The advantage of accuracy is even clearer if the target presented a narrow frontage. This was a rare occasion, to be sure, as most battles – in both Europe and East Asia – were fought in line formations to maximize firepower. But in cases like the Battle of Bladensberg in 1814, when the British infantry was crossing a river by a single narrow bridge, Korean musketeers would have fared far better than their American peers, who blazed away – with several hundred muskets within effective range – with no apparent effect. Compare this to the Battle of Kimhwa in 1636, when Korean musketeers lured Manchu cavalry into a narrow valley and volley-fired at close distance, with disastrous effect.[106]

The East Asian advantage was not just in accuracy. From the meticulous ramming and loading of their pieces, musketeers in the region could fire downwards from higher ground unlike their Western counterparts; if you recall, some Japanese even trained in *sageya* ("downward target practice"). Europeans, however, did not load their pieces faithfully, so their bullets were prone to fall out when the guns pointed downward. For this, we have a concrete example from a war fought between the Russians and a coalition of Qing and Korean forces. In 1654, the story goes, a Qing-Korean army found itself in a retreating situation during a riverine combat with a Russian fleet. The day was saved, however, by Korean musketeers who had stationed themselves on a hill overlooking the river, where they fired behind stockades and from higher ground. When annoyed

---

105 Firth, *Cromwell's Army*, 87.
106 Hughes, *Firepower*, 26. Kang, "Big Heads," 148–156.

by their volleys, the Russians – mostly armed with flintlocks – stormed the hill, but they were beaten back by effective counterfire, leaving behind a trail of death. Had the two switched positions, the outcome would have been drastically different: with Russian musketeers failing to fire downward at the climbing enemy. This happened at least once in European history: in the Battle of Wittstock (1636), the Imperial-Saxon army – ensconced atop a hill – paid a dear price for not loading their pieces correctly; to their detriment, when enemies approached from below, and they presented their pieces, the bullets fell out and left them defenseless.[107]

In other situations, however, European musketry held an edge: the sheer volume and speed of fire in Europe rendered it more efficacious against cavalry charges. In the famous Battle of Pavia (1525), French cavalrymen were shot off their horses by "volleys of deadly lead shot . . . [which] penetrated not only one man-at-arms, but often two, and two horses as well."[108] What is impressive is that the Spanish handgunners fought without field fortifications – unlike their East Asian counterparts – and in a dense fog. Despite the odds, their firepower left the field "covered with the pitiful carnage of dying noble knights as well as with heaps of dying horses."[109] Even after Pavia, many European armies continued to raise mounted troops, but musketry in Europe was such that by the mid-seventeenth century, the traditional head-on cavalry charge was avoided for its own sake.[110]

In contrast, East Asian musketry did not fare so well against cavalry. For every tale of musketry victory in East Asia, there seems to be another testimonial of a brutally effective cavalry charge. For instance, despite the oft-mentioned story of Nobunaga's musketeers at Nagashino (1575), Japanese musketeers, in most cases, could only stand their ground against even light horsemen.[111] Indeed, during the East Asian War of 1592 to 1598, Japanese musketeers were often outmaneuvered by Ming cavalry, who were proficient in Mongol-derived tactics.[112] Likewise, Korean musketeers left an ambivalent legacy against the formidable Manchu cavalry: when left in the open without the support of close-quarter units,

---

**107** Firth, *Cromwell's Army*, 90.
**108** Hall, *Weapons and Warfare in Renaissance Europe*, 182.
**109** Hall, *Weapons and Warfare in Renaissance Europe*, 182.
**110** Hall, *Weapons and Warfare in Renaissance Europe*, 179; Stephenson, *The Last Full Measure*, 77–81.
**111** Kubota, *Nippon no gunji kakumei*, 48.
**112** No, "16–17 segi choch'ong," 121.

they were easy targets.[113] A Korean observer at the Battle of Sarhū (1619) put it as clearly as possible:

> The musket is a craft that allows shooting from a great distance, but it is very slow to reload and fire. Without relying on a fortress or rough geography, its use is never to be tested against cavalry in the open field. Last year, our forces faced cavalry charges with only musketeers and before they could even finish reloading, the enemy cavalry dashed into the heart of our formation.[114]

After Sarhū, Koreans learned their lesson and even scored a few battles against the Manchu cavalry by exploiting terrain and fortifications, by sniping at the officer corps, and through mass firing at close range.[115] By and large, however, it seems that East Asian musketeers still preferred to fire behind defensive structures when countering horsemen, and often found themselves outmatched in a way that their European counterparts would not.

Last but not least, there was an unforeseen advantage in European musketry, which gave way to an actual military divergence eventually – and unexpectedly at that. In East Asia, the same infantrymen who trained in marksmanship were also volley-firers. Therefore, they were on average more lethal, well-rounded, and versatile than the regular European soldier in the line.[116] Yet, while this may have been an East Asian edge through the seventeenth and eighteenth centuries, it was no longer so in the nineteenth century. In Europe, because light infantry remained separate from the line infantry, there was a greater compass for specialization of sharpshooting. Initially, due to the neglect in Europe for marksmanship, this separation rendered a European "backwardness" in terms of accuracy. But as interest in light infantry tactics grew in the late eighteenth century, Europeans benefited from, if you will, an "advantage of backwardness," to borrow the words of Kenneth Pomeranz: when formally systematized in the military, sharpshooting units acquired their own institution of practice ammunition and target shooting – which was more generous than that of their average peers, and reaped windfall gains from the advanced rifling technologies of Europe.[117] This small, unforeseen

---

[113] Kang, "Big Heads," 148–156.
[114] Kang, "Big Heads," 149.
[115] Kang, "Big Heads," 148–156. Both the Koreans and Japanese increased the proportion of musketeers and drilled them harder after their qualified successes against cavalry. For more, see No, "16–17 segi choch'ong," 121.
[116] In a sense, they resembled the well-trained light infantry corps of nineteenth-century Europe, who operated both as specialist sharpshooting units and as regulars firing volleys in the line.
[117] Conversely, early military rifles – given their extremely slow reloading process – may have rendered them inutile for the East Asian musketeer who not only had to aim carefully but volley fire

(dis)advantage in European musketry, then, grew further in the nineteenth century when bouts of warfare brought standards of musketry accuracy in Europe up to those in East Asia and beyond. This moment of Western ascendancy was also the time when the Age of Parity finally gave way to the Great Military Divergence.[118]

## Conclusion

This chapter has appreciated, if you will, "difference in an age of parity," a conundrum represented by the disparate ways in which the military musket was used and produced in Europe and East Asia, as well as their nonetheless comparable levels of innovation in bringing this artifact to bear on early modern warfare. As demonstrated, speed took precedence in Europe and accuracy in East Asia. Yet, rather than being mutually exclusive, mass firing and marksmanship were also deployed simultaneously in both regions: European skirmishers aimed, and East Asian sharpshooters volley-fired. The difference in musketry, then, stemmed not from an incommensurable gap (or "divergence") in drill and training – as suggested by Kubota and Kim, but from a certain "interpretation" of technology that shaped (and was being shaped by) its surrounding context, social and cultural as well as political, economic, and aesthetic. In other words, a military and technological parity did prevail between Europe and Asia during the early modern period, but with intriguing differences that we have only begun to understand.

In closing, then, I present two further directions for investigation, with an eye for moving the field towards a truly comparative study of technology between Europe and East Asia. First, military historians should take to heart initial charges by their colleagues in the history of technology and much more. In the early responses to the theses by Roberts and Parker, Bert Hall and Kelly Devries challenged the field to consider the social shaping of technology, but the field of the history of technology has since further flourished, and we have ample sources to draw on. Rather than replacing technological determinism with social constructivism – arguably, two sides of the same coin, historian of military science and technology Antoine Bousquet, for instance, makes a compelling case to use "assemblage theory," which embeds technology in a "network" of relations between society, artifact, culture, politics, and so on.[119] Another voice in the field –

---

118 Andrade, *The Gunpowder Age*, 5.
119 Bousquet, "A Revolution in Military Affairs?"; Bousquet, "Welcome to the Machine."

and drawing on a case study of guns in Central Africa, Giacomo Macola argues instead for grasping the "interpretive flexibility" of technologies when they move between contexts: he draws attention to changes made in the physical alterations of the hardware, as well as the ideologies and values that shape its use.[120] We can move towards a truly comparative history of technology, only after building on these rich works. For the purposes of this paper, this will also help answer questions that remain unresolved: Why did European militaries give up on accuracy until the nineteenth century? What explains the difference in East Asia?

But historians of technology can also learn from global military historians. The global turn in military history has changed that field with a revisionist force that has yet to be seen in the history of technology. As exemplified by works by Tonio Andrade and others, the former has been broken open by new comparative data, and East Asia, for one, is as central to the study of the Military Revolution as Europe. Of course, this, too, has room to grow: as other global historians quip, we need to move beyond comparative studies that shuttle between the opposite ends of Eurasia and consider the space in between – or to grasp what Ander Gunder Frank has called "The Centrality of Central Asia."[121] A truly comparative history of technology, then, would immerse in globalist discourses and data – even when its main thrust may still be anchored in a particular locale, or a juxtaposition of Eurasian poles – and bring such expansive views to understand the different uses to which machines and techniques are put in different places. Again, rephrased in the context of this paper, we should revise the centrality of European data in the formation of our understanding about early modern firearms, and their technical parameters: if the East Asian data on marksmanship stands, then, what of the histories of technology that quote the laws of physics to deny possibilities of smoothbore accuracy? It is indeed time now to interrogate European history with non-European standards and to generate new comparative insights that change our understanding of both: How, for instance, should we rethink the rise of ballistic science in eighteenth-century Europe, in light of the differences reported in this study concerning East Asia?

---

120 Macola, *The Gun in Central Africa*.
121 Smith, "Nodes of Convergence," 16.

# Bibliography

Andrade, Tonio, Hyeok Hweon Kang, and Kirsten Cooper. "A Korean Military Revolution? Parallel Military Innovations in East Asia and Europe." *Journal of World History* 25.1 (2014): 51–84.

Andrade, Tonio. "The Arquebus Volley Technique in China, c. 1560: Evidence from the Writings of Qi Jiguang." *Journal of Chinese Military History* 4 (2015): 115–141.

Andrade, Tonio. *The Gunpowder Age: China, Military Innovation, and the Rise of the West in World History*. Princeton: Princeton University Press, 2016.

Anzai Minoru安斎實. *Hōjutsuka no seikatsu: zōho Ensei buki zuryaku Seiyō heigaku kunmō* 砲術家の生活: 増補遠西武器図略, 西洋兵学訓蒙. Tokyo: Yūzankaku Shuppan 雄山閣出版, 1989.

Paul Varley. "Oda Nobunaga, Guns, and Early Modern Warfare." In *Writing Histories in Japan: Texts and their Transformations from Ancient Times Through the Meiji Era*, edited by James C. Baxter, and Joshua A. Fogel, 105–125. Kyoto: International Research Center for Japanese Studies, 2007.

Bousquet, Antoine. "A Revolution in Military Affairs? Changing Technologies and Changing Practices of Warfare," In *Technology and World Politics: An Introduction*, edited by Daniel R McCarthy, 165–182. London: Routledge, 2017.

Bousquet, Antoine. "Welcome to the Machine: Rethinking Technology and Society Through Assemblage Theory," In *Reassembling International Theory*, edited by Michele Acuto, and Simon Curtis, 91–97. London: Palgrave Macmillan UK, 2014.

Börekçi, Günhan. "A Contribution to the Military Revolution Debate: The Janissaries Use of Volley Fire during the Long Ottoman-Habsburg War of 1593–1606 and the Problem of Origins." *Acta Orientalia* 59.4 (December 2006): 407–438.

Brown, Delmer M. "The Impact of Firearms on Japanese Warfare, 1543–98," *The Far Eastern Quarterly* 7.3 (May 1948): 236–253.

Carmen, W.Y. *A History of Firearms: From Earliest Times to 1914*. London: Routledge, 2016.

Chandler, David. *Dictionary of the Napoleonic Wars*. Hertfordshire: Wordsworth, 1999.

Chang Chŏngsu. "17 segi chŏnban Chosŏn ŭi p'osu yangsŏng kwa unyŏng" 17세기 전반鮮의 砲手 養成과 運用. Ph.D. dissertation, Korea University, 2010.

Charney, Michael. *Southeast Asian Warfare, 1300–1900*. Leiden: Brill, 2004.

Chase, Kenneth. *Firearms: A Global History*. Cambridge: Cambridge University Press, 2003.

Chŏng Yŏnsik 정연식. "Hwasŏng kongsimdon ŭi yurae wa kinŭng" 화성 공심돈의 유래와 기능. *Yŏksa Hakpo* 歷史學報169 (2001): 1–35.

Chŏng Yŏnsik 정연식. "Hwasŏng ŭi pangŏ sisŏl kwa ch'ongpo" 화성의 방어시설과 총포. *Chindan Hakpo* 震檀學會91 (2001): 133–161.

Douglas, Howard. *A Treatise on Naval Gunnery*. London: J. Murray, 4th ed., 1855.

Elting, John R. *Swords around a throne*. London: Weidenfield & Nicholson, 1999.

Equiluz, Martin de. *Milicia Discurso, y Regla Militar, del Capitan Martin de Eguiluz, Bizcayno*. Antwerp: Casa Pedro Bellero, 1595 [1586].

Firth, Charles H. *Cromwell's army: a history of the English soldier during the civil wars, the commonwealth, and the protectorate*. London: Greenhill Books, 1992 [1902].

Fissel, Mark Charles. "Military Revolutions." In *Oxford Bibliographies in Military History*, edited by Kaushik Roy. New York: Oxford University Press, https://www.oxfordbibliographies.com/view/document/obo-9780199791279/obo-9780199791279-0212.xml.

Gommans, Jos. *Mughal Warfare*. London: Routledge, 2002.
Grant, James. *The Cavaliers of Fortune, or, British Heroes in Foreign Wars*. London: Routledge, Warnes and Routledge, 1859.
Grossman, Dave. *On Killing: The Psychological Cost of Learning to Kill in War and Society*. Boston: Little, Brown, 1996.
Hall, Bert. *Weapons and Warfare in Renaissance Europe: Gunpowder, Technology, and Tactics*. Baltimore: Johns Hopkins University Press, 1997.
Hall, Bert S., and DeVries, Kelly. "The 'Military Revolution' Revisited." *Technology and Culture* 31.3 (July 1990): 500–507.
Hanger, George. *Reflections on the menaced invasion and the means of protecting the capital by preventing the enemy from landing in any part contiguous to it*. London: Printed for John Stockdale, 1804.
Haythornthwaite, Philip J. *Weapons and Equipment of the Napoleonic Wars*. Blandford Press, 1979.
Hŏ Sŏndo. *Chosŏn sidae hwayak pyŏnggisa yŏn'gu* 朝鮮時代火藥兵器史研究. Seoul: Ilchogak, 1994.
Hoffman, Philip T. *Why Did Europe Conquer the World?* Princeton: Princeton University Press, 2015.
Hora Tomio 洞富雄, *Teppo denrai to sono eikyo: Tanegashimaju zohoban* 鉄砲伝来とその影響: 種子島銃増補版. Azekura Shobo 校倉書房, 1959.
Hughes, B.P. *Firepower: Weapons Effectiveness on the Battlefield, 1630–1850*. New York: Sarpedon, 1997.
Ishioka Hisa 石岡久夫. *Naganumaryū heihō* 長沼流兵法. Tokyo: Jinbutsuōraisha 人物往来社, 1967.
Jacob, Frank, and Gilmar Visoni-Alonzo. *The Military Revolution in Early Modern Europe: A Revision*. London: Springer, 2016.
Jomini, Antoine H. *The Art of War: Restored Edition*. Kingston, Ont: Legacy Books Press, 2008.
Kang Hyeok Hweon. "Big Heads and Buddhist Demons: The Korean Musketry Revolution and the Northern Expeditions of 1654 and 1658." *Journal of Chinese Military History* 2 (2013): 127–189.
Kang Seok Hwa. "A Study on the Assimilation of Qing Military Techniques in Chosŏn during the Seventeenth and Eighteenth Centuries." In *Space and Location in the Circulation of Knowledge (1400–1800): Korea and Beyond (Research on Korea)*, edited by Marion Eggert, Felix Siegmund, Dennis Würthner. Peter Lang GmbH, Internationaler Verlag der Wissenschaften, 2014.
Kim Yŏngjung 김영중. "17 segi Chosŏn ŭi p'osu hullyŏn pangsik kwa kunyul" 17C 조선의 포수 훈련방식과 군율. *Kunsa* 92 (September 2014): 167–197.
Krenn, Peter, Paul Kalaus, and Bert Hall. "Material Culture and Military History: Test-Firing Early Modern Small Arms." *Material History Review* 42 (1995): 101–109.
Kuba Takashi久芳崇. *Higashi Ajia no heiki kakumei: jūrokuseiki Chūgoku ni watatta Nihon no teppō* 東アジアの兵器革命: 十六世紀中国に渡った日本の鉄砲. Tokyo: Yoshikawa Kōbunkan, 2010.
Kubota Masashi. 久保田正志. *Nippon no gunji kakumei*日本の軍事革命. Tokyo: Kinseisha, 2008.
Lavin, James D. "A Study of Spanish Firearms." Ph.D. diss., The Florida State University, 2017.
Lorge, Peter A. *The Asian Military Revolution: From Gunpowder to the Bomb*. Cambridge: Cambridge University Press, 2008.

Lugs, Jaroslav. *A History of Shooting: The Development of Target Guns, Shooting Ranges and Rifle Associations: A History of Duelling and Exhibiton Shooting: Magic and Superstition*. Feltham: Spring Books, 1968.
Macola, Giacomo. *The Gun in Central Africa: A History of Technology and Politics*. Athens: Ohio University Press, 2016.
Monck, George. *Observations upon Military & Political Affairs*. London: Printed by A.C. for Henry Mortlocke and James Collins, 1671.
Morillo, Stephen. "Guns and Government: A Comparative Study of Europe and Japan." *Journal of World History* 6.1 (1995): 95–100.
Muir, Rory. *Tactics and the Experience of Battle in the Age of Napoleon*. New Haven: Yale University Press, 1998.
Müller, Wilhelm. *The Elements of the Science of War: Containing the Modern, Established, and Approved Principles of the Theory and Practice of the Military Sciences . . . Illustrated by Seventy-five Plates, on Artillery, Fortification, &c. and Remarkable Battles Fought since the Year 1675, for the Use of Military Schools and Self-instruction*. Longman, Hurst, Rees, Orme and Etc., 1811.
No Yŏnggu 노영구. "Chosŏn hugi pyŏngsŏ wa chŏnbŏp ŭi yŏn'gu" 朝鮮後期兵書와 戰法의 研究. Ph.D. dissertation, Seoul National University, 2002.
No Yŏnggu 노영구. "16–17 segi choch'ong ŭi toip kwa Chosŏn ŭi kunsajŏk pyŏnhwa", 16–7 세기 조총의 도입과 조선의 군사적 변화. *Han'guk munha* 58 (November 2012): 111–137.
No Yŏnggu 노영구. "Kihoek nonmun: chŏnjaeng ŭi sidaejŏk yangsang; 'kunsa hyŏngmyŏngron (Military Revolution)' kwa 17~18 segi Chosŏn ŭi kunsajŏk pyŏnhwa" 기획논문 전쟁의 시대적 양상: '군사혁명론 (Military Revolution)'과 17~18 세기 조선의 군사적 변화. *Sŏyangsa yŏngu* 5.5 (2007): 39–43.
*On-saki teppo kakucho* [Score Sheets for Bakufu Musketeers]. 1849/41/5, 1850/5/10, 1853/ 5/ 19.MSS. H-60-12-88/89, H-60-12-77, H-60-12-227. Honda Family Documents. National Museum of Japanese History, Sakura, Chiba.
Parker, Geoffrey. *The Military Revolution: Military Innovation and the Rise of the West*. Cambridge: Cambridge University Press, 1996.
Peterson, Harold L., and Robert Elman. *The Great Guns*. London: Hamlyn Publishing Group, 1971.
Peterson, Harold L. *Encyclopedia of Firearms*. New York: Dutton, 1967.
Picard, Ernest, and Paul Azan. *La campagne de 1800 en Allemagne*. Paris: Chapelot, 1907.
Pye, Henry James. *The Sportsman's Dictionary; Containing Instructions for Various Methods to be Observed in Riding, Hunting, Fowling, Setting, Fishing. The Management of Dogs, Game. And the Manner of Curing Their Various Diseases and Accidents*. London: J. Stockdale, 1807.
*Pyŏnghak chinam chuhae* 兵學指南註解 [*Annotations on the Guide to the Military Arts*]. National Library of Korea, Seoul, South Korea.
Qi Jiguang 戚繼光. *Ji xiao xin shu: shi si juan ben* 紀效新書: 十四卷本. Edited and annotated by Fan Zhongyi 范中義. Beijing: Zhonghua shuju, 2001.
Quimby, Robert. *The Background of Napoleonic Warfare*. New York: Columbia University Press, 1957.
Raudzens, George. "War-Winning Weapons: The Measurement of Technological Determinism in Military History." *The Journal of Military History* 54.4 (Oct. 1990): 403–434.

Roberts, Michael. "The Military Revolution, 1560–1660." Inaugural Lecture Delivered before the Queen's University Belfast. Belfast: M. Bond, 1956, reprinted in *The Military Revolution Debate*, edited by Clifford J. Rogers, 13–35. Boulder: Westview Press, 1995.

Robins, Benjamin, and W.S. Curtis. *New Principles of Gunnery*. Richmond [Surrey]: Richmond Publishing Co. Ltd, 1972; originally published: London: J. Nourse, 1742.

Rogers, Clifford J., ed. *The Military Revolution Debate*. Colorado: Westview Press, 1995.

Rogers, John M. "The Development of the Military Profession in Tokugawa Japan." Ph.D. Diss., Harvard University, 1998.

Rothenberg, Gunther E. *The Art of Warfare in the Age of Napoleon*. Chalford: Spellmount, 2007.

Saito Tsutomu齋藤努, Takatsuka Hideharu高塚秀治, Udagawa Takehisa 宇田川武久. "Hihakai bunseki niyoru teppō jūshin no zaishitsu to seisaku gihō no kaiseki" 非破壊分析による鉄砲銃身の材質と製作技法の解析. Kokuritsu Rekishi Minzoku Hakubutsukan kenkyū hōkoku 国立歴史民俗博物館研究報告 136 (2007): 237–265.

Sasama Yoshihiko笹間良彦. *Kakyū bushi ashigaru no seikatsu*下級武士足軽の生活. Tokyo: Yūzankaku雄山閣, 1991.

Smith, Pamela H. "Nodes of Convergence, Material Complexes, and Entangled Itineraries." In *Entangled Itineraries: Materials, Practices, and Knowledges Across Eurasia*, edited by Pamela H. Smith, 5–24. Pittsburgh: University of Pittsburgh Press, 2019.

Smythe, Sir John. *Certain Discourses Military*, edited by J.R. Hale. Ithaca: Published for the Folger Shakespeare Library by Cornell University Press, 1964.

Sherer, Idan. *Warriors for a Living: The Experience of the Spanish Infantry in the Italian Wars, 1494–1559*. Leiden: Brill, 2017.

Siegmund, Felix. "Circulation of Military Knowledge and its Localization. Some notes on the Case of Military Techniques in Late Chosŏn Korea." In *Space and Location in the Circulation of Knowledge (1400–1800): Korea and Beyond*, edited by Marion Eggert; Dennis Wrthner. Frankfurt/Bern: Peter Lang GmbH, Internationaler Verlag der Wissenschaften, 2014.

Siegmund, Felix. "The Adaptation of Chinese military techniques to Chosŏn Korea, their validation, and the social dynamics thereof." In *Civil-Military Relations in Chinese History: From Ancient China to the Communist takeover*, edited by Kai Filipiak. London/New York: Routledge, Taylor & Francis Group, 2015.

Sin Yu 申瀏. *Pukchŏng ilgi*北征日記, translated by Pak Tae-gŭn 朴泰根. Kyŏnggi-do, Sŏngnam-si: Han'guk chŏngsin munhwa yŏn'guwŏn, 1980.

*Sŏnjo sillok* 宣祖實錄 [Veritable Records of Sŏnjo's Reign]. Edited by Kuksa p'yŏnch'an wiwŏnhoe. Kuksa p'yŏnch'an wiwŏnhoe, 1955–1958.

Stephenson, Michael. *The Last Full Measure: How Soldiers Die in Battle*. London: Duckworth Overlook, 2016.

Stravos, Matthew. "Military Revolution in Early Modern Japan" *Japanese Studies* 33.3 (2013): 243–261.

Sugawa Shigeo. *The Japanese Matchlock: A Story of the Tanegashima*. Published by the author in Tokyo; distributed by Kogei Shuppan, 1991.

Sun Laichen. "Military Technology Transfers from Ming China and the Emergence of Northern Mainland Southeast Asia (c. 1390–1527)." *Journal of Southeast Asia Studies* 34.3 (2003): 495–517.

Sung Ying-Hsing. *T'ien-kung K'ai-wu; Chinese Technology in the Seventeenth Century*, translated by Sun, E-Tu Zen, Sun, Shiou-Chuan. University Park: Pennsylvania State University, 1966.

Swope, Kenneth. *A Dragon's Head and a Serpent's Tail: Ming China and the First Great East Asian War, 1592–1598*. Norman: University of Oklahoma Press, 2009.

Udagawa Takehisa. *Edo no hōjutsu: Keishōsareru bugei* 江戶の炮術: 継承される武芸. Tokyo: Tōyō Shorin, 2000.

von Friedrich Wilhelm Ludolf Gerhard Augustin Steuben, Baron. *Regulations for the Order and Discipline of the Troops of the United States*. New Hampshire: Exeter, printed and sold by Henry Ranlet, 1794.

von Pivka, Otto. *Armies of the Napoleonic Era*. Newton Abbot: David & Charles, 1979.

von Scharnhorst, Gerhard Johann David. *Results of Artillery and Infantry Gun Trials*, translated by Bill Leeson. Hemel Hempstead: B. Leeson, 1992.

Wells, William R. *Shots that Hit: A Study of U.S. Coast Guard Marksmanship, 1790–1985*. Washington, D.C.: U.S. Coast Guard, Historian's Office, 1993.

Wilson, Guy M. "Some Important Snap Matchlock Guns," *Canadian Journal of Arms Collecting* 26.1 (1988): 3–10.

Wong, R. Bin. "Tonio Andrade. The Gunpowder Age: China, Military Innovation, and the Rise of the West in World History." *The American Historical Review* 122.2 (2017): 464–467.

Yerxa, Donald A., ed. *Recent Themes in Military History: Historians in Conversation*. Columbia: University of South Carolina Press, 2008.

Yi Chinho, *Chinpŏp ŏnhae chuhae* 陣法諺解 註解. Seoul: Hyŏngsŏl Chulpansa, 2011.

*Yungwŏn p'ilbi* 戎垣必備 [Essential Weapons for the Commander]. Special Manufactory, 1813, Toyo Bunko Collection, VII-3-127.

Zhmodikov, Alexander, and Yuri Zhmodikov. *Tactics of the Russian Army in the Napoleonic Wars*, Vol. 2. West Chester: The Nafziger Collection, 2003.

Aliaksandr Kazakou
# Gunpowder Revolution in the East of Europe and the Battle of Orsha, 1514

On 14 July, 1500, the Grand Duchy of Lithuania's army led by the Superior *Hetman* (the chief commander), Prince Konstantin Ostrozhsky, suffered a crushing defeat on the river Vedrosha while attempting to prevent yet another Muscovite invasion. Ostrozhsky marshalled 3,500 to 4,000 cavalry along with an unknown number of infantry. Considering that foot soldiers played no significant role in the battle, most likely their numbers were few. The victorious army under command of Daniil Shchenya likewise consisted mostly of cavalry, and outnumbered their enemy.[1] However, the Muscovites achieved success not only due to superior numbers and their skilful tactics, but also because of mistakes made by the Lithuanian army. The troops of the Great Duchy of Lithuania were almost entirely either killed or captured. The defeat in the battle of Vedrosha cost the *hetman* himself several years of freedom, until he managed to escape in 1507. Seven years later, heralded by the sounds of cannon and infantry fire, Ostrozhsky took revenge for his bitter defeat. This time his forces complied far better with the most advanced military practices of the time, and the *hetman* performed as a masterful tactician.

These events occurred within a crucial era for the development of warfare, both in and outside Europe. The end of the fifteenth century and the following decades experienced significant changes in the use of firearms, tactics, siegecraft and fortification, and military organization. Scholars of the history of warfare from the perspective of military revolution and its critics have discussed this period voluminously. Here I deliberately avoid a broad discussion of methodological problems within the concept of military revolution (revolutions) or of revolutions in military affairs because the relevant methodology and historiography are expounded upon elsewhere in this volume. I will only note that although this approach has been criticized by one group of scholars, or corrected by another group, or rejected by a third, it has endured and provided a powerful impulse to military history as a branch of historical science. In Western, or to be precise, Anglophone historiography (if such a term may be applied), this highly developed theory still constitutes a hard to ignore, perhaps even dominant, methodological paradigm.[2]

---

[1] Alekseev, "Kampaniya 1500 goda v svete noveyshikh issledovaniy" ['1500 Campaign'], 326.
[2] For one of the latest works touching on the questions of military revolution, see Garza, *The Mughal Empire at War*. For a thematic bibliography, see Fissel, "Military Revolutions."

The end of the fifteenth century witnessed the eruption of a series of conflicts known as the Italian Wars (1494–1559). At that time the Apennine peninsula became a field for experiments conducted by military leaders of France, Spain, the Holy Roman Empire, and the Italian city-states. Firearms laid the foundation for multiple innovations. Cannon which the king of France, Charles VIII, brought to Italy in 1494 greatly impressed contemporaries. The quality of the ordnance employed by the French was significantly different from that they had used to seize the English fortresses in the last phase of the Hundred Years' War. In Italy, the French fired iron balls from cast bronze cannon whose barrels could withstand a far more powerful gunpowder charge than cast iron ordnance. Their iron projectiles were three times as dense as those made of stone.[3] All this significantly increased the destructive power of artillery. Moreover, the French ordnance became more mobile due to the use of improved gun carriages, whilst cannon crews were better trained. These factors allowed the artillery to keep pace with the rest of the army, to quite rapidly deploy for battle, and (astonishing to Italian observers) – destroy stone fortifications with unprecedented speed. Significantly, the walls of the Neapolitan castle of Monte San Giovanni which had earlier withstood a seven-year siege were destroyed within eight hours.[4] Although the amount of French artillery and its success in that campaign were somewhat exaggerated,[5] still it demonstrated a new level of possibility. The Pyrenean peninsula was another theater of warfare during the end of the fifteenth century where cannon played a noticeable role. The "Catholic Monarchs" – King Ferdinand of Aragon and Queen Isabella of Castile – as a consequence of the war against the Emirate of Granada in 1481 to 1492 – finished the *Reconquista*. Gunpowder artillery became the primary means of seizing Muslim fortresses, although projectile weapons of medieval times were employed as well. This success has inspired discussion about the "cannon conquest" of Granada.[6]

There were two ways to resist the enhanced power of siege artillery: upgrading fortifications and equipping defensive positions with enough firearms of various calibers to enable counter-fire against the besiegers. Although the idea of lower sloped walls had originated in the middle of the fifteenth century, the famous *trace italienne*, to which Geoffrey Parker ascribes an important role in his concept of military revolution, started to spread in the 1530s.[7] Until bastioned forts made long lasting sieges more ubiquitous, defenders were well advised to

---

3 Hall, *Weapons and Warfare*, 93–94.
4 Duffy, *Siege Warfare*, 9.
5 Pepper, "The Face of the Siege," 36–40.
6 Cook, "The cannon conquest of Nasrid Spain and the end of the Reconquista," 43–70.
7 Parker, *The Military Revolution*.

defeat the enemy in the field before they could attack fortified positions. As a result, pitched battles erupted frequently during the Italian wars. The intensity of field battles promoted "cultural exchange" between the military systems of different countries and encouraged active tactical experimentation.[8] Although cannonry had already been successfully used in the field during the Hundred Years' War (for example, the battle of Castillon in 1453), during the Italian Wars era artillery played a more decisive and more noticeable role as field artillery. The most striking example is the battle of Ravenna (1512), when adversaries fired cannon at each other over a period of two hours, causing considerable damage to both sides. As Sir Charles Oman observes, the battle of Ravenna "was the first example of a new sort of tactics."[9]

Improvements in hand firearms in the second half of the fifteenth century led to the introduction of the harquebus. The word itself derived from the German *Hackenbüchse*, which described a small caliber weapon with a hook.[10] The hook was used to mount the weapon on city wall parapets or other fortifications in order to soften the recoil. German cities which sought to protect their walls but did not need massive cannon to attack their enemies' fortifications were particularly successful in spreading that type of firearm. While in the fifteenth century, harquebuses mostly remained a means to protect fortifications, the era of the Italian Wars witnessed the mass employment of harquebusiers in pitched battles.

The major turning points are considered to be the battle of Cerignola in 1503 and the battle of Pavia in 1525. While at Cerignola harquebusiers were employed under the cover of field fortifications for the first time, at Pavia they were already assuming an offensive role, using trees and bushes as cover.[11] By the middle of the sixteenth century, infantry armed with hand firearms became an essential element for the armies of numerous states, and their importance on the battlefield increased proportionally. Pierre Terrail, *seigneur de Bayard*, "the knight without fear and beyond reproach," slain by a bullet at the battle of river Sesia in 1524, personifies the change in this era. However, use of heavy cavalry did not decline for some time; moreover, the main reason for that decline was not so much the expansion of infantry pike and shot tactics, but the use of pistols as weapons of choice in close combat.[12] The directions of development outlined in Western

---

[8] Hall, *Weapons and Warfare*, 165.
[9] Oman, *A History of the Art of War in the Sixteenth Century*, 138.
[10] In the Slavic world, including Eastern Europe, such a type of firearm was called *hakownica*, corresponding to the English term *hackbut*; however, in Muscovy a different term *zatinnaya piszchel* was used.
[11] Hall, *Weapons and Warfare*, 167–169, 180–183.
[12] Hall, *Weapons and Warfare*, 195–197.

Europe in the late fifteenth and early sixteenth centuries became dominant in the years that followed.

What innovations occurred in the east of Europe in the meantime? This region has been neglected by Western historians studying warfare in the early modern era, including those who have participated in, and are still participating in, the discussion on military revolution. Jeremy Black avoids the "Western-centric" view, adopting some skepticism towards ideas of military revolution and technological determinism. In his opinion, neither the implementation of firearms per se became the cause of important changes in military affairs in the West, nor did it trigger a severe breakage of military culture elsewhere.[13] Researchers who peered into the east of Europe tended to focus on Muscovy, mostly during the period of the second half of the sixteenth century and later.[14] Two scholars in particular explored military history in the region: Robert Frost researched the Northern Wars (1558–1721)[15] and Brian L. Davies, who focused on the Black Sea region.[16] The latter also edited a volume of essays dedicated to the study of Eastern Europe.[17] These authors made notable observations on the development of military affairs in Eastern Europe, and encouraged further research.

Furthermore, a massive literature covering diverse aspects of the region's military history exists in the national languages of corresponding countries and within national historiographical traditions. These works share a common trait of almost entirely avoiding both the problematics of military revolution or of adopting a comparative approach which would address events in a pan-European (or broader) context.[18] This essay aims to fill that gap to a modest extent. It attempts to summarize our understanding of the tactical use of firearms by the troops of three countries: Poland, the Grand Duchy of Lithuania, and the Grand Duchy of Muscovy, from the late fifteenth century to the 1530s. It explores how successful the gunpowder revolution that took place there was, and how the tactical

---

**13** Black, *European Warfare*. Recent geographical spread of this research to Asia is worth mentioning. For instance, see Andrade, *The Gunpowder Age*; Andrade, Kang, and Cooper, "A Korean Military Revolution?" 51–84; Garza, *The Mughal Empire*.
**14** I will cite just a few works: Gustave, "Muscovite Military Reforms," 73–108; Hellie, "Warfare, changing military technology," 74–99; Poe, "The military revolution," 247–273; Lohr, and Poe, ed, *The Military and Society in Russia*.
**15** Frost, *The Northern Wars*.
**16** Davies, *State, Power and Community*; Davies, *Warfare, State and Society*; Davies, *Empire and Military Revolution*.
**17** Davies, ed., *Warfare in Eastern Europe*.
**18** The following works can be named as positive examples: Penskoy, *The Great Gunpowder Revolution*; Mleczek, "Armia koronna"; Gawron, "Wojskowość zachodnioeuropejska"; Bołdyrew, "Przemiany," 113–138.

use of firearms evolved in that geographical context. The essay also seeks to compare these developments with similar processes which took place simultaneously in Western Europe and the Near East. The analysis emphasizes the Grand Duchy of Lithuania which, unlike its more noticeable neighbors, historians tend to neglect, and the battle of Orsha in 1514, which brilliantly manifested the transformed role of firearms in the region.

## War and Armies in Eastern Europe

**Figure 1:** Eastern Europe c. 1522, by Dzmitry Vitsko.

By the end of the fifteenth century, Poland was a relatively formidable state on the eastern border of the Catholic world. The State of the Teutonic Order, one of Poland's major historical adversaries, had ceased to be a serious threat and in 1525 became a Polish vassal. By 1490, various representatives of the Jagiellonian dynasty ruled Hungary, Bohemia, Poland, and the Grand Duchy of Lithuania. In the southeast, Polish Jagiellonians struggled against Moldavia over *Pokucie* territory. The southern regions of the Polish kingdom in the first half of the fifteenth century had already suffered attacks by Crimean Tatars. Both Moldavia and Crimea became dependent upon the Ottomans and were backed by them. Toward the east, Polish foreign policy targeted the Grand Duchy of Lithuania. The Jagiellonian dynasty itself had been founded by the Lithuanian Grand Duke (also called *hospodar* – sovereign) Jogaila, who in 1385 became the king of Poland. The two states were unified from 1447 until 1492 to 1501 when Alexander (1461–1506) became the Grand Duke of Lithuania, and John Albert (1459–1501) ascended the throne of Poland. After the death of Alexander, his brother Sigismund I became both King of Poland and Grand Duke of Lithuania, reigning until 1544 in Lithuania and until 1548 in Poland. For the Polish elite, close political ties with their Eastern neighbor also meant sharing their foreign policy problems.

Until its first territorial losses in the wars with Muscovy in the late fifteenth century, the Grand Duchy of Lithuania was one of the largest states in Europe, incorporating modern day Lithuania, Belarus, the western part of Russia, a large portion of Ukraine, and part of Poland. The lion's share of its territory consisted of Ruthenian (Eastern Slavic) lands with predominately Slavic populations.[19] However, the population density and level of urbanization there were considerably lower than in Poland. English-language literature commonly describes the union of the two countries as the *Polish-Lithuanian Commonwealth*. However, one should not exaggerate their unity. In spite of the essentially Polish and European (to a large extent via Poland) influences and close ties, the Grand Duchy of Lithuania in its internal arrangement and political life, including foreign affairs, remained quite independent. Notably, the Lithuanian elites refused to join Poland in the campaign against Moldavia in 1509.[20] It was said that such a war was not in the interests of the Grand Duchy of Lithuania, which already had enough enemies. Apparently, Lithuania remained aloof from the Polish-Moldavian struggle over *Pokucie*. Relations between Poland and Lithuania remained quite complicated even following their union into *Rzeczpospolita* in 1569.

Two fundamental external threats to the Grand Duchy of Lithuania arose at the end of the fifteenth century. In the south, the Crimean Khanate began regularly harassing its northern neighbors with destructive raids. Complications were compounded after 1478, when the Crimean khans became vassals of powerful

---

**19** One must note that the southern territories close to the Tatars were in reality controlled rather formally.
**20** Halecki, *Dzieje unii Jagiellońskiej*, 49–50.

Ottoman sultans, and after 1480, khan Mengli Gerey became an ally of the Grand Prince of Moscow Ivan III. While Tatars' raids per se led to devastation, rulers of Muscovy claimed Ruthenian lands possessed by Jagiellonians and were ready to take them by violence. Frequent and prolonged conflicts with Muscovy made Lithuania more and more dependent upon Polish military aid, which exploited Lithuania's weakened position to impose Poland's political will on her.

The late fifteenth and early sixteenth centuries were crucial to the establishment of Muscovy dominance in the east of Europe. During the rule of two grand princes – Ivan III (1462–1505) and Vasily III (1505–1533) – the territory of the state expanded several times over. First, other Ruthenian duchies were conquered or subdued one by one, along with the Novgorod and the Pskov Republics. Subsequently most territories conquered were chiseled out of the Grand Duchy of Lithuania. The distance from Moscow to the western border in 1492, slightly above 100 kilometers, had by 1522 grown to 400 kilometers. Such success derived preponderantly from the imposition of an authoritarian model of political order that effectively mandated mobilizing resources in order to perform vital tasks. Moscow, not being an economically prosperous power, owed its rise, firstly, to skilful and determined implementation of far-sighted domestic and foreign policies by its rulers and, secondly, to the military service class.[21]

For Muscovy, as for Lithuania, the Tatar threat was prominent: originally in the form of the Golden Horde (until its final breakdown in 1502), and later in the guise of the Crimean Khanate, which broke its alliance with Muscovy after the death of Ivan III. With the annexation of Novgorod (1478) and Pskov (1510) Muscovite rulers acquired new adversaries in the north-west – the Livonian Order and Sweden. However, Moscow did not seek territorial expansion in that direction, so the border there remained relatively stable. Muscovite-Lithuanian conflicts, which involved Poland to a certain extent, were the focal point of military events in Eastern Europe.[22] During the half-century under study, five wars occurred: 1486 to 1494, 1500 to 1503, 1507 to 1508, 1512 to 1522, and 1534 to 1537. Lithuania and Muscovy were at war with each other approximately 27 years out of 50. Moreover, none of these conflicts ended with a lasting peace treaty; each of the sides (or at least Muscovy and its allies) understood that a formal secession of hostilities only meant a short break before the next war. The main reason for endemic conflict was Moscow's carrying out of its grand strategy – repossession

---

21 Zimin, *Vityaz at the Crossroads*, 191–211.
22 For some works that indicate the relations between the two countries at that time, see Temushev, *The First Muscovy-Lithuanian Frontier War*; Krom, *Between Ruthenia and Lithuania*.

of the lands that once formed the so-called *Kievan Rus* and were considered by the Moscow house of Rurikids to be their lawful possessions. For the Grand Duchy of Lithuania, the net results of these wars were great territorial losses in the east. Its leaders could not fashion an effective strategy to meet the challenge.

What was the military like in the three countries? Any state's army is largely defined by societal structure, by its culture in general, and by the character of the relationships between government and society. The medieval obligation of the local nobility (*szlachta*) to serve militarily when summoned by their monarch persisted both in Poland and in the Grand Duchy of Lithuania (under the respective terms of *pospolite ruszenie* and *zemskaya sluzhba*). Gradually, as limitations of a monarch's power in favor of *szlachta* evolved, the decision to summon *zemskaya sluzhba*, along with resolving other questions, became the responsibility of the *Sejm*. Feudal duty required each landowner to muster arms, on a good horse, leading a band of warriors (*pochet*) in numbers proportionate to the size of his dominions. Therefore, *zemskaya sluzhba* comprised cavalry, except for poor *szlachta* which could not afford horses and mustered as unmounted men. Although the performance of *szlachta* militia was inferior to professional soldiers, the system endured in Poland; moreover, in the Grand Duchy of Lithuania the system's influence remained prominent even longer.[23] It must be acknowledged, however, that early in the sixteenth century, *szlachta* could perform quite well, as demonstrated in the victorious battle against the Tatars at Kletsk in 1506.

Throughout the fifteenth century, mercenaries steadily participated in Polish wars. In cavalry, the tendency was to gradually increase the substitution of foreign soldiers of fortune for locals, while for the infantry the conscription ratio remained 1:1. However, already in the first half of the sixteenth century, Poles prevailed among the foot soldiers.[24] As with armies elsewhere in the region, cavalry dominated both in terms of quantity and impact on the battlefield. Unlike in Western Europe, in Poland during the first half of the 1500s, infantry continued to play a secondary role,[25] although in some campaigns and battles the infantry's contribution was substantial. For instance, during the last war with the Teutonic Order from 1519 to 1521, foot soldiers were quite numerous and actively employed by the Polish command. Admittedly, that war was dominated by sieges.[26]

In the fifteenth century, the necessity of protecting the southern borders against Tatar invasions led to the creation of *obrona potoczna*. These salaried

---

23 Łopatecki, *Organizacja*, 11.
24 Grabarczyk, *Jazda*, 14–20; Grabarczyk, *Piechota*, 78; Bołdyrew, *Piechota*, 122.
25 Bołdyrew, *Piechota*, 336.
26 Plewczyński, *Wojny*, 32; Bołdyrew, *Piechota*, 312–319.

soldiers, unlike those hired for a certain campaign, were permanently employed. Due to shortages of ready money, *obrona potoczna* remained small and did not have an established number of men.[27] In 1501, King John Albert increased the unit's size from several hundred men to 2,150, and in 1502 to 1506 it comprised at most a thousand mounted soldiers. From 1513 to 1522, Sigismund kept *obrona potoczna* numbers at the level of 2,000 horsemen. When additional funds became available, the unit's size could significantly increase. For instance, in the Obertyn campaign of 1531, during the war with Moldavia *obrona potoczna* boasted 4,452 cavalry and 1,167 infantry.[28]

Before the beginning of the intense struggle against Tatar invasions and Muscovy's aggression at the end of the fifteenth century, the Grand Duchy of Lithuania, unlike Poland, enjoyed several decades of peace. The last prominent Lithuanian military event of the fifteenth century was the battle of Vilkomir in 1435. A prolonged period of peace is rarely good for the readiness of an army and does not strengthen the marketability of professional soldiers. Lithuania's first recruitment of a sizeable mercenary contingent is dated at 1480, when 6,000 to 8,000 cavalry were hired in Poland for the abortive campaign against Ivan III.[29] Later, the government of the Grand Duchy of Lithuania would consistently hire foreign soldiers to carry out military actions in the east of the country. The weakness of their own professional service market, along with Poland's geographical proximity and the two countries' close political ties, naturally led Lithuania to draw upon their neighbor's human resources, although mercenaries from other countries were also hired, mostly Czech, Hungarian, and German.

The formidable expense of paying foreign professionals, and the intrinsic difficulties with their mobilization caused by physical remoteness, forced the Grand Duchy of Lithuania to recruit internal sources of manpower. According to Gediminas Lesmaitis, the Grand Duchy of Lithuania initiated its professional army early in the sixteenth century. By 1544 it was already well established. In that year the *Sejm* agreed to send 4,000 mounted soldiers to succour the Poles. Paid military service of mounted "courtiers" became one important factor in this development.[30] In 1509, more than 476 were mustered along with their military retainers.[31] In contrast to the soldiers of the *obrona potoczna*, "courtiers" were

---

[27] In my opinion, there is no reason to see a certain analog of a standing army in *obrona potoczna*: Gawron, "Wojskowość," 47.
[28] Plewczyński, *Wojny*, 27–28.
[29] Bokhan, *Military Affairs*, 334; Lesmaitis, *Wojsko*, 71.
[30] Lesmaitis, *Wojsko*, 42–52.
[31] Kazakou, "Muscovites among the courtiers of the Lithuanian grand duke Sigismund the Old," 65.

personal servants of the *hospodar*. In peacetime, they functioned as his representatives in various parts of the country and carried out official tasks. Based on their genesis and typology *hospodar* courtiers possessed some resemblance (even if remote and culturally dissonant) to Ottoman *kapukulu* cavalry, and also to the English king's "spears."[32] During the campaign of 1514, their number reached 1,081, and in 1525 it was planned to send 2,000 of them to defend the southern frontiers.[33] Unquestionably, paying "courtiers" for their service powerfully impacted the prospects for the establishment of genuine mercenary armies. However, in my opinion, the military service of Sigismund's "courtiers" should instead be considered an attempt to create a standing army. During wartime, "courtiers" formed a cohesive military unit; in 1514, for example, every courtier mobilized for war garnered payment for his service.[34] Moreover, during peacetime, courtiers who directly depended on the *hospodar* also received allowance in the form of money, goods, and land. In return, they were expected to provide for themselves and their combat servants. A willing acceptance of immigrants into contingents that were relatively quick to muster could be explained by the system's utility and the need for such a force; in 1509 natives of the Grand Duchy of Muscovy constituted about 21 percent of courtiers.[35] As a detailed analysis of attempts to create a standing army in the Grand Duchy of Lithuania is beyond the scope of this chapter, it should be noted that this topic deserves special study.

## Firearms

In Eastern Europe, firearms appeared more or less simultaneously. The earliest documented mentions date to 1383 in Poland and 1382 in the Grand Duchy of Lithuania.[36] There is information about firearms production in Krakow since at least 1391; however, prior to the creation of a large royal foundry (*ludwisarnia*) in 1516, only individual craftsmen supplied firearms. By 1519, production of bronze cannon began in Gdańsk. Lviv also traditionally functioned as a large center for artillery production. Throughout the fifteenth century, cannons spread across Poland becoming both an inseparable part of castles' arsenals and common in field armies.[37]

---

32 Murphey, *Ottoman Warfare*, 36; Raymond, *Henry VIII's Military Revolution*, 145.
33 Lesmaitis, *Wojsko*, 51; Lulewicz, "Jerzy Mikołajewicz Radziwiłł," 24.
34 Krom, ed., *Acts of Radzivill*, 28–33.
35 Kazakou, "Muscovites," 65.
36 Szymczak, *Początki broni palnej w Polsce*, 13–14; Bokhan, *Military Affairs*, 238.
37 See below on the implementation of firearms.

Notably, toward the end of 1530s, castles exchanged outdated models for modern weaponry, and the number of handheld firearms increased dramatically. The construction of the Royal Arsenal in Krakow, completed in 1533, was a distinctive physical institutional manifestation of this development; a parallel organizational culmination was the appearance of an artillery commander in a field army as early as 1520.[38]

The proliferation of firearms in Poland occurred under a powerful German influence, manifested in the importing of both technology and qualified specialists. Out of 155 artillerists renowned in Poland before the sixteenth century, 72 were German, eight Czech and Moravian, and two Italian.[39] However, in this respect the Polish kingdom resembled a broad range of other states. The "gunpowder empire" of the Ottomans owed a huge debt to Christian military technology transfer, from guns to armor. French, German and Italian specialists contributed greatly to the "cannon conquest" of Granada. Frenchmen in English royal service developed the technology of cast iron cannon production.[40] As shown below, Muscovy also serves as an excellent example of dependence on imported technology.

Along with the larger calibers, there were also *hakownica* and *rusznica* guns in Polish castles' arsenals. *Rusznica* denotes a weapon consisting of a barrel, a wooden butt with gunstock, and a matchlock. It was intended for hand (*ruka*) shooting, hence the name. From the 1530s, the word "harquebus" along with *rusznica* appears in Poland, though it remains unclear exactly what is meant by the term which in Europe applied to the gun known as the *rusznica*.[41] Clear evidence indicates that a harquebus powder charge was twice that of a *rusznica*. According to Alexander Bołdyrew, in Polish documents harquebus refers to a gun with a wheellock and thicker barrel walls than the *rusznica*, allowing it to withstand the explosion of a larger powder charge. As a result, shooting range and accuracy increased.[42] Curiously, at the same time in Italy, a heavier version of the *rusznica* was known by the name of harquebus (*archibugio*), and later became known as a musket. The French in this case could have used the term *arcuebuse à croc*.[43]

---

**38** Szymczak, *Początki*, 17, 87–88, 179–191, 331–332.
**39** Szymczak, *Początki*, 191–196.
**40** Ágoston, *Guns for the Sultan*, 42–48; Cook, "The cannon conquest," 49; Raymond, *Henry VIII's Military Revolution*, 32–33.
**41** Later I will use the term *rusznica* while referring to Eastern Europe, and *harquebus* when referring to the West.
**42** Bołdyrew, *Piechota*, 223–225, 233–235.
**43** Taylor, *The Art of War in Italy*, 40–41; Hall, *Weapons and Warfare*, 176–177.

Polish mercenary infantry were mostly shooters. Typically, a company consisted of crossbowmen and a small number of soldiers with *pavises* who stood in the first row and shielded the whole unit. While other countries increasingly employed pikemen to protect marksmen from attacks in close combat (or, on the other hand, marksmen with firearms supported pike formations), in Poland cavalry carried out this function. In Italy in the fifteenth century, a slightly different pattern developed. Along with marksmen there were assault companies of swordsmen or spearmen, which gave infantry detachments more tactical flexibility.[44] At the turn of the sixteenth century, literally within a few years, Polish infantrymen almost totally abandoned crossbows in favor of the *rusznica*, as evidenced in the registers of infantry companies. Unfortunately, the documentary evidence does not survive in full, preventing us from clearly tracing the process of rearmament. Available sources are sufficient, however, to outline a general picture. Firearms, in the form of a very few *hakownicas* and *piszczels*, are registered in the hands of infantrymen for the first time in 1471.[45] The *piszczel*, a primitive hand weapon without a butt-end and a matchlock, approximately correlates to the English term handgun.[46] Oddly, in 1474 and 1477 *hakownicas* and *piszczels* were absent in the infantry.[47] We do not know what armaments those companies used between 1477 to 1495 because no registers from those years survive. However, by 1496 perhaps 29 percent of shooters had *rusznicas*, a reasonable proportion. In the following years, the numbers increase: 1497 – 54 percent, 1498 – 60.5 percent, 1500 – 83 percent.[48] No data exists for 1501 to 1521, but from 1522 to 1547 of 119 companies only four crossbows were registered, moreover two were in the hands of cavalrymen.[49]

Although the firing speed of the crossbow exceeded that of the *rusznica*, the choice was made in favor of the latter.[50] Among the advantages of the *rusznica*

---

[44] Mallett, and Hale, *Military Organization of a Renaissance State*, 76. On marksmanship, see also the preceding essay by H.H. Kang in this volume.
[45] In a census of companies in 1471, less than one percent of combatants had firearms; Grabarczyk, *Piechota*, 143–144.
[46] Szymczak, *Początki*, 36–41.
[47] There is only one, somewhat unclear, mention of a *bombard* among weapons lost by Bartolomeo's company in 1474. Taking into account an insignificant compensation, it could not have been a cannon; Grabarczyk, *Piechota*, 145.
[48] Grabarczyk, *Piechota*, 146–149.
[49] Bołdyrew, *Piechota*, 219–221. In infantry companies, a small number of horses can be found registered, with one of them ridden by the captain.
[50] In the Polish authors' estimation, charging and firing a *rusznica* could take ten or more minutes, which is bewildering; Szymczak, *Początki*, 45, 303; Grabarczyk, *Piechota*, 152; Plewczyński, *Daj nam*, 52; Plewczyński, *Wojny*, 375. A period of half a minute for a trained soldier is

over the crossbow, Tadeusz Grabarczyck identifies the following reasons: it cost less, was less prone to damage, and its operation required less physical strength. Moreover, firearms could frighten the enemy's horses. The slow firing speed did not play a crucial role. During combat the infantry often had time to fire only one round. If a possibility to fire several times occurred, the shooters needed protection from enemy attack. Possibly this defensive requirement is linked to the appearance of spearmen in infantry companies during the last decade of the fifteenth century, which chronologically coincides with the change of weapons in favor of the *rusznica*. On average, from 1496 to 1500, a company consisted of 10 percent *pavisiers*, 10 percent spearmen, and 80 percent handgunners.[51] In the first half of the sixteenth century, this ratio remained almost the same.[52]

In terms of structure, a company was divided into units of ten, which most frequently consisted of eight soldiers managed by a "headman" (almost always a spearman). Arrayed for battle, *pavisiers* steadied the first line, then there were pikemen, and behind them six files of shooters. A unit of ten lined up into one file, one soldier after another. According to another version, spearmen constituted the first line. The arrangement may have been dictated by the nature of the weaponry of their adversaries.

Prescribed firing methods are unknown. According to the famous Polish commander Jan Tarnowski's work *Consilium rationis bellicae* (1558), the first line fired a volley and knelt down, then the second, and so on. According to another version, the lines fired in the reverse order. The whole company except for the last line knelt down; the last line fired first then the next line got up and fired, and so on. Spearmen and *pavisiers* probably flanked the formation during volley fire. It is difficult to imagine a scenario in which one volley was fired by the whole company while, except for the first one, all the lines of shooters were firing over the heads of their comrades. There is only one source providing evidence of such a way of firing – a painting 'The Battle of Orsha,' discussed below.[53]

Among the cavalry, firearms were known from the end of the fifteenth century, though even in the first quarter of the sixteenth century they were rare. Crossbows – a predominant weapon of mounted shooters of the 1400s – gradually

---

far closer to the reality; Hall, *Weapons and Warfare*, 149. Of course, a unit's real firing speed in combat conditions was not reduced to just recharging personal weapons (because it also depended at least on collective actions) and could have taken up to two to three minutes; Rogers, "Tactics and the Face of Battle," 213.

51 Grabarczyk, *Piechota*, 148–153, 180–186.
52 Bołdyrew, *Piechota*, 65–68.
53 Bołdyrew, *Piechota*, 338–343; Wimmer, *Historia*, 103.

gave way to bows, and then completely disappeared from cavalry armament in the 1520s.[54]

The development of firearms in the Grand Duchy of Lithuania was strongly impacted by foreign influences. Because the country did not deploy any native infantry, inventories of Lithuanian gunpowder weapons involved mostly artillery. Prior to 1500, the types of firearms documented were cannon (increasingly called *delo*), *piszczels, and hakownicas*. Gradually the name *piszczel* began to disappear and *rusznica* became mentioned more frequently. One significant shift in technology was the displacement of older specimens of ordnance that fired stone projectiles by "modern" bronze artillery using iron cannonballs. In general, the number of guns stockpiled in arsenals increased throughout the period under discussion.

Castle inventories (unfortunately only fragmentarily preserved) documented artillery, firearms, ammunition and accoutrements such as carriages. Artillery of predominantly medium and small calibers was prevalent. For example, in 1471 in Vinnitsya there were one cannon and two *piszczels*, in Chudov three cannon and two *piszczels*, in Zhytomyr four big and five small cannon. In 1480, the Kremenets arsenal consisted of four cannon, nine medium caliber weapons of various types, and one *piszczel*. Viciebsk castle in 1508 retained just two cannon out of six that had been sent there a year earlier.[55] Charsvyaty's comparatively rich store of ordnance in 1523 boasted 24 medium caliber guns and 65 *hakownicas* (while lacking even a single large caliber weapon). This artillery was complemented by ample supplies of gunpowder and cannonballs.[56] Kyiv's fine arsenal in 1535 held 34 guns and 59 *hakownicas* (11 classified as outdated).[57] In some instances, direct evidence of artillery being present is unavailable for a given castle but mention of resident gunners suggests cannon were maintained there. For instance, two gunners served at Putivl in 1499.[58] Fragmentary data also survive regarding deliveries of artillery and ammunition to fortresses.[59]

The government monitored the defensive capabilities of the Grand Duke's strongholds and provided them with garrisons and weapons. Cannon were cast in Poland by royal commission and then delivered to the castles. In the Grand Duchy of Lithuania local artillery manufacturing had not yet developed. Sigismund apparently assumed manufacture in Lithuania would not be cost effective

---

54 Grabarczyk, *Jazda*, 83–85; Plewczyński, *Wojny*, 49–57.
55 Bokhan, *Military Affairs*, 241–242; *Lietuvos Metrika* [subsequently *LM*], Kn. 6, 330.
56 Bokhan, *Military Affairs*, 243–244; *LM*, Kn. 11, 146.
57 Bokhan, *Military Affairs*, p. 244; *LM*, Kn. 15, 209.
58 *LM*, Kn. 6, 246.
59 For example, in 1514 *hakownicas*, bullets for them, gunpowder and saltpeter were sent to Smolensk, Polatsk and Krychaū: *LM*, Kn.7, 643–644.

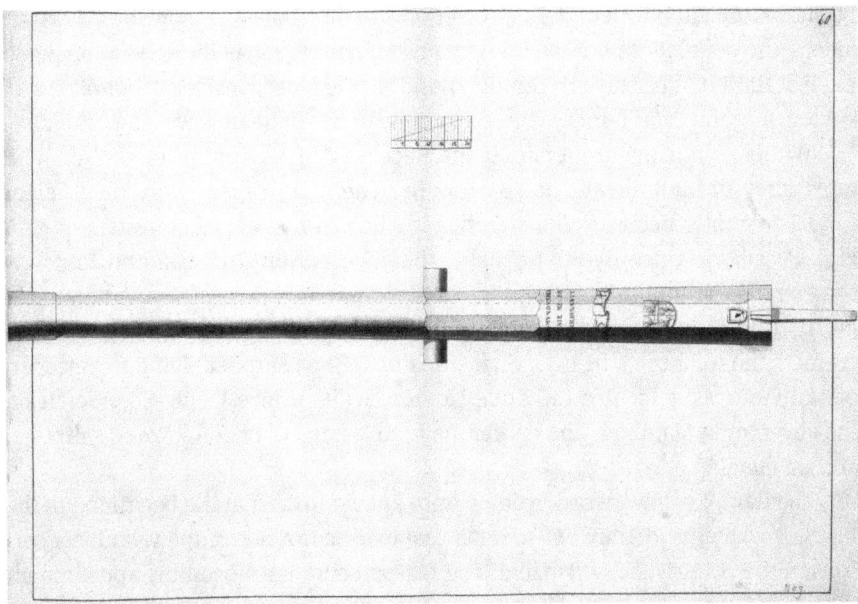

**Figure 2:** View of an artillery piece of 1514 from the collection of drawings of Anna Maria and Philip Jacob Thelott, Armémuseum, Stockholm. The coat of arms indicates that this specimen was cast by the order of the Polish king and Lithuanian grand duke. It may well have been present at Orsha. In 1705, it was decided to make drawings of the multitude of cannons captured by Swedes before melting them down. Anna Marie Thelott and her brother Philipp Jacob made around 1,300 drawings including the three cannons from 1514, in Armémuseum Ritningssamlingen J 3, volume 45, Part 1, 5377, 113; Part 2, 5378, 95; Part 3, 5379, 60. All the drawings are available at the digitaltmuseum.se. *Image by Kjell Hedberg, Armémuseum in Stockholm.*

when Polish foundries could satisfy the demand for ordnance in both of his states. Central government also provided castles with ammunition and gunpowder. The latter could also be produced on the spot by gunners whose main responsibility was maintenance of artillery and commanding cannon fire during rebuffs of enemy attacks.[60] Sometimes documents mention *rusznica* gunpowder in the arsenals, but not *rusznicas*. Evidently, this gunpowder was intended for the mercenary foot soldiers who formed the garrison, or for town citizens who participated in the collective defense. According to a description of the Zhytomyr castle in 1545: "[A]ll landed gentry and town citizens have *rusznicas* and can shoot well, but do

---

60 For example, a Kyiv gunner in 1520 must have produced saltpeter and gunpowder and fired the cannons; *LM, Kn. 10*, 62–63.

not have any gunpowder."⁶¹ This record vividly illustrates the spread of firearms among the population; however, it remains difficult to judge the state of affairs in the first third of the sixteenth century. Besides, one must consider the proximity of Zhytomyr to the Tatar border, where an ability to wield firearms was important.

While authorities understood the necessity of equipping fortresses with guns, they did not always do so properly. The main causes were insufficient funds, corrupt officials, and a significantly flawed management system in general. Possibly, Prince Ilya Ostrozhsky, the governor of Vinnytsia and Bratslav, exaggerated writing to the *hospodar* circa 1538, that he expected a Tatar attack, but had nothing to protect the castles entrusted to him with: no cannon, no gunners, no infantry.[62] In 1533, Sigismund expressed surprise that an inspection of Mahilioū castle located only one cannon, while prince Vasily Solomeretsky, the governor of Mahilioū, had earlier reported lacking only gunpowder, saltpeter, and tin metal.[63]

The first privately owned artillery were known to exist at the beginning of the 1530s. Documents identify two cannon cast to order for one of the wealthiest persons in the country, Yury Radzivil. The reason could have been his appointment to the position of Superior *Hetman*. According to German terminology, the cast weapons appeared to be long-barrelled *shlange* guns.[64] It is possible that Radzivil ordered more guns. In 1535, his *pochet* consisted of 400 mounted soldiers and 200 footmen with 20 cannon, seven described as "large." The same year his ordnance was utilized in the successful siege of Homiel.[65]

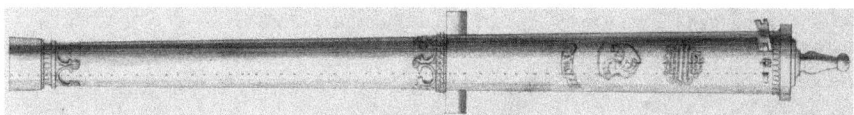

**Figure 3:** Cannon of 1532 from the collection of drawings of Anna Maria and Philip Jacob Thelott, Armémuseum, in Stockholm. The coat of arms "Horns" indicates that this artifact belonged to Yury Radzivil, the superior *hetman* of the Grand Duchy of Lithuania. It is likely that this gun was used in the siege of Homiel in 1535. *Image by Kjell Hedberg, Armémuseum in Stockholm.*

---

61 *Lietuvos Metrika. Knyga 561*, 251.
62 Archiwum Główne Akt Dawnych, Archiwum Radziwiłłów, vol. 2, 3296.
63 Metryka Vialikaha Kniastva Litouskaha, no. 226: 70–72.
64 Volkau, *Niasviž Castle Artillery*, 26.
65 Lesmaitis, ed, *Popisy wojskowe*, 50; Krom, *Acts of Radzivill*, 103; no. 73: 162–163.

What were the armed forces of the Grand Duchy of Muscovy like? Here cavalry absolutely dominated, which was defined by the social foundation of the army. By the late fifteenth century and later on, landowners were still obligated to go to war if so ordered by the Grand Duke. However, *pomestnaya* cavalry, established by Ivan III, assumed ever greater importance. Such horsemen were mustered by *deti boyarskie* – minor landowners who received manors (*pomestie*) from the sovereign with no right of ownership, their tenure wholly dependent on the will of the Grand Duke. The holder of each manor, the *pomeshchik*, had to serve in the military, leading a number of combat servants defined by the size of his landholdings. This arrangement resembles the *timar* system of the Ottomans. In general, according to Vitaly Penskoy, the Muscovy military organization by the middle of the sixteenth century could be tentatively labelled "Ottoman" based on certain shared traits.[66] By the period under discussion, the Muscovite cavalry had adopted Tatar-style armaments and tactics. In the first half of the sixteenth century, a typical horseman carried a *sagaydak* (a bow and quiver with arrows) and a sabre as his main weapons, and wore limited armor. In combat, such cavalry strove to weaken the enemy by launching arrows before the shock of a sabre attack and avoided direct confrontation with heavily armed horsemen. Firearms were not widely used by cavalry then because of prohibitive costs, a low rate of fire, and the inherent difficulty of use.[67]

Two periods can be nominally marked out in the development of artillery in the Grand Duchy of Muscovy. The first mention of firearms there occurred in 1382, a reference to cannon and *tyufyaki* (a handgun similar to a *hakownica*, but firing grapeshot).[68] In the first half of the fifteenth century, the Tver Duchy's artillery dominated; however, in the second half of the century the balance fundamentally and forever shifted in favor of Moscow.[69] The turning point came in 1475, when Aristotele Fioravanti, an architect and founder from Italy, arrived to supervise the resident craftsmen. He established a large casting factory, first mentioned in 1478 as a *pushechnaya izba* (cannon foundry), and began casting guns of various calibers. Such centralized manufacture existed neither in the lands of the Jagiellonian dynasty, nor in the territories of Moscow's other enemies. Meanwhile, the diplomacy of Ivan III and Vasily III continued to recruit foreign specialists. From the second decade of the sixteenth century, German craftsmen gradually replaced Italians.[70] One can only imagine how tempting an

---

66 Penskoy, *The Great Gunpowder Revolution*, 295.
67 Penskoy, "The Development of Russian Military Affairs," 214–223; Ostrowski, "The replacement."
68 Kirpichnikov, *Military Affairs*, 77.
69 Penskoy, *The Great Gunpowder Revolution*, 299–300.
70 Lobin, "Russian artillery," 107–114.

offer must have been to convince Italian and German craftsmen to travel as far as Muscovy.[71] The diplomats' efforts, and undoubtedly a great amount of money were not wasted. Muscovy governors acquired qualitatively and quantitatively superior ordnance compared to any neighboring power.[72] Artillery became a constant factor in the larger campaigns of Muscovy troops, at least of those aiming to seize strongholds.[73] In this example, military technology transfer (central to military revolution scholarship) is laid bare.

Data on hand firearms remains less clear. The dearth of a causal line of development can be explained by the fact that in Eastern Rus a weapon of any caliber with a smooth bore that shot iron projectiles could be called a *piszczel*. Furthermore, it is impossible to distinguish whether the term identifies small cannon, *hakownicas* or *rusznicas*.[74] Contacts with the Livonian Order and the Grand Duchy of Lithuania, including those on the battlefield, seem to have promoted the proliferation and improvement of handheld firearms, and *pishchalniki* (literally, those who fire *piszczels*) seem to have appeared in the North West before other regions. In my opinion, it would be too optimistic to anticipate infantry with firearms in Ivan III's order to his Pskov governor in 1478 to prepare for the campaign against Novgorod "with cannons and *piszczels* and crossbows."[75] If such a prospect were justified it would have been the only explicit reference to infantry with handheld firearms at that time.[76] *Pishchalniki* referred to those who produced artillery and fired them. They originated in the first decade of the sixteenth century, with their first mention in the Grand Duchy of Muscovy dating back to 1500. In 1508, *pishchalniki* were mobilized for the war in the West. Two years later the chronicle mentions 'kazionnye pishchalniki' (literally, *pishchalniki* paid by the Treasury) indicating the Grand Duke kept them on a permanent basis. In 1512, *pishchalniki* among other troops were stationed on the bank of the river Ugra in order to hold off an anticipated Tatar attack, and in the winter

---

71 In 1509, young Pyotr Fryazin, who had earlier escaped to the Grand Duchy of Lithuania, asked to re-enroll in the service of Muscovy. Lobin, "Russian artillery," 111. He perhaps returned to Moscow due to the difference in the salaries of cannon-founders.
72 Lobin, "Russian artillery"; Kirpichnikov, *Military Affairs*, 80–81.
73 Penskoy, *The Great Gunpowder Revolution*, 300–301.
74 For example, in 1514 the chronicles talk about Polish infantrymen armed with *rusznicas*, "zholniers with pizchels" *Collection of Chronicles called the Patriarch's or Nikonovskaya Chronicle, Polnoe Sobranie Russkih Letopisei* [subsequently *PSRL*], vol. 13, 23.
75 *Sofiyskaya Vtoraya Letopis*, PSRL 6.2: 260; Penskoy, *The Great Gunpowder Revolution*, 303.
76 Penskoy, *The Great Gunpowder Revolution*, 303–304.

1512 to 1513 they participated in an unsuccessful Smolensk campaign. In 1535, they were also mentioned in the garrison of Homiel, attacked by Lithuanian forces.[77]

Oddly, we have no solid evidence on the character of weapons *pishchalniki* carried. The same can be said of the Muscovite infantry in general. The ambassador of the German Emperor Sigismund von Herberstein declared that Vasily III had 1,500 infantry consisting of "the Lithuanians and various ragtag."[78] Foot soldiers participated in repulsing the Tatar attack near Tula in 1517 and were "killing" (which might mean both close combat and skirmishing) Tatars "in forest"; unfortunately, it remains unknown what kind of weapons they carried.[79] According to Anatoly Kirpichnikov, until the middle of the sixteenth century the main responsibility of *pishchalniki* was gun maintenance, including small caliber weapons. These were employed either during sieges (both as besiegers and defenders), or on the banks of rivers, along with other types of troops, as a shield against Tatar invasions. In 1556, *pishchalniki* in Nevel were supposed to "be at the city at the ordnance, cannons and piszczels relentlessly." The same year in Ivangorod *pishchalniki* and gunners are mentioned, but only their total number is listed.[80] Evidently, officials saw no need to mention them separately, as their duties were roughly similar.

The first unequivocal mention of *pishchalniki* bearing handheld firearms dates to 1545 when Ivan IV demanded that Novgorod muster them for the Kazan campaign, 1,000 infantry and 1,000 cavalry, each soldier armed with a *ruchnaya* (handheld) *piszczcl*.[81] It may be noted that even mounted shooters were mentioned, but these probably used horses for transportation and fought dismounted as a sort of prototypical dragoon or light cavalry. It is also clear that these were militia rather than a standing army. The first corps of foot soldiers carrying firearms that the sovereign kept on a permanent basis were *streltsy* (literally, "shooters"), whose origin dates back to the 1550s.

Therefore, the question of the Muscovy infantry's make-up and armament in the first third of the sixteenth century remains unresolved.[82] One can only assume that a certain part of the *pishchalniki* could be formations of foot soldiers armed with *rusznicas*. It is evident that the army needed infantry, while the internal

---

[77] *The Great Gunpowder Revolution*, 305; Lobin, *Bitva pod Orshey*, 43–48; Pahomov, "Pishchalniki of Vasily III," 6–9.
[78] ". . . ex Lithvuanis, variaque hominum colluvie, mille & quingentos fere pedites"; Herberstein, *Notes on Muscovite Affairs*, 244–245; Ostrowski, "The replacement," 516, note 14.
[79] PSRL 13.2: 26; Ostrowski, "The replacement," 516, note 14.
[80] Kirpichnikov, *Military Affairs*, 92–93. *Dopolneniya k aktam istoricheskim*, 141, 144.
[81] Kirpichnikov, *Military Affairs*, 93.
[82] For a brief review of opinions on *pishchalniki*, see Pakhomov, "Pishchalniki," 6.

manpower sources were scarce and the mercenary markets far away. Vasily III tried to hire infantry from German territories, but transportation was a long and laborious process; moreover, frequent conflicts with Poland and the Great Duchy of Lithuania closed direct access routes through those countries.[83] An incident in 1505 illustrates implicitly the dearth of native infantry. In order to repulse a Nogay attack on Nizhny Novgorod, 300 "Lithuanian" *zholners* captured during the battle at Vedrosha in 1500 were set free and recruited.[84] Only their marksmanship made it possible to defend the city.[85] One might recall von Herberstein's account, indicating that Vasily III's patchwork infantry consisted partially of Lithuanians who likely had been captured at some point, and had consented to serve Muscovy's sovereign.[86]

## Firearms on the Battlefield

Consideration of battlefield implementation of firearms in Eastern Europe invites comparison with other regions of the continent and the Near East. First, let us address the Polish experience, leaving for later the Poles' participation in wars against Muscovy in the eastern part of the Grand Duchy of Lithuania. As far back as the Thirteen Years' War against the Teutonic Order (1454–1466) artillery played no major role, remaining a weapon of low importance even in siegecraft.[87] The same is true for the less significant conflicts from 1470 into the 1480s. In his 1497 campaign against Moldavia, John I Albert deployed impressive ordnance. During

---

[83] Evidently, Vasily III had the funds for mercenaries. In 1520 he sent money to Albrecht, the master of a Teutonic Order, for the hiring of 1,000 infantrymen for service against Poland and also promised to send funds for 10,000 foot and 2,000 mounted soldiers when the master had demonstrated significant military achievements, Ordensbriefarchiv, No 23709, 23891, 23985.
[84] *Zhelnyr/zholner* is a borrowing from Polish (*żołnierz*), which means soldier, a term that got into the Russian language perhaps through Old Belarusian. There is agreement in Russian literature that in Muscovy only the infantry was implied under the term *zholnery*, although this is doubtful. For instance, according to the Nikonovskaya chronicle, in 1513 *zholners* along with other people *rode out* from Smolensk to combat the avant-garde of the Muscovite army; PSRL 13.2: 16. In the Grand Duchy of Lithuania, *zholnery* implied both foot and mounted soldiers.
[85] Penskoy, *The Development*, 234–235; Pakhomov, "Pishchalniki," 8; *Istoriya o Kazanskom Tsarstve*, PSRL 19: 24–25.
[86] It might seem to invite speculation about native infantry in the Grand Duchy of Lithuania. But one should be cautious because under the term "Lithuanian," Muscovites might have meant Polish infantrymen in Lithuanian service.
[87] Szymczak, *Początki*, 249–263.

the siege of Suchava's stone fortress his cannon performed well, breaching enemy fortifications. The defenders, however, managed to seal the breaches overnight and counterattacked with retaliatory fire. Their strenuous defense successfully repulsed the Poles for about a month. Encountering difficulties with supplies and receiving reports about a possible relief force, John I Albert decided to retreat.[88]

The final war against the Teutonic Knights from 1519 to 1521 aptly shows the increasing impact of artillery. The Poles deployed 10 to 12 large siege guns against Kwidzyn castle. Throughout the day of 12 March, 1520, they pummelled fortifications at various points forcing the garrison to capitulate on 14 March. Delivered before Pasłek castle, those cannon then bombarded that stronghold on 20 April, damaging it considerably. Pasłek castle's garrison capitulated on 29 April. Interestingly, that fortress had been besieged in February of the same year but unsuccessfully due to the lack of ordnance.[89]

Instances of the successful use of artillery to defend fortifications are also worth mentioning. The Poles' besiegement of Braniewo in the autumn of 1520 failed because the defenders' cannon managed to suppress Polish artillery. Gdańsk's successful defense in November 1520 against German mercenaries who had come to the aid of the Teutonic Order owed much to cannon. The Germans fielded 19 guns, which on 8 November fired 400 rounds; the citizens of Gdańsk, however, answered with 800 shots. On the following day, the defenders killed one of the Germans' foremost cannoneers, one among many factors which persuaded them to lift their siege.[90] The Polish-Teutonic war of 1519 to 1521 illustrates the disrupted "balance" between means of fortification and means of siege. Efforts to improve old fortifications generally failed if the enemy managed to deploy guns of adequate quality and quantity, and to provide them with sufficient ammunition and professional gunners. The development of the ability to destroy fortifications surpassed developments in the art of fortification. However, the scales balanced when the besieged could bring to bear adequate retaliatory fire. In the conflict of 1519 to 1521 between Poland and the Order, mostly consisting of sieges and "small war," cannon played a crucial role.

Weapons with enormous potential also demonstrated significant logistical limitations. The abundance of rivers, in particular, became a serious hindrance in transporting cannon. For instance, during a passage of the Polish army across the river Świeża in May 1520, only small caliber guns succeeded in arriving on

---

88 Szymczak, *Początki*, 267–268.
89 Szymczak, *Początki*, 267–268.
90 Szymczak, *Początki*, 282–283.

the opposite bank. The dearth of large caliber cannon took its toll during the siege of Königsberg by Polish troops later same year, which ended in failure.[91]

As mentioned above, in the 1519 to 1521 war against the Teutonic Order, infantry constituted a substantial part of the active Polish army, evidenced by surviving muster rolls, *listy przypowiedne*, and contemporaries' accounts. In certain campaigns infantry even outnumbered cavalry.[92] No vivid examples exist of the infantry's firearm capabilities during that 1519 to 1521 conflict, unlike during the campaign against Moldavia in 1531, the pinnacle of which was the battle at Obertyn on 22 August. In this battle both sides employed cannon. The crown *Hetman* Jan Tarnowski deployed about 6,000 men including 1,000 infantry and 12 smaller cannons. Moldavian *hospodar* Peter Raresh confronted them with an army of 17,000, predominantly light cavalry carrying 50 small cannon. Facing a superior enemy, the Polish *Hetman* placed his troops on the elevated plateau and encircled them with their carts. *Tabor* or *wagenburg* provided the Poles with solid cover. However, the Moldavian quadruple superiority in artillery did not allow the Poles to rely on purely defensive tactics.

Apparently, the Polish cannons' projectiles caused the enemy few significant losses. Infantrymen fired from behind the carts, but it was hard to judge the effectiveness of that shooting. Besides, the foot soldiers performed two significant maneuvers in the battle. Initially, 800 men left the *tabor* cover to fire to distract the enemy's attention; in the battle's final stage the infantry secured the cannon seized from the enemy.[93] If we analyze this battle in terms of the tactical application of gunpowder weaponry, the most important element was the Moldavians' superiority in artillery. As Moldavian cannons fired repeatedly with increasing precision, Tarnowski had to maneuver to avoid higher losses and to make the enemy relocate their cannon. One crucial miscalculation occurred when Raresh moved his artillery to another position, which ended with its complete loss. Tarnowski's maneuvers also caused the enemy to overextend their front, and enabled cavalry attacks from the *tabor* in the right places at the right moments. These secured a Polish victory.

Cannon and handheld firearms deployed under cover of carts were not new: Hussites often practised this tactic, which later became widespread in Central

---

**91** Szymczak, *Początki*, 281.
**92** For example, in October 1520, 1,800 foot and 935 mounted soldiers were hired, and in July 1520 Braniewo was besieged by 2,418 infantry and 2,102 horsemen, and that occurred after a successful outing of the enemy during which the foot soldiers suffered losses: Bołdyrew, *Piechota*, 313–315.
**93** Plewczyński, *Obertyn*, 169–218; Szymczak, *Początki*, 284–287; Bołdyrew, *Piechota*, 324–326.

Europe and the Near East.⁹⁴ Firearms were hardly used during active defense of the southern frontier, by either the Polish or the Lithuanian sides. When battling nomads, guns and *ruznicas* proved useful defending fortifications if the enemy attacked. Massed infantry armed with firearms could be, under certain circumstances, quite effective against Tatars in the open field. However, disparity in mobility between infantry and Tatar cavalry made it very unlikely for them to meet on the battlefield.⁹⁵ Fast mobilization of forces and rapid advancement against the enemy in order to overtake and destroy them was the basis of anti-Tatar tactics. Such an approach, however, excluded or made radically complicated the utilization of artillery and infantry. Perhaps the only large battle against Tatars involving both artillery and infantry occurred in 1512 near Vishnevets (also known as the battle of Lopushno).⁹⁶ The Polish-Lithuanian side deployed two light cannon and an infantry company of 300 men who apparently just happened to be handy. Neither played a significant role in that victory.⁹⁷

Now let us analyze the Muscovites' tactical implementation of gunpowder weapons. During the reign of Ivan III, cannons participated in several campaigns, however the only example of their direct and successful use occurred at the siege of the Livonian fortress of Fellin in 1481. There the artillery breached a stone wall of the fortifications' outer lodgement.⁹⁸ The 1482 campaign against Kazan featured the first large long-distance march involving cannon. They were shipped to Kazan by river under the command of Aristotle Fioravanti. Access to a large navigable river was crucial when transporting siege artillery over long distances characterized by undeveloped road networks. Fioravanti also supervised ordnance in the 1485 expedition against Tver. However, he never fired the cannon under his command.⁹⁹

In the first third of the sixteenth century, Muscovite armies constantly deployed artillery, but rarely employed them successfully. Frequently, either they simply never got round to using their ordnance, or the enemy prevented them from doing so. On the other hand, surviving documentation verifies the Muscovy artillery's considerable power and greater capacity. The most notable instance

---

94 Davies, "Guliai-gorod," 99–100.
95 The clash at Slutsk in 1508 when an infantry company defeated a unit of Tatars illustrates this; Bołdyrew, *Piechota*, 332–333.
96 The opinion that at the battle of Kletsk in 1506 infantry participated from the Lithuanian side lacks evidence: Szymczak, *Początki*, 271. There is merely a mention of cannon (evidently not very numerous) and just a few *rusznicas* in a 6,000 soldiers strong army; *Kroniki Bernarda Wapowskiego*, 67; Stryjkowski, *Kronika Polska*, 332, 334. Evidently showers of arrows from numerous bows had an incomparably larger effect than fire from a few firearms.
97 Szymczak, *Początki*, 274.
98 Alef, "Muscovite military reforms," 80–81; *Ioasafovskaya Chronicle*, 123.
99 Alekseev, *Russian military campaigns under Ivan III*, 270, 278.

occurred during the Smolensk campaigns of Vasily III. An initial attempt to penetrate the city occurred in 1502, when Ivan III's son Dmitry besieged Smolensk until approaching relief forces compelled his retreat. Smolensk's citizens joined with the garrison and defended themselves using firearms, sortied out several times, and repelled a major assault. According to the King of Poland and Lithuanian Grand Duke Alexander, the gunners in Smolensk managed to kill one of the enemy artillerists and disable a large cannon which earlier had damaged the fortifications.[100] Vasily III attempted three expeditions against Smolensk, during the winter of 1512 to 1513, the summer of 1513, and in spring/summer of 1514. The evolution of siege tactics is noteworthy. During the first siege, after a powerful artillery barrage, drunken *pishchalniki* stormed at night, and were defeated whilst suffering great casualties.[101] During the second siege a major assault was not attempted. Vasily sent letters to inhabitants of Smolensk, trying to persuade them to surrender while his cannon bombarded the city walls. The Pskov chronicle hints that the artillery fire was intended to destroy the fortifications, but the defenders managed to repair them: "[W]hatever they broke during the day was fixed at night."[102]

In the spring/summer of 1514, the largest and best prepared Smolensk campaign ended successfully. This time Muscovite artillery apparently focused on bombarding the interior of Smolensk rather than aiming at fortifications. That tactic determined the siege's outcome. Evidence indicates that even more weapons had been gathered for the third expedition, a factor that also must have played its role. Evidently, the Smolensk garrison commander Yury Salahub abandoned any hope of outside help. He capitulated under pressure from citizens desperate to save their lives, homes and belongings.[103]

No reliable data reports the numbers and types of guns used during the siege of Smolensk. Records from the Polish-Lithuanian side mention three and even four digit numbers. Up to 200 such weapons may be close to accurate, if including guns of all calibers.[104] At this point some aspects of siegecraft and fortifications in Eastern Europe should be clarified. Very few towns of the region had stone or brick walls, namely on the eastern frontiers of the Grand Duchy of Lithuania only Vitebsk and partially Orsha. Earth and wood, always abundantly available, traditionally served as materials for erecting fortifications. From the

---

100 Szymczak, *Początki*, 270–271.
101 Lobin, *Bitva pod Orshey*, 46–47; *Pskovskie Chronicles*. Second issue, 259.
102 *Pskovskie Chronicles*. Second issue, 259; Lobin, *Bitva pod Orshey*, 52–53.
103 *Ustyuzhskie I Vologodskie Chronicles of XVI–XVIII centuries.*, PSRL 37: 100. In Vilno it was classified as a betrayal, which is quite convenient if one needs to justify their failure in organizing the country's defense.
104 Lobin, *Bitva pod Orshey*, 45–46.

tenth through the thirteenth centuries, it was common practice to erect wooden structures and fill them with earth. Over time, as a result of wood erosion and destruction, those structures gradually turned into earth ramparts, on top of which wooden fortifications were erected. Such fortifications, comparatively inexpensive, were quickly and relatively easily made. Earthen ramparts' capacity to absorb the shock of cannonballs may have been accidentally discovered by the citizens of Pisa in 1500 but was in fact well known in Eastern Europe.[105] Wooden fortifications, while easily restored when damaged by enemy fire, also possessed a major drawback, flammability, which was partially remedied by layering clay on the timber. Obviously, cannons' ability to rapidly destroy medieval stone fortifications could not be fully implemented in the east of Europe. Even if assault troops managed to compromise defenders' artillery positions and wooden fortifications, the assailants often still had to cross a trench and climb a high rampart.

At Smolensk in particular, various sources describe this fortress as quite formidable. Its hilltop location was surrounded by the Dnepr river and marshes. Artificial impediments consisted of trenches, the earthen rampart, and walls fashioned from rows of logs, the spaces between filled with earth and stones. According to one anonymous observer, cannonballs could not penetrate these barriers.[106] Nevertheless, in 1514, the Smolensk stronghold – the eastern gateway to the Grand Duchy of Lithuania – succumbed to Muscovite artillery.

Apart from their skills in siegecraft, Muscovites consistently deployed ordnance to protect their southern frontier from Tatar incursions. In the fifteenth century they developed a relatively effective system of defense, deploying troops, including artillery of different calibers, at river crossings accessible to the Tatars. Cannon and *piszczels* could fire at an enemy on the opposite bank while remaining safe; a sudden attack was not a threat. One famous example of mass implementation of firepower is the battle of the Ugra River in 1480.[107] The period under study offers not a single example of utilization of firearms by the Muscovites in an open field battle. Cannon were employed in defense of river crossings, and to attack and defend strongholds.

In reviewing the Grand Duchy of Lithuania's applications of gunpowder weapons, one should note changes in the composition of its army at the turn of the sixteenth century. Significantly, data from narrative sources is supported by exact figures, though not complete, from the account books of the Grand Duke Alexander.[108]

---

105 McNeill, *The Pursuit of Power*, 90.
106 Lobin, *Bitva pod Orshey*, 43–44.
107 Penskoy, *The Great Gunpowder Revolution*, 300–301.
108 *Lietuvos didžiojo*.

When the war against Muscovy began in 1500, the court *chorągiew* (not to be confused with the courtier corps in Sigismund's reign) increased to 460 cavalrymen. Late in 1500 and early in 1501, Alexander hired several thousand Poles, Czechs, and Germans. Sources mention two commanders of foreign mercenaries, Jan Polak Karnkowski and a Czech, Piotr Czernin.[109] Mercenaries had already served the Grand Duchy of Lithuania in 1500, as mentioned above; foot soldiers were among prisoners taken at the battle of Vedrosha. In the summer of 1501, the Grand Duchy of Lithuania planned to deploy 4,000 infantry, 1,000 cavalry, and a small troop of Polatsk nobility for joint actions with the Livonians near Pskov.[110]

It appears mercenary troops served in two major areas, near Polatsk and in the vicinity of Smolensk. At the end of 1502, six infantry companies received pay for their service in Smolensk, numbering 803 infantrymen and 23 mounted soldiers.[111] Obviously, these troops defended Smolensk from Muscovy's army in the same year. In 1503, 279 mounted men of the court *chorągiew* were registered under Chernin's command along with 810 infantrymen and 56 horses in eight infantry companies. At the same time, Jan Polak commanded 300 cavalry, and 4,076 infantry with 147 horses in 22 companies. According to Krzysztof Pietkiewicz, by the end of the war in 1503 the mercenary army consisted of 1,163 cavalry and 5,952 infantry, and 237 mounted men in infantry companies.[112] Historians considering these figures do not mention that for the first time ever massive numbers of infantry appeared in the Grand Duchy of Lithuania's army.

Moreover, among mercenaries, infantry outnumbered cavalry by several times, which reflected the predilections of their employer. One should keep in mind that by the beginning of the sixteenth century Polish infantry companies mostly used firearms. While *zemskaya sluzhba* allowed no options, in hiring professionals emphasis on infantry represents an intentional choice. Terming this an "infantry revolution" would be an exaggeration, but the data testify to the high demand for infantry at this time and in this context. It is impossible to account for all the operating forces, but even considering the mounted nature of the *zemskaya sluzhba*, infantry must have constituted quite a prominent part of the army. As demonstrated by plans for the joint Polish-Livonian campaign against

---

[109] Г. Лясмайціс, 'Ад Вядрошы да Оршы: змены ў сістэме камплектавання войска ВКЛ', у *ARCHE*, 12 (2014): 137; Pietkiewicz, "Dwor," 100–101.
[110] Papee, *Aleksander Jagiellończyk*, 43–44.
[111] Pietkiewicz, "Dwor," 103; *Lietuvos didžiojo*, 84.
[112] Pietkiewicz, "Dwor," 102–104; *Lietuvos didžiojo*, 88–90, 96–98. Of course, there is no guarantee the data is complete. It certainly does not take fortresses' permanent garrisons into account.

Pskov in 1501, in certain operations infantry could outnumber cavalry several times over. Besides, many castles had permanent infantry garrisons which might enlarge their forces when anticipating combat. For example, in May of 1500, 500 florins were expended to cover salary to foot soldiers in Kyiv.[113] As an average infantryman earned two florins per quarter, the Kyiv garrison must have been 250 men strong. Accounts show city garrisons receiving pay subsequently.[114] There is also surviving data about gunners in Lithuanian service. By the end of the war in 1503 they were as many as 11.[115] In 1500, five gunners arrived to buttress the defensive capabilities of Smolensk. Unlike its Eastern neighbor, the Grand Duchy of Lithuania seldom initiated campaigns to capture fortresses; their gunners were mostly sedentary, servicing ordnance in castles and upon city walls. The battle of Orsha was a striking exception, as we shall see below. One notable example of successful employment of guns in siege warfare occurred at Homiel in 1535.[116] The day after arriving at the city, Yury Radzivil's artillery attacked. The royal artillery train arrived that evening, fired all night and most of the next day, until the defenders capitulated. The ordnance performed as expected: rapid deployment and intensive fire. Two days had been sufficient to force Homiel's garrison to surrender.

## The Battle of Orsha, 1514

In subsequent actions against Moscow, infantry became a regular component of the army. However, we lack reliable evidence on numbers. Also, prior to the battle of Orsha, no surviving records describe in detail the actions of infantry in large numbers in an open field battle.[117] Unlike the Italian Wars, or even the campaigns of the Ottomans in the Near East in the 1510s, in Eastern Europe in spite of the intensity of wars in general major battles rarely occurred. Hence, there were few opportunities for infantry or artillery to demonstrate their potential as a field force. At Vedrosha, in 1500, the Lithuanian army deployed 3,500 to 4,000 men, not counting infantry; at Kletsk in 1506 about 6,000; at Vishnevets in 1512, joint Polish-Lithuanian forces numbered 6,000; at Obertyn in 1531,

---

113 *Lietuvos didžiojo*, 62.
114 For example, см. Лясмайціс, 'Ад Вядрошы', 135–136.
115 Pietkiewicz, "Dwor," 102–104.
116 Krom, *The Acts of Radzivill*, 162–163.
117 There is a record of a skirmish between an infantry company numbering 200 men and a superior Tatar detachment near Sluck in 1508. The Tatars could not use their numerical advantage as the enemy encircled their position with carts, Bołdyrew, *Piechota*, 332–333.

the Poles also had about 6,000 men. Against such a background, Orsha stands out for its scale, one of the major reasons contemporaries long remembered it as "the great battle." Orsha attracted the attention of many historians, but controversial and unexplored aspects still remain. Even the sole monograph, the most sound although not an indisputable study, omits analysis of the tactics regarding firearms implementation.[118]

What events led combatants to a field near Orsha? Learning of the Muscovite siege of Smolensk in the spring of 1514, Sigismund started to muster troops, but his response came too late. When the fortress fell, his army was encamped near Minsk. Nevertheless, he moved to the east, at that point aiming to recapture Smolensk from the Muscovites. Vasily III, knowing of Sigismund's preparations, advanced his army, which was later reinforced with reserves. Not long before the battle, Ivan Andreevich Chelyadnin took charge, his main task being to block the enemy's advance towards Smolensk for which the large river Dnepr served as a convenient frontier. Sigismund, then formally the army's leader, moved no further than Barysau and resigned his command to Ostrozhsky. In the first days of September, the army reached Orsha, lying on the right bank of Dnepr. The Muscovite forces located on the other bank controlled the crossing.

Ostrozhsky had different types of troops under his command including infantry. Understanding the makeup and ratio of foot soldiers requires some calculations.[119] Without taking into consideration clearly exaggerated numbers, sources agree that the army amounted to between 16,000 and 17,000 men.[120] It is somewhat more difficult to know the sizes of its separate military contingents. There are detailed payment rolls for the courtiers corps and a record of mercenary force numbers but no reliable evidence about the size of *zemskaya sluzhba*. The other difficulty is that the entire army did not participate in the battle.

Calculating the number of mercenaries is rather easier. In April to May of 1514, Lithuanian authorities hired about 6,663 soldiers, 3,000 of these infantry.[121] A 94 person company of Rascians and 200 cavalrymen should be added to the above-mentioned figures.[122] Also, before mid-April, Jan Spergaldt's infantry unit

---

[118] Lobin, *Bitva pod Orshey*. Tomasz Mleczek has lightly touched this topic: Mleczek, "Armia koronna."
[119] Numerous historians have tried to do it. Aleksey Lobin is the closest to being objective, however even his deductions need to be reviewed.
[120] Lobin, *Bitva pod Orshey*, 132–133.
[121] Lobin, *Bitva pod Orshey*, 112–118.
[122] Krom, *The Acts of Radzivill*, 27.

of 500 arrived at Orsha.[123] One should remember that the actual number of men in mercenary companies was always lower than the nominal figures. Also, the mercenaries may have lost men in combat before Orsha. Thus, it would not be inappropriate to lower their numbers by approximately 20 percent, leaving about 3,200 cavalry and 2,800 infantry.

However, not all these mercenaries accompanied the *hetman* to Orsha. A unit of both mounted and foot mercenary soldiers remained with Sigismund in a camp near Barysau.[124] The *ablegate* Jacob Piso, in Vilnius during these events, estimated its strengths as 4,000. Numerous authors up to the present have repeated those numbers. However, Piso in general demonstrated an inaccurate reckoning of numbers, and his estimates must not be taken as reliable. He himself originated the misstatement that the king's army constituted 35,000 men.[125]

In estimating the size of the detachment left in Barysau one must consider two points. Firstly 4,000 constitutes around 67 percent of all the mercenaries, the most efficient unit. A prudent commander would have deployed them against the enemy, not left them out of action. Second and more important, as shown below, infantry played a critical role in the battle. That would have been unlikely if only 33 percent out of 2,800, that is under a thousand soldiers, had been fighting at Orsha.

In my opinion, a couple of infantry companies must have remained with the king, perhaps 300 men, and a few mounted companies, under a thousand in all. In this case, the calculation of mercenary infantry at Orsha would decrease to 2,500 and the mercenary cavalry to 2,200. Hence, the total manpower of the army that fought at Orsha should be reduced from between 16,000 and 17,000 to between 14,700 and 15,700.

The Polish court *choragiew* was deployed; while there is no indication of the unit's size, we know that in the battle of Kletsk in 1506 it was 300 men strong, and one should be guided by this number.[126] In general, the Polish court *choragiew* may have contained several hundred mounted men. Polish *szlachta* volunteers also sought chivalrous glory in the East. In the absence of even approximate data, any estimation of their numbers remains speculative. Assuming few would willingly risk their lives a thousand kilometers from home (and at their own expense!) I would estimate *szlachta* fighters and their *pochets*

---

123 *Kronika Polska Marcina Bielskiego*, 971.
124 *Acta Tomiciana*, vol. III (Posnaniae, 1853), no. CCXXVII; 162; Górski, "O rozwoju sztuki wojennej w Polsce w wieku XV," 417.
125 *Acta Tomiciana*, vol. III, no. CCXLIV: 203–204.
126 Stryjkowski, *Kronika*, 335.

making up to 500 men.[127] A unit of *hospodar* "courtiers" was approximately a thousand men strong. With no evidence on *zemskaya sluzhba*, one must calculate by process of elimination. The 2,200 mercenary cavalry, 2,500 infantry, 500 volunteers, 300 court *choragiew* combatants, and 1,000 *hospodar* courtiers total 6,500 men. Subtracting these 6,500 from the 14,700 to 15,700 leaves around 8,200 to 9,200 men of the *zemskaya sluzhba*. The ratio of Polish to Lithuanian troops was slightly different, with around 9,200 to 10,200 Lithuanians, compared to 5,500 Poles.

Considering the total number of Ostrozhsky's army as 14,700 to 15,700, the infantry constituted 16 to 17 percent. Sixty to 80 percent of Polish infantrymen were marksmen, 83 percent of whom were armed with *rusznicas* by 1500. One may venture that by 1514 the crossbow had been displaced completely. Sigismund's requirements for the armament of mercenaries to be hired in spring 1514 for war against Muscovites support this conclusion, as he insisted that foot soldiers must bear "pixides bonas manuales" ("good handheld guns").[128]

These figures do not take into the account the presence of a unit of engineers under the command of Jan Baszta, as well as gunners. We have no reliable information on the number and composition of the Polish-Lithuanian artillery. In the famous painting "Battle of Orsha," an anonymous artist portrays 14 guns, 13 of which could be *schlange*, according to the German classification.[129] However, while being a wonderful artistic representation, a painting created in the 1530s or even 1540s remains an unreliable source for reconstructing the composition of troops and the course of the battle.[130]

The Muscovite troops were completely mounted and lacked ordnance. No reliable sources on headcount survives. Alexey Lobin estimated the size of Muscovy's army by his own methods, none of which, in my opinion, yielded a valid result.[131] The extreme scarcity of accounts renders quite problematic the size of Muscovite armed forces in the first half of the sixteenth century in general, and

---

**127** A. Lobin, *Bitva pod Orshey*, 120–121, suggests that there could have been up to 2,000 volunteers, based on Stanisław Sarnicki's report: "polaków dobrze zbrojnych 2000" (Sarnicki, *Księgi hetmańskie*, 362). However, below Sarnicki writes that those 2,000 Poles were commanded by Janusz Świerczowski. According to other sources it is well known that Swierczowski commanded the mercenaries.
**128** *Acta Tomiciana*, vol. III, no. CXXXI: 99.
**129** Lobin, *Bitva pod Orshey*, 136–138.
**130** A. Kazakou, "Bitwa pod Orszą 1514 r," 128–132. For the most recent attempt at dating the painting, see Janicki, "Obraz *Bitwa pod Orszą*," 173–225.
**131** Lobin, *Bitva pod Orshey*, 83–105. For a critique of methods and results, see Kazakou, "Hunting the Orsha myth," 88–97; Kazakou, "The Historiography of the Battle of Orsha, 1514," 20–24.

**Figure 4:** "The battle of Orsha," executed *circa* 1530s to 1540s, preserved in the National Museum in Warsaw. This is the only known work of its kind depicting a battle fought in Eastern Europe in the early sixteenth century. The artist remains unknown as all attempts to attribute it to a specific painter lack solid, factual evidence. The representation was long believed to be an historically accurate source regarding battlefield topography and the unfolding events of the battle. These views rested upon the conviction that the painting was created not long after the battle had been fought, and that the artist had personally witnessed the event. According to recent research, "The Battle of Orsha" should be dated as late as the 1530s, perhaps even the 1540s. Dendrochronological analysis reveals that the wood upon which it was painted was still a growing tree in 1524. Nevertheless, this masterpiece incorporates the salient turning points in the battle – the crossing of the river by Lithuanian and Polish forces, intense hand-to-hand combat, artillery fire from a hidden position, and the flight of the Muscovites. *Image: National Museum in Warsaw.*

at the battle of Orsha in particular. We can say Ostrozhsky was unlikely to have enjoyed numerical superiority over his enemy.[132]

Quite a large river divided the opposing sides. By locating a clandestine fording point where Chelyadnin did not anticipate it, Ostrozhsky transferred his army to the other bank by a pontoon bridge built under Jan Baszta's supervision. The location of the crossing, as of the battlefield itself, remains debatable.[133]

---

[132] For a detailed analysis of the chronicle's reports, see Kazakou. "Opravdyvaya porazhenie," 315–334.
[133] Kazakou, "Bitwa."

Primary sources on the battle are regrettably scarce. The absence of reports by firsthand participants (with two exceptions mentioned below) compels us to content ourselves with accounts by contemporaries and more recent authors. However, we can reconstruct the picture of the events of 8 September in broad strokes.

According to the majority of sources, Ostrozhsky placed the Poles on the left wing and the Lithuanians on the right, stationing his infantry and artillery "in convenient places." He concealed some cannon and a part of his infantry in a small forest, most likely toward the center of his dispositions. The Muscovites initiated the battle by attempting to outflank the Ostrozhsky's positions on their left, a manuever checked by a counterattack of the court *choragiew*.[134] Then, or soon after, the struggle commenced on the left flank and in the center. In the battle's climatic turning point, from the forest ambush the infantry and artillery opened fire at the clustered Muscovites, probably on the left flank. According to Sigismund von Herberstein, the bombardment aimed at the crowded rear rows of horsemen which had not expected the attack. The fire from *rusznicas* and cannon sowed confusion among the Muscovites, and the Lithuanian cavalry assault along the front broke them decisively. After that, Ostrozhsky ordered all his troops to attack which resulted in a crushing defeat of the enemy. Part of the Muscovite army pressed to the Krapivenka river and was put to the sword or captured.

The extent of Muscovy's losses is another debatable question.[135] Death and capture of the majority of the commanders serves as an indirect, yet eloquent, testimony to total defeat. The number of Muscovite captives presented to Sigismund and later imprisoned in castles (apart from the mass of people taken by *szlachta*) was so great that for the first time authorities maintained centralized accounts of them.

Now let us consider the actions of infantry and artillery in greater detail. Most sources agree that infantry were divided up and deployed in different locations.[136] One may suppose that a significant number of foot soldiers lay in ambush, otherwise their fire would have had no serious effect. Several authors mention infantry actions, yet only Stanisław Sarnicki devotes undiverted attention to them (more about this below). Stanisław Górski merely remarks that infantry fired at the enemy. Ludwik Decjusz states that infantrymen with *rusznicas* proved very useful and caused great damage to the enemy.[137] Von Herberstein writes about ordnance

---

[134] According to another version, some Muscovy forces attacked the enemy's rear.
[135] Lobin, in *Bitva pod Orshey*, tends to minimize them (166–170).
[136] *Acta Tomiciana*, vol. III: 5; *Kroniki Bernarda*, 120; Stryjkowski, *Kronika*, 32. 382.
[137] Decius, *De Sigismundi regis temporibus liber 1521*, 78.

**Figure 5:** "The battle of Orsha," a detail. Artillery and infantry repel the attack of Muscovite cavalry. Three lines of harquebusiers are shielded by a line of *pavisiers* and a line of spearmen wearing plate armor. *Image: National Museum in Warsaw.*

and infantry laid in ambush discharging against packed enemy ranks.[138] Jacob Piso, describing the muster of the royal troops, mentions infantry who "carry the main burden of the war." However, he ignores their participation in the battle.[139] Marcin Bielski, as well as Maciej Strykowski, also neglect the infantry but ascribe a notable role to artillery. Bernard Wapowski was the only one to report an interesting detail, of cannon located on the high points which struck the rear ranks of the enemy.[140]

Muscovite chronicles yield little information, apart from remarks about infantry who "were killing" Muscovite horsemen "from the forest," and a characterization of the place where the fusillade occurred as "tight." The artillery report is

---

[138] Decius, *De Sigismundi regis temporibus liber 1521*, 382. Herberstein, *The Notes*, vol. 1, 84–85. In the Latin version of his work, von Herberstein mentions only cannons, and adds infantry only in a later German edition.
[139] *Acta Tomiciana*, vol. III: p. 203.
[140] *Kroniki Bernarda*, 120.

**Figure 6:** "The battle of Orsha," a detail. Two cannons hidden in a spruce grove on high ground and firing upon the enemy. The artist follows the (erroneous) popular belief that artillery exclusively, not infantry, were involved in this phase of the battle. *Image: National Museum in Warsaw.*

reduced to a reference to the death of one of the army commanders from a cannonball.[141]

A military treatise deserves a mention, Stanisław Sarnicki's *Hetman Books* written sometime in the late 1570s or early 1580s. Besides his theory of the art of war, Sarnicki describes a series of battles, including Orsha. Unlike many primary sources cited here, Sarnicki wrote long after the conflict, and his description contains evident mistakes. Nevertheless, his account is underestimated. It is quite important that Sarnicki incorporated sources which have not survived. He twice refers to the words of Ostrozhsky (in addition to a manufactured allocution encouraging his troops), and this likely represents the only extant testimony from a participant in the battle, even if it is paraphrased. Both citations are quite short, but one is crucial: according to the *hetman*, the battle's success depended on the infantry. Significantly, Sarnicki devotes a good half of his description to the actions of infantry that add up to the following. Infantrymen, located behind cavalry formations, repulsed a surprise rear attack and a flank attack, apparently one from the left. Discharging their *rusznicas* they caused considerable loss to enemy soldiers, "confined to a tight place," finishing them off with steel.[142] Unfortunately, Sarnicki does not specify what the "tight" place was.

One more factor that underscores the importance of infantry units for the *hetman* should also be mentioned. Wapowsky states that Ostrozhsky and Janusz Świerczowski, commander of the Polish contingent, convened a war council

---

**141** PSRL 13: 1, 23.
**142** Sarnicki, *Księgi*, 365–366.

together with the "foot" shortly before the battle.[143] Who commanded the infantry remains unknown. Sarnicki mentions Połobsza, though this name goes unreported in other sources.

Now let us analyze this data taking into consideration what we know about the technical capacity of firearms at the time. Combining Sarnicki's narrative with the other authors' accounts it seems infantrymen distinguished themselves at two major points of the battle: by unleashing a fusillade from ambush, and by repulsing rear and flank Muscovite attacks. In both cases, the location is some tight space, a critical circumstance. As the effective range of *rusznicas* hardly reached a hundred meters, cavalry, especially light cavalry, could easily and quickly avoid being exposed to bullets if given space for maneuver. Apparently, some particular place on the battlefield allowed them no such opportunity. Infantrymen, in contrast, fired from a position protecting them from enemy attack – the woodline. In another location, while repulsing rear and flank attacks, they exploited additional natural obstacles, for example "gulleys" (presumably low spots in an undulating topography). Even if that was not possible, rows of pikemen and *pavisiers* could provide quite reasonable protection from a light cavalry sabre attack. However, it is important that the infantry's adversaries in both cases were confused, preventing an organized onslaught. A mob of cavalry lacking organization and explicit orders provided an exceptional target for an enfilade. Considering that reloading a *ruznica* took one to two minutes, barrages from hundreds of men at a proximate distance lasting, say, 20 minutes and aimed at lightly to moderately armed cavalry could be remarkably punishing. One should recall that wounds from firearms led to higher mortality rates than wounds from edge weapons.[144]

It is not surprising that some contemporaries and Ostrozhsky himself considered the infantry's role in the battle to be outstanding. Nonetheless, one may have doubts, remembering that foot soldiers constituted only 16 to 17 per cent of the troops. The army deployed two or more lines deep, each line consisting of several files. Infantry occupying relatively little space along the front line could mass on the line of contact at the right time. Unfortunately, we do not know the exact location of the battlefield and the nature of the landscape there 500 years ago. However, references to "tight" places suggest that the front line may have been quite limited. If so, in the percentage ratios infantry could contribute far more to the line of contact than could the army in general.

---

143 *Kroniki Bernarda*, 120.
144 In 1575, Luis de Requesens, a Spanish governor-general of the Netherlands, reported to Phillip II that soldiers wounded by cold weapons would recover soon, whilst those wounded by firearms would die; Hale, *War and society in Renaissance Europe*, 121.

**Figure 7:** "The battle of Orsha," a woodcut, *Ad divum Sigismundum Polonie regem et magnum ducem Lithuanie semper invictum, post partam de Moskis victoriam Andree Krziczki inclite coniugis sue cancellarii carmen*, unknown artist, 1515. This is the earliest iconographic work depicting the battle. Foot soldiers are present and occupy the central position. Infantry cross a river via a bridge. A cannon barrel is depicted, ahead of the advancing troops.

As evident in accounts of the engagement, ordnance proved far less useful, which is not surprising considering their low rate of fire. At the beginning of the sixteenth century, a cannon could fire four rounds an hour.[145] In addition, at Orsha artillery had to perform against an entirely mounted and mobile enemy, while ordnance performed best when firing at targets like serried infantry formations. Wapowsky's report of cannon firing from high ground at the rear rows of the enemy over the heads of front rows is unconfirmed by other sources, although one cannot

---

145 Potter, *Renaissance France at War*, 154.

completely reject such a possibility. In this case, several hours of bombardment could have caused considerable damage. Cannon, along with infantry, firing from the forest edge would most likely be limited to one or two rounds. That would have had more of a psychological than a physical effect. In any case, even such limited participation of artillery in a field battle marked something new in Eastern Europe. The cannon did enough to justify their presence at the battlefield by killing one of the enemy's commanders. Although infantry played a larger role at Orsha, ascribing success to artillery is popular.[146] Something similar happened with the common picture of the battle of Mohács in 1526.

The victory achieved at Orsha depended on two major factors. The first is the combined deployment of various troop units (infantry, artillery, light and heavy cavalry), whose capabilities under the circumstances were accurately reckoned by the *hetman*. Understanding the inherent limits of infantry disposition, he positioned them in locations that maximized their capabilities. Infantry and heavy cavalry, no matter how good, could not under ordinary conditions have inflicted upon the enemy such a stunning defeat unilaterally. Muscovite troops suffered heavy losses during the pursuit, likely accomplished by light and middle cavalry. The second factor was effective coordination between different categories of troops in different places on the front line. Achieving this synchronization was, tactically, the most difficult task any commander faced, and Ostrozhsky rose to the occasion.

## Firearms and New Tactics: A Broad Context

During this era, in the fields of Italy and beyond, the role of firearms was changing. In terms of infantry firearm tactics, the first key event occurred in 1503 at the battle of Cerignola. Spanish commander Gonsalo de Cordoba placed 2,000 harquebusiers (apparently Germans sent by the Emperor Maximilian) behind a bulwark garnished with cannon. These field fortifications surprised the French cavalry and infantry who, unable to overcome the bulwark, were riddled with harquebus fire. A French commander, the Duke of Nemours, was among the casualties.[147]

Tactical implementation of harquebusiers included firing from city walls. When deployed in the field harquebusiers needed protection from the skirmishing of the enemy, especially by cavalry. Field fortifications in the form of trenches and ramparts provided shelter for foot soldiers firing and reloading as well as for

---

146 For example, see Hellie, "Warfare," 77.
147 Hall, *Weapons and Warfare*, 167–168.

artillerists. Technically, exchanging gunfire and discharging ordnance from atop a rampart was roughly akin to sharpshooting from a city wall.[148] This tactic was also employed later on; however, it did not always produce the expected results. One drawback of a fixed position (particularly for heavier field pieces) is that the enemy can renew the attack, carry out a turning movement, and force the defenders to abandon their position. The presence of significant numbers of cannon also permitted another option – artillery fire at a fortified position, as at Ravenna in 1512.

Pedro Navarro, who commanded the gunners and artillery at Cerignola, now led the whole army. He fortified his position while awaiting the French attack, but they initiated the battle with a cannonade. The Spanish ordnance responded, and for two hours the opposing sides bombarded each other. The French artillery surpassed that of the Spanish in numbers and quality, which in the end proved decisive. The Spanish cavalry broke first, after launching an unauthorized attack. The ensuing battle afforded little opportunity for firearms, and the French carried the day due to their advantage in cavalry. Clearly, at Ravenna artillery proved its ability to force the enemy to abandon an advantageous fortified position. In order for such tactics to be successful, superiority over the enemy in quantity and/or quality of ordnance was needed. Apparently, casualties inflicted by cannon fire were, for the era, horrendous.[149]

Artillery also played its part in the battle of Marignano in 1515, although there the French achieved victory over the Swiss infantry primarily through greater tactical flexibility and coordination of infantry (including some harquebusiers), cavalry, and artillery. At Bicocca (1522), the Spanish utilized the same tactics as earlier: deepening a sunken road to create a bulwark, concealing cannon and infantry. The Viscount of Lautrec, commanding the French, planned to repeat the Ravenna scenario, but the Swiss who constituted the bulk of his army attacked prematurely with unfortunate results.[150]

The next step in the development of European handgun tactics was its deployment beyond field fortifications. In the 1524 battle of the Sesia River, harquebusiers using horses for transportation attacked the retreating French army, firing from the flanks and the rear. During the attempts of the French to counterstrike, these gunmen mounted and retreated. Harquebusiers also employed new tactics at the famous battle of Pavia in 1525, which cannot be termed a traditional field battle. About 20,000 Frenchmen besieging Pavia were attacked by the

---

[148] Hall, *Weapons and Warfare*, 170.
[149] Hall, *Weapons and Warfare*, 171–173.
[150] Hall, *Weapons and Warfare*, 174–175.

Spanish Imperial troops consisting of approximately 17,000 men. The engagement began in dense early morning mist, and disintegrated into chaos, as reflected in contradictory accounts of the conflict. However, it is clear that 1,000 to 1,500 of Spanish harquebusiers exploited undulating terrain as ad hoc cover, mauling the enemy's heavy cavalry commanded by King Francis himself. The horsemen's moment of indecisiveness rendered them palpable targets.[151]

As one can note, the gunners' behavior in the battles discussed above resembles that of the infantry at Orsha, though the Polish infantry relied on its *pavise* men and pikemen instead of barricades. Like the gunners at Pavia nine years later, the infantry at Orsha exploited natural cover to pummel the enemy's cavalry, who failed to manuever. Significantly, in both cases the proportion of harquebusiers in the army was relatively low.

While tracing evolving tactics employed by firearm-bearing infantry and artillery, it is interesting to look for the parallels in other regions, first in the Balkans and the Near East. Towards the beginning of the sixteenth century, the Ottomans, in terms of developing and in particular manufacturing their own firearms, were a progressive power. During the rule of Selim I (1512–1520) the use of cannons and infantry with firearms (janissaries) in the open field became common practice. The Ottomans used *tabor* to protect their firing positions, as in major battles against the Safavids at Çaldıran in 1514 and against the Mamelukes at Marj Dabiq in 1516. In the former case, a mounted Safavid army recklessly attacked a superior force and was defeated. In the latter case, the enemy of the Ottomans had both firearm infantry and artillery but in far less quantity; the same can be said about the force ratio in general. For Mamelukes, the betrayal of one of the commanders and the death of the sultan Al-Ghawri in mid-battle complicated the situation.[152] For Ottoman infantry and artillery, *tabor* provided the same cover that earth bulwarks had for the Spanish infantry and artillery.

The 1517 battle of Raidaniya against the Mamelukes is of a special interest because circumstances forced the Ottomans to abandon their tactics and improvise on the spot. Moreover, the armies were equal in numbers, and the Mamelukes gathered all the firearms they could, though their 200 artillery pieces were likely to have been of a small caliber, and the number of harquebusiers was reduced to 200. The new sultan Tumanbay had learned the lesson of Marj Dabiq. As the Ottomans had then, he used *tabor* and provided earthwork to protect the artillery. After observing these preparations, the Ottomans began outflanking the Mamelukes from the right without approaching within range of effective fire. Meanwhile,

---

151 Hall, *Weapons and Warfare*, 179–183; Oman, *A History*, 194–205.
152 Ágoston, "War-winning weapons?," 134–138.

field artillery and janissaries-harquebusiers advanced to the outer left flank. Thus, the Mamelukes were compelled to abandon their fortified positions, making them vulnerable to the Ottoman fire. Ottoman cavalry charged and forced the major part of the enemy artillery out of action. Combining cavalry attacks with cannon and janissary fire, the Ottomans achieved final victory.[153] Notably, Ottoman artillery and infantry with firearms did not occupy fixed positions but took the offensive. Compared to European practice, it was an example of advanced tactics, even if it was spontaneous.

In the battle of Mohács in 1526, Ottoman cannons and harquebusiers acted in a somewhat different manner. They chained together about 150 to 200 field pieces and placed entanglements in front of them, creating a barrier against cavalry. About 4,000 janissaries-harquebusiers were deployed behind the row of cannons. Neither Hungarian cavalry nor infantry penetrated the barriers, and the janissaries standing behind them decimated the enemy from a short distance. It is interesting that Ottoman artillery failed to perform adequately; later in the historical literature its role was exaggerated.[154]

Quite similar changes in tactical usage of artillery and handheld firearms occurring simultaneously in different regions raise the question of borrowing and influence (i.e., technology transfer). Unlike with imports of firearm production technology and craftsmen, tracing the spread of tactics is quite difficult, if possible at all. For example, gunmakers in the Ottoman service bearing Christian names is a clear indication of the employment of foreign specialists by Ottoman rulers. Regarding tactics, we need strong evidence like a reliable account from a competent contemporary in order to prove adaptation from a different military culture.

Returning to Sarnicki's description of the battle of Orsha, the infantry tactics resembled those of the Ottoman janissaries. Sarnicki goes further, suggesting that military actions during the Ottoman Empire's internal struggle from 1511 to 1512 offered a definite model from which to borrow. The fact that the commander of the Polish contingent at Orsha, Janusz Świerczowski, visited Istanbul on a diplomatic mission in 1511 is even more intriguing.[155] Ambassadors routinely studied the martial capabilities of countries where they served, and many combined a military career with one in diplomacy. Such borrowing must remain theoretical until supported by evidence. Reflections on potential borrowing can be geographically widened to Western Europe, where Poland and the Grand Duchy

---

153 Ágoston, "War-winning weapons?," 138–140.
154 Ágoston, "War-winning weapons?," 140–142; Ágoston, *Guns*, 24.
155 Archiwum Główne Akt Dawnych, Archiwum skarbu koronnego, 1: 44.

of Lithuania had various contacts. For example, prince Mikhail Glinski, a famous commander and adventurer, who defeated the Tatars at Kletsk in 1506, earlier had spent several years in Western Europe and participated in military actions there.[156]

At the same time, there is every reason to believe that tactics employed by the harquebusiers at Orsha, at Pavia, and at Raidania resulted from innovations by talented and resourceful commanders, who relied on their own experiences and those of their predecessors. Sometimes, novel tactics were developed via experimentation and exposure to new types of military equipment and thus found it difficult to gain acceptance in a military committed to traditional modes of warfare. Perhaps firearms represent one of the most vivid examples. It was associated with devilish powers, mocked by the aristocrats; reportedly, the hands of captured gunners were chopped off.[157] Nevertheless, as firearms evolved through different stages of development, they proved effective time after time. One can agree with the skeptics who deny they played a decisive role in certain battles and who caution against technological determinism in general.[158] Indeed, one cannot confidently assert that harquebuses won the battle of Pavia.[159]

According to Gábor Ágoston, the Ottomans owed their victories instead to a highly developed system for mobilizing troops, good campaign management, intelligence, discipline, soldiers' courage, and numerical superiority more so than to firearms.[160] Yes, out of 60,000 men at Mohács only 4,000 were harquebusiers; moreover, the Hungarians had about the same number. Certainly, no weapon by itself, whether catapult or atomic bomb, can guarantee victory. Superior weapons can serve their purpose only when skillfully used by trained and motivated troops in coordination with other units, provided with enough ammunition of adequate quality, delivered on time to the appropriate place, and sufficiently protected from enemy attack. Even when all of these conditions are fulfilled is success possible, but never guaranteed.

As Bert Hall showed, from the 1540s firearm modernization stalled for an extensive period, and, if one omits the spread of the "pistol gun," even from the 1520s. However, tactical development with firearms implementation did not halt. In the eighteenth century, firearms became the principal armament of the modern army. Muskets were used to defeat the enemy from close range, cannons from long range. Cavalry could rarely succeed without support from infantry and artillery, while infantry with enough guns could win a battle even without cavalry.

---

156 Pocieha, "Glinski Michał." 65.
157 Chase, *Firearms*, 59; Mallett, and Hale, *Military Organization*, 84, 143, note 186.
158 DeVries, "Catapults are not atomic bombs," 454–470.
159 Black, *European Warfare*, 77.
160 Ágoston, "War-winning weapons," 140–143.

In the late fifteenth and early sixteenth centuries, crucial changes in warfare were just emerging. In Western Europe, coordinated pike and shot tactics began to dominate much later, when building armed forces according to a single model would be a necessary condition to establish any state's serious political claims. But before that, the competition between various military systems reminds the keen observer, Bert Hall, of a rock-paper-scissors game.[161] The Swiss fought on foot in dense columns with no cavalry and very few firearms. The French, famous for their heavy cavalry and artillery, failed to create native infantry qualitatively equal to the Germans or Spanish. The English, known for their conservatism, remained committed to the long bow and lacked heavy cavalry. They made some efforts to create modern ordnance; in 1523 their cannon could level medieval fortifications on a par with the French artillery.[162] Nothing, however, indicated serious developments in warfare at Flodden in 1513 when English bowmen defeated Scottish pikemen in a medieval style battle. The Scots, after some success in developing firearms under James IV, showed little interest in further innovation.[163]

Evolution of warfare in Eastern Europe went its own way. Technical innovations tended to be applied rather quickly, but only when in demand and demonstrably effective. Employing artillery to defend and besiege fortresses rapidly became standard military practice. Handheld firearms, predominately infantry weapons, proved of little use fighting nomads on the steppes.[164] One clear illustration is the contrasting composition of Polish troops opposing Tatar raids in the south and those combating the Teutonic Order in the north. The sovereigns of Muscovy, needing satisfactory ordnance for their conquests, centralized production of cannon earlier than any other polity in the region. The battle of Orsha proved that infantry with firearms in a forest area could defeat a substantial contingent of mounted bowmen. If large battles had occurred there as frequently as in Western Europe, more such examples would exist. However, the infantry's increasing role in the Grand Duchy of Lithuania's warfare in the early sixteenth century did not precipitate long-term changes. The *szlachta* culture required fighting while mounted and Lithuanian authorities lacked sufficient funds to afford the necessary numbers of infantry, or mercenaries in general. During the Livonian War (1554–1582) attempts to force landowners to muster foot soldiers

---

161 Hall, *Weapons and Warfare*, 157–158.
162 Fissel, *English Warfare*, 7–8.
163 Phillips, "Scotland in the age of the military revolution," 186–190.
164 Chase, *Firearms*. Chase argues that firearms proved to be of a little use fighting nomads. He thus explains why gunpowder weapons were developed and used widely in the West and not in those parts of the world threatened by nomads. His concept has received some criticism with regard to China; see Andrade, *The Gunpowder Age*, 112–113.

ended in failure. *Szlachta*, fearing royal power becoming stronger, opposed any move to create a solid permanent army in the Polish-Lithuanian Commonwealth. Eventually, military weakness mirroring political weakness was one factor in the country's downfall.

# Bibliography

*Acta Tomiciana*, vol. 3. Posnaniae, 1853.
Ágoston, Gábor. "War–winning weapons? On the decisiveness of Ottoman firearms from the siege of Constantinople (1453) to the battle of Mohács (1526)." *Journal of Turkish Studies* 39 (2013): 129–143.
Ágoston, Gábor. *Guns for the Sultan: Military Power and the Weapons Industry in the Ottoman Empire.* New York: Cambridge University Press, 2005.
Alekseev, Yuriy. "Kampaniya 1500 goda v svete noveyshikh issledovaniy" [1500 Campaign in the Light of the Newest Research]. In *Po liubvi, v pravdu, bez vsiakie hitrosti. Druziya I kollegi k 80-letiyu Andreya Vladimirovicha Kuchkina. Sbornik statey* [Lovingly, in truth, with no tricks. Friends and colleagues on the 80th anniversary of Andrey Vladimirovich Kuchkin. A collection of articles], edited by Boris Floria, 317–330. Moscow: Indrik, 2014.
Alekseev, Yuriy. *Pokhody russkikh voysk pri Ivane III* [Russian Troops' Campaigns under Ivan III]. Saint Petersburg: Izdatel'sto Sankt-Peterburzhskogo universiteta, 2009.
Andrade, Tonio, Hyeok Hweon Kang, and Kirsten Cooper. "A Korean military revolution? Parallel military innovations in East Asia and Europe." *Journal of World History* 25.1 (2014): 51–84.
Andrade, Tonio. *The Gunpowder Age: China, Military Innovation, and the Rise of the West in World History.* Princeton: Princeton University Press, 2016.
Archiwum Główne akt dawnych w Warszawie. Archiwum skarbu koronnego, vol. 1, Rachunki królewskie.
Archiwum Główne Akt Dawnych w Warszawie, Archiwum Radziwiłłów, vol. 2, 3296.
Bielski, Marcin. *Kronika Polska Marcina Bielskiego.* Sanok: K. Pollak, 1856.
Black, Jeremy. *European Warfare, 1494–1660.* London/New York: Routledge, 2002.
Bokhan, Yury. *Vayskovaya sprava u Vialikim Kniastvie Litouskim u druhoy palove XIV – kancy XVI stahoddzia* [Military affairs in the Grand Duchy of Lithuania in the second half of the 14th – to the end of the 16th century]. Minsk: Belaruskaya Navuka, 2008.
Bołdyrew, Aleksander. "Przemiany uzbrojenia wojska polskiego na przełomie średniowiecz i nowożytności (1454–1572) jako przejaw (r)ewolucji militarnej." *Roczniki dziejów społecznych i gospodarczych* 80 (2019): 113–138.
Bołdyrew, Aleksander. *Piechota zaciężna w Polsce w pierwszej połowie XVI wieku.* Warszawa: Neriton, 2011.
Chase, Kenneth. *Firearms: A Global History to 1700.* Cambridge/New York: Cambridge University Press, 2003.
Cook, Weston F. "The cannon conquest of Nasrid Spain and the end of the Reconquista." *Journal of Military History* 57.1 (January 1993): 43–70.

Davies, Brian L. "Guliai-gorod, wagenburg and tabor tactics in 16$^{th}$–17$^{th}$ century Muscovy and Eastern Europe." In *Warfare in Eastern Europe, 1500–1800*, edited by Brian L. Davies, 93–108. Leiden/Boston: Brill, 2012.
Davies, Brian L. *Empire and Military Revolution in Eastern Europe: Russia's Turkish Wars in the Eighteenth Century*. London/New York: Continuum, 2011.
Davies, Brian L. *State, Power and Community in Early Modern Russia: The Case of Kozlov, 1635–1649*. Houndmills, Basingstoke: Palgrave Macmillan, 2004.
Davies, Brian L. *Warfare, State and Society on the Black Sea Steppe, 1500–1700*. London, New York: Routledge, 2007.
Davies, Brian L., ed. *Warfare in Eastern Europe, 1500–1800*. Leiden/Boston: Brill, 2012.
Decius, Iodocus L. *De Sigismundi regis temporibus liber 1521*. Kraków: Akademia umiejętności, 1901.
DeVries, Kelly. "Catapults are not atomic bombs: towards a redefinition of effectiveness in premodern military technology." *War in History* 4.4 (1997): 454–470.
Dopolneniya k aktam istoricheskim [Additions to historical acts], vol. 1, edited by Yakov Berednikov. Saint-Petersburg, 1846.
Duffy, Christopher. *Siege Warfare: the Fortress in the Early Modern World, 1494–1660*. London: Routledge, 1979.
Fissel, Mark Charles. "Military Revolutions." In *Oxford Bibliographies in Military History*, edited by Kaushik Roy. New York: Oxford University Press, https://www.oxfordbibliogra phies.com/view/document/obo-9780199791279/obo-9780199791279-0212.xml.
Fissel, Mark Charles. *English Warfare, 1511–1642*. London/New York: Routledge, 2001.
Frost, Robert. *The Northern Wars: War, State and Society in Northeastern Europe, 1558–1721*. Harlow, New York: Pearson Education Limited, 2000.
Garza, Andrew de la. *The Mughal Empire at War: Babur, Akbar and the Indian Military Revolution, 1500–1605*. London/New York: Routledge, 2016.
Gawron, Przemysław. "Wojskowość zachodnioeuropejska w dobie bitwy pod Orszą i jej związki z litewską–polską sztuką wojenną," *Biblioteka epoki nowożytnej* 3 (2015): 37–52.
Górski, Konstanty. "O rozwoju sztuki wojennej w Polsce w wieku XV." *Biblioteka Warszawska* 198.2 (1890): 197–215, 407–422.
Grabarczyk, Tadeusz. *Piechota zaciężna Królestwa Polskiego w XV wieku*. Łódź: Ibidem, 2000.
Grabarczyk, Tadeusz. *Jazda zaciężna Królestwa Polskiego w XV wieku*. Łódź: Blue Note, 2015.
Gustave, Alef. "Muscovite Military Reforms in the second-half of the Fifteenth Century." *Forschungen zur Osteuropäischen Geschichte* 18 (1973): 73–108.
Hale, John R. *War and Society in Renaissance Europe, 1450–1620*. Baltimore: Johns Hopkins University Press, 1985.
Halecki, Oskar. *Dzieje unii Jagiellońskiej*, vol. 2, *W XVI wieku*. Kraków: Akademia Umiejętności, 1920.
Hall, Bert S. *Weapons and Warfare in Renaissance Europe*. Baltimore: Johns Hopkins University Press, 1996.
Hellie, Richard. "Warfare, changing military technology and the evolution of Muscovite society." In *Tools of War. Instruments, Ideas, and Institutions of Warfare, 1445–1871*, edited by John A. Lynn, 74–99. Chicago: University of Illinois Press, 1990.
Herberstein, Sigismund von. *Zapiski o Moskovii* [Notes on Muscovite Affairs], vol. 1, edited by Anna Khoroshkevich. Moscow: 2008.
*Ioasafovskaya letopis'* [Ioasafovskaya Chronicle], edited by Aleksandr Zimin. Moscow: Izdatelstvo Akademii nauk SSSR, 1957.

Janicki, Marek. 'Obraz *Bitwa pod Orszą* – geneza, datowanie, wzory graficzne a obraz bitwy 'na Kropiwnej' i inne przedstawienia batalistyczne w wileńskim pałacu Radziwiłłów', *Biblioteka Epoki Nowożytnej* 3 (2015): 173–225.

Kazakou, Aliaksandr. "Histaryyahrafiya Arshanskay bitvy 1514 h.: dasiahnenni i prablemy" [The Historiography of the Battle of Orša, 1514: achievements and problems]. *Belarusian Historical Review* 22 (2015): 20–24.

Kazakou, Aliaksandr. "Muscovites among the courtiers of the Lithuanian grand duke Sigismund the Old: Evidence from the Census of 1509." *The Fifth International Congress of Belarusian Studies. Working papers*, vol. 5, 63–66. Kaunas: Vytautas Magnus University Press, 2016.

Kazakou, Aliaksandr. "Opravdyvaya porazhenie: Orshanskaya bitva 1514 g. glazami letopiscev" [Excusing the defeat: the Battle of Orsha, 1514 in the eyes of chroniclers]. *Ukrayina v Centralno-Shidnoy Evropi* 18 (2018): 315–334.

Kazakou, Aliaksandr. "Paliavannie na arshanski mif" [Hunting the Orsha myth]. *ARCHE* 5 (May 2012): 88–97.

Kazakou, Aliaksandr. "Bitwa pod Orszą 1514 r.: kwestia lokalizacji," *Biblioteka epoki nowożytnej* 3 (2015): 127–141.

Kirpichnikov, Anatoliy. *Voennoe delo na Rusi v XIII – XV vv.* [Military affairs in Rus' from XIII to XV centuries]. Leningrad: "Nauka," 1976.

Krom, Mikhail, ed. *Radzivillovskie akty iz sobraniya rosiyskoy nacionalnoy biblioteki* [Acts of Radziwill from Russian National Library Collection], Pamiatniki Istorii Vostochnoy Evropy [Historical Heritage of Eastern Europe series], vol. 6. Moscow/Warsaw: Drevlekhranilishche, 2002.

Krom, Mikhail. *Mezh Rus'yu I Litvoy. Pogranichnye zemli v sisteme russko-litovskikh otnosheniy v konce XV – pervoy treti XVI veka* [Between Ruthenia and Lithuania. Borderlands in the system of Russian–Lithuanian relations in the end of the 15th – the first third of the 16th century]. Moscow: Obyedinennaya Redakcyia MVD Rossii, 2008.

*Kroniki Bernarda Wapowskiego (1480–1535).* Scriptores Rerum Polonicarum, vol. 2. Cracoviae, 1874.

Lesmaitis, Gediminas, ed. *Popisy wojskowe pospolitego ruszenia Wielkiego Księstwa Litewskiego, 1524–1566.* Białystok: Instytut Badań nad Dziedzictwem Kulturowym Europy, 2016.

Lesmaitis, Gediminas. "Ad Viadroshy da Orshy: zmeny u sistemie kamplektavannia voyska VKL" [From Viadrosha to Orsha: changes in the composition of the GDL's army]. *ARCHE* 12 (2014): 121–139.

Lesmaitis, Gediminas. *Wojsko zaciężne w Wielkim Księstwie Litewskim w końcu XV – drugiej połowie XVI wieku.* Warszawa: Neriton, 2013.

*Lietuvos didžiojo nkunigaikščio Aleksandro Jogailaičio dvaro sąskaitų knygos, 1494–1504,* Vilnius: "Lietuvos pilys", 2007, edited by Darius Antanavičius and Rimvydas Petrauskas.

*Lietuvos Metrika. Knyga 15: (1528–1538). Užrašymų knyga 15,* edited by Artūras Dubonis. Vilnius: Žara, 2002.

*Lietuvos Metrika. Knyga 7: (1506–1539). Užrašymų knyga 7,* edited by Laimontas Karalius and Darius Antanavičius. Vilnius: Lietuvos istorijos instituto leidykla, 2011.

*Lietuvos Metrika. Knyga 10: (1440–1523). Užrašymų knyga 10,* edited by Egidijus Banionis, Algirdas Baliulis. Vilnius: Mokslo ir enciklopedijų leidybos institutas, 1997.

*Lietuvos Metrika. Knyga 11: (1518–1523). Užrašymų knyga 11,* edited by Artūras Dubonis. Vilnius: Lietuvos istorijos instituto leidykla, 2003.

*Lietuvos Metrika. Knyga 6: (1496–1506). Užrašymų knyga 6,* edited by Algirdas Baliulis. Vilnius: Lietuvos istorijos instituto leidykla, 2007.

*Litovs'ka metrika. Knyga 561. Reviziya ukrayins'kyh zamkiv 1545 hodu* [Lithuanian Metrica. Book 561. Inspection of the Ukranian Castles in 1545], edited by Volodymyr Kravchenko. Kyiv: Nacionalna akademiya nauk Ukrayiny, 2005.
Lobin, Aleksey. "Russian artillery made on the Italian models in 1480–1520s." *Armi Antiche* (2015): 107–130.
Lobin, Aleksey. *Bitva pod Orshey 8 sentyabrya 1514 goda* [The Battle of Orsha on 8 September, 1514]. Saint-Petersburg: Obshchestvo pamyati Igumenii Taisii, 2011.
Lohr, Eric, and Marshall Poe, eds. *The Military and Society in Russia, 1450–1917*. Leiden/Boston/Köln: Brill, 2002.
Łopatecki, Karol. *Organizacja, prawo i dyscyplina w polskim i litewskim pospolitym ruszeniu (do połowy XVII wieku)*. Białystok: Instytut Badań nad Dziedzictwem Kulturowym Europy, 2013.
Lulewicz, Henryk. "Jerzy Mikołajewicz Radziwiłł, starosta grodzieński, hetman dworzański (1514–1522/23?)." In *Radziwiłłowie w służbie Marsa*, edited by Mirosław Nagielski and Karol Żojdź, 13–32. Warszawa: DiG, 2017.
Mallett, Michael E., and John R. Hale. *Military Organization of a Renaissance State: Venice, c. 1400 to 1617*. Cambridge: Cambridge University Press, 1984.
McNeill, William H. *The Pursuit of Power: Technology, Armed Force and Society Since A.D. 1000*. Chicago: University of Chicago Press, 1982.
Metryka Vialikaha Kniastva Litouskaha (KMF–18) [Metrica of the Grand Duchy of Lithuania (KMF–18)]. Nacyyanalny Histarychny Arkhiu Belarusi [National Historical Archive of Belarus].
Mleczek, Tomasz. "Armia koronna w pierwszej połowie XVI w. na tle przeobrażeń w sztuce wojennej doby wojen włoskich." PhD diss., Warsaw University, 2012.
Murphey, Rhoads. *Ottoman Warfare, 1500–1700*. London: UCL Press, 1999.
Oman, Charles. *A History of the Art of War in the Sixteenth Century*. London: Greenhill Books, 1999.
Ordensbriefarchiv. Geheimes Staatsarchiv Preußischer Kulturbesitz.
Ostrowski, Donald. "The replacement of the composite reflex bow by firearms in the Muscovite cavalry." *Kritika: Explorations in Russian and Eurasian History* 11.3 (2010): 513–534.
Pakhomov, Igor. "Pishchalniki Vasiliya III" [Pishchalniki of Vasily III]. *Tseyhgaus* 20 (2002): 6–9.
Papee, Fryderyk. *Aleksander Jagiellończyk*. Kraków: Universitas, 1991.
Parker, Geoffrey. *The Military Revolution: Military Innovation and the Rise of the West, 1500–1800*. Cambridge: Cambridge University Press, 1988.
Penskoy, Vitaliy. *Velikaya ognestrelnaya revolutsiya* [The great gunpowder revolution]. Moscow: Eksmo, Yauza, 2010.
Penskoy, Vitaly. "Razvitie vooruzhennykh sil Rossii i voennaya revolutsiya v Zapadnoy Evrope vo vtoroy polovine XV – XVIII vv.: sravnitelno-istoricheskiy analiz" [The development of Russian armed forces and the Military Revolution in Western Europe in the second half of the XV – XVIII centuries: comparative historical analysis]. Habilitation diss. Moscow University for Humanities, 2004.
Pepper, Simon. "The face of the siege: fortifications, tactics and strategy in early Italian wars." In *Italy and the European Powers: The Impact of War, 1500–1530*, edited by Christine Shaw, 33–56. Leiden: Brill, 2006.

Phillips, Gervase. "Scotland in the age of the military revolution." In *Military History of Scotland*, edited by Edward M. Spiers, Jeremy Crang, and Matthew Strickland, 182–208. Edinburgh: Edinburgh University Press, 2012.
Pietkiewicz, Krzysztof. "Dwor litewski wielkiego księcia Aleksandra Jagiellończyka (1492–1506)." In *Lietuvos valstybe XII–XVIII a.*, edited by Zigmantas Kiaupa, Artuars Mickievicius, and Jolita Sarceviciene, 75–125. Vilnius: Lietuvos Istorijos institutas, 1997.
Plewczyński, Marek. *Daj nam, boże, sto lat wojny*. Warszawa: Bellona, 1997.
Plewczyński, Marek. *Obertyn, 1531*. Warszawa: Bellona, 2008.
Plewczyński, Marek. *Wojny i wojskowość polska w XVI wieku*, vol. 1, *Lata 1500–1548*. Zabrze: Inforteditions, 2011.
Pocieha, Władysław. "Glinski Michał." In *Polski słownik biograficzny*, vol. 8, 65–69. Wrocław, Kraków, Warszawa: Zakład narodowy imienie Ossolińskich, Wydawnictwo Polskiej Akademii Nauk, 1959–1960.
Poe, Marshall. "The military revolution, administrative development, and cultural change in early modern Russia." *Journal of Early Modern History* 2.3 (1998): 247–273.
*Polnoe sobranie russkikh letopisey* [The full collection of Russian chronicles], vol. 13, *Letopisnyy sbornik imenuemyy Patriarshey ili Nikonovskoyu letopis'yu* [Collection of Chronicles called the Patriarch's or Nikonovskaya Chronicle], part 1, edited by Sergey Platonov. Saint-Petersburg, 1904.
*Polnoe sobranie russkikh letopisey* [The full collection of Russian chronicles], vol. 37, *Ustyuzhskie and Vologodskie letopisi XVI–XVIII vekov* [Ustyuzhskie and Vologodskie Chronicles of XVI-XVIII centuries], edited by Ksenya Serbina. Leningrad: "Nauka", 1982.
*Polnoe sobranie russkikh letopisey* [The full collection of Russian chronicles], vol. 19, *Istoriya o Kazanskom Tsarstve* [A story about Khazan' tsardom], edited by Georgiy Kuntsevich. Saint-Petersburg, 1903.
*Polnoe sobranie russkikh letopisey* [The full collection of Russian chronicles], vol. 6, *Sofiyskaya Vtoraya Letopis* [Sofiyskaya second chronicle], issue. 2, edited by Sergey Kisterev, and Liudmila Timoshina. Moscow: Yazyki russkoy kultury, 2001.
Potter, David. *Renaissance France at War: Armies, Culture and Society, c. 1480–1560*. Woodbridge/Rochester: Boydell & Brewer, 2008.
*Pskovskie letopisi* [Pskovskie Chronicles]. Second issue, edited by Arseniy Nasonov. Moscow: Izdatelstvo Akademii nauk SSSR, 1955.
Raymond, James. *Henry VIII's Military Revolution: The Armies of Sixteenth-Century Britain and Europe*. London: Tauris, 2007.
Ritningssamlingen, J 3, volume 45, part 1, 5377, 113; part 2, 5378, 95; part 3, 5379, 60. Armémuseum in Stockholm.
Rogers, Clifford J. "Tactics and the face of battle". In *European Warfare, 1350–1750*, edited by Frank Tallett, and David J. B. Trim, 203–236. New York: Cambridge University Press, 2010.
Sarnicki, Stanisław. *Księgi hetmańskie*. Krakow: "Historia Iagellonica", 2015.
Stryjkowski, Maciej. *Kronika Polska, Litewska, Żmódzka i wszystkiej Rusi*, vol. 2. Warszawa, 1846.
Szymczak, Jan. *Początki broni palnej w Polsce (1383–1533)*. Łódź: Wydawnictwo Uniwersytetu Łódzkiego, 2004.
Taylor, Frederick L. *The Art of War in Italy, 1494–1529*. Cambridge: Cambridge University Press, 1921.
Temushev, Viktor. *Pervaya moskovsko-litovska pogranichnaya voyna 1486–1494* [The First Muscovy–Lithuanian frontier war, 1486–1494]. Moscow: Kvadriga, 2013.

Volkau, Mikola. *Artyleryya Niasvizhskaha zamka* [Niasvizh castle artillery]. Minsk: Janushkevich, 2015.

Wimmer, Jan. *Historia piechoty polskiej do roku 1864*. Warszawa: Wydawnictwo Ministerstwa Obrony Narodowej, 1978.

Zimin, Aleksandr. *Vityaz na rasput'e: feodalnaya voyna v Rossii* XV veka [Vityaz at the crossroads: The feudal war of the 15th century in Russia]. Moscow: Mysl', 1991.

Wayne E. Lee
# To Stop a Cannonball: Ottoman Fortress Design and Comparing Military Revolutions, 1350–1730

Studying a military revolution or a "revolution in military affairs" inevitably requires comparative work.[1] Even within a single broad theater of combat like Europe, China, or the modern globalized Pacific, one must know the evidence and literature for the before and the after to accurately assess whether any changes have been revolutionary. Furthermore, in most circumstances claims for a military revolution or an RMA have been proposed as both revolutionary in time, and also within a comparative context of different societies, cultures, or polities. Geoffrey Parker's paradigmatic study of the early modern military revolution was predicated on the idea that certain key changes in European military practice conferred crucial advantages on Europeans vis-à-vis other combatants around the world, although they were not yet sufficient to overwhelm China or Japan.[2] His critics include both those interested in the details of intra-European chronology and its implications, and those more concerned with the extra-European comparison.[3] The complexity of the problem is compounded when the change in question is actually a plural. Cannon, firearms, drill, fortifications, and ships are all central to Parker's model and all are individually complex technical subjects,

---

[1] Thanks to Clifford Rogers and Kelly DeVries for citations, advance copies of papers, and discussions on European developments. Virginia Aksan helped prevent me making key mistakes and gave me some additional sources as well. Thanks also to David Nixon for sending me his unpublished paper. This paper was greatly shaped by Kahraman Sakul, who took me to Çanakkale and Kilidülbahir, and provided some key citations and some of his own unpublished work on Ottoman fortifications, especially in the later period. Also thanks to Rebecca Seifried for helping survey the forts at Kelefa and Passavas and sharing Ottoman defter data, and to Sylvia Galaty for helping with some Albanian. And as always, I am grateful to Rhonda Lee, my extremely capable in-house editor. The great variety of alphabets, languages, and consequently placenames in the Balkans complicates having a consistent way to present them. I have opted for those versions of placenames that appear most frequently in the literature that deals with them, and even then that is a judgment call. No political statement is implied by any of my choices. I have mostly chosen to avoid listing all the alternative names.
[2] Parker, *The Military Revolution*.
[3] The literature is enormous and is referenced elsewhere in this volume. But for one example of the former (intra-European chronology) see Rogers, "The Military Revolutions of the Hundred Years War," 55–94. For the latter (global comparison), see Black, *Beyond the Military Revolution*. Most recently, see Andrade, *The Gunpowder Age*.

https://doi.org/10.1515/9783110661415-004

made more so in combination. As a result, many students of the subject since Parker have drilled into ever narrower niches, seeking to perfect the chronology, clarify which cause for which effect, and typically assert more evolutionary than revolutionary change.

Such detailed studies clarify what actually happened, but they often lose their comparative usefulness. Studying military history from a world historical perspective allows us to see patterns of action, reaction, adoption, and change among warring combatants over long periods of time, all generally indicating that they often responded as much to each other as to changes in technology or other conditions. On one hand, Jeremy Black has quite reasonably complained that we should not measure world militaries against an essentially arbitrary European standard.[4] Armies in China, or wherever, had different problems to solve. On the other hand, John Lynn, although he was writing about intra-European conflict, argued that paradigmatic armies forced neighboring others to copy aspects of them in order to survive.[5] And indeed, there is a fair amount of evidence in military history for convergence in military systems (if not ideologies) among competing entities. Alex Roland's short work on war and technology further suggests a specific variant of convergence, arguing that prior to about 1980, within contexts of state-based combatants, military technologies often tended toward symmetrical "gigantism." Ships of war enlarged (whether powered by oars, sails, or steam); medieval aristocratic armor got heavier; stable-fed horses got bigger; tanks swelled in weight and firepower; fortresses became more extensive and complex; and one can even think of ever-more-rapid-firing firearms as a functional equivalent of "gigantism." In contrast, a combatant who lacked the productive capacity to engage in a gigantism race pursued different, "asymmetric" modes of combat.[6]

These questions of revolutionary change, convergence, or even gigantism have animated the relatively smaller body of literature comparing western European developments to those in eastern Europe and the Ottoman Empire. The Ottomans have long seemed a valuable comparison because they were culturally "different" but still operated within the same general level of technological development (unlike, for example, early modern sub-saharan Africa or the New World.) While I was working on my own effort to place early modern European military developments into a comparative context, I stumbled into a seemingly understudied problem: Ottoman fortress design in the gunpowder era. This

---
4 Black, *War in the World*, 214–218.
5 Lynn, "The Evolution of Army Style in the Modern West, 800–2000," 505–545.
6 Roland, *War and Technology*, 45, 47–48, 52, 73, 115–117. I develop other examples of this kind of symmetrical gigantism in Lee, *Waging War*.

seems to me a particularly fruitful case study: The combatants were hierarchical, sedentary, agricultural states, engaged in conflict over a very long period, and definitely experiencing convergence (and gigantism) on some fronts in both guns and ships (at least), something widely acknowledged in the literature on the Ottoman-European confrontation.[7] But something different, or at least something less clear, happened in the design of fortresses. Furthermore, as in my earlier work on Native American fortifications, the Ottoman case has promise for doing world military history in a comparative mode when the textual sources are limited. The forts are artifacts that in some ways tell their own story, and surprisingly, we will find that they force us to examine deeper questions than simple functionality. Forts are not just walls for keeping out enemies. They are also the incarnation of cost-benefit calculations, strategic orientation, instruments of domestic social control, and messages about power and legitimacy.

The history of Ottoman fortifications highlights key differences between Ottoman and western and central European designs. At an early stage, those differences reflected the complexity of solving a problem that was itself a moving target. The interactions of guns, gunpowder, metallurgy, and existing and emerging fort design from roughly 1400 to 1500 was a difficult and shifting puzzle. Early Ottoman solutions were as innovative and as effective as their western (primarily Italian and French) counterparts. Over time, however, a trajectory of difference was established that widened the gap between European designs and Ottoman ones, and that trajectory favored western systems. We must ask *why* they diverged? As we will see, one explanation for the supposed Ottoman general disinterest in Italian fortification styles – the so-called *trace italienne* – was simply their strategic orientation (at least through the end of the seventeenth century). It was not a matter of backwardness, nor close-mindedness – the Ottomans were extremely open to new ideas, and technicians and thinkers moved freely in and out of the empire – but instead we find them making a different set of cost-benefit analyses in accord with their vision of the role and future of the empire. In this sense, Ottoman fortification systems provide another example of how a "revolutionary" new technology or method may provide a substantial advantage only within a specific context, and be irrelevant to those operating with a different definition of victory.[8] Even beyond strategic orientation, however, we will

---

7 Murphey, *Ottoman Warfare, 1500–1700*, 107. Although it is also true that since the Ottomans faced other opponents to the east and north with different fighting styles, they did not completely "converge." Chase, *Firearms: A Global History to 1700*.

8 This is far from the last word on Ottoman fortification. It proposes a series of ideas and suggestions based on wide reading in regional studies of Ottoman military architecture and my

also track the effects of printing, of bureaucracy, and even the chemistry of the mortar holding the stones together. Design does not just follow function.

## The Italian Style Artillery Fortress in Europe

Most readers will be familiar with the basic outlines of the emergence of the *trace italienne*. But some key details and some chronological categorization is necessary to sustain the comparison to developments within the Ottoman Empire. Throughout Europe and the Mediterranean, changes in fortification technique were driven by the arrival and improvement of gunpowder artillery, and so the story must begin with guns.

While there are legitimate claims that gunpowder artillery accompanied almost every siege in northwestern Europe as early as the end of the fourteenth century, the usual and still prevailing argument is that truly effective siege artillery had to await the middle of the fifteenth century.[9] The uncertainty derives from the rapidity of change in guns and gunpowder in this early period. Too often historians lump the 1325 to 1450 period together as a single "early artillery" phase. In reality, as Clifford Rogers points out, "gunpowder artillery changed as much in *each* twenty-five year period between 1325 and 1500 as it did over the following three centuries." In the earliest phases, essentially most of the fourteenth century, gunpowder weapons were mostly used to fire *over* walls rather than *at* walls, much the way trebuchets had been used.[10] Even these early cannon, however, compared to bows or crossbows, or even compared to standard medieval siege weapons like springalds or trebuchets, could deliver projectiles over much greater range and with much more penetrating power. They also grew rapidly in size over the course of the fourteenth and early fifteenth centuries, from firing projectiles of well under 100 kilograms in the 1370s, to balls of 1,700 kilograms by around 1415 (these figures represent the largest then available

---

own observations of fortresses in Greece, Hungary, Albania, and western Turkey. I acknowledge a great debt to the work of Simon Pepper and David Nicolle, as well as other specialists.

**9** DeVries, "The Impact of Gunpowder Weaponry on Siege Warfare in the Hundred Years War," 227–244, esp. 229. For summaries of the more conventional view, see the up-to-date examination in Rogers, "Gunpowder Artillery in Europe, 1326–1500: Innovation and Impact," 39–71. For older but still extremely useful works on key details, see Pepper and Adams, *Firearms and Fortifications*, esp. 3–17 for the early changes in guns; Parker, *Military Revolution*; Hall, *Weapons and Warfare in Renaissance Europe*. There are many other studies, some cited below.

**10** Rogers, "Gunpowder Artillery in Europe"; this latter point is also made in Hale, "The Early Development of the Bastion: An Italian Chronology, c. 1450–1534," 474.

cannon rather than an average, but the shift is indicative).[11] It was this marked increase in power that, as Kelly DeVries argues, made guns a standard part of a siege by the end of the fourteenth century.

Even so, for most of the early fifteenth century, strong medieval stone fortifications remained resistant to bombardment with stone balls, however large. Change continued apace over the course of the century, however, as barrels lengthened, ratios of powder to ball weight changed, powder chemistry shifted, and the number of cannon in a typical siege train increased. In Rogers' estimation: "relative to a major siege of around 1400, one conducted around 1450 might involve something on the order of 8 times as many great bombards, each delivering 7.5 times as much kinetic energy per shot, and shooting 3 times as frequently, for a total of around one hundred and eighty times as much destructive power deliverable per day."[12] From mid-century to about 1500 another series of design changes took hold, eventually settling into a standard pattern that would remain very similar for the next three centuries: barrels lengthened further, cast-iron balls replaced stone – allowing smaller caliber cannon to produce equivalent or even improved penetration; and cast bronze generally replaced wrought iron hoop-and-stave bombards. The combination of longer barrels, iron balls, and corned gunpowder meant that wall-breaking guns no longer needed to be monstrous. In addition, guns were increasingly standardized into categories by size of ball, which simplified and amplified the effects of the swelling number of cannon in a siege train. And not least, the new bronze cannon were cast with trunnions to make changing the elevation of the gun a simple process, and they were mounted on two-wheeled carriages to increase their strategic and tactical mobility. In 1494, the French king Charles VIII invaded Italy with a mobile artillery siege train featuring all of these changes. His army progressed down and back up the length of Italy, taking cities seemingly at will. The contemporary Italian historian Francesco Guicciardini wrote that the new French artillery moved so quickly that it

> almost kept Pace with the Army. They were planted against the Walls of a Town with such speed, the Space between the Shots was so little, and the Balls flew so quick, and were impelled with such Force, that as much Execution was done in a few Hours, as formerly, in Italy, in the like Number of Days.[13]

In reality, there were a variety of factors beyond just Charles's artillery that rendered Italian cities vulnerable at that time, but the contemporary perception of

---

[11] See the chart in Rogers, "Gunpowder Artillery in Europe," 50.
[12] Rogers, "Gunpowder Artillery in Europe," 67–68.
[13] Gucciardini, *The History of Italy*, 147–149.

the decisiveness of his artillery siege train hastened already ongoing changes in fortification.[14]

Effective siege cannon from the 1420s and especially by the 1490s increasingly had rendered vertical medieval curtain walls and tall towers vulnerable. Early design adjustments in that period focused on reinforcing or altering existing walls to increase their resistance to shot. Some of those adjustments, such as removing overhanging machicolations and crenellations – which were likely to splinter under fire – made the walls more vulnerable to escalade. As a consequence, new systems of embrasures in walls and especially in projecting towers were created to generate close-range, mostly anti-personnel fire, designed particularly to cover the "dead space" at the foot of walls and towers formerly protected by machicolations. In the longer term, as walls got thicker and proportionally lower, the platforms originally designed to increase close-in defensive fire also became strong enough and expansive enough to mount heavy cannon capable of longer-range offensive firepower, designed to keep the enemy at a distance from the walls. Those platforms eventually became the angular bastions that would be the signature component of the *trace italienne* artillery fortress. But the dominance of the "offensive" bastion design emerged only from the 1520s.[15] We should consider first the variety of fifteenth-century responses to artillery – what might be called a transitional period in fortress design between the high round or square towers of the medieval period and the interlocking long- and short-range fire of low profile, linked, angular bastions.[16]

The transitional period produced a variety of solutions within western Europe as different designs were tried out and as designers sought economy in the face of enormous costs and had to deal with existing walls. To simplify: one can alter walls to make them resist cannon fire, or one can alter walls to support cannon that fire back, or one can try to do both. Firing back can be done either defensively at short-range to prevent escalade, or offensively at long-

---

**14** Potter, *Renaissance France at War*, 153–154. The 1494 invasion has often been portrayed as catalytic to sudden changes in fortification design in Italy. More detailed work, some summarized below, emphasizes that many changes in fortification design preceded the invasion, all in response to the pre-1494 changes in guns. For one introduction, see Pepper and Hughes, "Fortification in Late 15th-Century Italy: The Treatise of Francesco di Giorgio Martini," 541–562.
**15** For this defensive/offensive characterization, see Hale, "Bastion," 475.
**16** Useful sources on changing fort design in this period include Hughes, *Military Architecture*; Pepper and Hughes, "Fortification in Late 15th-Century Italy," 541–542. For the pre-1494 development of bastions, see Hale, "Early Development." Other key summaries of the literature are: Parker, *Military Revolution*; Duffy, *Siege Warfare*. Other more specialized sources are cited throughout.

range in the form of counter-battery fire directed against the besieging artillery and its protective infantry screen.

In early defensive wall redesign, one improvisation was to pile up earth mounds behind existing walls.[17] Another more permanent fix was to "scarp" the lower third or so of otherwise vertical walls (scarping thickened and angled the lower part of the wall). Scarping – sometimes called battering – appears in various locations around Italy as early as the 1440s, soon accompanied by the removal of vulnerable machicolations and crenellations, and then by the reduction of towers to the same height as the wall. All of these changes could be made to existing walls but were incorporated into new designs as well (see Figure 1).[18]

**Figure 1:** Scarped walls at Salses, built in 1498 on the French-Spanish frontier. https://upload.wikimedia.org/wikipedia/commons/8/83/Castell_de_Salses_-_Forteresse_de_Salses_01.JPG. Accessed 3/4/2020.

These passive or defensive changes to the walls proceeded in parallel with changes designed to increase firepower from the walls. One early attempted solution was to add an outer ground-level layer to what was often already a concentric system of walls. Medieval castles had long relied on walls within walls to increase their strength, usually anchored by a central keep, each taller than

---

**17** DeVries, "Impact of Gunpowder Weaponry," 238.
**18** Hughes, *Military Architecture,* 75; Hale, "Early Development," 478; Pepper and Adams, *Firearms and Fortifications,* 17–18.

the next. The walls at Constantinople, for example, although lacking a central keep, were three layers deep. Adding another circuit wall at ground level with embrasures for cannon was a relatively simple way to enhance existing fortifications. The fact that it was at ground level meant that it was a simple matter to include a battery of heavy, offensive-oriented cannon.[19]

Another early improvisation to increase firepower, common by the late fourteenth century and continuing in France and elsewhere through the late fifteenth century, was to add gunports (embrasures) for cannon within medieval towers, or in towers newly built for that purpose (although still architecturally resembling a medieval round tower). In general medieval towers were not suited to supporting the weight of large bombards, and so these early artillery towers mounted mostly small caliber weapons for close range defensive fire against escalade.[20] This was quickly followed by the purpose-built, multi-level "artillery tower" (Pepper calls the early version the "Multi-Gun Tower"): a round, thickly walled tower with cannon pointing in all directions, and with embrasures at different levels – the round design here was sometimes exaggerated to create tight curves that would deflect cannon balls.[21] The "Cow Tower" at Norwich, built in 1398, combines all these characteristics: a multi-level artillery platform, tight curves in still vertical walls, and primarily designed to mount small caliber artillery (see Figure 2a). Eventually the artillery tower design would become thicker and squatter, followed soon thereafter by "water lily" designs of linked artillery towers, a design that would persist for a long time in parts of Europe, especially in England, where Henry VIII would build a variety of coastal artillery tower complexes on that principle in the 1540s (although some were likely designed by a German architect) including at Camber, Deal, and St. Mawes (see Figure 2b).[22] In general, artillery towers were designed to increase offensive firepower, to make the towers resistant to shot (through curvature, and eventually through squat thickness), but they did not solve the problems of dead space.

The thicker walls and shorter towers improved the forts' defenses against artillery, but additional changes were needed that *both* increased offensive firepower and eliminated dead space at the foot of the walls and towers in the

---

[19] For adding concentric outer walls in Italy in the mid-fifteenth century, see Pepper and Adams, *Firearms and Fortifications*, 18–22. For similar works called *boulevards* in France, see the nineteenth-century work by Eugene-Emmanuel Viollet-le-Duc, in a modern English translation as *Castles and Warfare in the Middle Ages*, 215–216.
[20] DeVries, "Impact," 233–236; Hale, "Early Development," 475 (Hale also calls these early towers "basically defensive," vis à vis the "aggressive" later bastion design).
[21] Pepper and Hughes, *Firearms and Fortifications*, 545; DeVries, "Impact," 239–241.
[22] Hughes, *Military Architecture*, 97–100.

**Figure 2a:** The Norwich "Cow Tower" (ca. 1398). Author photo.

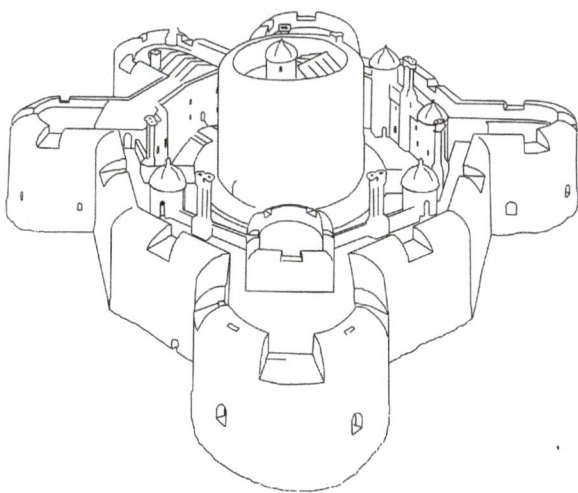

**Figure 2b:** Camber (1542). Edmund Potter, based on aerial photography.

event of an infantry assault. In the decade before the invasion of Charles VIII in 1494, Italian architects were beginning to converge on the squat, triangular bastion to solve both problems.[23] Many archaisms persisted, most notably the desire for a central keep within the walls, but after the 1494 invasion, the triangular bastion system spread rapidly in Italy. The design appealed to the mathematical bent of Italian Renaissance artists – often also military engineers – and their "cult of harmony, proportion and symmetry," but it also solved a host of problems.[24] Its squat, earth-filled, stone-faced walls absorbed the blows of artillery better than high vertical stone walls. Its low profile behind deep ditches further protected it from artillery fire, while the ditch provided the first line of protection against escalade. The triangular bastions anchoring the curtain walls or detached angular ravelins (usually protecting gates or bridges) provided firepower into and along the line of the ditch, and at least some cannon could be preserved in the recessed angles (eventually behind curved *orillons* or "ears") of the bastion for fire against the final infantry assault on the wall. The angled walls of the bastions deflected shot, and the wide platform of the bastion allowed heavier cannons to fire heavier shot over greater distances. Creating a full circumference of such bastions that accommodated local topography was a complex exercise in surveying and geometry, and was not mastered all at once, but this basic system spread throughout western Europe between 1520 and 1700, proliferating rapidly in Italy, France, the Low Countries, and on the Hungarian frontier, coming later to Germany (mostly after 1600), and even more slowly to England, Ireland, and northeastern Europe.[25]

We might summarize all these changes in European fortification as having progressed through three overlapping phases for the period 1350 to 1650: 1. Early Transitional – a longish period of improvisational responses to rapidly changing cannon, roughly 1350 to 1500, in which military architects all around Europe (and, as we will see, within the Ottoman empire) struggled to weigh the costs and benefits of different designs, the value of existing fortifications, how to modify them, and what the likely role of cannon would be for both defending and attacking such a fortress. 2. Proto-*trace italienne* – during which Italian engineers began experimenting with geometric designs based on angular bastions,

---

[23] Hale, "Early Development," 481.
[24] Hale, "Early Development," 473 (quote).
[25] Parker, *Military Revolution*, and Duffy, *Siege Warfare* trace these movements within Europe. Regarding England, David Nixon's study of Tudor fortresses built in northern France in the 1540s suggests that some English military engineers, in this case John Rogers, were well versed in the *trace italienne*'s primary features earlier than heretofore acknowledged. Nixon, "Tudor Coastal Forts in Northern France."

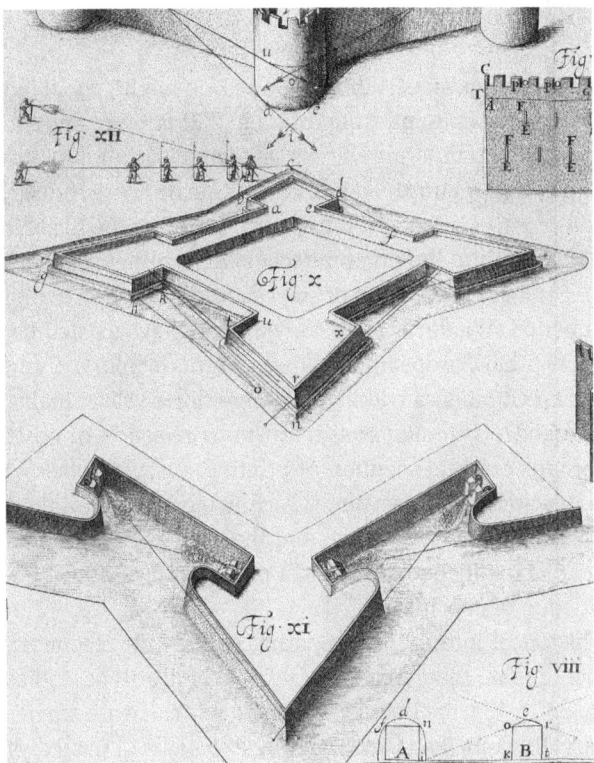

**Figure 3:** A rendering of a standard bastion as of 1647, showing the ditch, gun platform, and the recessed cannon behind the "ears" of the bastion.
Dögen, *Architectura Militaris Moderna*, 12.

roughly 1460 to 1500, often still in combination with older features like round towers. 3. The matured *trace italienne* artillery fortress, as the Italian designs stabilized into fairly standard patterns (always responsive to topography) and as siege artillery also standardized and became more predictable. The matured system saw continuous and incremental improvements generating ever more complex designs, ultimately perfected by Vauban in the late seventeenth century including his modifications to enclose artillery within casemates to protect against mortar fire.[26] These labels are simplistic and inevitably overlap, but they provide a shorthand for discussing Ottoman designs.

---

26 Hale, "Early Development," 470 calls the whole 1450 to 1530 period "transitional." But as his own article indicates, this period can be subdivided. Pepper and Hughes call the 1350 to 1500 period "transitional."

## The Ottomans and Fortifications

We must begin with some crucial caveats. Whatever else one might say about Ottoman fortification design, there is no question of their skill at besieging European forts of whatever era. Contemporary witnesses were clear about Ottoman skill in siegecraft, and they regularly captured European forts. Indeed, Ottoman successes in taking castles in eastern Europe from 1521 to 1566 likely is what prompted the Habsburgs to hire Italian architects to design their new fortifications.[27] Furthermore, Ottoman production of siege cannon matched that of Europeans for this whole period. Whatever technical gap may have existed between the Ottomans and any other European power was slight, temporary, and often counter-balanced by an Ottoman advantage in some other year. Finally, there is no issue of a generalized Ottoman conservatism or rejection of other technologies (with the perhaps crucial exception of printing, discussed below) or of being isolated from developments elsewhere. Technicians and specialists followed employment around the Mediterranean, and the empire often offered the most profitable venue, making it "one of the most porous and receptive environments for the introduction of new ideas."[28]

Despite the growing historical interest in comparing military developments in Europe and the Ottoman Empire, there remains a relative of dearth of studies of fortifications, and the majority of what does exist is regional. There are studies of Ottoman forts in Hungary, in the Balkans, of castles of all periods in Greece, of the Ottoman forts in and around Istanbul and on the straits, of those protecting the pilgrimage route to Mecca and Medina, and even one on Yemen.[29] But there are many regions of the empire whose forts remain relatively unknown, and there are only two attempts at a general survey of Ottoman fort design in the gunpowder era. In 2000, Simon Pepper, an expert in early European fortifications, published a fine overview of Ottoman fortifications in Europe, but his study covers developments only up to about 1550.[30] In 2010, as part of the popular Osprey press series, David Nicolle prepared a more extensive overview of Ottoman fortifications around the Mediterranean from the late medieval period to 1710.[31] He

---

27 Ágoston, "Empires and Warfare in East-Central Europe, 1550–1750," 118–119.
28 For a recent re-statement of this issue from an Ottomanist perspective, see Şakul, "Military Engineering in the Ottoman Empire," 179–200 (note that Şakul has complained that some of this text was changed significantly and without his knowledge); Murphey, *Ottoman Warfare, 1500–1700*, 106 (quote), 111–115. On Ottoman cannon, see Agoston, *Guns for the Sultan*.
29 Each of these studies will be cited below.
30 Pepper, "Ottoman Military Architecture in the Early Gunpowder Era," 282–316.
31 Nicolle, *Ottoman Fortifications, 1300–1710*.

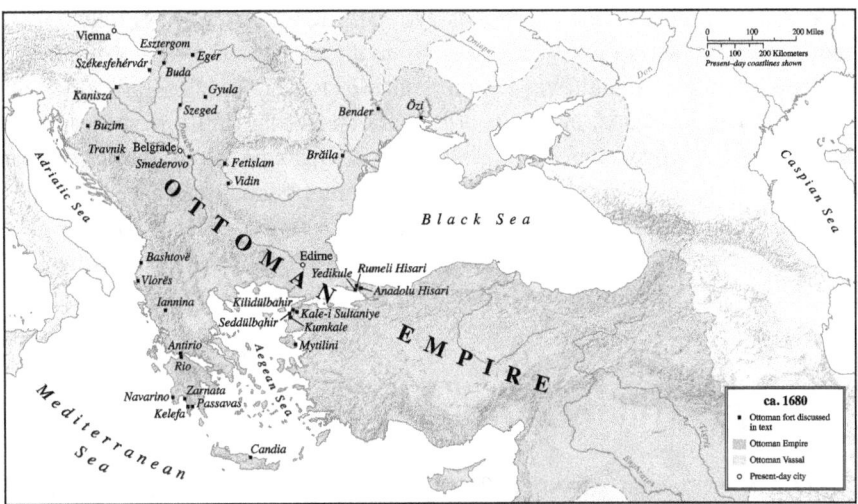

**Map 1:** The Ottoman Empire around 1680, showing fortresses mentioned in the text. This is not a complete list of sites fortified by the Ottomans. Map drawn by Justin Morrill, TheMapFactory.com

draws few comparative conclusions, but it is extremely useful, not least because of its many photos and artist's reconstructions. The present essay supplements and expands on both of these works, while also incorporating other regional studies, some unpublished data, and my own site surveys in Greece and Turkey (see Map 1).

Ultimately, we must make some suggestions about why the Ottomans built forts the way they did. The usual assessment, including judgments by contemporary Europeans, has been that the Ottomans ignored, or at best only poorly understood, the *trace italienne*, at least until about 1650, if not beyond. The famous seventeenth-century Ottoman traveler Evliya Çelebi (1611–1682) rejected such criticism, remarking that those men clearly had not seen "the Segedin (Szeged) castle at the Eger frontier, or Bender castle by the Dniester, or Özi," or Avlonya (Vlorës). If they had, he remarked, they would "realize how the House of Osman build[s] castles."[32] We will encounter each of these forts later in this essay and can then better assess Çelebi's protest, but later historians have tended to agree with those early western critics. Even Ottoman specialist Colin Imber concludes that "there is, as yet, no evidence to suggest that the Ottomans adopted the new

---

32 Çelebi, *Evliya Çelebi Seyahatnamesi*, 312. Thanks to Kahraman Şakul for the translation. He discusses this quote briefly in Şakul, "Siege Warfare in Verse And Prose" 205–242.

style of fortification."[33] Some recent assessments agree, but blame Ottoman budgetary calculations, arguing that the Ottomans "were well aware of the latest technology in fortress-building . . . [but] such forts were only sporadically built in certain key locations, mainly on the border with the Habsburgs. Budgetary constraints were a constant limit . . . [so the Ottomans preferred] the restoration and adaptation of earlier structures."[34] On the other hand, Simon Pepper argues for a brilliant efflorescence of creativity in design from the 1450s to the 1520s, succeeded by a more practical and budget-conscious approach to fortification that did not indeed partake much of the Italian style.[35]

## The First Artillery Fortresses

The timing of Ottoman expansion and of their eventual capture of Constantinople may help explain the nature of early gunpowder fortress construction within the empire. Pepper is surely correct in pointing out that in some ways the Ottomans in this period were actually anticipating western designs.[36] But why? Recall that truly effective siege artillery was emerging in France in the early to mid-fifteenth century. Those developments were quickly copied in eastern Europe and among the Ottomans, famously playing a role in the capture of Constantinople in 1453.[37] It was the context of that capture, however, that may have played a key role in Ottoman fortress design in this early period. Ottoman forces had long since crossed the straits from Anatolia into Europe. They had established a European capital in Adrianople (Edirne) in the 1360s, and gained final control of Thessaloniki in 1430. In between they destroyed the Serbians in 1389 and defeated a force of western European crusaders at Nicopolis in 1396. Their most dangerous regional enemies at the time may have been the maritime Venetians and Genoese, who continued to threaten Ottoman communications across the Dardanelles and the Bosphorus. As the Ottomans slowly closed the circle around Constantinople,

---

33 Imber, "Ibrahim Peçevi on War," 7–22.
34 Peacock, "Introduction: The Ottoman Empire and its Frontiers," 20.
35 Pepper, "Ottoman Military Architecture," 309.
36 Pepper, "Ottoman Military Architecture," 308.
37 Ágoston, "Empires and Warfare," 116–117; Ágoston, *Guns for the Sultan*, 1, 17–18, 20–21, 28–29; Ágoston, "Firearms and Military Adaptation," 110–111; Imber, "Ibrahim Peçevi on War," 8–9. Kelly DeVries argues that the role of cannon in the siege may have been overplayed in "Gunpowder Weapons at the Siege of Constantinople, 1453," 343–362. Simon Pepper agrees that other factors such as mining and especially Ottoman numbers were most important. Pepper, "Ottoman Military Architecture in the Early Gunpowder Era," 293–294.

maritime fortifications played a crucial role, and it was those forts which precociously incorporated new gunpowder-related characteristics. In contrast, the continental frontier remained fluid, and there the Ottomans were generally on the offensive. Furthermore, the final capture of the great city demanded not only consolidation by further building, but also symbolic celebration of Mehmed II's historic achievement. Meanwhile, the architects themselves continued to be influenced, as were their European contemporaries, by medieval precedents.

This combination of precedent, celebratory public works, and a focus on maritime control produced several sophisticated cutting-edge fortress complexes in and around the new capital and its key maritime connections, and in some other key straits, but designs in more landward-oriented contexts remained more conservative. Intriguingly, the new Ottoman designs from about 1452 to 1510 were more offensive in their use of artillery than their European equivalents, and were much less concerned with their defensive qualities, especially against attack from the land. The new forts were designed instead to project fire across open waters to deny maritime access.

Consider the two fortresses that Mehmed built to close the Bosphorus north of Constantinople, intended to prevent the Genoese fleets in the Black Sea from interfering with operations against the Byzantine capital. On the Asian side, the Castle of Anatolia (Anadolu Hisari) had been originally built ca. 1390 in a traditionally medieval style, with high curtain walls, towers, and a central keep, and was primarily designed to protect the crossing point to the European side. In 1452 to 1453, however, a much wider circuit wall was added, with large, ground-level artillery embrasures facing the strait. The fort's overall defensive role shifted from a passive one of protecting a ferry embarkation site to one of projecting firepower offensively to close the straits to enemy ships. The design reflected that intention: It incorporated no new passive defenses against cannon fire (see Figure 4a). In 1452, Anadolu Hisari was paired with Rumeli Hisari directly across the strait (see Figure 4b). Here too, there is little evidence that the castle's design incorporated fears of bombardment by artillery, except perhaps in the thickness of the tower walls (7.5m at the base), although such thick walls were hardly unknown in the medieval world. The fort was relatively easy to dominate from nearby hills, the towers were originally roofed with cupolas, and there is no evidence of embrasures for artillery within them. But like its counterpart directly across the strait, it incorporated a massive ground-level water battery to fire against enemy ships in the strait.[38]

---

[38] Nicolle, *Ottoman Fortifications*, 8–11; Pepper, "Ottoman Military Architecture in the Early Gunpowder Era," 288–289, 292; Ćurčić, *Architecture in the Balkans*, 767–768.

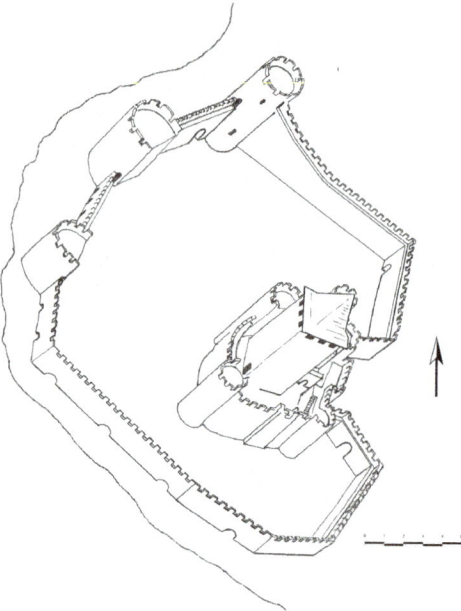

**Figure 4a:** Anadolu Hisari (1390, expanded 1452) The water battery embrasures are those in the curtain wall facing southwest. Edmund Potter, based on satellite imagery, and Nicolle, *Ottoman Fortifications, 1300–1710*, 10.

**Figure 4b:** Rumeli Hisari (1452). Author photo.

From 1459 to 1462, Mehmed also built a pair of forts to dominate the Dardanelles strait, Kilidülbahir on the European side and Kale-i Sultaniye on the Asian side (see Figures 5a and b). Both are enceintes with corner towers built around a massive central keep. Both fortresses seem intended to fulfill a similar role as the Bosphorus forts, essentially protecting a ground-level water battery, that, as the contemporary Kritovoulos noted, made "entirely impossible the navigation up or down for any who might wish it . . . For as soon as ships approached they were immediately sunk and demolished by the immense stone balls fired from the cannon."[39] (Note that modern modifications have removed the majority of the water batteries at both sites; note too that the large almost freestanding tower in the lower left of Figure 5a at Kilidülbahir was added later, see below). But unlike the Bosphorus fortresses, these slighter newer complexes incorporated several design aspects to enhance their passive defenses against attack by enemy artillery. The central keep in both forts uses a thick curved banquette instead of thin vertical crenellations to deflect cannon balls, and the smaller towers in the curtain walls are the same height as the walls, facilitating the movement of defensive cannon around the curtain wall. Kale-i Sultaniye follows a traditional rectangular design, but Kilidülbahir is remarkably creative, with three concentric layers, placing a triangular keep/artillery platform within a thick cloverleaf intermediate wall, and then a final smaller outer wall that also incorporated the ground level water battery. The cloverleaf structure created separate internal courtyards that could be dominated by the triangular keep, while the cloverleaf walls supported additional defensive artillery oriented to create interlocking fire outside the enceinte. Both the curvature of the cloverleaf and the triangular tower seem designed to deflect shot as much as possible. Conservative elements remain, however, including the thin vertical walls of the outer line and its traditional crenellations, while the fort's placement renders it vulnerable to fire from nearby higher hills. For both forts, however, the primary function of offensive fire into the straits dictated their otherwise vulnerable location.[40]

Meanwhile, in Constantinople itself Mehmed II built a remarkable and seemingly forward-looking towered gate complex as an addition to the Byzantine walls known as Yedikule (literally, seven towers). Pepper argues that the complex added substantially to the defensive strength of that part of the Byzantine walls (where the royal road into the city passed through the walls), and that its formal pentagular shape pre-dated later Italian angular designs. On the

---

**39** Kritovoulos quoted in Thys-Şenocak, *Ottoman Women Builders*, 132–133.
**40** Pepper, "Ottoman Military Architecture in the Early Gunpowder Era," 301–304; Nicolle, *Ottoman Fortifications*, 11–12, 51. To clarify, this is the *Kale-i Sultaniye* built in what is now the town of Çanakkale.

**Figure 5a:** Kilidülbahir on the European side of the Dardanelles. The water battery wall is now missing and is reconstructed here based on older drawings. Edmund Potter, based on author photographs.

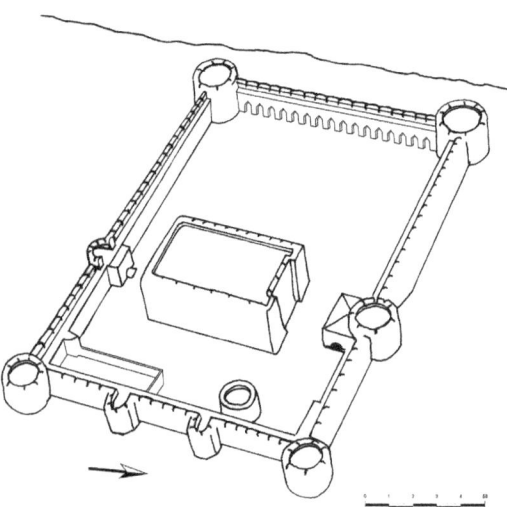

**Figure 5b:** Kale-i Sultaniye on the Asian side, seen as if approaching from the landward side – the long water battery shown from the inside is reconstructed from older drawings.
Edmund Potter from satellite imagery, assorted historic photographs, and Pepper, "Ottoman Military Architecture in the Early Gunpowder Era," 303.

other hand, although it strengthens the defenses of the gate, it does so only after they have been breached. The entire new complex is *behind* the existing wall and gate, and it therefore does not provide any sort of fire covering the main forward walls. Furthermore its towers rise above the associated curtain wall, are almost strictly vertical (unscarped), provide very little flanking fire protection along the walls, and are equipped with standard medieval crenellations rather than curved banquettes. Worse, the curtain walls are punctuated by small triangular projections (also present at Kilidülbahir) that could not have mounted any substantial guns, and partially obscure the line of fire of the larger towers, see Figure 6a.) Finally, unlike the Ottoman maritime forts discussed thus far, it has almost no "offensive" fire capability. It provides no platforms for mounting substantial cannon, and only a few platforms for smaller guns. The complex was indeed geometrically complex, but its formality probably speaks to its status as a palace, an archive, and a treasury (preceding the later palace at Topkapı) rather than to its sophistication as a fortress (see Figure 6b).[41]

This assortment of forts, built over a remarkably short time span (1452–1462), and all by the same ruler, has inspired substantial commentary and interpretation. Lucienne Thys-Şenocak believes that the variety of designs at Kilidülbahir, Kale-i Sultaniye, and Yedikule may have been deliberate political showmanship by Mehmed the Conqueror, who thus displayed his knowledge of military architecture at the time.[42] A more functionalist argument might simply suggest that each fort had different purposes, and design followed function: Yedikule was mostly a palace, while the other two were designed to project a large volume of fire on a skipping trajectory out over the water, thus leading to extensive curtains walls for the water batteries, with central keeps to protect them. Slobodan Ćurčić argues that the uniqueness of these designs within the empire suggests imported architectural talent, but Pepper favors simply accepting that this time period saw a "lively spirit of improvisation" rather than a reliance on imported designers or even different designers.[43]

Pepper's argument for Ottoman design creativity and even repetition in this period is supported by the strong resemblances between the cloverleaf towers at Kilidülbahir and the artillery towers that anchor the fortress complexes at Rio and Antirio in Greece (the Castle of the Morea and the Castle of Roumeli respectively). Both were built ca. 1500, and like the paired forts already discussed on either side of the Bosphorus and the Dardanelles, these were paired to close the entrance to the Corinthian Gulf near modern Patras. The original

---

41 Pepper, "Ottoman Military Architecture in the Early Gunpowder Era," 294–300.
42 Thys-Şenocak, *Ottoman Women Builders*, 173 n. 150.
43 Ćurčić, *Architecture in the Balkans*, 768; Pepper, "Ottoman Military Architecture in the Early Gunpowder Era," 301 (quote).

**Figure 6a:** Yedikule, view from easternmost tower to northeast tower (note the intervening triangular projection). Author photo.

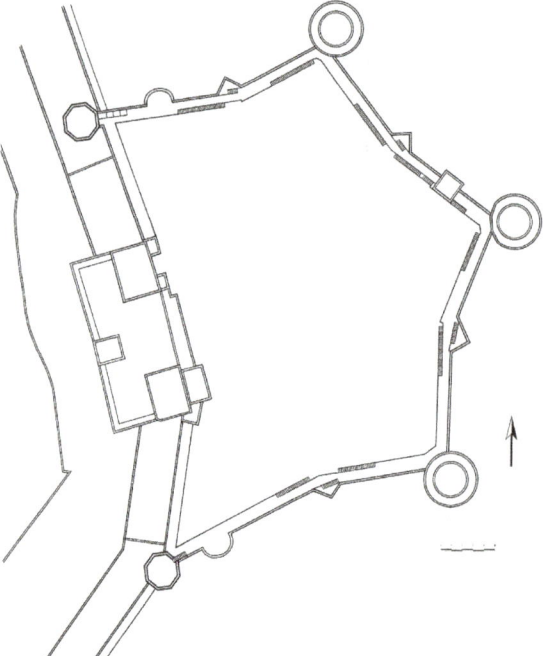

**Figure 6b:** Yedikule, aerial view. The square towers in the center left are the original Byzantine gate towers.
Edmund Potter based on aerial photography, seventeenth-century travelers' drawings, and Pepper "Ottoman Military Architecture in the Early Gunpowder Era," 291.

design combined Kilidülbahir-like artillery towers at its heart, now supplemented with aspects of angled bastion designs, while also retaining some conservative medieval qualities. Rio, the most well preserved of the two, was built in 1499 during the Ottoman campaign against the Venetians in Messenia.[44] In Pepper's analysis, this was an extremely modern design for its time, with full-sized artillery platforms on the trefoil towers behind a curved banquette, strongly curved tower walls to deflect shot, a separate powder magazine behind the towers, and sea-level embrasures for a water battery (see Figure 7a).[45] Here again we see an early design emphasizing offensive firepower, but now also presented in a triangular form that paralleled proto-*trace italienne* shifts in western European designs (see Figure 7b). It as yet lacked the full emphasis on corner tower bastions that could fire down the length of the wall, but it did have embrasures in the gate complex and in the main seaward tower designed for that purpose. In the early eighteenth century the Venetians added *trace italienne* outworks to protect the gate complex and extend the defenses. Even so, in its Ottoman manifestation in 1500, Rio was an advanced, though idiosyncratic design.[46]

As the Ottoman conquest progressed through the Balkans in the fifteenth century, one finds them also working through the problem of how to upgrade existing medieval fortifications. At Buzim in Bosnia-Herzegovina, the Ottomans found an existing medieval square curtain anchored by four round towers. In the 1490s, in much the same way that Italian architects in the transitional period were adding low, outer, concentric walls to existing keeps, the Ottomans added a concentric outer wall, much lower than the inner keep walls, anchored by polygonal towers designed to support artillery (see Figure 8a).[47] At Travnik, also in Bosnia Herzegovina and captured in the 1460s, the Ottomans found a

---

**44** See Andrews, *Castles of the Morea*, 130–134, and plate XXX. Key recent work on Rio was presented by Koumousi, "Rion Fortress: Restoration Works and New Archaeological Data (2006–2016)," Paper presented at "Fortifications of the Ottoman Period in the Aegean," conference held at Mytilene, 30–31 Oct. 2018. Proceedings are forthcoming; personal correspondence with Anastasia Koumousi, Nov. 2018.

**45** Pepper, "Ottoman Military Architecture in the Early Gunpowder Era," 308. Very little of Antirio's design has been preserved and Pepper's comments on it remain speculative.

**46** Koumousi (see note 44), has discovered an earlier sixteenth-century Ottoman phase of extensive reconstruction after Andrea Doria's brief capture of the fort in 1532 and Machiel Kiel has discovered Ottoman documents ordering repairs at Rio after the defeat at Lepanto in 1571. Kiel, "Construction of the Ottoman Castle of Anavarin-i Cedid," 266. It is possible that one of these repair episodes included building the curved banquette at the top of the artillery towers. But this is speculation based on the strong resemblance they bear to other Ottoman mid-sixteenth-century artillery towers discussed below.

**47** For Buzim, see Ćurčić, *Architecture in the Balkans*, 772.

**Figure 7a:** Rio castle, seaward facing side. Note the water battery embrasures at sea level. http://www.kastramoria.gr/en/content/rio accessed 3/3/2020.

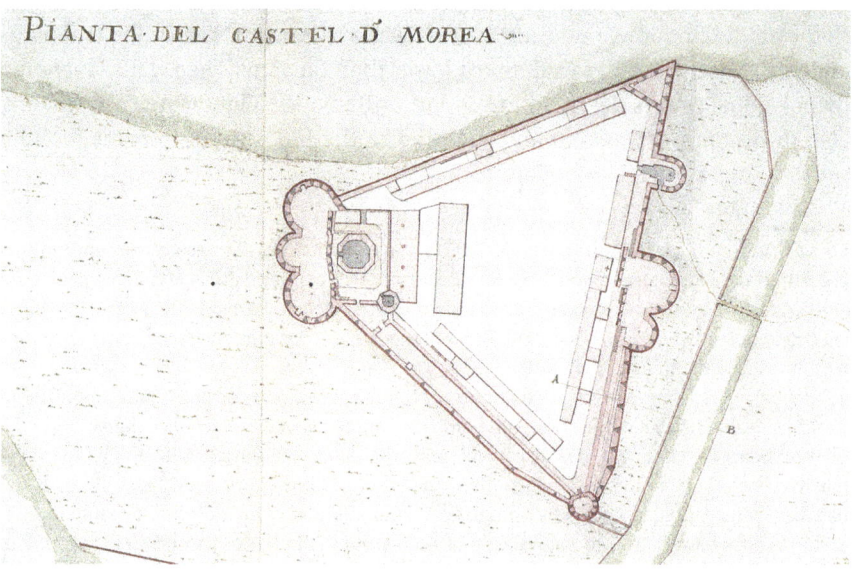

**Figure 7b:** A Venetian plan of Rio, drawn, ca. 1700. Digital image provided by the Gennadius Library (also available in Andrews, *Castles,* Plate XXX).

castle on a precipitous slope and were therefore unable to add a new external concentric wall. Instead, the Ottomans repaired the existing walls and built two thick polygonal towers on the foundations of older medieval towers, but now redesigned as artillery platforms (see Figure 8b).[48]

Regarding this early period of what we might oxymoronically call "medieval artillery fortresses," Simon Pepper's "lively spirit of experimentation" seems a reasonable assessment.[49] Whether it was partly for political display or not is irrelevant. The Ottomans were interested and engaged in the problem of combining cannons and walls, and indeed were on the forefront of design through 1500 or so, at least. Furthermore, some of their early designs precociously emphasized creating a platform for long-range, large-caliber artillery fire, rather than merely enhancing the defensive strength of the walls by increasing crossing fire and eliminating dead space. This is in part because the most innovative of the forts reviewed thus far were almost entirely designed to fire out to sea, and in fact were not especially defensible against attack from land, most being dominated by nearby high ground.

On the other hand, Pepper may have been a bit optimistic about Ottoman inventiveness, even in this early period. There are other forts from this early period that betray a distinct conservatism, or at best a lack of concern for the likelihood of an enemy arriving equipped with siege cannon. The fort at Bashtovë, Albania, built from the ground up between 1467 and 1469, for example, was almost strictly medieval in design: a square curtain, with vertical crenellations and a mix of square and round projecting towers (see Figure 9a).[50] The fort at Fetislam, raised in 1524 on the banks of the Danube in modern Kladovo, Serbia, is also severely conservative. Although it incorporates aspects of concentric design, it also hearkens back to design elements of Kale-i Sultaniye and Yedikule, with a rectangular layout, a keep inside the enceinte, and the curtain anchored by round corner towers. Its design makes only the barest concessions to the existence of artillery (see Figure 9b).[51] The inner citadel at Bender, along the Dniester River in northern Moldova, rebuilt on possibly Genoese foundation by the Ottomans in 1538, although cited in the 1660s by Evliya Çelebi as particularly strong, was actually quite similar to Bashtovë or Fetislam, essentially rectangular, with a mix of artillery towers, generally higher than the curtain. Like

---

**48** Travnik needs further investigation and publication. These comments are based on various tourism-related websites in Bosnia. For example, http://www.ahlanbosna.com/en/travnik-old-fort/, accessed 22 September, 2017.
**49** Pepper, "Ottoman Military Architecture," 301 (quote).
**50** Ćurčić, *Architecture in the Balkans*, 770.
**51** Ćurčić, *Architecture in the Balkans*, 774.

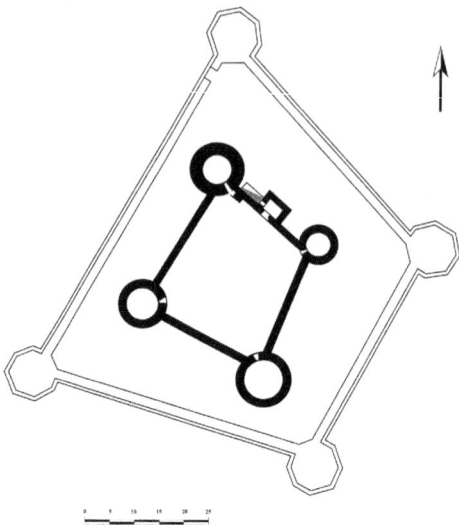

**Figure 8a:** Buzim, outer walls added by the Ottomans in the 1490s. Edmund Potter, based on Ćurčić, 2010, 772.

**Figure 8b:** Travnik, with the two added artillery towers from the 1460s. Edmund Potter, based on http://www.ahlanbosna.com/en/travnik-old-fort.

Fetislam, there is a lower outer wall around the citadel with platform bastions, but that may have been added later. It is possible that the fort Çelebi praised was a much enhanced and expanded edifice, reflecting Ottoman improvements made at the beginning of the seventeenth century. Those works may even have included an expansive *trace italienne* system of bastions surrounding the citadel, but it is also possible that the bastion line dates to further Turkish improvements in the early eighteenth century.[52]

Finally, Vlorës (Avlonya), also built from the ground up in 1537, and also praised by Çelebi for its strength, was almost entirely medieval in conception and construction, being a moated curtain wall punctuated by towers (see Figure 9c). Lacking even battered walls, its only concession to artillery was the provision of gun platforms on the curtain wall towers, the larger ones apparently held four to five guns, and the smaller ones one each. In addition there was a massive internal tower, not unlike the White Tower of Thessaloniki (an observation first made by Evliya Çelebi himself), capable of mounting more cannon and firing over the outer walls or against an enemy who had breached the walls (all around embrasures in the central tower are visible in a later Venetian plan of the tower).[53] What Vlorës lacked, as did Bashtovë, Fetislam, Travnik, and Buzim, were the complex layered set of defenses associated with the *trace italienne* that defended against artillery siege and ultimately against assault into a breach.[54] The walls in all of these places

---

[52] There are some cursory observations in an old survey of European fortifications which also reprints a 1740 map of the fort: Ebhardt, *Der Wehrbau Europas im Mittelalter*, 610–611. A more modern study in Russian by Georgiĭ Osipovich Astvat͡saturov, *Benderskai͡a krepost'* (Bendery: Poligrafist, Benderskai͡a tipografii͡a, 2007) was recently made available partly in English at a website dedicated to the fort: https://bendery-fortress.com/en/construction-phase-of-the-bendery-fortress-written-by-georgy-astvatsaturov-2007, accessed 21 October, 2021.

[53] Ćurčić, *Architecture in the Balkans*, 770, 774–775; Bace, "Kalaja e Vlores," 43–57; Çelebi, *Evliya Çelebi in Albania and Adjacent Regions*, 135–137; Kiel, "The Building Accounts of the Castle of Vlorë/Avlonya (S. Albania) 1537–1539, 3–31, esp. 14; Kiel, *Ottoman Architecture in Albania (1385–1912)*, 269–275. The fortress did not survive; this reconstruction is based on a series of drawings of the fort analyzed in Bace. The 1537 date is from Kiel, "Building Accounts," which documents the original expense accounts associated with construction, but which unfortunately had no design discussion.

[54] The Venetians added a pair of ravelins to protect the seaward facing gate after 1690. Smederovo would also seem to fit into this category, captured in 1459, and progressively redesigned. The Ottomans appear to have added an outer partially concentric system to what was originally a distinctly Byzantine wall design. The Ottoman additions included the now-characteristic polygonal towers, lower and thicker, and clearly designed as artillery platforms, but the original medieval design dominated the works. More research is needed on the Ottoman period of the fort. Ćurčić claims the Ottoman towers have scarped walls but that does not appear to be the case. Ćurčić, *Architecture in the Balkans*, 771.

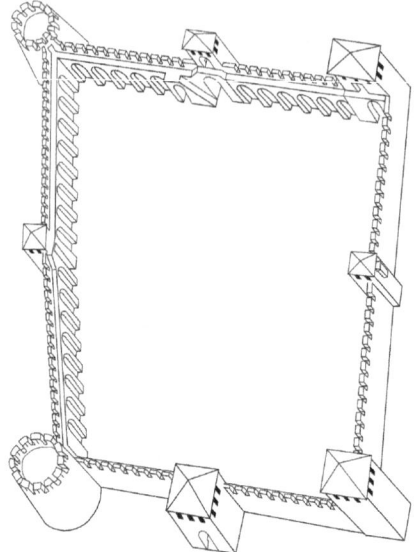

**Figure 9a:** Bashtovë (ca. 1468). Edmund Potter, based on Ćurčić 2010, 880.

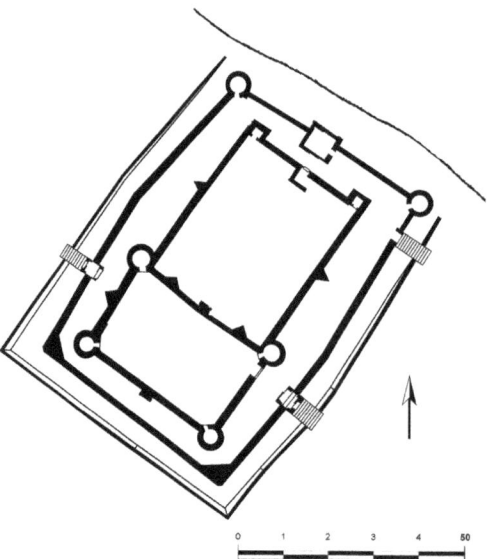

**Figure 9b:** Fethülislam (Fetislam) (1524). Edmund Potter, based on Ćurčić 2010, 774.

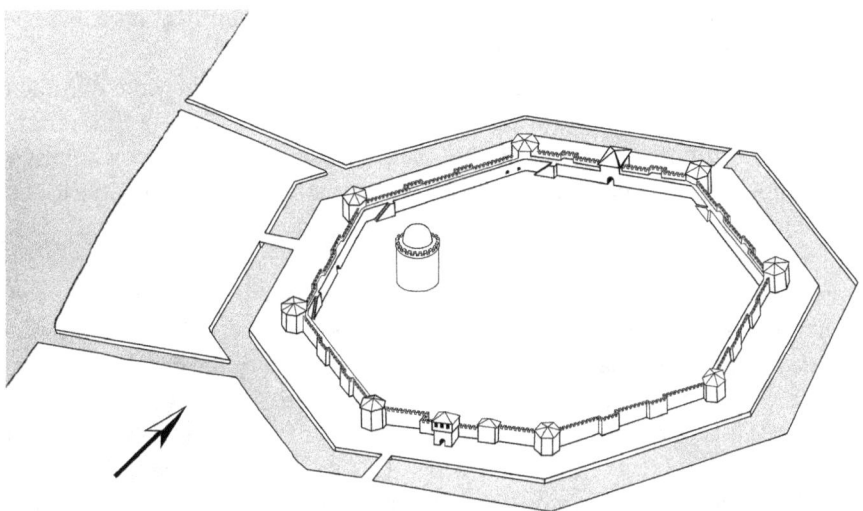

**Figure 9c:** Vlorës (1531). Edmund Potter, based on Bace "Kalaja e Vlores," 46.

were thick, even impressively so in the design of the central towers, but they seemed to presume that no long siege was likely, perhaps in the belief that an Ottoman relief army would never be far away. The projecting towers along the curtain wall provided only minimal coverage of each other, they did not eliminate dead space at the foot of the walls and towers, and they remained more vulnerable to enemy artillery fire than the lower thicker, angled, or at least scarped walls of their Italian competitors.

A critical component of most of these designs, and one that was shared with early European responses to the problem of artillery, was the artillery tower – a relatively low, thick, generally polygonal or round tower, with an open platform on top. The more advanced versions incorporated embrasures through a curved banquette to protect the guns and gunners, and possibly with some lower embrasures to protect surrounding works. Not necessarily standalone towers, they were often part of a new wall built around an existing set of walls, extending the circuit, and sometimes with one or two towers dramatically pushed out to command the surrounding grounds or waters. They could also anchor new construction, however, as at Rio. Several examples have already been mentioned – Travnik's new outer tower could fit in this category for example, or the central tower at Vlorës – but we return to them here because the idea seemed to have quite a long life.[55]

It persisted well into the seventeenth century and even beyond, and their persistence supports the idea of Ottoman economic pragmatism in their approach to fortification design. Rather than tear down or redesign whole systems, they instead added one or a series of artillery towers to prolong the useful life of an aging design, although these towers did not usually provide the type of interlocking fire that eliminated dead space in the manner of the *trace italienne*. The towers at Rio were built around 1500, and they may have had antecedents in the wall towers of Kale-i Sultaniye built in the 1450s and 60s (see Figure 10a). The Ottoman addition of four outer towers to the distinctly medieval fortress at Smederovo in Serbia in the 1480s shared some of these characteristics, but it is unclear if they originally had the curved banquette that would later be characteristic.[56] The tower added to Kilidülbahir by Sultan Süleyman the Magnificent apparently in 1551 is typical of the type, now with the lower part of the tower slightly scarped (see Figure 10b). One can see the design evolution clearly when the new tower is compared to the original towers at that site.[57] The North round bastion or "Earth" tower at Esztergom, built in the late sixteenth century also meets this description, as do quite a few round, sometimes, multi-level embrasured tower bastions built by the Ottomans during this period in Hungary.[58] Some time between 1600 and 1650 the Ottomans built a very similar set of towers (also mostly scarped) to reinforce the older fortress when they captured Mytilini, on the island of Lesbos (see Figures 10c and 11).[59] The type persisted into the eighteenth century, including in combination with more European-style angled bastions, and can be seen in a particularly massive form in the southwestern round tower at Ioannina, built by Ali Pasha in the 1790s.[60]

---

55 Machiel Kiel has also noted this Ottoman tendency to rely on a "single solid tower as pivot of a defence system," which he briefly tracks from 1445 to the late sixteenth century at several additional sites not discussed here. *Ottoman Architecture*, 272.
56 Ćurčić, *Architecture in the Balkans*, 771.
57 Thys-Senocak, *Ottoman Women Builders*, 130; Nicolle, *Ottoman Fortifications*, 11, 51.
58 Horváth, "Ottoman Military Construction in Esztergom," 79–80; Bencze, "Recent Research into Ottoman-period Remains in Buda," 55 (tower at Víziváros, built 1565–1566); Irás-Melis, "Pest During the Ottoman Era," 166, 169 (tower bastions in Pest built 1684).
59 McGilchrist, *Blue Guide Greece: The Aegean Islands*, 506; Nicolle, *Ottoman Fortifications*, 28–29.
60 Brooks, *Castles of Northwest Greece*, 248, 254. Brooks describes similar artillery towers/ bastions added by the Ottomans to earlier fortifications at Naupactus as well, although their dating is less clear (although likely post-1571); see 26, 30.

**Figure 10a:** Artillery towers at Rio (1499). Edmund Potter, based on http://romeartlover.tripod.com/Rio.html (accessed 3/4/2020).

**Figure 10b:** At Kilidülbahir (1551). Author photo.

**Figure 10c:** At Mytilini (ca. 1630s). Edmund Potter based on Nicolle, *Ottoman Fortifications*, 29.

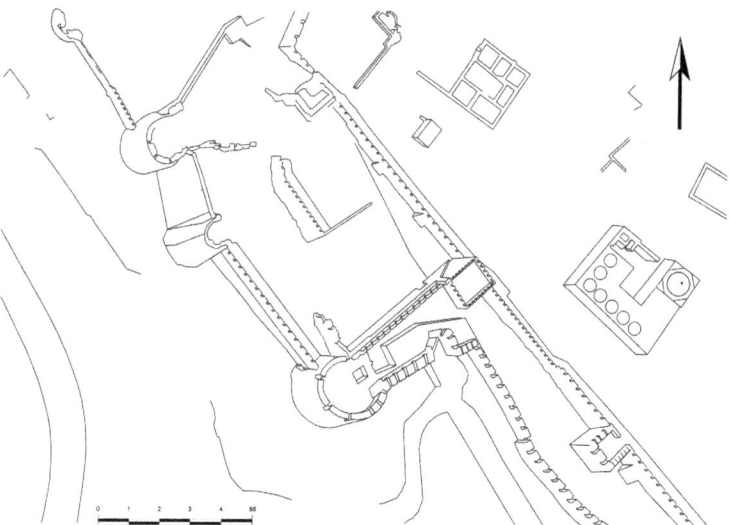

**Figure 11:** Ottoman towers added to Mytilini's fortifications in the seventeenth century. They appear to be scarped artillery towers, mixing round and polygonal construction, with the walls and the towers mostly at the same height. Edmund Potter, based on satellite imagery.

In all of these responses to the problem of artillery fire, interacting with the problem of incorporating existing older fortifications, we certainly see experimentation and effective innovation, but Ćurčić argues that the "experimentation" in the early period (1460–1540) was more improvisation than experimentation. This may be a distinction without a difference, but it does seem that the Ottomans preferred *not* to build forts, and assumed that if they did, it would only serve a temporary purpose en route to further imperial expansion.[61] Many of the most innovative forts considered thus far were exceptional in their permanent strategic value defending key straits or even the entrance to Constantinople itself.

## Ottoman Forts in the Sixteenth and Seventeenth Centuries – Stagnation?

On the other hand, this period of experimentation appears to have stopped or slowed, and there is very little evidence that Ottoman fortresses built in the late sixteenth and seventeenth centuries adopted the *trace italienne* system of angled bastions, or even represented significant refinements of their own techniques. The preceding discussion of artillery towers should hint at this basic conservatism, with the only change being the appearance of scarping on the later versions. Thus far the evidence here has mostly buttressed Pepper's argument for creative, receptive, and adaptive responses to gunpowder artillery in Ottoman fortification design, but as one enters the mid-sixteenth century and beyond, the historiography provides very little help. The actual sites that Pepper considers in his reevaluation of Ottoman forts were built no later than about 1500, although parts of his analysis proceed beyond that. So let us return to existing interpretations and what they can tell us before turning again to specific forts.

Already mentioned is Ćurčić's conclusion, based on his examination of Ottoman forts in the Balkans, that "the Ottomans apparently invested in new fortifications seldom and did so only under particularly demanding circumstances."[62] Geoffrey Parker's work on the military revolution dwells at some length on Ottoman infantry and artillery, but does not mention fortification at all.[63] Oddly, Rhoads Murphey's highly detailed study of Ottoman military systems from 1500 to 1700 makes almost no comment on Ottoman fortifications, except to note the

---

61 Ćurčić, *Architecture in the Balkans*, 773, 775.
62 Ćurčić, *Architecture in the Balkans*, 767.
63 Parker, *Military Revolution*, 125–130.

role of *khorasan* masonry in possibly making otherwise traditional masonry "nearly cannon-proof" (about which see below). He instead argues that the vast size of the empire's periphery rendered it financially unfeasible to fortify in any systematic way (setting aside the concomitant cost of garrisons, something Murphey also documents). This of course became more true as the empire stopped expanding. Local forts were built by local resources (often in the explicitly temporary palanka style, more below), and only a few key forts were kept up and/or modernized and then garrisoned (in Europe key forts included Eger, Buda, Belgrade, and Esztergom).[64]

Christopher Duffy's expansive overview of worldwide fortifications from 1494 to 1660 notes how Ottoman expansion provoked "an epidemic of fortress-building all over the Christian Mediterranean."[65] And, given the timing of Ottoman expansion, many of those new forts (especially the maritime ones) were built in accord with emergent Italian artillery fortress designs. The Ottomans nonetheless proved masters at besieging and often taking such fortresses, and in the process they no doubt learned the nature of the design and how the bastioned artillery fortress worked in practice. The walls surrounding Nicosia in Cyprus, for example, although newly built in the fully modern late sixteenth-century *trace italienne*, were taken by storm by the Ottomans in 1570 (the other modern fort on Cyprus at Famagusta took somewhat longer to take). Ottoman siegecraft showed every awareness of the power of overlapping fields of fire from the projecting bastions, and once they had captured it, the Ottomans promptly repaired the walls.[66] Like Murphey, Duffy also notes how the Ottomans held certain key European forts once captured, including for example Eger, Esztergom, Kanizsa, and Székesfehérvár, captured from the Habsburgs in the early seventeenth century.[67] As for Ottoman fortification techniques, he devotes only a few pages to the subject, relying on contemporary western authors to conclude that they rarely fortified places, and when they did, they built works "with high masonry ramparts in the medieval or Byzantine style." Duffy sees the situation changing with Austrian offensives at the end of the seventeenth century leading to some fortresses "suddenly sprout[ing] retrenchments and bastioned works that were clearly designed by renegade Western architects." It is true that western observers were dismissive of Ottoman fortifications: one, writing in the 1820s, complained that "they have no idea of a regular system

---

64 Murphey, *Ottoman Warfare*, 18–19, 113–14.
65 Duffy, *Siege Warfare*, 194.
66 Duffy, *Siege Warfare*, 195–196.
67 Duffy, *Siege Warfare*, 202.

either of bastions or of lines, or outworks or covered ways, nor of conforming the height of the works to the nature of the ground in front."[68]

Jeremy Black's more recent survey noted that the Ottomans failed to fully "reevaluate" their fortification systems, and argues that the Ottomans were "not under attack" in the period up to 1600. They were instead on the offensive, with a strategic system that emphasized "field forces and mobility." He notes, however, that they were not averse to building forts, and points to the forts protecting the pilgrim trail to Mecca, and a late sixteenth-century chain of forts on the Red Sea coast "from Suakin to Massawa" to "consolidate its recapture from Ethiopia." He also explains the fortifications in Yemen as necessitated by the province's distance from the imperial center. He concludes that in much of the empire they "had the resources and logistical capabilities to overwhelm their opponents," and fortification during the phases of great Ottoman power was simply a low priority.[69]

David Nicolle's overview is mostly descriptive, but he too opens his study by recounting the long-held view of Ottoman deficiency in fortress architecture. He also suggests, however, that this difference or deficiency only became problematic under Austrian assault in the late seventeenth and eighteenth centuries. He then oddly excuses it as resulting from their "limited interest" in Western European designs, presumably because of their expense. This seems to dodge the question.[70]

I may have missed other examples in the literature, especially in Turkish, but Ottoman forts on the frontiers after 1550 is not a well-studied field. What historians seem to agree on is that in general the Ottomans relied on whatever fortifications they found, destroyed most, kept some and altered them slightly, but rarely built wholly new systems. They also did not generally establish new urban communities needing fortifications, although new forts would eventually attract communities. Statistically speaking, this should be most apparent, or not, on the Hungarian frontier, where the Ottomans and the Hungarians/Habsburgs invested substantial effort fortifying a region that saw much back-and-forth warfare over two hundred years and more.

After a long period of border warfare from the late fifteenth century through 1521, the Ottomans swept into the Kingdom of Hungary after their victory at Mohács in 1526. Over the next forty years they took control of nearly all of Hungary.[71]

---

68 Duffy, *Siege Warfare*, 215–216 (Valentini, 1828 quoted on 216).
69 Black, *War in the World: A Comparative History, 1450–1600*, 174–175.
70 Nicolle, *Ottoman Fortifications*, 4–5.
71 A good summary is in Ágoston, "Habsburgs and Ottomans: Defense, Military Change and Shifts in Power," 126–141.

Gábor Ágoston estimates that between 1521 and 1566 only thirteen Hungarian fortresses resisted for more than ten days, and only nine for more than twenty days. Only four held out on a more permanent basis. Many of those forts had remained essentially medieval in design, but some of the more important had received artillery-related improvements. This stinging series of defeats led the Habsburgs to turn to Italian architects to update their remaining fortresses and to build a vast fortified defence line extending over 1,000 kilometers, with crucial sites or cities receiving the full complex of the *trace italienne* artillery fortress. This modernization was well under way, although not yet complete in many cities, when a new war, the so-called "Long War," broke out in 1593.[72]

Meanwhile the Ottomans had settled into their own system of garrisoning and fortifying parts of that frontier. Unfortunately the detailed history of Ottoman fortification in the region has yet to be written.[73] We can begin by noting that as late as 1577, only the Hungarian fortresses at Eger, Kassa, Komárom, and Győr were equipped with cannon heavy enough to fulfill the offensive function of the artillery fortress design: that is, to keep besieging cannon at a distance with long range fire from gun platforms on bastions. The many other fortresses of that era were equipped with small guns only, designed to defend against the final assault on the walls. To be fair, such a design might be thought adequate for the strategic purpose of such smaller forts, as they were intended to resist only raids, not large armies.[74] The implication here is that such forts when captured by the Ottomans prior to the Long War, barring reconstruction or major renovation, would still have had that limitation. Newly constructed Ottoman forts built prior to 1593 that can be documented via garrison payrolls suggests a similar purpose for the fortresses the Ottomans built in the course of their sixteenth-century expansion in Hungary. That is, they protected river transit, supported the collection of tolls, and constituted a zone of fortifications around the regional (*vilayet*) capital at Buda (Budun). Key forts in that zone included former Hungarian fortresses at Csókakő and Gesztes, but with new Ottoman constructions at Zsámbek (1549), Vál (1550), Battyán (1568), and

---

72 Ágoston, "Habsburgs and Ottomans," esp. 129; Ágoston, *Guns for the Sultan*, 35, 195 (see his Figure 4 for an example of the partly modernized fort at Szigetvár under siege in 1556); Ágoston, "Empires and Warfare in East-Central Europe, 1550–1750," 118–119; Szakály, "The Hungarian-Croatian Border Defense System and its Collapse," 141–160 (esp. 142, 148, 149–150). For one example of hurried Italian-designed upgrades in the 1540s, see Horváth, "Ottoman Military Construction," 78; for Habsburg updates after the 1570s, see Ágoston, "Firearms," 116.

73 Stein, *Guarding the Frontier* might seem a likely candidate to expand our knowledge, but it is primarily about the problem of garrisons and pay, and not about fortress design or construction.

74 Kelenik, "The Military Revolution in Hungary," 137–138.

Kızılhisar (early 1570s). There were other similar processes of combining Hungarian fortifications with new Ottoman ones around key sites at Esztergom and in the Danube bend.[75] In contrast, in line with the notion of Ottoman practicality and frugality, the Hungarian fortress at Gyula, partially modernized by the Hungarians with Italian help in the 1560s, was captured by the Ottomans and held by them without major change for 130 years. Again, to be fair, it was not seen as a key site, although they did maintain its palisade over the years.[76]

Unfortunately, the documentary record identifying all these forts tells us little about the nature of their construction, although some travelers' accounts and place names make it clear that much of the construction was the palanka type, while other garrisons were simply housed in existing unfortified buildings. Palankas were expedient, generally simple, square outline, earth-and-timber constructions, ditched and gated with a gate tower, but otherwise unimpressive (see Figure 12). They were built by both sides on the Hungarian frontier, with perhaps somewhat greater frequency by the Ottomans. Connected by key roads or waterways to create a fortified corridor, they were designed to defend against small raiding parties or to slow down the advance of larger forces. They were not intended to hold out against a serious siege, much less to project offensive firepower against a besieger.[77]

Klára Hegyi's study of the documentary record indicates that many of the garrison sites attested by payrolls were in fact palankas, and Hungarian local histories sometimes provide other illuminating details. As an example, the first fort that Hegyi documents at Cankurtaran, built in the 1540s, was originally known as Cankurtaran Palankası,[78] while the garrison at Canfeda (now Jászberény), apparently occupied a ruined Franciscan abbey.[79] These are hardly definitive examples, but a more wide-ranging survey of the architecture sheltering Ottoman garrisons in Hungary and on the Danube is needed. During the Long War of 1593 to 1606 frontier forts changed hands and the frontiers shifted substantially, followed by endemic fighting over much of the second half of the seventeenth century and the continued exchange of fortresses. This constant exchange encouraged the continued use of cheaply built palankas (and on sites where the forts were more permanent, that same exchange has made their architectural history complex). And although, as Hegyi notes, "hardly any new fortresses were built by the

---

75 Hegyi, "The Ottoman Network of Fortresses in Hungary," 166–167.
76 Gerelyes, "Ottoman Architecture in the Town of Gyula," 177.
77 Özgüven, "The Palanka," 113.
78 Özgüven, "The Palanka," 4.
79 https://en.wikipedia.org/wiki/J%C3%A1szber%C3%A9ny (accessed 28 September, 2017); Hegyi, "Ottoman Network," 168–169.

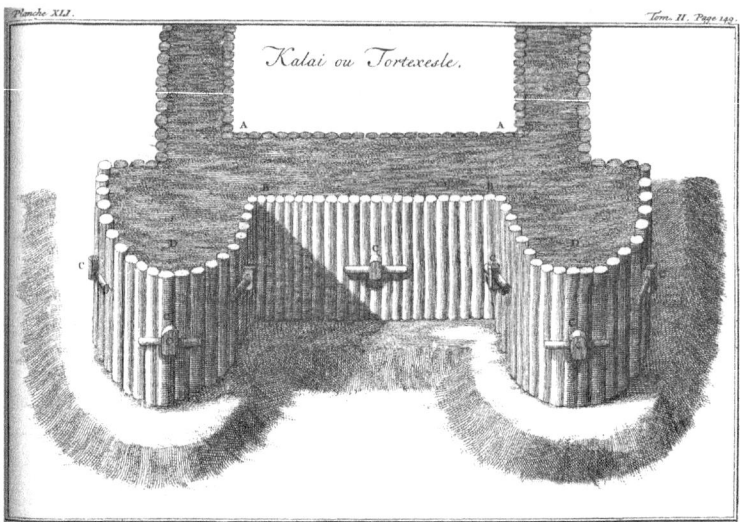

**Figure 12:** A drawing of one side of a palanka, by Marsigli, early eighteenth century. Marsigli, *Stato Militaire dell' Imperio Ottomanno*, 149 (facing).

Ottomans in the seventeenth century," except along key waterways, palankas continued to be built well into the eighteenth century, initially as a response to the new frontiers created by the 1699 Treaty of Karlowitz.[80]

The rest of this essay examines this whole concept of a late sixteenth- and seventeenth-century stagnation in fort design, noting its reality, likely exceptions, and possible explanations. It is a difficult generalization to prove, much less explain, in part because although Ottoman bureaucrats were effusive on the problems of pay, costs, garrions, and so on, there are few to no Ottoman records of fortress design, nor are there official plans of their forts from the time of their building.[81] Furthermore, there was no equivalent of the European tradition of printed manuals of fort design. Pepper partly attributes the success of the Italian method and its spread around Europe to the "parallel development of a technical literature on fortification (most of it written in Italian) in which advances were reported, new ideas advertised, canons of sound practice established and deviations denounced."[82]

---

[80] Hegyi, "Ottoman Network," 171 (quote); Özgüven, "Palanka Forts and Construction Activity in the Late Ottoman Balkans," 171–188.
[81] Stein, *Guarding the Frontier*, 48.
[82] Pepper and Hughes, "Fortification in Late 15th-Century Italy," 541.

In contrast, there was essentially no Ottoman theoretical literature on architecture or engineering, and such as there was would have circulated in manuscript only, greatly limiting its reach.[83] There was a centralized Corps of Royal Architects in Istanbul, and there is evidence of an Ottoman tradition of architectural drawings and models, but much remains unclear about how plans were created or transmitted from the Corps of Royal Architects to builders in the field. Furthermore, one architectural historian argues that although deeply trained in geometry and architectural proportion, "Ottoman architects appear to have been bound by the conventions of miniature painting, which was strictly two-dimensional . . . unlike their European contemporaries who were trained in perspective drawing and scientific methods of orthogonal projection."[84] He concludes that although overhead plans were likely available from the central administration and likely were sent to the provinces, elevations depended heavily on local interpretation.

In addition to the still unclear question of central control over design and its transmission, there was the perhaps more fundamental problem of the lack of commercial printing presses to make design ideas cheaper, more widely available, and subject to mutual critique. The first Arabic letter-printing press in the Ottoman empire was not established until 1727 (earlier presses existed in the empire, but most were run by Greeks and other ethnicities, and printed works in their own languages).[85] Furthermore, the bare existence of a press did not necessarily mean it was used to print military advice manuals, as was happening on a vast scale in Europe from the mid- to late-seventeenth century. A study of Ottoman books on military subjects from 1299 to 1923 has found only 70 military works from before the eighteenth century, and those were almost entirely focused on pre-gunpowder subjects like archery. There is in fact only a single Ottoman-authored text on firearms from before the eighteenth century, and virtually everything written before the end of the seventeenth century was in manuscript only – not printed. Even in the eighteenth-century Ottoman presses printed only two military books. To be sure, in the later eighteenth century the Ottomans began actively copying (and eventually printing) European military texts, something they did with increasing rapidity in the nineteenth century. Interestingly, however, even over the whole period covered by that survey, that is, right up to 1923, only 3.9 percent of the books are categorized as "military engineering"

---

83 Sakul, "Military Engineering," 188.
84 Thys-Şenocak, *Ottoman Women Builders*, 162; Necipoğlu-Kafadar, "Plans and Models in 15th- and 16th-Century Ottoman Architectural Practice," 224–243 (esp. 224, 231, 236 [quote], 243).
85 Ágoston and Masters, eds. *Encyclopedia of the Ottoman* Empire, sv "Printing"; Hanioğlu, *A Brief History of the Late Ottoman Empire*, 38–41.

(much of which is about trench construction) with a further 1.9% about "fortification and strategy." The nature of fortress design plans, and the ease with which they can spread in printed books, suggests that this difference in print culture may have played a marked role in differentiating European and Ottoman fort design.[86] Ottoman design remained idiosyncratic and local, even as it was partly beholden to and perhaps handcuffed by an imperial corps of architects, while European designs tended to become standardized based on a competitive pan-European literature.

This combination of Ottoman centralization without standardization and local idiosyncrasy can be seen in the well-studied and possible exception to apparent Ottoman conservatism at New Navarino (Pylos, or Anavarin-i Cedid), built in 1573 (See Figure 13).[87] Its construction also tells us something about the role of the royal architect corps in ordering and designing forts. New Navarino was another maritime fort, designed to control the southern entrance to the bay of Navarino, supplementing an earlier medieval construction ("Old Navarino" or Anavarin-i Atik) on the north side of the bay. Here we have a fort built from the ground up by the Ottomans, late in the sixteenth century, and clearly with the Venetian threat in mind – all factors that would seemingly encourage building it in the most modern and effective style possible. The result was a mixture of *trace italienne*-like features and rectangular bastions otherwise seemingly unique in design, but intended to project offensive firepower to sea, much like the other Ottoman maritime forts from a century earlier. In this case, the star-shaped citadel replicates many aspects of contemporary Italian designs, although its pentagonal bastions lacked orillons to protect the cannons that provided flanking fire along the wall. The large enceinte, however, was defended with far less sophistication. The curtain walls were only weakly defended by a few circular bastions and thin vertical walls, neither scarped nor ditched. The primary function of the complex would seem to be the two nearly square bastions on the northwest and southwest

---

[86] Ihsanoğlu, et al., eds., *Osmanlı askerlik literatürü tarihi*, xiii–xix; Black, *War in the World*, 111–113, 216; Black, *War in the Eighteenth-Century World*, 69. For compendia of European military publications, see Jähns, *Geschichte der Kriegswissenschaften*; Cockle, *A Bibliography of English Military Books Up to 1642*. Systematic comparison to the recent bibliography of Ottoman military works is needed; Ágoston, "The Ottoman Empire and the Technological Dialogue between Europe and Asia", 27–39.

[87] It must be said that despite New Navarino's publication in Andrews's well-known volume on castles in Greece and its remarkable state of preservation, neither Nicolle nor Pepper discuss it. There are intriguing similarities between this design and the design of the Ottoman fort at Porto Kagio in the southern Peloponnese, the 1570 siege map of which was recently released online from the map collection of King George III. https://militarymaps.rct.uk/other-16th-century-conflicts/maina-cape-matapan-and-porto-kagio-1570fortezza (accessed 3/4/2020).

corners, both situated nearly at sea level facing the harbor or its entrance. A letter from 1576 (see below) suggests that the curtain wall was built after the southwest bastion and the star-shaped citadel. The whole complex likely began with the water battery in the southwest bastion, with the citadel designed to protect it from fire from the nearby height on which it sits (unlike at Rumeli Hisari or Kilidülbahir, both of which were dominated by nearby heights). The two water battery bastions conform neither to the Italian style nor to the common Ottoman circular or polygonal artillery tower design discussed previously.[88]

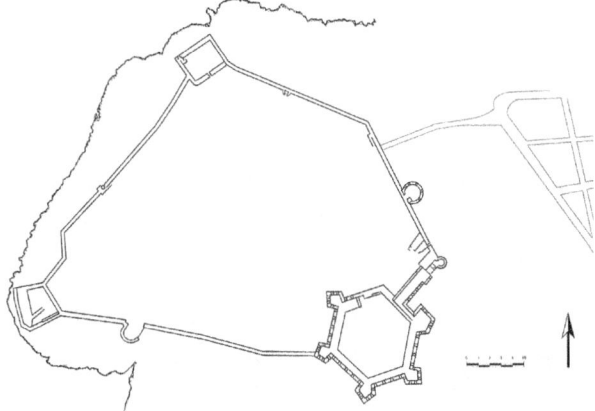

**Figure 13:** New Navarino (1573), internal buildings omitted. Edmund Potter based on Καρποδίνη – Δημητριάδη, *Κάστρα*, 193 and satellite imagery.

Unusually, we know something about the building history of this fort from the publication of a series of documents by Machiel Kiel. In the wake of their defeat at Lepanto in 1571, the Ottomans reassessed their defenses all along the western coast of Greece, including repairs to Rio and Naupactus (discussed above). The local commander at Old Navarino called for additional troops, cannon, and help repairing the fortress. Some initial orders were directed at the old fort, but by February 1573 the emphasis had shifted to building a new fort at the southern entrance to the harbor. In a letter from September of that year we learn that "the plan of the castle to be built has been previously dispatched with the messenger," that local revenues had been allocated to cover costs, and that construction should begin immediately. Meanwhile the admiral of the fleet had been ordered to "leave behind [in Anavarin] the architect who is with him." Much correspondence

---

**88** Andrews, *Castles of the Morea*, 49–57, plates XI and XII.

then ensued over the next several years about costs and labor supply, including one detailed letter from the chief architect in Istanbul to the provincial governor specifying the thickness of the curtain wall, the distance between towers, and the use of internal arches (the arches were characteristic of Ottoman curtain walls in the period). The towers were ordered to be sufficient to support large guns of the "*kolomborna*" type. The governor was also ordered to build a battery at sea level. It is clear from these letters that traveling representatives of the Corps of Royal Architects were expected to work from a centrally provided plan – presumably much adjusted for conditions on the ground. At the same time, however, the central architect ordered that the number of towers be determined by the length of the wall using a standard figure for the distance between towers, potentially ignoring the lay of the land.[89] In this circumstance, one wonders if the star-shaped citadel was a centrally provided idea, while the harbor-facing bastions were designed on the spot? But this is only speculation. We know that the Venetians, having captured the fort from the Ottomans in 1686, made plans in the early eighteenth-century to greatly expand the landward-facing fortifications, protecting that whole side of the fortress with *trace italienne* ravelins and hornworks (although those works were never carried out).[90]

The well-preserved walls at Navarino, the existence of Venetian plans to alter them, and the preservation of construction records all make it unique. What we lack for the Ottomans here, as elsewhere, are their detailed design documents, or even their plans of their own forts – something quite common among the Venetians and other European powers. The lack of either plans or design manuals undermines tracking changes over time or even assigning construction phases to specific periods. This difficulty is one reason that scholars have not delved deeply into Ottoman fort design, and Ottoman designs are even less well known for forts built after the mid- to late-seventeenth century.[91] Ironically, this was the exact period when Ottoman offensive power was ebbing and one might expect a burst of new forts designed to consolidate control.

There is in fact substantial evidence for a renewed (or new?) interest in fortification in this period, especially in improving those places the Ottomans already held. They rebuilt the citadel at Candia in Crete (modern Heraklion) after they finally captured it in 1669 (after a 21-year siege), but the details of modifications

---

**89** Kiel, "Construction of the Ottoman Castle," 266–267, 272.
**90** Andrews, *Castles*, 56–57.
**91** Another reason is a nationalist historiography in numerous Balkan countries that has ignored or glossed over Ottoman components of forts within their modern boundaries. In some cases, obviously Ottoman components (a tower or bastion) are simply ignored in the description of the fort or is glossed with a single sentence referring to Ottoman additions.

are elusive, and Candia was already a fully realized *trace italienne* complex built by the Venetians. The Ottomans improved key forts on the Danube frontier at Eger, Esztergom, Kaniszsa, Székesfehérvár – all mentioned above – and elsewhere. The late seventeenth and early eighteenth century also found them improving their fortifications along the Black Sea coasts. It is clear that some of the forts on the southern shore of the Black Sea were strengthened or rebuilt in that period, but some of those sites remain in Turkish military hands and are thus barely studied (although some work is ongoing).[92] The Ottoman forts on the northern and western shores of the Black Sea are also not well known, but a recent study of the fort at Özi (or Ochakov) supports the overall notion of Ottoman military construction as being fit to purpose as opposed to fit to a standard design.

Originally, Özi and the other late fifteenth- and early sixteenth-century fortifications along the north shore of the Black Sea and in the Crimea were intended to exert influence on the steppe and to resist steppe raiders. As the Russian empire began expanding into the region in the early eighteenth century, however, bringing artillery with it, the Ottomans invested in updating the fortifications. Caroline Finkel and Victor Ostapchuk's study of Özi charts its evolution in several phases (see Figure 14). It began as a small medieval style fort that morphed into a larger but still fundamentally medieval castle in the early seventeenth century, capable of mounting small cannon against Cossack naval assaults.[93] Even then, however, the Ottoman traveler Evliya Çelebi could describe the fort in 1657 as resembling "the high stone walls of medieval Europe," and they had remained that way "because there was no need for them to be otherwise." The Cossacks, although great diggers, lacked siege artillery. Russian pressure around the turn of the century, however, led to a new circuit wall surrounding the older complex, in a design substantially influenced by the *trace italienne*. Finkel and Ostapchuk are careful to point out, however, that only extreme need led to this modernization, and that it was limited in scope. The remoteness of the site, and the lack of local trees, stone, and other building materials meant that the decision to build a full European-style artillery fortress was a major commitment of resources.[94]

Fortunately, although the fort does not survive above ground, the extensive correspondence Finkel and Ostapchuk review, and a map of the fort from the

---

[92] Dorter and Thys-Şenocak, "Ottoman Fortifications on the Upper Bosphorus," (abstract available at http://www.eaa2014istanbul.org/pdf/abstracts.pdf).
[93] The strategic situation created by Cossack raids is described briefly in Ágoston's "Military Transformation in the Ottoman Empire and Russia, 1500–1800," 289.
[94] Finkel and Ostapchuk, "Outpost of Empire," 150–188, esp. 150–153, 159, 163–164 (quote). Çelebi's negative assessment here is at odds with his elsewhere holding up Özi as a model fortress (see above).

1730s, reveals some of the design considerations that the Ottomans struggled with. Most of the correspondence deals with materials, labor, cost, and the geological problems of maintaining a ditch in that region, but there was also some discussion of the actual design of the bastions, their angles, functions, and so on. In 1695 three "bastions" of uncertain design had been built, and then in 1710 we learn from Isma'il Pasha, the new warden of the fort, that the 1695 renovations had added a ditch around the entire complex and some necessary bastions (presumably the aforementioned three), but the ongoing building of a new ditch and seven new bastions were rendering them obsolete. He thus suggested filling in the old ditch and using its stone to revet the new one. He also suggested cutting down the 13 stone towers of the original castle to the level of the walls, which could then serve as bastions for light cannons. A separate anonymous treatise from the same period complained that the current seven bastions (presumably those built in the 1710 renovations) were too tall and were designed for musketeer use, not for cannon fire. This treatise recommended lowering them and making them cannon platforms. The author also noted, however, that they would only be "big enough for two cannons each, and thus incapable of doing what bastions 'usually' do, which is to mount four or five cannons to provide fire 'scissor-wise.'"[95] From this evidence, Finkel and Ostapchuk argue for a real awareness of the basic requirements of the *trace italienne*, here being applied with difficulty to an aging fortification in an expensive and resource-poor zone. Some of the modifications, such as cutting down towers to the height of the walls to support cannon, date back to the fifteenth century in Italy, but on this steppe frontier that change previously had not been necessary. Finkel and Ostapchuk conclude that this awareness of the value of the *trace italienne* and knowledge of its basic characteristics in the 1695 to 1710 period contradicts the common assumption that the Ottomans remained ignorant of the European science of fortification until the Baron de Tott educated them at the end of the eighteenth century.

There is further evidence to support this claim of Ottoman interest in fortress design in the late seventeenth century. The War of the Holy League (1683–1699) combined with Russian pressures led to a surge in renewal and repair of fortifications in the late seventeenth century accompanied by a growing interest in European designs that is especially evident from early in the eighteenth century when the 1699 Treaty of Karlowitz set the empire to ordering its frontiers. For one thing, we can see the Ottoman vocabulary for fortification expand. For centuries the two basic words for a fort or a bastion, *hisar* or *burç*, were interchangeable,

---

95 Finkel and Ostapchuk, "Outpost of Empire," 173 (quote).

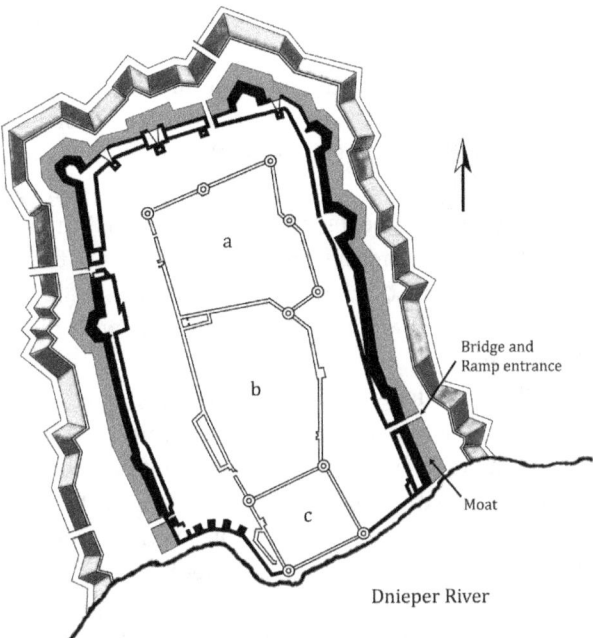

**Figure 14:** Özi. Internal castles a, b, and c represent the original late fifteenth-century fort (a) and its later expansions through the mid-seventeenth century (b and c). The outer wall complex with angular bastions was built in the 1710s.
Edmund Potter based on Finkel and Ostapchuk, Figure 3. An Ottoman plan from 1737 published in the same article indicates an additional outer line of bastions, but the authors do not make clear whether those bastions existed in the 1710s.

modified by shape (round, square, and so on). *Kale* was also used generically for castle or tower, and Ottoman bureaucrats categorized castles as either big (*büyük kale*) or small (*küçük kale*) for financial reasons. The former tended to be key border fortresses or regional capitals, usually identified as royal castles. Finally, *palanqa* referred to the earth and wood fortresses already discussed. Then, in the mid- to late-eighteenth century a new terminology began to emerge to identify those forts "built in line with the principles of geometry /theoretical design drawn prior to the beginning of the construction" (*hendese üzere inşa*). A fort built per the *trace italienne* would be referred to as "*hendese üzere hisar/sur/kale.*" Other words also emerged to describe European bastions, including *yıldız* for star-shaped constructions.[96] Further evidence for an intellectual shift in approach to fortification

---

**96** Additionally, *Kistel* could be used for castle. Personal communication with Kahraman Şakul, November 7, 2017.

design comes from two different Ottoman manuscripts now in archives in Vienna, both of which sought to transmit European designs to an Ottoman audience. One is just a single page within a geographical handbook rendering the hornwork of an unknown European fortress. The other, however, is a partial translation and presentation of Raimondo Montecuccoli's (1609–1680) and Allain Manesson Mallet's (1630–1706) works on fortifications, including detailed renderings of complex *trace italienne* designs with extensive commentary in Turkish. The latter was produced in 1786, and the former in 1783. With respect to the linguistic shift just discussed, it is interesting to note that notations on folios 4a and 5b of the latter manuscript compare older style fortresses with those "built on the principles of geometry."[97]

Beyond this increased intellectual interest in European designs, one can see even in the early eighteenth century a marked interest in using *trace italienne* designs as part of defending an empire under siege. The renovations at Özi under Russian pressure from 1695 to 1710 are only one example. The War of the Holy League led to renewed emphasis on the Danube frontier fortifications as well, the details of which need further work. There are hints of a greater interest in European designs at Vidin and Brăila, for example. Vidin, in modern Bulgaria, had long been an important regional city on the Danube, but was well within the empire. It was considered so secure that most of its medieval fortifications were actually dismantled in the late sixteenth century. In 1689, however, the Habsburgs captured it and held it for about ten months. During that time they initiated a massive refortification program following Vauban's principles. They were unable to finish them before the Ottomans regained the city, but the Ottomans continued the construction along those same lines, and had completed them by 1722 to 1723.[98] A late nineteenth-century German traveler to the city claimed that the improvements at Vidin, as well as at Belgrade, Oršova, Kladova, and Niš had all been built or at least initiated by the Holy Roman Emperor, and that those were now the most important places in the Ottoman northern defensive system.[99] That

---

97  Duda, ed. *Islamische Handschriften II*, 173–178, 197–198, and *Tafelband*: abb. 316–324, Farbtafel X (Turkish translations courtesy of Kahraman Şakul, to whom I also owe finding this source).
98  Gradeva, "Between Hinterland and Frontier," 331–336; Gradeva, "War and Peace Along the Danube," 154; Bearman, et al, eds., *Encyclopedia of Islam* 2nd ed., s.v. "Widin." Consulted online on 9 November 2017, https://referenceworks.brillonline.com/browse/encyclopaedia-of-islam-2. Details of what the Ottomans built versus the Austrians is not clear, although the Bulgarian literature may be more helpful.
99  Kanitz, *Donau-Bulgarien und der Balkan*, 14–15, 18.

may simply be European condescension about Ottoman forts, but it is also clear that the Ottomans were invested in completing those European designs at Vidin. At Brăila (further down the Danube from Vidin, in modern Romania) the details of the fortress's development is less clear, but a 1711 European map clearly shows *trace italienne* design features, although it was also hybridized with some semicircular bastions.[100]

A more thorough survey of Ottoman defensive constructions in the seventeenth and eighteenth centuries must await more specialist studies, but in addition to the detailed examination of turn-of-the-century Özi, way out on the frontier, we now also have a solid analysis of two additional forts, Seddülbahir and Kumkale, built at the heart of the empire in the mid-seventeenth century, from the ground up, and with royal patronage. Presumably this represented an opportunity to build to the most up-to-date designs and with a full understanding of likely enemy artillery technology. The two forts were built across from each other, as part of reinforcing control over the entrance to the Dardanelles, and like their earlier brethren on the straits they sought to project offensive firepower out over the water, and they were not built around population centers. Lucien Thys-Şenocak has used the Ottoman documentary record to establish the foundation dates and has examined the financial records and all the available graphic data from the two sites (both of which were heavily disturbed by refortification and bombardment during World War I).[101] Seddülbahir is on the European side, and its partner, Kumkale, is 4,140 meters away on the Asian side of the straits (the latter is often referred to by its more formal name Kale-i Sultaniye; it is here called Kumkale, following Thys-Şenocak, to distinguish it from the Kale-i Sultaniye discussed earlier and built in the fifteenth century at nearby Çannakale). Both forts were under construction by 1658 via the patronage of Turhan Sultan, mother of Sultan Mehmed IV.

Despite the symbolic and strategic significance of the location, and despite the threat of a resurgent Venice whose fleets were slipping into the Dardanelles past

---

**100** Candea, "Braila 1711," 201, Muzeul Brailei-Editura Istros, Braila, 2011, reproduced in Engin, "1787–1792 Osmanli-Rus," 165; Kiel, "Building Accounts," 14–15 also discusses the Ottoman interest in newer designs at Niš, Serbia, and Vidin.
**101** Much of what follows on Seddülbahir and Kumkale is from Thys-Şenocak, *Ottoman Women Builders*, 107–180, esp. 111, 113–114, 136, 147, 153, 155, 161, 167–175. See also the project website at www.seddulbahir-kumkale.org (accessed 13 October, 2017). Both sites remained under Turkish military control and were unsurveyed until recently (Thys-Şenocak, 135). Also useful is http://www.etudmimarlik.com/en/seddulbahir-fortress-documentation-and-onservation-project.php (accessed 22 June, 2018). Christopher Duffy mistakenly identifies the works at this time as both being on the European side, and including *Kilidülbahir*, rather than *Kumkale*. Duffy, *Siege Warfare*, 197.

the older fortifications there, and even despite the largesse of a royal patroness, the designs of both forts represent only minor departures from the other Ottoman forts on the Dardanelles built two centuries earlier. During a ceremony at their completion in 1659, both forts demonstrated the efficacy of their water batteries by skimming *stone* cannonballs across the width of the strait (although the cannon at Kumkale failed to reach the far side).[102] Seddülbahir is rectangular in plan, with separate walls enclosing an upper and lower section, the latter providing a sea-level water battery firing out over the strait in a design similar to the fifteenth-century fortifications at Kilidülbahir or even Rumeli Hisari (see Figure 15). The water battery wall (the lower southeast curtain) is 6.5 meters high, and pierced by eighteen vaulted cannon embrasures and supports a parapet that connects the two round corner towers, each roughly 9 meters tall. In this respect alone the design is archaic, in that towers taller than the curtain were almost unheard of by the seventeenth century. They were simply too vulnerable. The inner southeast wall paralleled the water battery but higher up on the slope, and it mounted a second battery, doubling the firepower extending out over the strait – although it is likely that these were somewhat smaller caliber guns. The batteries facing the straits were thus built to suit a very specific purpose and although their design might appear archaic, they can also be said to have been built to purpose. The walls facing the landward side, however, are strictly rectangular, with one surviving polygonal projecting tower at the corner and a square tower in the center of the curtain (other corner towers and the center tower on the northeast wall and possibly one on the southwest are now missing). An early Venetian drawing shows crenellations throughout the complex (also archaic by this time), and shows the landward corner towers as polygonal with scarped bases. Across the strait, Kumkale followed a similar simple rectangular plan with round landward corner towers and mid-curtain square or polygonal towers. The wall facing the straits no longer survives, but nineteenth-century diagrams show two massive square tower batteries (cut to the height of the wall with embrasures at sea-level), and a late seventeenth-century traveler described the two seaward towers as square, with a combined forty sea-level embrasures in the tower and the seaward curtain wall.[103]

---

**102** The Ottomans' continued use of stone-firing cannon, which persisted in this maritime role into the nineteenth century, probably reflected this desire to skip balls along the surface. Stone balls had less mass relative to surface area and would skip further and longer than iron balls. Evliya Çelebi's description of the occasion is available in English in Karateke, ed. and trans., *Evliya Çelebi's Journey from Bursa to the Dardanelles and Edirne*, 111–112.
**103** Thys-Şenocak, *Ottoman Women Builders*, 111, 113–14, 136, 147, 153, 155, 161, 167–175; Choiseul-Gouffier, *Voyage pittoresque de la Grèce*, plate 48, facing 440).

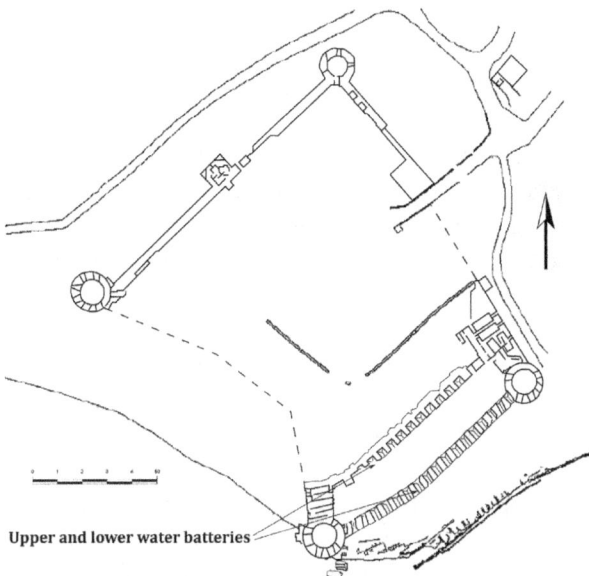

**Figure 15:** Seddülbahir (1658). Note the two batteries facing the straits on the lower right. This diagram shows the current state of the fort and some of the walls and curtain are missing. Edmund Potter based on project website, www.seddulbahir-kumkale.org.

It is hard not to critique these forts, intended to be a major display of royal investment in the defense of the empire, for their archaic designs. One could argue that the architects of both forts simply did not fear a major attack from land, or even that the necessity of siting the fort on such low ground to fire across the straits meant that any major land attack would dominate the high ground further inland and thereby rendered expensive fortifications irrelevant. Even so, this would be to display a remarkable confidence, since such an attack, even one by a small party, was the usual way to disable coastal batteries like this one – as would be true right up through the allied landing at Gallipoli in World War I. In the event, however, that confidence proved correct! The fort was not in fact assaulted from land until 1916.

## Expedience and Economy?

Ottoman builders clearly gauged the purpose of a fort and the most economical methods for satisfying that intent. There seemed to be any number of occasions when they deemed it unnecessary, or not worth the cost, to build bastioned artillery forts – much like we saw at Özi in its pre-1709 incarnations. Ćurčić reached this conclusion about many of the forts built during the Ottoman offensive in Serbia and Hungary from 1460 to 1500. Ottoman modifications to the Hungarian castle at Smederovo and their building of a new small frontier fort at Ram in Serbia, suggested to him a medieval-mindedness on the part of the masonry crew, who were being advised how to make the forts accommodate cannon as an offensive aspect of the fort. But the overriding strategic intent was fundamentally temporary: these forts were mere pauses on the long road of conquest and did not merit the more extensive investment required by *trace italienne* designs.[104]

Even in later periods, when "temporary" was clearly no longer the primary design criterion, the Ottomans built severely conservative fortifications, some of which they retained and repaired for long periods of time. Such forts, like Özi, continued to fulfill their purpose, however. They worked. Three examples in Greece include Kelefa, Passavas, and Zarnata, all of which seem primarily designed for local population control. Each may have been originally pre-gunpowder castles, but it was their later history that is most revealing. The Venetians captured all three forts in the 1680s, and the Venetian governor's dismissive description in 1708 conveyed the archaic nature of the designs but also hinted at their real strategic intent: "[Kelefa and Zarnata are] little forts, weak and outmoded in construction, built by the Turks more to control that contumacious population than to serve as bulwark to assaults from overseas."[105] As if to prove both points, the Ottomans recaptured Kelefa almost without a fight in 1715, and then continued to find it useful for controlling the local population for another 65 years. Kelefa itself is a simple, nearly rectangular fort with round corner towers, square towers in the center of the curtain to protect the gates, and vertical crenellated walls. It should be clear by now that such a design conforms in many ways to standard Ottoman practice: it compares to Bashtovë (built in late 1460s) and with Kumkale and Seddülbahir (both in the 1650s). At Kelefa at least the walls and towers appear to have been the same height. The dating of Kelefa remains controversial. But even if, as Malcolm Wagstaff argues, it was originally a medieval fortress, there is little doubt that the Ottomans later occupied it and they do not appear to have modified it in any substantial way.[106] Furthermore, there is documentary evidence that the Ottomans dispatched troops to Kelefa in 1670/71 with orders to "rebuild it." They may not have completely razed the site, but this means that the drawing by Vincenzo

---

104 Ćurčić, *Architecture in the Balkans*, 770–771.
105 Quoted in Andrews, *Castles*, 36.
106 Wagstaff, "Evliya Çelebi, the Mani and the Fortress of Kelefa," 16–20, 36–40 and Καρποδίνη – Δημητριάδη, *Κάστρα της Πελοποννήσου*, 132–139.

Coronelli from the 1680s (published ca. 1708, see Figure 16) – which shows a very conservative design – likely reflects those late seventeenth-century works.[107]

The fort at Passavas probably makes the point even more clearly. It lies on the eastern slopes of the Mani peninsula and is perfectly situated on a steep hill overlooking the road to one of the few passes that traverses the peninsula heading west (Kelefa lies on the other end of that road, on the western slopes of the peninsula). Ramsay Traquair suggested that the site had been fortified since antiquity, and indeed some of the lower stone courses on one side have been credited to the Bronze Age (based on the so-called and still visible "cyclopean" masonry).[108] The Franks (re)fortified the site in 1254, after which it changed hands several times in a still medieval configuration, only to be abandoned – according to Greek historical traditions – and then repaired and garrisoned by the Ottomans in 1670.[109] Recent research in Ottoman bureaucratic records, however, makes it clear that Passavas had an Ottoman garrison in 1583, although there was a period of abandonment after that date, leading to the rebuilding episode in 1670.[110] Traquair's suggestion that the Venetians destroyed it as useless after their conquest of the Peloponnese in 1684 seems unlikely, since nearly all of its wall circuit remains standing to its full height even today, and other sources attest to a new Ottoman garrison in the eighteenth century.

The fort itself is remarkably simple and remarkably antiquated. The curtain wall, essentially rectangular in plan except where it conforms to the line of a sheer cliff on the north, is vertical and quite thin, topped by crenellations, many of which have been equipped with a gun slit in the center of the merlon. There are round towers on three corners, but none of them were large enough to support much in the way of cannons, and they are also crenellated as if for infantry. Indeed, the main entrance on the eastern side does not appear large enough to admit a cannon mounted on wheels. The seeming bastion in Coronelli's drawing (see Figure 16) on the northern corner above the cliff is not a bastion at all, but

---

**107** Çelebi, *Evliya Çelebi in Albania*, 4 n. 5 (translation by Kahraman Şakul).
**108** Traquair, "Laconia: I. Medieval Fortresses," 274–`275; Greenhalgh and Eliopoulos, *Deep into the Mani*, 49; Παν. Σταμ. Κατσαφάδος, *Τα Κάστρα της Μαϊνης*, 185, 278.
**109** Traquair, "Laconia: I," 275; Greenhalgh and Eliopoulos, *Deep into the Mani*, 50; Καρποδίνη – Δημητριάδη, *Κάστρα*, 244–245.
**110** Defter *tapu tahrir* (TT) 677, an *icmal defteri* from *hijri* 991 (1583 CE), shows the tax paid by villages designated for the support of the garrison of the "*kal'a-ı İmanya der liva-i Mizistre*" (fortress of Imanya in the district of Mystras). A later scribal note, dated 17–26 August, 1671, refers to this fortress as "Pasava," and also mentions the ensuing period of abandonment due to rebellions in the Mani. Elias Kolovos provided the translation of the defter, as fully discussed in Seifried, "Community Organization and Imperial Expansion in a Rural Landscape," 124–125 and fig. 23.

more of an observation platform. There is also no evidence outside the curtain walls for the angular ditch-and-bastion outworks that Coronelli's drawing indicates (they might, as at Navarino, have been suggestions for improvement).[111]

In terms of its design and function, however, Passavas, like Kelefa, was intended to serve as a "power centre for an army of occupation." The Ottomans seem to have assumed that it would be threatened only by local revolts – not by any major enemy army with a siege train (see Figures 17a and b). The fort was deemed adequate for that purpose, or so it seemed. When the Venetians in fact invaded and captured the fort in 1684, they were allied with the local Maniates, and since the fort would therefore not be needed for local control, the Venetians preferred to destroy it (although again, the destruction appears to have been fairly limited).[112]

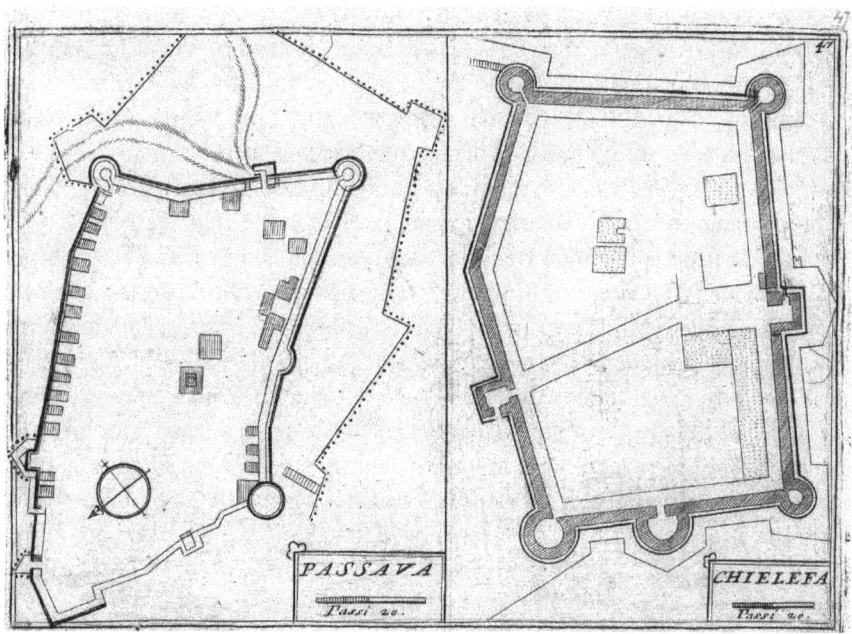

**Figure 16:** Coronelli's drawings of Passavas (left) and Kelefa made in the 1680s (published ca. 1708). The indications of outer angular bastions on the south side of Passavas do not appear justified by site survey. Coronelli 1708. Digital image provided by the Gennadius Library.

---

**111** These observations are based on Coronelli's drawing and close personal survey of the site. Coronelli, *Morea, Negroponte & Adiacenze*. Images from this volume are archived on the Travelogues website, http://eng.travelogues.gr/item.php?view=55026 (accessed 13 October, 2017).
**112** Greenhalgh and Eliopoulos, *Deep into the Mani*, 50–51 (quote).

**Figure 17a:** Passavas. Interior of north wall, showing crenellations, the parapet (wall walk), and occasional gun slits. Note thinness of the wall. Author photo.

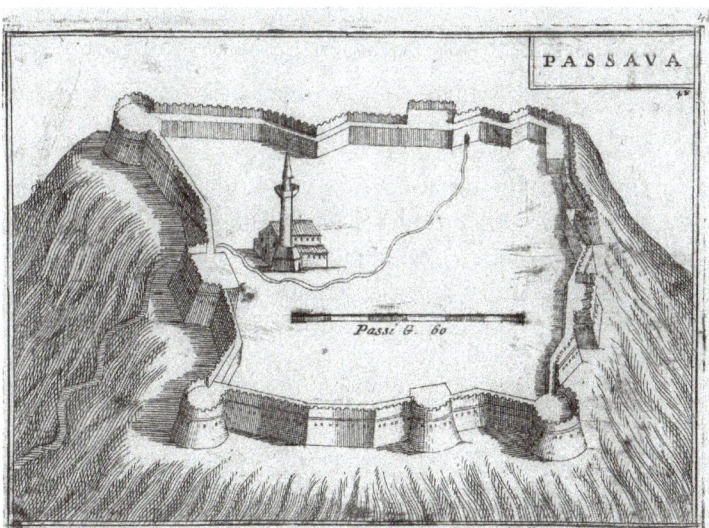

**Figure 17b:** Coronelli's elevation of Passavas. It is not nearly as formidable as this drawing makes it appear. Coronelli 1708. Digital image provided by the Gennadius Library.

The Ottomans reconquered the Peloponnese in 1715, and they re-garrisoned Passavas at some point, something we know only because that garrison is reported as fleeing during the Orlov rebellion in 1770. What is all the more remarkable is that Ottoman repairs to the fort after the Orlov rebellion appear not to have modernized it much at all, and it ironically proved vulnerable to the locals after all, who stormed the fort and killed the garrison in 1780, after which it was left abandoned as it is now.[113]

Passavas and Kelefa suggest a kind of functional economy in fort design, and there are other examples of the Ottomans building forts around the empire with an eye to economy, weighing likely threat versus certain costs, and building to suit their purpose. Some were clearly intended as semi-temporary measures, such as bases built while enroute to another offensive. Mehmed II, for example, the very ruler who ordered the building of the innovative Yedikule and Kilidülbahir (among others), opted for a very medieval upgrade to the ancient fort at Elbasan to support his campaign there.[114] In other places, strategic requirements may have suggested more permanent designs, but the locations were deemed as unthreatened by artillery. The pilgrims' route to Mecca and several coastal locations in Yemen were fortified seemingly without concern for enemy siege trains – a consideration that also may have dictated design decisions at places like Passavas, Kelefa, and the first iterations of work at Özi.[115]

# Conclusion

This study has only scratched the surface of Ottoman fortifications. A short exemplary list of the sites *not* covered includes major locations like Cairo, Algiers, Mosul, Erivan, and Tabriz.[116] And even within the studied regions, any number of forts have been elided, sometimes because we know so little about them. The

---

113 Greenhalgh and Eliopoulos, *Deep into the Mani*, 51–52.
114 Ćurčić, *Architecture in the Balkans*, 770.
115 For the pilgrim forts, see Petersen, "The Ottoman Conquest of Arabia and the Syrian Hajj Route," 81–94. For Yemen, see Farah, *The Ottoman Forts and Castles of Yemen*.
116 Examples: In 1583 at Erivan, the Ottomans built a new outer wall of 43 towers and 1,726 embrasures, and an inner defense with eight towers and 725 embrasures. Similarly impressive works were built around Tabriz in 1585. Unfortunately, we do not know much about their design. They also updated the defences at Mosul in 1631 at great cost. Murphey, *Ottoman Warfare*, 217, n. 23, n. 24; Murphey, "Construction of a Fortress at Mosul in 1631," 163–178; https://en.wikipedia.org/wiki/Erivan_Fortress. Nicolle, *Ottoman Fortifications*, briefly examines some of these locations, including Cairo, Algiers, and Baghdad.

Habsburg frontier had some 85 to 90 fortresses in the seventeenth century, while the Black Sea coast boasted at least 14 major Ottoman forts, and there were many more in the interior of Serbia, Croatia, Albania, and Greece.[117] Some of these were merely captured and garrisoned by the Ottomans, but others they either built new or extensively renovated. There is also more work to do studying the ongoing changes in fortification within the empire in the late eighteenth and nineteenth centuries, especially as local lords or *ayans*, increasingly acted independently of the central government, and ensconced and fortified themselves to maintain their independence. Ali Pasha of Ioannina, for example, massively re-built the fortifications there after 1786, although the designs seem to have been a hybrid of various theories and possibilities limited by the existing walls.[118]

This vast array of forts, many of them garrisoned during this period by centrally paid regular troops (the janissaries) creates a conundrum within the terms of the military revolution debate. A key part of Parker's original thesis was that the proliferation of large expensive artillery fortresses in Europe, and the need to garrison them, generated "a major reorganization of government in most Western states in which the inherited administrative system (based on the household) gave way to a more complex bureaucratic edifice."[119] The irony, as Rhoads Murphey has pointed out, is that the Ottoman empire of the sixteenth century was substantially in advance of their western European counterparts with respect to central standing armies, imperial tax administration, and center-to-periphery financial and military power projection.[120] It was that sophistication which enabled them to garrison as many locations around the empire as they did. In contrast, Portuguese, Spanish, Dutch, French, and then English colonial power projection relied much more on private enterprise activity throughout this period, sometimes in partnership with the central government. To grossly simplify: western European empire building was a process of centralizing what started out as private enterprise ventures; the Ottoman empire began as highly centralized

---

117 Murphey, *Ottoman Warfare*, 113; Finkel and Ostapchuk, "Outpost of Empire," 151.
118 Brooks, *Castles of Northwest Greece*, 248, 257–258. In her study of Ottoman Preveza, Emily Neumeier argues that Ali Pasha deliberately constructed his fortresses to appear more imposing than they were and for that purpose he persuaded European architects to design some of the many forts that he (re)constructed. Neumeier, "Ottoman Fortifications and the 'image value' of European Military Engineering," cited with permission.
119 Parker, *Military Revolution*, 24–26, 147.
120 Much of the point of Murphey's *Ottoman Warfare* is to show the level of sophistication in Ottoman central administration and logistical capability through 1700. Ágoston's "Military Transformation" also tracks an Ottoman advantage vis-à-vis Russia lasting until the eighteenth century (see 282).

patrimonial state that in time progressively decentralized as late eighteenth-century regional lords (*ayans*) asserted autonomy.[121] The Russian empire's trajectory differed from both, but its military power also began to outpace the Ottomans by the late seventeenth century.[122] Ottoman decline in military capability is now generally held to have begun at the earliest in the 1680s, and primarily due to their deteriorating logistical capability in the face of wars on numerous fronts. That decline in central power projection then allowed regional lords to slip the reins of the sultan. For some scholars, however, even this eighteenth-century capability gap was temporary. They argue that the Ottomans recognized the problem, actively tracked developments in Europe, and initiated reforms that kept them nearly current.[123]

Still, despite clear innovation in fortification technique and design in the first half of our period, it seems hard to deny that Ottoman fortifications within the empire from ca. 1550 to ca. 1750 lagged well behind European standards, and this at a time when the Ottomans were at least current in other military technologies and were actively engaged in the wider exchange of information and technology. Ottoman military capability overall had not yet lagged, but their design of fortifications was certainly different, and arguably less effective. If we return to Çelebi's list of model fortifications at Avlonya, Szeged, Bender, and Özi, we can say that at the time he was writing in the late 1660s at least two of those forts (details of the design at Szeged are obscure) were quite antiquated and therefore vulnerable, and Bender was likely antiquated as well. Was the lag economic calculation? strategic orientation? cultural resistance? A different path of innovation? These are hard questions and no single answer suggests itself.

The older claim that the Ottomans were generally uninterested in proliferating fortresses simply will not do. There is clear and ever mounting evidence that the Ottomans did build fortified garrisons and even fortified frontier systems, especially if reckoned on an imperial scale rather than a regional one. This claim of disinterest persists even for the seventeenth and eighteenth centuries, even as

---

[121] The larger issue of the decline of Ottoman patrimonial control over regional lords and/or generalized economic decline is somewhat distinct from the issue of its relative military power within the region. See a brief overview of this literature in Ágoston's "Military Transformation," 286–288.

[122] Ágoston's "Military Transformation," passim.

[123] Ágoston, *Guns for the Sultan*, 201–202, for diplomatic reconfiguration, see Inalcık, "Military and Fiscal Transformation in the Ottoman Empire, 1600–1700," 283–337; Ágoston, "Ottoman Empire and the Technological Dialogue"; Şakul, "General Observations on the Ottoman Military Industry, 1774–1839," 41–56; Şakul, "The Evolution of Ottoman Military Logistical Systems in the Later Eighteenth Century," 307–328.

the Ottomans clearly adopted a more defensive strategic posture. The explanation for this supposed disinterest has been that the expansionist ideology of the Ottomans mitigated against permanent frontier fortifications. The problem is that Ottoman religious and ideological claims to a universal empire did not stop them from adopting a more pragmatic approach in practice, and they had established clearly demarcated frontiers well before 1700.[124] Furthermore, as we have seen, they invested heavily to protect key "heartland" locations that could be threatened by naval forces. The early forts in such places were on the cutting edge of artillery fortress innovation (like Kilidülbahir or Rio). Later forts, however, equally "heartland," like Seddülbahir and Kumkale, were built in an outmoded and vulnerable fashion. Perhaps most interestingly, although the Habsburgs on the Danube frontier quickly imported Italian architects in the mid-sixteenth century to update their fortresses, and although those fortresses impressed the Ottomans, the Ottomans nevertheless managed to capture many of them and then showed little interest in matching them.[125]

Rather than argue general disinterest, therefore, we might do better to argue for a different strategic calculation. The Ottomans did long sustain an ideological and strategic posture favoring the offensive. As a result, for the early part of the period covered here, large field armies were often in being, and there was, unusually for the sixteenth century, a large standing army distributed around the empire, with a strong core in Istanbul, rising from at least 29,175 in 1574 to 84,675 in 1670. This standing army was then supplemented by the military service of landed timariots who might add an average of 50,000 men to a campaign army.[126] In contrast, the French monarchy through the period from roughly 1445 to the 1630s maintained a peacetime standing army of about 10,000, with occasional jumps to 25,000. French wartime forces in that same period rose to between 30,000 and 60,000.[127] Standing forces available to smaller states like sixteenth-century Venice or Hungary were correspondingly smaller. Campaign armies of course could be much larger, and they are often taken as a point of comparison. The Ottomans' large and readily accessible standing force, however, allowed them to assume a strong probability of being able to arrive in time to lift a siege –

---

**124** Kolodziejczyk, "Between Universalistic Claims and Reality," 205–219.
**125** For arrival of Italian architects, see Ágoston, "Empires and Warfare," 118–119; Ágoston, *Guns for the Sultan*, 194–195; Murphey, *Ottoman Warfare*, 111. For Ottoman impressions of Habsburg fortress at Uyvar, see Murphey, *Ottoman Warfare*, 113.
**126** Murphey, *Ottoman Warfare*, 45, 40, 49 (note that I have added in the provincial janissary garrisons to the 1670 figure. They amounted to 14,379 men, in addition to the janissaries in the central army). Ágoston, "Firearms," Tables 3 and 7 show comparable, if slightly different numbers.
**127** Lynn, "The *trace italienne* and the Growth of Armies," 170.

at least within large sections of the empire. Any given fort, no matter how well designed, was never believed to be invulnerable. The key variable was whether it could hold out long enough to receive help from a relief army or exhaust the supplies of the army besieging it. Even as sprawling as the Ottoman empire was, the empire's standing army could be deployed relatively quickly to a threatened fort or frontier. Transportation by sea in the Aegean, Black, and Adriatic seas was fast and efficient, and only somewhat less so up the Nile, Danube, Dnieper, and Dniester Rivers.[128] Other outposts of the Ottoman empire in Baghdad, Algiers, Ethiopia, Yemen, or the interior of places like Albania were more problematic, and indeed some of those places spun out of centralized control in the Age of Ayans in the late eighteenth century.

Separate from the problem of supposed interest or disinterest in fortifying in general is the question of design. In one sense this is an argument over whether the Ottoman military architects were "ignorant" of changes in the west, or simply expedient and perhaps financially conservative.[129] The documents regarding Özi clearly show the imperial treasury's bureaucrats making calculations about the costs of moving materials to such a distant location, and Rhoads Murphey has documented the steep labor costs for the repairs of the fort at Mosul.[130] Furthermore, the design of forts like Passavas, Kelefa, the updates to Smederovo, and even the common pattern of simply adding an artillery tower to an existing fortification, all suggest a choice of economy in light of expected function. It has long been an article of faith among historians that the expense of the *trace italienne* forts in Europe was enormous.[131] For the Ottomans, these costs may have dominated design considerations. Thys-Şenocak specifically wonders if the acute strain on the Ottoman treasury during the building of Seddülbahir and Kumkale limited architectural options. Even if that were so,

---

[128] Murphey emphasizes the cost of sea transport and the overall difficulty given the size of the empire, but within the narrower sphere outlined by the Adriatic, Black, and Aegean Seas, and up the Danube, transport could be rapid.

[129] Thys-Senocak, *Ottoman Women Builders*, 172–173, esp. n. 148, 174–175.

[130] Murphey, "Construction of a Fortress."

[131] Parker, *Military Revolution*, 12. Contrarily, Azar Gat argues that fortifications themselves comprised only a small percentage of a state's military expenditure, often only 5 to 10 percent, and reaching 17 percent only during the radical program of fortification conducted in late seventeenth-century France by Sébastien Le Prestre de Vauban. The main cost for state armies was personnel – salary and logistics. Gat, *War in Human Civilization*, 466–471, 474–476. Murphey's analysis of the Ottoman military budget supports Gat: he finds payments to the Janissaries alone ranging from 37 percent to 77 percent of Ottoman royal expenditures in the sixteenth and seventeenth centuries (not every year can be reconstructed). Murphey, *Ottoman Warfare*, 44.

however, there are aspects of their design that reflect more design conservatism than cost savings. Consider, for example, the corner towers at Seddülbahir being higher than the curtain: there is no expedience or cost savings in such a plan, but it did make them markedly more vulnerable. The same is true regarding the persistence of crenellations, which were actively dangerous to the defenders as sources of stone shrapnel when hit by artillery fire. And unlike Passavas, Seddülbahir was distinctly likely to be attacked by heavy ship-borne artillery.

Thus despite successful efforts by historians to rehabilitate Ottoman military technology, tactics, and suitability to purpose in the early modern period, especially with regard to firearms, cannon, and drill, there may yet be a case for a persistent conservatism in military architecture. The lack of a printing press certainly played a role in limiting the spread of specific techniques, and it may have played a role in the lack of standardization. But a similarly low output of printed Ottoman books regarding cannon or guns did not hold back the empire on those fronts. Is it possible that some of the problem lay with the centrally controlled guild of architects? Italian military architects of the late fifteenth and sixteenth centuries (if not beyond) moved freely between competing cities, copied each other's work, and published competing manuals. And what we know of military engineers in other periods and places in early modern western Europe also suggests the rapid movement of experts and their ideas from country to country, always in competition with each other as experts for hire – a competition separate from that between states.[132] We know from other studies of the empire that experts moved in and out of the Ottoman sphere as well, but if ideas about buildings had to be filtered through an imperial guild, which also approved expenditures (as the Navarino documents suggest), that process may have fostered conservatism in design. This institutional conservatism may ultimately have been broken by the increased emphasis on imperial defense in the eighteenth century, as we saw with the interest in Montecuccoli, the changes in technical vocabulary, and the designs of the additions at Özi and Vidin. And from the late eighteenth century it is clear that the Ottomans moved decisively to catch up, and were happy to use whatever expertise was available to do so.

From a completely different angle, there is an essentially untested technical claim that the Ottoman use of *khorasan* or *horasani* mortar, in which the cement

---

**132** Hale, "Early Development," 473 for mobility of Italian designers; Ágoston, "Habsburgs and Ottomans," 132–133 discusses the role of Italian, Spanish, Austrian, and German designers working on the late sixteenth-century defenses in Hungary; Nixon, "Tudor Coastal Forts," discusses the movement of Italian engineers and ideas into France and into English designs in France as well.

was blended with crushed brick to produce a stronger, rose-colored mortar, rendered their masonry forts – regardless of design – nearly impervious to early modern cannon. Rhoads Murphey echoes this claim based on a seventeenth-century account, and claims that "only fire at very close range was capable of penetrating walls constructed using this technique."[133] If true, this passive wall strength certainly would have incentivized the offensive design of Ottoman forts, that is, redesigning the walls and towers to mount cannon, while deemphasizing the defensive aspects – lowering and reshaping the walls, and preparing defenses against a breach and assault. Nevertheless, this claim seems improbable on its face, given the number of Ottoman forts breached and captured by artillery over the course of the early modern era. It is worth pointing out, however, that modern tests on two samples of historic khorasan masonry found one to be 2.2 and the other 3.9 times stronger than a standard stone masonry wall.[134] That would be a substantial improvement, but it is not clear that even this greater strength would render a vertical masonry wall cannon-proof.

It is difficult to place all this limited evidence into a single interpretive framework, but it does seem that the Ottomans carefully tailored their investments in fortifications to fit their strategic intentions and threat projections. Unfortunately for them, those projections were sometimes inaccurate, and antiquated forts sometimes faced well-equipped siege trains. Nevertheless, on the whole this economical attitude toward design does not appear to have significantly undermined Ottoman frontier security in and of itself. After all, as work on the development of the *trace italienne* makes clear, much of the architectural detail was fundamentally both tactical and *defensive*. The early steps on the path to the mature *trace italienne* system in the West were intended to solve problems in wall re-design that had been made to deal with early siege cannon. The first guns had led to the loss of tall towers, crenellations, machicolations, and so on. All of those features had been designed to repel a direct infantry assault on the walls. With them removed, fort designers looked for new ways to deal with that infantry assault, using crossing fire from different angles to eliminate dead space at the foot of the walls. Those efforts ultimately evolved into the bastioned artillery fortress. In contrast, those early Ottoman fortresses whose designs appeared most innovative were intended to project *offensive* firepower further and further away – primarily in a maritime access control role. The designers were less worried defending against assault because of their confidence in the availability of a land-based

---

**133** Murphey, *Ottoman Warfare*, 113–114.
**134** Akman, et al., "The History and Properties of Khorasan Mortar and Concrete," 5 (paper available at https://www.researchgate.net/publication/265167126); Bağbanci, Özcan and Bağbanci, "Characteriztion of Materials Used in the Fourteenth Century," 301–312.

relief army. Neither the Italian nor the Ottoman initial design "trajectory" was permanent, or even singular at any one time (for example, so-called defensive design features easily and often overlapped with offensive design elements – the triangular bastion increased defensive strength but also provided a platform for heavy offensive artillery fire). But this difference of emphasis in different cultures at different times suggests the complexity of the problem with which actors at the time were struggling. There was no obvious pre-ordained solution. In the longer term, however, possibly due to excessively centralized design control, Ottoman fortifications lagged in ways that made them distinctly vulnerable. There are subtle lessons here about the military revolutions of the present and the future. Technological interactions are difficult to predict; the "correct" trajectory is not always apparent; and rigid bureaucratic systems may not change rapidly enough with the times.

# Bibliography

Ágoston, Gábor. "Gunpowder for the Sultan's Army: New Sources on the Supply of Gunpowder to the Ottoman Army in the Hungarian Campaigns of the 16th and 17th Centuries." *Turcica* 25 (1993): 15–48.

Ágoston, Gábor. "Ottoman Artillery and European Military Technology in the Fifteenth and Seventeenth Centuries." *Acta Orientalia Academiae Scientiarum Hungaricae* 47 (1994): 75–96.

Ágoston, Gábor. "Habsburgs and Ottomans: Defense, Military Change and Shifts in Power." *Bulletin of the Turkish Studies Association* 22.1 (1998): 126–141.

Ágoston, Gábor. "Ottoman Warfare in Europe: 1453–1826." In *War in the Early Modern World*, edited by Jeremy Black, 118–144. Boulder: Westview Press, 1999.

Ágoston, Gábor. *Guns for the Sultan: Military Power and the Weapons Industry in the Ottoman Empire*. Cambridge: Cambridge University Press, 2005.

Ágoston, Gábor. "Military Transformation in the Ottoman Empire and Russia, 1500–1800." *Kritika: Explorations in Russian and Eurasian History* 12.2 (2011): 281–319.

Ágoston, Gábor. "The Ottoman Empire and the Technological Dialogue between Europe and Asia: The Case of Military Technology and Know-How in the Gunpowder Age." In *Science Between Europe and Asia: Historical Studies on the Transmission, Adoption and Adaptation of Knowledge*, edited by Feza Günergun, and Dhruv Raina, 27–39. New York: Springer, 2011.

Ágoston, Gábor. "Firearms and Military Adaptation: The Ottomans and the European Military Revolution, 1450–1800." *Journal of World History* 25.1 (2014): 85–124.

Ágoston, Gábor, and Bruce Masters, eds. *Encyclopedia of the Ottoman Empire*. New York: Facts on File, 2009.

Akman, M.S., et al. "The History and Properties of Khorasan Mortar and Concrete," paper presented at the Uluslararası Türk-İslim Bilim ve Teknoloji Tarihi Kongresi, İ.T.Ü., 28 Nisan-2 Mayıs 1986 [2nd International Congress on the History of Science and

Technology in Turko-Islamic Era], 1986. Paper available at https://www.researchgate. net/publication/265167126.

Andrade, Tonio. *The Gunpowder Age: China, Military Innovation, and the Rise of the West in World History*. Princeton: Princeton University Press, 2016.

Andrews, Kevin. *Castles of the Morea*. Revised edition. Princeton: American School of Classical Studies at Athens, 2006.

Bace, Apollon. "Kalaja e Vlores." *Monumentet* 5–6 (1973): 43–57.

Bağbanci, M. Bilal, Reşat Özcan, and Özlem Köprülü Bağbanci. "Characteriztion of Materials Used in the Fourteenth Century: The Early Ottoman Ördekli Bath, Bursa, Turkey." *Studies in Conservation* 55.24 (2010): 301–312.

Bearman, P., et al, eds. *Encyclopedia of Islam* 2$^{nd}$ (online) ed. Leiden: Brill, 2012. Accessed November 9, 2017, https://referenceworks.brillonline.com/browse/encyclopaedia-of-islam-2

Bencze, Zoltán. "Recent Research into Ottoman-Period Remains in Buda." In *Archaeology of the Ottoman Period in Hungary*, edited by Ibolya Gerelyes, and Gyöngyi Kovács, 55–62. Budapest: Hungarian National Museum, 2003.

Black, Jeremy. *Beyond the Military Revolution: War in the Seventeenth-Century World*. New York: Palgrave Macmillan, 2011.

Black, Jeremy. *War in the World: A Comparative History, 1450–1600*. New York: Palgrave Macmillan, 2011.

Black, Jeremy. *War in the Eighteenth-Century World*. New York: Palgrave Macmillan, 2013.

Brooks, Allan. *Castles of Northwest Greece*. Huddersfield: Aetos Press, 2013.

Çelebi, Evliya. *Evliya Çelebi in Albania and Adjacent Regions*. Edited and translated by Robert Dankoff, and Robert Elsie. Leiden: Brill, 2000.

Çelebi, Evliya, *Evliya Çelebi Seyahatnamesi*, vol. 8. Edited by Y. Dağlı, S.A. Kahraman, and R. Dankoff. Istanbul: Yapi Kredi Yayinlari, 2003.

Ćurčić, Slobodan, *Architecture in the Balkans: From Diocletian to Süleyman the Magnificent*. New Haven: Yale University Press, 2010.

Chase, Kenneth Warren. *Firearms: A Global History to 1700*. Cambridge: Cambridge University Press, 2003.

Choiseul-Gouffier, Gabriel Florent Auguste de. *Voyage pittoresque de la Grèce*. Paris: J.-J. Blaise, 1822.

Cockle, J. D. *A Bibliography of English Military Books Up to 1642 and of Contemporary Foreign Books*. London: Simpkin, Marshall, Hamilton, Kent & Co., 1900.

Coronelli, Vincenzo Maria. *Morea, Negroponte & Adiacenze*. Venice: n.p., ca. 1708.

DeVries, Kelly. "Gunpowder Weapons at the Siege of Constantinople, 1453." In *Guns and Men in Medieval Europe, 1200–1500*, edited by Kelly DeVries, 343–361. Burlington: Ashgate, 2002.

DeVries, Kelly. "The Impact of Gunpowder Weaponry on Siege Warfare in the Hundred Years War." In *The Medieval City Under Siege*, edited by I.A. Corfis, and M. Wood, 227–244. Woodbridge: Boydell, 1994.

Dögen, Matthias. *Architectura Militaris Moderna*. Amsterdam, Louis Elzevier,1647.

Dorter, Gizem, and Lucienne Thys-Şenocak. "Ottoman Fortifications on the Upper Bosphorus." Paper presented at the 20th Annual Meeting of the European Association of Archaeologists, Istanbul, 10–14 Sept. 2014. Abstract available at http://www.eaa2014istanbul.org/pdf/abstracts.pdf.

Duda, Dorothea, ed. *Islamische Handschriften II, Teil 2: Die Handschriften in Türkischer Sprache*. Wien: Verlag der Österreichischen Akademie der Wissenschaften, 2008.

Duffy, Christopher. *Siege Warfare: The Fortress in the Early Modern World, 1494–1660*. London: Routledge & Kegan Paul, 1979.

Ebhardt, Bodo. *Der Wehrbau Europas im Mittelalter*. Bd. 2, Teil II. Stollhamm: Helmut Rauschenbusch Verlag, 1958.

Engin, Hakan. "1787–1792 Osmanli-Rus, Avusturya Harpeleri Sirasinda Ibrail Kalesi." M.A. thesis, Trakya Üniversitesi, Edirne, 2013.

Farah, Caesar E. *The Ottoman Forts and Castles of Yemen: A Photographic and Architectural Analysis*. Lewiston: Edwin Mellen Press, 2010.

Finkel, Caroline, and Victor Ostapchuk. "Outpost of Empire: An Appraisal of Ottoman Building Registers as Sources for the Archeology and Construction History of the Black Sea Fortress of Özi." *Muqarnas* 22 (2005): 150–188.

Gat, Azar. *War in Human Civilization*. Oxford: Oxford University Press, 2006.

Gerelyes, Ibolya. "Ottoman Architecture in the Town of Gyula." In *Archaeology of the Ottoman Period in Hungary*, edited by Ibolya Gerelyes, and Gyöngyi Kovács, 173–180. Budapest: Hungarian National Museum, 2003.

Gradeva, Rossitsa. "War and Peace Along the Danube: Vidin at the End of the Seventeenth Century." *Oriente Moderno* Nuova serie 20 (2001): 149–175.

Gradeva, Rossitsa. "Between Hinterland and Frontier: Ottoman Vidin, Fifteenth to Eighteenth Centuries." In *The Frontiers of the Ottoman World*, edited by A.C.S. Peacock, 331–351. Oxford: The British Academy, 2009.

Greenhalgh, Peter, and Edward Eliopoulos. *Deep into the Mani: Journey to the Southern Tip of Greece*. London: Faber and Faber, 1985.

Gucciardini, Francesco. *The History of Italy*. 3$^{rd}$ ed. Translated by Austin Parke Goddard. London: Z. Stuart, 1763.

Hale, J.R. "The Early Development of the Bastion: An Italian Chronology, C. 1450–1534." In *Europe in the Late Middle Ages*, edited by J.R. Hale, J.R.L. Highfield, and B. Smalley, 466–494. Evanston: Northwestern University Press, 1965.

Hall, Bert S. *Weapons and Warfare in Renaissance Europe: Gunpowder, Technology, and Tactics*. Baltimore: Johns Hopkins University Press, 1997.

Hanioğlu, M. Şükrü. *A Brief History of the Late Ottoman Empire*. Princeton: Princeton University Press, 2008.

Hegyi, Klára. "The Ottoman Network of Fortresses in Hungary." In *Ottomans, Hungarians, and Habsburgs in Central Europe: The Military Confines in the Era of Ottoman Conquest*, edited by Gézá Dávid, and Pál Fodor, 163–193. Leiden: Brill, 2000.

Horváth, István. "Ottoman Military Construction in Esztergom." In *Archaeology of the Ottoman Period in Hungary*, edited by Ibolya Gerelyes, and Gyöngyi Kovács, 75–88. Budapest: Hungarian National Museum, 2003.

Hughes, Quentin. *Military Architecture*. London: Hugh Evelyn, 1974.

Ihsanoğlu, Ekmeleddin, et al., ed. *Osmanlı Askerlik Literatürü Tarihi = History of Military Art and Science Literature During the Ottoman Period*. 2 vols. Istanbul: IRCICA, 2004.

Imber, Colin. "Ibrahim Peçevi on War: A Note on the 'European Military Revolution.'" In *Frontiers of Ottoman Studies: State, Province, and the West, Vol. II*, edited by Colin Imber, Keiko Kiyotaki, and Rhoads Murphey, 7–22. London: I. B. Tauris, 2005.

İnalcık, Halil. "Military and Fiscal Transformation in the Ottoman Empire, 1600–1700." *Archivum Ottomanicum* 6 (1980): 283–337.

Irás-Melis, Katalin. "Pest During the Ottoman Era." In *Archaeology of the Ottoman Period in Hungary*, edited by Ibolya Gerelyes, and Gyöngyi Kovács, 161–172. Budapest: Hungarian National Museum, 2003.

Jähns, Max. *Geschichte Der Kriegswissenschaften: Vornehmlich in Deutschland*. 2 vols. Munich: Oldenbourg, 1889.

Kanitz, Felix Philipp. *Donau-Bulgarien und der Balkan: Historisch-geographisch-ethnographische Reisestudien aus den Jahren 1860–1879*. Leipzig: Renger'sche Buchhandlung, 1882.

Karateke, Hakan T., ed. and trans. *Evliya Çelebi's Journey from Bursa to the Dardanelles and Edirne: From the Fifth Book of the Seyhatatname*. Leiden: Brill, 2013.

Καρποδίνη – Δημητριάδη, Έφη. *Κάστρα της Πελοποννήσου*. Athens: Adam Editions, 1993.

Κατσαφάδος, Παν. Σταμ. *Τα Κάστρα της Μαΐνης*. Athens: Σπανός – Βιβλιοφιλία, 1992.

Kelenik, József. "The Military Revolution in Hungary." In *Ottomans, Hungarians, and Habsburgs in Central Europe: The Military Confines in the Era of Ottoman Conquest*, edited by Gézá Dávid, and Pál Fodor, 117–162. Leiden: Brill, 2000.

Kiel, Machiel. "The Building Accounts of the Castle of Vlorë/Avlonya (S. Albania) 1537–1539." In *Proceedings of the Second International Symposium on Islamic Civilisation in the Balkans (Tirana, Albania, 4–7 December 2003)*, edited by Ali Çaksu, 3–31. Istanbul: IRCICA, 2006.

Kiel, Machiel. "Construction of the Ottoman Castle of Anavarin-i Cedid." In *A Historical and Economic Geography of Ottoman Greece: The Southwestern Morea in the 18th Century*, edited by Fariba Zarinebaf, John Bennet, and Jack L. Davis, 265–281. Princeton: The American School of Classical Studies at Athens, 2005.

Kiel, Machiel. *Ottoman Architecture in Albania (1385–1912)*. Istanbul: Research Centre for Islamic History, Art and Culture, 1990.

Kolodziejczyk, Dariusz. "Between Universalistic Claims and Reality: Ottoman Frontiers in the Early Modern Period." In *The Ottoman World*, edited by Christine Woodhead, 205–219. London: Routledge, 2012.

Koumousi, Anastasia. "Rion Fortress: Restoration Works and New Archaeological Data (2006–2016)." Paper presented at conference on "Fortifications of the Ottoman Period in the Aegean," Mytilene, 30–31 Oct, 2018. Proceedings are forthcoming.

Lee, Wayne E. *Waging War: Conflict, Culture, and Innovation in World History*. New York: Oxford University Press, 2016.

Lynn, John A. "The *Trace Italienne* and the Growth of Armies." *The Military Revolution Debate: Readings on the Transformation of Early Modern Europe*, edited by Clifford J. Rogers, 169–200. Boulder: Westview Press, 1995.

Lynn, John A. "The Evolution of Army Style in the Modern West, 800–2000." *International History Review* 18.3 (August 1996): 505–545.

Marsigli, Luigi Ferdinando. *Stato Militaire dell' Imperio Ottomanno*. Amsterdam: Pierre Gosse & Jean Neaulme, 1732.

McGilchrist, Nigel. *Blue Guide Greece: The Aegean Islands*. London: Blue Guides, 2010.

Murphey, Rhoads. "Construction of a Fortress at Mosul." In *Social and Economic History of Turkey, 1071–1920*, edited by Osman Okyar and Halil İnalcık, 163–177. Ankara: Meteksan Şirketi, 1980.

Murphey, Rhoads. *Ottoman Warfare, 1500–1700*. New Brunswick: Rutgers University Press, 1999.

Necipoğlu-Kafadar, Gülru. "Plans and Models in 15th- and 16th-Century Ottoman Architectural Practice." *Journal of the Society of Architectural Historians* 45.3 (1986): 224–243.

Neumeier, Emily. "Ottoman Fortifications and the 'image value' of European Military Engineering." Paper presented at the Middle East Studies Association, Boston, November 17–20, 2016.

Nicolle, David. *Ottoman Fortifications, 1300–1710*. Oxford, New York: Osprey Publishing, 2010.

Nixon, David. "Tudor Coastal Forts in Northern France." Paper presented at the International Medieval Congress, Leeds, July 4, 2017.

Özgüven, Burcu. "The Palanka: A Characteristic Building Type of the Ottoman Fortification Network in Hungary." *Electronic Journal of Oriental Studies* 4 (2001): 1–12. Accessed February 10, 2021, http://web.archive.org/web/20041203232633fw_/http://www2.let.uu.nl/Solis/anpt/ejos/pdf4/34Ozguven.pdf.

Özgüven, Burcu. "Palanka Forts and Construction Activity in the Late Ottoman Balkans." In *The Frontiers of the Ottoman World*, edited by A. C. S. Peacock, 171–188. Oxford: The British Academy, 2009.

Parker, Geoffrey. *The Military Revolution: Military Innovation and the Rise of the West, 1500–1800*. 2nd ed. Cambridge: Cambridge University Press, 1996.

Peacock, A.C.S. "Introduction: The Ottoman Empire and Its Frontiers." In *The Frontiers of the Ottoman World*, edited by A.C.S. Peacock, 1–28. Oxford: Oxford University Press, 2009.

Pepper, Simon. "Ottoman Military Architecture in the Early Gunpowder Era: A Reassessment." In *City Walls: The Urban Enceinte in Global Perspective*, edited by James D. Tracy, 282–316. Cambridge: Cambridge University Press, 2000.

Pepper, Simon, and Nicholas Adams. *Firearms and Fortifications: Military Architecture and Siege Warfare in Sixteenth-Century Siena*. Chicago: University of Chicago Press, 1986.

Pepper, Simon, and Quentin Hughes. "Fortification in Late 15th-Century Italy: The Treatise of Francesco Di Giorgio Martini." In *Papers in Italian Archaeology I: The Lancasater Seminar, Part Ii.*, edited by H. McK. Blake, T.W. Potter, and D.B. Whitehouse, 541–562. Oxford: British Archaeological Reports, 1978.

Petersen, Andrew. "The Ottoman Conquest of Arabia and the Syrian Hajj Route." In *The Frontiers of the Ottoman World*, edited by A.C.S. Peacock, 81–94. Oxford: The British Academy, 2009.

Potter, David. *Renaissance France at War: Armies, Culture and Society, C. 1480–1560*. Woodbridge: Boydell & Brewer, 2008.

Rogers, Clifford J. "The Military Revolutions of the Hundred Years War." In *The Military Revolution Debate: Readings on the Transformation of Early Modern Europe*, edited by Clifford J. Rogers, 55–94. Boulder: Westview Press, 1995.

Rogers, Clifford J. "Gunpowder Artillery in Europe, 1326–1500: Innovation and Impact." In *Technology, Violence and War. Essays in Honor of John F. Guilmartin, Jr.*, edited by Sarah Douglas, and Robert Ehlers, 39–71. Leiden: Brill, 2019.

Roland, Alex. *War and Technology: A Very Short Introduction*. New York: Oxford University Press, 2016.

Şakul, Kahraman. "General Observations on the Ottoman Military Industry, 1774–1839: Problems of Organization and Standardization." In *In Science Between Europe and Asia: Historical Studies on the Transmission, Adoption and Adaptation of Knowledge*, edited by Feza Günergun, and Dhruv Raina, 41–56. New York: Springer, 2011.

Şakul, Kahraman. "Military Engineering in the Ottoman Empire." In *Military Engineers and the Development of the Early-Modern European State*, edited by Bruce P. Lenman, 179–200. Dundee: University of Dundee Press, 2013.

Şakul, Kahraman. "The Evolution of Ottoman Military Logistical Systems in the Later Eighteenth Century: The Rise of a New Class of Military Entrepreneur." In *War, Entrepreneurs, and the State in Europe and the Mediterranean, 1300–1800*, edited by Jeff Fynn-Paul, 207–228. Leiden: Brill, 2014.

Şakul, Kahraman, "Siege Warfare in Verse And Prose: The Ottoman Conquest Of Kamianets-Podilsky (Kamaniçe), 1672." In *The World of the Siege*, edited by Anke Fischer-Kattner, and Jamel Ostwald, 205–242. Leiden: Brill, 2019.

Seifried, Rebecca M. "Community Organization and Imperial Expansion in a Rural Landscape: The Mani Peninsula (AD 1000–1821)." Ph.D. diss., University of Illinois at Chicago, 2016.

Stein, Mark L. *Guarding the Frontier: Ottoman Border Forts and Garrisons in Europe*. London: Tauris Academic Studies, 2007.

Szakály, Ferenc. "The Hungarian-Croatian Border Defense System and Its Collapse." In *From Hunyadi to Rákóczi: War and Society in Late Medieval and Early Modern Hungary*, edited by János M. Bak, and Béla K. Kirjly, 141–160. New York: Columbia University Press, 1982.

Thys-Şenocak, Lucienne. *Ottoman Women Builders: The Architectural Patronage of Hadice Turhan Sultan*. Burlington, VT: Ashgate, 2006.

Traquair, Ramsay. "Laconia: I. Medieval Fortresses." *The Annual of the British School at Athens* 12 (1905–1906): 259–76.

Viollet-le-Duc, Eugene-Emmanuel. *Castles and Warfare in the Middle Ages*. Translated by M. Macdermott. Mineola: Dover Publications, 2005.

Wagstaff, Malcolm. "Evliya Çelebi, the Mani and the Fortress of Kelefa." In *The Frontiers of the Ottoman World*, edited by A.C.S. Peacock, 113–136. Oxford: The British Academy, 2009.

James O'Neill
# Firearms and Fieldworks: Military Transformation and the End of Gaelic Ireland

The modern contention over the extent or even existence of the military revolution has been fractious. Michael Roberts instigated the debate in a lecture given at Queen's University Belfast in 1955.[1] Roberts stated that in the century after 1560, European warfare was transformed by a revolution in tactics, army size and operational strategy, with a corresponding accentuation of the impact of warfare on society.[2] Since then the discussion that grew around these assertions became vigorous and contentious.[3] Geoffrey Parker contended the revolution stemmed from the fifteenth century, when the angular *trace italienne* structures supplanted medieval masonry in response to the threat of mobile artillery.[4] The "revolution" of the sixteenth century with its bastioned fortresses, disciplined firepower-centric armies, and ocean-going broadside sailing ships, was the beginning of military dominance of Europeans over non-Europeans.[5] Clifford Rogers countered with his punctuated equilibrium model with the origins of the changes set in the early fourteenth century, whereas Jeremy Black argued for transformation concentrated the fifteenth century, but more emphatically in the late seventeenth century.[6] A new facet of the argument has opened up with work on the gunpowder revolution in Asia by Tonio Andrade, Hyeok Hweon Kang and Kirsten Cooper.[7] Despite Bert Hall's request for the debate to move beyond the concept of a distinct revolutionary event occupying a singular moment in historical

---

[1] Clifford Rogers demonstrated that the idea of a military revolution associated with the introduction of gunpowder long predated Roberts, but Roberts certainly ignited the current debate. See Rogers, "The Idea of Military Revolutions in Eighteenth and Nineteenth-century Texts," 395–415. Many thanks to Mark Fissel for bringing this to my attention.
[2] Roberts, "The Military Revolution," 13–35, Parker, "The 'military revolution' 1955–2005," 205–206.
[3] Thoroughly summarized by Rogers, "The Military Revolution in History and Historiography," 1–10. Fissel, "Military Revolutions."
[4] Later published as Parker, "The 'military revolution'1560–1660-a myth?," 196–214.
[5] Parker, *The Military Revolution*.
[6] Rogers, "The Military Revolutions of the Hundred Years' War," 94.
[7] Andrade, Kang and Cooper, "A Korean Military Revolution?," 51–84; Andrade, *The Gunpowder Age*.

time, claim, counter-claim and reworkings of opinion and chronology are likely to rumble on for some time yet.[8]

How does all this debate relate to Ireland? Sitting on the western fringe of Europe it is easy to believe that the innovations forged in the fires of war wracking continental Europe were slow to influence the Irish. Indeed, John Brewer suggested that the military decline of England during the sixteenth and seventeenth centuries ensured that that country "was not a major participant in the so-called 'Military Revolution' and only played a larger role in the military affairs of Europe from 1688."[9] This was strongly disputed by James Raymond, who claimed there was little evidence to suggest England fell behind their continental counterparts during the sixteenth century, and called for the idea of English military stagnation to be rejected.[10] If the influence of modernizing trends in England was disputed, then surely England's western neighbor was less likely to be effected by military changes in Europe. Moreover, how could a process of modernization take root in a society described by A.L. Rowse as "part medieval, in part pre-medieval . . . in a stage of rapid social decomposition"?[11]

Rolf Loeber and Geoffrey Parker intervened by examining the military revolution's relevance to the wars in Ireland during the 1640s. They itemized four key developments: artillery fortifications, naval broadsides, primacy of firepower in combat, and strategies requiring simultaneous deployment of multiple armies. Contrary to Brewer's position, they claimed "all four appeared in Ireland during the 1640s, and transformed the nature of the conflict."[12] James Scott Wheeler also noted that English financial, logistic and military achievements of the 1640s suggested that the transformations of military revolution were active by at least the mid seventeenth century.[13] This still left the conflict known as the Nine Years War or Tyrone's Rebellion out in the historical cold, as Loeber and Parker unambiguously stated "the Military Revolution scarcely touched Ireland directly."[14]

In his 2002 O'Donnell lecture, Thomas Bartlett raised the issue of the marginalization by Irish historians of the effects and experience of warfare in Ireland. He proposed part of the reason for this may have been the perception of Irish war as primitive, with little to relate to the events of the wider military revolution.[15]

---

8  Hall, *Weapons and Warfare*, 210.
9  Brewer, *The Sinews of Power*, 7.
10  Raymond, *Henry VIII's Military Revolution*, 190–193.
11  A. L. Rowse, *The expansion of Elizabethan England* (London, 1973), pp. 18–9.
12  Loeber and Parker, "The Military Revolution," 67.
13  Wheeler, "The Logistics of Conquest," 205.
14  Loeber and Parker, "The Military Revolution," 71.
15  Bartlett, *The Academy of Warre*, 9.

Bartlett argued that a study of early modern Irish warfare may well have something to contribute to the military revolution debate. He highlighted the paradoxical appearance of skilful use of pike and shot formations in battle, and the construction of sophisticated field fortifications by the Irish, when the image (almost iconic) of the uncouth and primitive Irish warrior abounds in the historiography. Nevertheless, warfare in Ireland was dramatically transformed in the 1590s. In the space of less than ten years, pike and shot replaced traditional hosts of galloglass and kerne. This was linked to the rise of Hugh O'Neill, second earl of Tyrone. Thought by many at the time to be a creature of the English state, Tyrone built a confederation of Irish lords which succeeded in bringing the power of the crown in Ireland to the point of collapse.[16] Yet even the influence of Tyrone has been contested.

M. Ó Báille claimed modernizing trends in the late sixteenth-century Irish military were due to a new politico-religious idealism imported from counter-reformation Europe coupled with Tudor aggression that stimulated the transformation; Tyrone was more of an opportunist than innovator.[17] Cyril Falls accorded "Tyrone's genius" as a factor in the creation of the modern soldiers in Ireland, but was convinced Tyrone's modernized troops were a minority within traditional Gaelic forces.[18] G.A. Hayes-McCoy claimed Tyrone deliberately and systematically built a modern army, armed with the latest weaponry and utilizing contemporary methods of pike and shot warfare fought "in the accepted manner of the age."[19] Furthermore, he believed Tyrone's new soldiers were "formed for combat like the English."[20]

Mark Fissel's *English Warfare, 1511–1642* addresses Hibernian warfare covering the period 1558 to 1601.[21] His principal focus (understandably given the title) is on the response of English arms to the threat of modern Irish troops. Fissel draws on the works of James Hill. Unfortunately, Hill had crafted a model of Irish shock tactics which is not borne out by the evidence. Hill believed that Tyrone's army's power was predicated upon shock action or "the same primitive strategy, tactics, logistics and weapons as their forebears."[22] However Hill's thesis was

---

[16] Tyrone's rise to power is detailed in Morgan, *Tyrone's Rebellion*.
[17] M. O'Báille, "The Buannadha," 57.
[18] Falls, *Elizabeth's Irish wars*, 72–74.
[19] Hayes-McCoy, "The army of Ulster," 105–117; Hayes-McCoy, "Strategy and Tactics in Irish Warfare," 261–262.
[20] McCoy, "The Army of Ulster," 105.
[21] Fissel, *English Warfare*, 207–235.
[22] Hill, *Celtic Warfare*, 3–4.

entirely misguided as will be seen later, as shock action and *melee* combat were rarely used by Tyrone's men.[23]

Certain key elements of the debated military revolution paradigm can be discounted, or at least relegated to a secondary role. The first is the effect of ocean-going broadside armed ships. The Irish confederate never possessed any effective offensive naval capacity. They did use small riverine craft and had access to coastal ships such as galleys, which were deployed to carry troops, supplies and trade with Scotland. However, apart from piracy in the Shannon estuary and small-scale interdiction of English naval traffic between Dublin and Carrickfergus, the Irish had nothing which could challenge even the lightly armed pinnaces or cromsters sent by the crown to secure Irish coastal waters. During the few occasions when English ships managed to come to grips with Irish (or Scots) ships, the firepower of English guns greatly overmatched their enemies.[24]

Artillery had a limited impact on the conduct of the war. Cannon were difficult to move due to the condition, or in some areas, total absence of roads. The roads, that were little more than rutted tracks, soon became boggy quagmires in wet weather. Restricted mobility and an elusive enemy meant that English heavy guns had little practical use if they could not be transported by sea or along rivers. In general, the Irish confederate forces did not use field artillery. Their operations relied on mobility, which would have been compromised by the physical and logistical burdens imposed by cannon. The relatively minor role of cannon and the generally mobile nature of the conflict meant that sieges, which were so prevalent in continental warfare, were of no real importance for most of the war in Ireland.

While the transformation of Irish warfare was sudden and the success of Irish infantry against the English unprecedented, access to technology was not a determining factor. Firearms had long been available to native Irish soldiers. Gunpowder weapons had been available in Ireland since the fourteenth century, though the earliest reference to one being fired was in 1487, when Brian O'Rourke was killed "by the shot of a ball."[25] As hand-held firearms proliferated in the English army during the mid-sixteenth century, Irish soldiers in English pay received training and arms. Indeed, Irish kerne equipped with firearms proved

---

[23] The idea of the Celtic charge has remained stubbornly entrenched in the current literature.
[24] Captain George Thornton to Sir Henry Bagenal, 27 July, 1595 (T.N.A., SP 63/182, f. 196); Glasgow, "The Elizabethan Navy in Ireland," 302.
[25] de hÓir, "Guns in Medieval and Tudor Ireland," 80–81; O'Donovan, ed. and trans., *Annals of the kingdom of Ireland by the Four Masters*, iv, 1145–1147 (henceforth cited *A.F.M.*).

effective, prompting Sir Anthony St. Leger to recommend them to King Henry VIII for service in Scotland, where armed with handguns they would "do your highness great service; for as for gunners, there be no better in no land than they be, for the number they have, which be more than I would wish they had, unless it was to serve your majesty."[26]

Clearly firearms were no mystery to the Irish and their easy access to modern weaponry caused some consternation, but there was no indication that technology was altering the tactics of native Irish soldiers. Firearms were present at the battle of Knockdoe in 1504, but the only reference to their use was when a gun was used to club a horseman's brains out.[27] From the 1550s, the traditional combination of arms deployed by English infantry developed from the bow and bill to the more modern *battalia* of pikes and shot, with shot composed of lighter arquebus/calivers and the heavier but more powerful muskets. However, for the traditional hosts of galloglass and kerne deployed by Irish lords, firearms supplemented other ranged weapons such as bows and darts (short javelins), but made little impact on deployment or tactics.

Shane O'Neill was one of the foremost thorns in the side of the English administration in Dublin. In a break with tradition, O'Neill armed churls, husbandmen and farm laborers, enabling him to field 4,000 foot and 700 horse.[28] Moreover, he was quick to equip many of his troops with firearms. This made sense as recruits could be quickly trained in their use, but O'Neill made no effort to maximize their effectiveness. During two major engagement, Glentaisie (1565) and Farsetmore (1567), shot played no significant, or at least noteworthy role during either engagement; traditional Irish modes of war still held sway, with the brutal melee between heavy infantry armed with axe and sword deciding the outcomes.[29] This did not mean that the Irish could not put their weapons to good use. Thirteen years after Farsetmore, Feagh MacHugh O'Byrne's shot hammered Lord Grey de Wilton's army of 2,000 in Glenmalure, Co. Wicklow. Drawn into the valley, the English column was assailed by volleys of Irish fire from shot placed on the rocky slopes on both flanks. Retreat turned into rout, with the English leaving anywhere from 500 to 800 dead.[30] Duplicity, terrain and stunningly bad judgement on the part of Lord Grey all played their part in the defeat. Firearms were an effective element in the Irish forces arrayed against the English but were not essential to

---

26 Quoted in White, "Henry VIII's Irish Kerne in France and Scotland," 213.
27 Hayes-McCoy, "The Early History of Guns in Ireland," 51.
28 Hayes-McCoy, *Scots Mercenary Forces*, 78; Sir Henry Sidney to the privy council, 12 Nov. 1566 (*Cal. S. P. Ire., 1509–1573*, 318).
29 Hill, "Shane O'Neill's Campaign," 137; Hayes-McCoy, *Irish Battles*, 68–86.
30 O'Byrne, "The Battle of Glenmalure," 173–179.

Irish victory. It took outside intervention, albeit unexpected, for the seeds of Irish the transformation to be sown.

The dramatic arrival of the Spanish Armada along the north and west coasts of Ireland during September and October 1588, and the subsequent wreck of 26 ships, left thousands of experienced officers and soldier scattered throughout Connacht and Ulster. The English moved quickly to neutralize the potentially destabilizing effect of the new arrivals.[31] In concert with their Irish allies, English officers such as Sir Richard Bingham, hunted down and executed the Spanish wherever they could be found.[32] Possibly 1,500 were killed or executed, but around 750 survivors were transported into Scotland by Irish and Scottish chiefs.[33] A much smaller number stayed in Ireland to train and command the retinues of local Irish lords. This was most evident in Connacht, where the windfall of military expertise and hardware enabled some to retrain and rearm their soldiers in line with continental methods.

While the O'Flaherties were active in handing Spanish prisoners over for execution, not all found their way into English custody.[34] Several Spanish captains remained to train their household troops. When engaged by crown forces in March 1589, they deployed pike and shot infantry led by a Spanish officer.[35] The Burkes in Mayo also retained Spanish soldiers, therefore when Morrough O'Flaherty joined them in rebellion, the council in Dublin reported that in addition to 20 Spaniards, their combined forces had 2,000 men armed and equipped with arms and ordnance recovered from the Armada wrecks.[36] When Sir Richard Bingham engaged the O'Rourkes in April 1589, he noted that they too had Spanish officers and equipment.[37] The arms and training had profoundly altered the manner and carriage of the O'Flaherties in battle, as their traditional host of galloglass and kerne were now pike and shot. When the crown moved to neutralize Tadhg O'Flaherty in March 1589, Captain Edward Bermingham found them in "battle array . . . there was a volley of shots on both sides. They came to the push of the pike with great courage."[38] Despite their weapons, modern tactics and Spanish officers, the English had little difficulty in defeating the modernized

---

31 Morgan, *Tyrone's Rebellion*, 105.
32 Connolly, *Contested Island*, 222, 225–226.
33 De Yturriga, "Attitudes in Ireland," 245.
34 Discourse of the Spanish wrecks, Sep. 1588 (T.N.A., SP 63/136, f. 236).
35 Captain Edward Bermingham to Sir Lucas Dillon, 31 Mar. 1589 (*Cal. S. P. Ire. 1588–1592*, 146–147); Sir Richard Bingham to Burghley, 6 Apr. 1589 (T.N.A., SP 63/143, f. 148).
36 "Act entered in the council," 5 Apr. 1589 (T.N.A., SP 63/143, f. 41).
37 Bingham to the Fitzwilliam, 24 Apr. 1589 (T.N.A., SP 63/151, f. 228).
38 Captain Edward Bermingham to Sir Lucas Dillon, 31 Mar. 1589 (*Cal. S. P. Ire. 1588–1592*, 146–147).

Irish and exhibited no real concern for their use of pike and shot. Though it was new for the Irish to utilize pike and shot in *battalia*, English troops had been engaging with and defeating these types of infantry formations while on service in the Low Countries and France. Simply reproducing the arms and methods of the English or Spanish (as Falls and Hayes-MacCoy suggested) could not translate into Irish primacy on the battlefield for the first seven years of Tyrone's Rebellion, nor did it. Tyrone created something altogether new in Ireland.

From his early years the young Hugh O'Neill's prospects appeared limited, as his father was murdered on the orders of Shane O'Neill and his older brother, Brian, killed by Turlough Luineach O'Neill. Brought up a ward of the state and fostered by the Hovendens in the Pale, O'Neill's Anglicisation and service to the crown in Ulster with the earl of Essex in the 1570s, Munster in the 1580s and his cultivation by the state as a counterbalance to the power of Turlough Luineach suggested that, to the English, O'Neill was very much "a creature of our own."[39] Ennobled in 1585, the new earl of Tyrone promoted ownership and training with firearms within his lordship. Peter Lombard described how the earl "gradually accustomed the natives of the whole of Tyrone to handle those kinds of arms [firearms]."[40] Shooting was encouraged and guns were given out as gifts, whereby Tyrone's people "became so proficient in a short time in this practice that one could scarcely believe how skilfully all here even the farmers, ploughmen, swineherds, shepherds, and very boys, have learned to use this weapon."[41]

The crown permitted Tyrone to retain a force of horse and infantry which were paid by the state and trained by English officers; the so-called "butter captains." In 1597, an anonymous author, possibly Nicholas Dawtry, wrote that Tyrone had been allowed 50 horse and 100 foot, but under the cover of this arrangement that the earl trained "all the lusty bodied men of the province . . . in a soldier like manner of service."[42] Fynes Moryson later repeated this, claiming that Tyrone processed recruits through the ranks, in order to establish a large reserve of trained men without raising suspicion from Dublin.[43] Furthermore, Tyrone had more than just English captains training his men. Tyrone played an active part in assisting the crown round up the Spanish stragglers in 1588, and it was his close associates who slaughtered large numbers of Armada survivors in Innishowen.[44] Despite his sanguinary display of outward loyalty,

---

39 As referred to by Queen Elizabeth. Quoted in Morgan, *Tyrone's Rebellion*, 85.
40 Lombard, *The Irish War*, 31–33.
41 Lombard, *The Irish War*, 31–33.
42 Dawtry, "A Booke of Questions and Answers," 92–93 (henceforth cited Q & A).
43 Moryson, *Itinerary*, ii, 189.
44 Statement of Juan de Nova, Jan. 1589 (*Cal. S. P. Spain, 1587–1603*, 506–510).

Tyrone protected several Spanish officers and common soldiers, some of whom entered into the earl's service.[45]

English infantry at home and in Ireland at this time has been well documented, therefore a brief summary would suffice.[46] By the 1590s English infantry were composed of pike and shot deployed in companies of approximately 100 men. The ratio of shot to pike was one or two-to-one.[47] The shot deployed a mixture of matchlock calivers and muskets. Muskets were heavy slow-firing weapons, whose ponderous weight (sometimes ten kilogrammes or more) required a forked rest to fire. They were powerful weapons which could penetrate armour and engage targets up to 100 metres away. However, most of the English shot carried calivers; possibly up to 75 percent.[48] These were much lighter weapons firing a smaller bullet and were suitable for skirmishing, but they had little chance in hitting a target over 50 metres. Shot were generally lightly armoured, either with a helmet or jack of plates, but the pikemen were heavily armoured with helmet, corselet and occasionally tassets to cover the thighs. They were armed with a five meter pike which could weigh approximately four to five kilogrammes. In formation or *battalia*, the pike formed a close order central core surrounded by the musketeers, while the cailvermen deployed on the flanks, front and rear. The pike provided protection to the shot who were vulnerable to cavalry while reloading, but the English were also prone to using their pike in shock action; charging into close combat to break enemy units. In 1598, Robert Barret described the combined arms teams of pike and shot as "the armed pike is the strength; the shot the fury, one without the other is weakened."[49] As seen above, the Irish in Mayo deployed a native analogue of these pike and shot units and were easily overthrown by Bermingham. When war came in 1593, the Irish under Tyrone did not make the same mistake.

Tyrone did not openly break with the crown at the start of the war. During 1593 to 1595, the earl fought a proxy war, by reinforcing Hugh Maguire with troops led by Tyrone's brother Cormac MacBaron O'Neill. While distracted with an ostensibly local rebellion in south-west Ulster, Tyrone (while still claiming

---

[45] See Memorials or petitions mentioned by Alonso Cobos, Oct. 1596 (*Cal. S. P. Spain, 1589–1603*, 641); Mountjoy to Cecil, 30 May 1603 (H.H.A., C.P. 100/59).
[46] Cruickshank, *Elizabeth's Army*; Webb, *Elizabethan Military Science*; Eltis, *The Military Revolution in Sixteenth-Century Europe*, 99–135; Fissel, *English Warfare*; Hammer, *Elizabeth's Wars*.
[47] Cruickshank, *Elizabeth's Army*, 114–115.
[48] A minute of a letter describing reinforcements for Ireland, 27 August, 1598 (A.P.C., XXIX, 94).
[49] Barret, *The theorike and Practike of Moderne warre*, 69.

loyalty to the queen) eliminated the crown's Irish allies in Ulster or forced them to swear fealty to him.[50] Even after the defeat of Sir Henry Duke's column on its way to relieve the garrison of Enniskillen in August 1594, the English authorities in Dublin failed to notice the growing threat and increasing sophistication of the Irish.[51] Only when Tyrone attacked the English field army at Clontibret in May 1595, did the English realize how far the Irish had come. Incessant warfare enabled the English to become accustomed to the traditional Irish strategies. The Irish would openly engage each other in battle with the heavily armed and armoured galloglass and light kerne, but even the most powerful Irish lords regularly withdrew or avoided contact with punitive English attacks.[52] At Clontibret, the English learned that the Irish led by Tyrone had broken with the past: pike and shot had supplanted the axe, dart and bow.

Tyrone had blockaded Monaghan, but reinforcement from England steeled English resolve to relieve the garrison. Sir Henry Bagenal took charge of 19 companies of foot and six troops of horse totalling 1,750 men. From the outset the army was at a disadvantage due to a relative lack of ammunition issued to the English shot.[53] A protracted firefight with the lead elements of Tyrone force, with some 700 to 800 shot on 26 May, did not bode well, but Bagenal's force made it to Monaghan later that day.[54]

Bagenal chose to take a more southerly route for his return to Newry, past the old church at Clontibret. Marching out on 27 May, Bagenal expected trouble therefore the English infantry formed into two *battalia*; the van and the rear. As the English approached a pass just beyond the church, Tyrone launched his attack. The Irish opened fire on the van and on both flanks of the column as Tyrone pressed his attack on the rear. Irish shot supported by horse and stands of pike forced the loose sleeves of English shot back to the main body of the column.

Sir Ralph Lane noted the skill of the Irish, writing "he never saw more ready or perfect shot, by estimation two thousand, nor a skirmish or rather fight . . . continually maintained [and] carried in better order."[55] The Irish assailed the English army from close range, who after four hours had moved less than a quarter of a mile. Moreover, shortages of ammunition forced repeated English pike

---

50 O'Neill, *The Nine Years War*, 40–42.
51 O'Neill and Logue, "The Battle of the Ford of the Biscuits," 913–922.
52 Hammer, *Elizabeth's Wars*, 74–75. Galloglass were typically clad in mail and carried a long-handled axe or large sword, whereas kerne were unarmoured and carried a sword, bow or small javelins known as darts. See Falls, *Elizabeth's Irish Wars*, 69–71.
53 Perrot, *The Chronicle of Ireland*, 94.
54 Captain Francis Stafford to Sir Geffrey Fenton, 4 June 1595 (T.N.A., SP 63/180, f. 66).
55 Sir Ralph Lane to Burghley, 9 June 1595 (T.N.A., SP 63/180, f. 82).

charges to push back Tyrone's shot.[56] After eight hours however the Irish fire began to slacken as they exhausted their supplies of gunpowder.[57] Tyrone had to withdraw and Bagenal reached the relative safety of Newry the following day. The English claimed they lost 31 dead and 103 wounded, but the muster-master later declared that there were many more killed and wounded than was "thought fit to be given forth upon the first advertisement."[58]

The shock was profound as English myopia to Irish military advances finally began to disappear. Though he was not present at Clontibret, the veteran officer Sir John Norreys noted "the state of these northern rebels is far different from that was wont to be; their number greater, their arms better, and munition more plenty."[59] Even the lord deputy conceded that "their arms and weapons, their skill and practice therein far exceeding their wonted usage, having not only great force of pikes and muskets, but also many trained and experienced leaders as appeared by their manner of coming to the fight, and their orderly carriage therein."[60]

The near disaster at Clontibret set the pattern for the next five years of the war as successive viceroy's vainly struggled to bring the Irish lords to heel. Moreover, the engagements were not a hit-and-run style of warfare which typified previous English experience with the Irish wars. The Irish repeatedly held the field or forced the English to break contact. Sir John Norreys found out first hand as his supply convoy to Armagh struggled to return to Newry. His men were attacked near Mullabrack (County Armagh) by Irish fighting on open ground where there was "neither bog nor bush."[61] Norreys and many other officers were wounded but his men broke through to Newry.[62] Norreys later wrote that "all those that have formerly seen the wars in this country do confess that they have not accustomly been acquainted with the like fight."[63]

As English influence in Ireland was rolled back by successive defeats, Sir Henry Wallop opined that the Irish were "well furnished with all the habiliments of war, and have so trained their men, as in sundry encounters that they have had with our men, they seem to be other enemies, and not those that in times past were wont never to attempt Her Majesty's forces in the plain field, but in

---

56  Lane to Burghley, 28 June, 1595 (T.N.A., SP 63/180, f. 179).
57  Lieutenant Tucher [Perkins], 1 June, 1595 (*Cal. Carew MSS, 1589–1600*, 109–110).
58  Lane to Burghley, 28 June, 1595 (T.N.A., SP 63/180, f. 179).
59  Sir John Norreys to Burghley, 4 June, 1595 (T.N.A., SP 63/180, f. 40).
60  Lord Deputy Russell and council to the privy council, 4 June, 1595 (T.N.A., SP 63/180, f. 9).
61  Fenton to Burghley, 7 September, 1595 (T.N.A., SP 63/183, f. 39).
62  McGleenon, "The battle of Mullabrack 5 September 1595," 90–101.
63  Sir John Norrey to Burghley, 8 September, 1595 (T.N.A., SP 63/183, f. 51).

some pass or strait."⁶⁴ Frequent ceasefires punctuated the war, but the series of Irish victories culminated in the devastating defeat of Sir Henry Bagenal's army at the Battle of the Yellow Ford on 14 August, 1598, when Tyrone routed the English main field army of 4,500 infantry and horse. Their commander dead, roughly half of the English army returned south without their arms, money or munitions.⁶⁵

Clearly the English were not encountering simple Irish copies of their own formations. The melee-focused combat of the galloglass and kerne was nowhere to be seen as Sir John Dowdall reported that the Irish "have drawn the greatest part of their kerne to be musketeers, and their galloglass pikes."⁶⁶ However, this oversimplified the changes occurring. Superficially the Irish troops may have resembled those of the crown (and there were several occasions when English officers could not tell the difference), but the composition and tactics of the Irish were radically different.

## O'Neill's Reforms

Tyrone had ordered his infantry into companies of approximately 100 men and had retained the custom of "dead pays." This allowed a captain to draw pay for more troops than were actually within the unit, with one source claiming Tyrone allowed 12 dead pays per infantry company.⁶⁷ However, Tyrone's troops had a larger proportion of officers than English companies, with "40, to 20, and to 10 of their foot, for the fitness of their service of the passes."⁶⁸ This enabled greater tactical flexibility than English infantry who were limited to a captain, a lieutenant, an ensign and two sergeants per company.⁶⁹

The new Irish formations were built around pike and shot as the English, but the ratio was different. Tyrone predicated the combat power of his infantry on firepower, whereas the English had one or two shot for every pike, Tyrone deployed four or five shot for each pike.⁷⁰ Unlike the English shot, Tyrone's

---

64 Sir Henry Wallop to Sir Robert Cecil, 9 February, 1596 (*Cal. S. P. Ire., 1592–1596*, 468–469).
65 McGurk, "The Battle of the Yellow Ford," 34–55.
66 Sir John Dowdall to Burghley, 9 March, 1596 (*Cal. S. P. Ire., 1592–1596*, 484–488).
67 O'Donovan, ed, "Proclamation, in the Irish language," 63.
68 "An advertisement of the earl of Tyrone's forces," 11 November, 1594 (*Cal. Carew MSS, 1589–1600*, 101–102).
69 Falls, *Elizabeth's Irish wars*, 36.
70 Russell to Cecil, 26 February, 1595 (T.N.A., SP 63/178, f. 133); "A design upon Tyrone upon the landing of the army at Lough Foyle," 12 March, 1599 (H.H.A., C.P. 139/54). Though the ratio

were almost exclusively armed with calivers. Though sacrificing firepower and range, the caliver's light weight suited the fast-paced mobile warfare preferred by the Irish. Sir Roger Williams preferred muskets over calivers, but conceded the lighter weapons were best suited "to make great marches . . . to surprise towns, fortresses or passages . . . to lie in ambush a far off, to cut off convoys, passengers, and such services."[71] They were therefore ideally suited to Tyrone's fluid tactics, as speed and operational mobility were advantages Tyrone was keen to maintain. Swift Irish marches enabled the Irish to outpace English units, allowing them to pick when and where to engage, and to break contact if English resistance was too strong. Dawtry reported the Irish "would not hazard their whole estate, and quarrel, upon a few desperate men engaged upon any indifferent ground."[72] The firepower of Tyrone's swiftly moving shot became the mainstay of Irish combat power.

Intense skirmishing was central to Tyrone's infantry tactics. From the earliest engagement of the war, Irish shot were reported skirmishing with English forces, often while the latter were marching in column, but also when the crown forces had formed into *battalia*. At the Ford of the Biscuits (1594), Irish shot halted Sir Henry Duke's relief army at the River Arney, while Cormac MacBaron O'Neill's skirmishers poured withering fire along the flanks and rear of the English column. When Irish fire disrupted the rear, pike led by O'Neill charged, smashing the cohesion of two thirds of the English force.[73] In 1597, Captain Francis Croft noted the Irish shot was of "as good discretion as ever I saw for loose skirmishes."[74] Tyrone's loose order shot were shown in the illustrated of the Battle of the Yellow Ford (1598) to the left and right of Bagenal's army.[75] Mountjoy described fighting in the Moyry Pass as "one of the greatest skirmishes and best maintained in all places, that hath been in this kingdom."[76] Skirmishing was not haphazard or random (years of Hollywood films may have fostered this image), but ordered and disciplined. The method of well-maintained skirmishing was described by Sir John

---

of pike to shot consistently fell during the seventeenth century, it took until the 1690s for pike to fall to the proportions fielded by Tyrone. See Black, *Beyond the Military Revolution*, 140.
71 Evans, ed., *The Works of Sir Roger Williams*, 37–38.
72 *Q & A*, 94; Moryson, *The Irish Sections*, 70–71.
73 O'Sullivan Beare *Ireland under Elizabeth*, 80–81.
74 Captain Francis Croft to Sir Robert Cecil, 30 June, 1597 (T.N.A., SP 63/199, f. 298).
75 "The description of the army which was defeated by the earl of Tyrone," 14 August, 1598 (T.C.D., MS 1209/35).
76 Lord Deputy Mountjoy and council in the camp to the lord chancellor and rest of the council in Dublin, 7 October, 1600 (*Cal. S. P. Ire.*, *1600*, 472–473).

Smythe.[77] He detailed how groups of three or four shot dispersed themselves in the field. One soldier advanced to fire then withdrew behind his comrades to reload, to be replaced by one of his group. Further, alternate "societies" of three or four shot would advance, fire then withdraw to reload. Clearly, Tyrone's skirmishers were nothing like loose mobs of Irish shot darting through the bushes and skipping over bogs conjured in popular (and in some cases academic) imagination.

Tyrone's shot may have retained some habits from their Spanish drillmasters. Often after engagements English officers reported the Irish firing from well beyond the effective range of their weapons. Norreys experienced this in 1595, so too did Essex in 1599 and Mountjoy during 1600 in the Moyry Pass and at the Blackwater in 1601.[78] Ineffectual volleys of fire appeared peculiar given the frequent references to the good conduct and discipline of the Irish shot. The Irish may have been replicating a Spanish firing method described by Sir John Smythe in 1590. He reported how the Spanish would fire "a very few shots at their enemy with no other intent but to abuse and procure them to give their volleys with all fury, that thereby they may spend their powder and bullets, heat their pieces and work no effect, whereby they, still keeping the force of their shot, may after give full volleys to their enemies . . .. And for that effect the Spanish do use this phrase: *disparese de lejos, para atraher, y engañar bobos*."[79]

Despite the Irish dependence on firearms, Tyrone's shot were vulnerable in the open against cavalry. Though light by European standards, the English horse could easily overthrow unprotected Irish infantry. Carrying lance and pistols, and armoured with helmet, cuirass and possibly tassets and vabraces, the English horse were the most powerful individual element deployed by either side during the war. Where possible terrain was exploited to neutralize the English horse. At the Battle of the Ford of the Biscuits (1594), Duke's horse could play no part as marshy ground forced the English horsemen to dismount.[80] Tyrone made elaborate earthworks and modified the terrain at the Battles of the Yellow Ford (1598) and the Moyry Pass (1600) to limit their movements, but the Irish could not always depend on terrain and fortifications to protect their shot, therefore pike units were trained to provide a defensive hedge of pikes behind which shot

---

77 Smythe, *Instructions, observations, and orders mylitarie*, 133–141.
78 Essex was forced to order his men to ceasefire in 1599 to spare their powder. See "A journal of the Lord Lieutenant's proceedings from the 22nd of June to the 1st of July 1599" (Lambeth, MS 621, f. 136).
79 Translated as "shoot from afar, to attract and deceive the fools." See Smythe, *Certain Discourses Military*, 63.
80 O'Sullivan Beare, *Ireland Under Elizabeth*, 80.

could shelter. Unlike their English counterparts, the Irish pike were unarmoured apart from a helmet, and were deployed in loose order. Thomas Douglas referred to this in Dungannon during 1601, describing how Spanish officers only trained the Irish "to discharge their pikes and to run up and down." Douglas wondered why after so long "have your soldiers not better trained in order, to divide themselves after their quantity and number in ranks, files and squadrons. He answered by reason of the country order could avail nothing."[81]

The rugged terrain of Ireland was unsuited to manoeuvring closely packed formations of infantry. De la Vega was unambiguous when he warned of the dangers of deploying pikes in constricted ground, as it was "not convenient to the companies of Pikemen, nor yet may not be served of them, as well for mountains, straight passages, and other difficult places, because pikes are troublesome and cannot be in such places so manuable, as speed and agility require."[82] This sentiment was echoed by Sutcliffe, who wrote in 1593 that "in woods and shrubby or bushy grounds, these kinds of long weapons are unprofitable, and unwieldy . . .. In straits when soldier come to lay hands, and have prize each other long pikes cannot be managed."[83]

Notwithstanding their limited use in woods, pike were still required to provide defensive screens for the Irish shot. Given the speed of Tyrone's calivermen, his pike had to eschew armour and deploy in loose order to keep pace, possibly along the lines of the Spanish *pica seca* or light pikemen. The need to maintain a high level of mobility in the Irish pike meant that significant compromises undermined their utility in combat. Unarmoured and in loose order, their role was primarily defensive. Furthermore, they had no real shock value and were useless against the close-ordered armoured pike deployed by the crown, which could "overthrow, disorder and break them [loose order pike] with as great facility, as if they were but a flock of geese."[84]

Though pike provided a tactical solution to the threat from the English cavalry, Tyrone also deployed significant numbers of targeteers or swordsmen to defend his shot. This was very much in accordance with Spanish practice. Sir Roger Williams believed them well-suited for use in breaches, trenches and "to cover shot that skirmishes in straights."[85] This was a clear departure from the English infantry in Ireland, as the crown had stopped deploying targeteers in Ireland by

---

81 "The progress of my services since I arrived in Ireland on 17 March," [1601] (T.N.A., SP 63/210, f. 192).
82 de la Vega, *A compendious treatise entitled, De re militari*, 7.
83 Sutcliffe, *The practice and proceedings and laws of armes*, 185–186.
84 Smythe, *Instructions, observations, and orders mylitarie*, 25–26.
85 Evans, ed., *Sir Roger Williams*, 37–38.

the start of the war in 1593.[86] Armed with a sword and light shield (there was little use for the heavy steel targets used in Europe), Irish swordsmen provided close protection to the Irish shot and when the opportunity arose, could prove deadly against stands of pike without supporting shot: both were seen at the Battle of the Yellow Ford (1598).[87] Though Irish targeteers were armed with melee weapons, they only followed up the success of the shot, disruption and disordering of English formations was only achieved through firepower.

Tyrone created a hybridized army, by retaining Irish advantages in tactical and operational mobility, while grafting the benefits of the gunpowder revolution in Europe. Where the shock action of the galloglass had once been the decisive stroke of Irish armies, Tyrone's reforms had fully embraced the power of firearms in battle. Firepower was the key element on which the combat power of Irish infantry now rested. Contrary to the claims by Hill that charges into close combat were key to Irish success, Irish infantry were now almost wholly dependent on their firearms to achieve victory. At the Battle of Clontibret (1595), Bagenal's men managed to break through because the Irish had exhausted their supply of gunpowder; some 14 barrels.[88] Bagenal's army had been badly mauled, but the battered stands of English infantry remained intact, therefore the Irish withdrew. This led Sir Robert Salesbury to suggest that it would have been a much bloodier affair if Tyrone's pike had advanced like his shot.[89] At Sir Conyers Clifford's retreat from Ballyshannon in 1597, rain extinguished the match of both English and Irish shot, causing the Irish to break off the attack; steady English pike deterred any further assault.[90] Even when Tyrone was in the midst of gaining his greatest victory at the Battle of the Yellow Ford, a shortage of gunpowder revealed his pikemen's reticence to closing with the enemy. Captain Richard Cuny's regiment had spent their ammunition and had only their pikes left to hold off the Irish, but no attack came. Cuney later commented that if the Irish had charged as his regiment pulled back to Armagh, none of his men would have survived.[91]

---

**86** Webb, *Elizabethan Military Science*, 89–90. Officers were occasionally illustrated using swords and shields, but they were not issued to the common soldiers. Captains Lee and Dowdall can be seen leading the charge at the Battle of the Erne Fords (1593). See The Battle of the Erne Fords, 10 October, 1593 (British Library, Cottonian MS Augustus I/ii, f. 38).
**87** "The description of the army which was defeated by the earl of Tyrone," 14 August, 1598 (T.C.D., MS 1209/35).
**88** A report of the service done by Sir Henry Bagenal in the relieving of Monaghan by Lieutenant Tucher, 1 June, 1595 (*Cal. Carew MSS, 1589–1600*, 109–110).
**89** Sir Robert Salesbury to Burghley, 3 June, 1595 (T.N.A., SP 63/180, f. 5).
**90** Clifford to Burgh and council, 9 August, 1597 (*Cal. S. P. Ire., 1596–1597*, 373–377).
**91** Captain Richard Cuney to Essex, 28 October, 1598 (H.H.A., C.P. 177/131).

Tactical flexibility did not on its own result in the new-found Irish strength and success against the English. Tyrone's use of sophisticated operations requiring multi-regional coordination of forces, turning tactical superiority into operational dominance across the island. A feature of Robert's thesis was a "revolution in strategy."[92] Parker and Loeber referred to strategies which relied on utilization of several armies at once and in concert, as one of the four key developments of the military revolution which appeared in Ireland in the 1640s "and transformed the nature of the conflict."[93] However, Tyrone and his allies developed unprecedented levels of cooperation and synchronization.

Inter-factional fighting and localism was a recurring feature of native Irish society, and alliances even at a local level could be tenuous and rarely long-lasting. Yet at the start of the war Tyrone had built a confederation of powerful Irish lords stretching from Tirconnell in the north to Wicklow on Dublin's doorstep; Red Hugh O'Donnell, Hugh Maguire, Brian O'Rourke and Feagh MacHugh O'Byrne and many other lesser lords. During 1593 to 1594, Maguire and O'Donnell fought the crown in Fermanagh, which was in effect a large strategic feint, drawing the crown's attention (and military strength) into west, while Tyrone neutralized the crown's Irish allies and assets in the rest of Ulster.[94] This set the stage for operations by different Irish armies acting in concert to displace or divert English forces, to give tactical and operational advantage in widely separated regions.

At the start of 1595, Feagh MacHugh asked Tyrone to burn and spoil to draw the lord deputy out of O'Byrne's Wicklow fastness.[95] Within a month Tyrone attacked and took the English garrison at the Blackwater forcing Russell to refocus his attention on Ulster. O'Byrne reciprocated the following year by provoking Russell to send troops into Wicklow, leaving "the heart of the English Pale, as in most of the other parts of Leinster" unprotected and open to attack from the north.[96] Sir John Norreys fell for Tyrone's ruse at the end of 1596. Norreys had made progress against the Irish in Connacht, but a renewed blockade against the English garrison in Armagh forced the English to concentrate forces on the borders of counties Armagh and Louth, which included Norreys and his men in Connacht; the full strength of the army would be needed to resupply

---

92 Parker, "The 'Military Revolution–1560-1660' –A myth?," 38.
93 Loeber and Parker, "The Military Revolution in Seventeenth-century Ireland," 67.
94 O'Neill, "Maguire's revolt but Tyrone's war," 43–68.
95 Confession of James Fitzmorris Fitzgerald brother to Walter, 13 and 17 February, 1595 (T.N.A., SP 63/178, f. 124).
96 Fenton to Burghley, 25 November, 1596 (T.N.A., SP 63/195, f. 101); Norreys to the council, 28 November, 1596 (*Cal. S. P. Ire., 1596–1597*, 177–178).

the fort at Armagh. Tyrone relented and allowed the fort to be revictualled, but with the English concentrated in Ulster O'Donnell's forces swept into Connacht unopposed, undoing much of Norreys' work from the previous year. Only then did it become apparent that the blockade of Armagh was a ploy to draw Norreys away to clear the way for O'Donnell's advance.[97]

The crown's 1597 summer campaign met a similar fate. The English attacked into Ulster in a two-pronged thrust in the east and west. Lord Deputy Burgh had engaged Tyrone on the river Blackwater and Sir Conyers Clifford besieged Ballyshannon. After Tyrone checked Burgh on the Blackwater, raiding in Leinster by Captain Richard Tyrrell (one of Tyrone's most experienced and trusted captains) caused Burgh to pull his men back to defend the Pale. This released Tyrone's men who streamed west to engage Clifford. Greatly outnumbered, Clifford was forced into a precipitous withdrawal. Assaulted along its flanks and rear, only a fortuitous rainstorm saved the English army in the west from annihilation.[98]

Tyrone's strategy rarely sought open combat with crown forces as an end in and of itself. Disease and desertion could damage the English armies more efficiently (and inexpensively) than Irish guns, therefore when the earl provoked confrontations at Clontibret (1595), Mullaghbrack (1595), the blockade of Armagh in 1596 and Yellow Ford (1598), it was to facilitate military and political advances elsewhere in Ireland. While the queen's officers fretted over the fate of the Blackwater garrison in 1598, Tyrone extended his control of the countryside to the south coast of Ireland and initiated the collapse of the Munster plantation.[99] Tyrone's domination of the country caused Sir Richard Bingham to write that "the countries are possessed by the rebels, and few holds left for us to defend."[100] A report at the start of 1600 identified five distinct Irish field forces. Tyrone was in Durrow (County Offaly) with 3,000 foot and 300 to 400 horse; O'Cahan and James MacDonnell guarded the north coast; Cormac MacBaron was in Monaghan; Brian MacArt O'Neill and the Magennises held Antrim and Down; Red Hugh O'Donnell was in Connacht.[101] This level of control required more than just the troops

---

[97] Russell and council to the privy council, 26 January, 1597 (T.N.A., SP 63/197, f. 182).
[98] Clifford to Burgh and the council, 9 August, 1597 (*Cal. S. P. Ire., 1596–7*, pp. 373–7); Ó Clérigh. *The life of Aodh Ruadh O' Domhnaill*, Part 1, 159–61; *A.F.M.*, vi, 2033–2037.
[99] Sheehan, "The overthrow of the Plantation of Munster in October 1598."
[100] Bingham to Cecil, 5 December, 1598 (*Cal. S. P. Ire., 1598–1599*, 392–393).
[101] Intelligences sent to Fenton, 6 February. [1600] (T.N.A., SP 63/203, f. 80). This document was misdated in the archives to 1598–1599, however the intelligencer talks of Essex's lieutenancy of 1599 in the past tense.

commanded by the earl in Ulster; it needed an expansion of military manpower which dwarfed all previous native Irish armies.

The number of troops available to an Irish lord was limited by population and economic capacity of the country. Irish armies tended to be relatively small, but local alliances meant that larger forces could be concentrated for limited times, as seen at the battle of Knockdoe when possibly 10,000 men were deployed in total.[102] Shane O'Neill managed to deploy 4,000 foot and 700 horse and Turlough Luineach could call on 3,000 Scots mercenaries. But Tyrone's armies were not temporary seasonal forces or dismissed when fighting died down,[103] as they were maintained in pay throughout the course of the war. The largest single concentration of Irish troops was around 6,000 to 8,000 men at the battles of the Yellow Ford and Kinsale, but the total number across Ireland was much greater. At the start of the war, Tyrone was thought to have 6,000 men, but as the conflict escalated and the earl's confederation spread throughout Ireland these numbers rapidly grew. Estimates of the Irish confederate forces in 1599 put Tyrone's forces at 20,000. Sir George Carew considered the Irish forces in Munster to be 7,000 and a report in the same year believed the Irish armies to total 23,000 foot and 1,550 horse.[104] The crown responded in kind, eventually raising the English army in Ireland to 19,000 horse and foot; the greatest number ever maintained in Ireland during the sixteenth century.[105]

The battle of Cerignola (1503) demonstrated the potential of firearms to dominate the battlefield, when arquebusiers protected by a trench and palisades devastated the French armoured cavalry.[106] Shot within fieldworks slaughtered the Swiss pikemen at the Battle of Bicocca (1522). The power of shot behind earthworks was formidable, but in Ireland field fortifications were rarely, if ever, a feature in warfare for most of the sixteenth century. The Irish had little use for trenches, banks or palisades, as axe-wielding galloglass and swordsmen could do little to an enemy on the opposite side of a rampart. However, when firearms sup-

---

**102** Ó Domhnaill, "Warfare in Sixteenth-Century Ireland," 31.
**103** Hayes-McCoy, *Scots Mercenary Forces in Ireland*, 78; Sir Henry Sidney to the privy council, 12 November, 1566 (*Cal. S. P. Ire., 1509–1573*, 318); Fitzwilliam to Cecil, 12 September, 1569 (*Cal. S. P. Ire., 1508–1573*, 420).
**104** Carew to Cecil, 2 May, 1600 (*Cal. S. P. Ire., 1600*, 141–146); "The main strengths of the rebels in Ireland," 1600 (U.L.C., Ms. Kk.1. 15, no. 147, f. 503).
**105** Sir George Carey to Henry Cuffe, 3 Apr. 1600 (*Cal. S. P. Ire., 1600*, p. 74); Bruckhurst to Cecil, 13 August, 1600 (*Cal. S. P. Ire., 1600*, 346–349).
**106** Turnbull, *The Art of Renaissance Warfare*, 67–68.

planted *melee* weapons as the principal strength of the Tyrone's Irish infantry, the changes were paralleled by the widespread use of defensive fieldworks.[107]

In 1593, when Lord Deputy Fitzwilliam dispatched Sir Henry Bagenal into Fermanagh to engage Hugh Maguire, the English advance was stopped cold in its tracks by the Irish shot protected by earthworks. Bagenal would not dare to force the position "without artillery or other implements it was not passable for us."[108] Fortified river crossings and passes were a regular feature of the conflict as Tyrone and his allies used earthwork defenses as a force multiplier for their shot. Sconces and trenches were raised in the passes between Athlone and Roscommon and the road to Philipstown in Offaly was blocked by ten "half-moons."[109] In response to Mountjoy's march north in September 1600, Tyrone fortified the south-Ulster borderlands, where it was reported that "Tyrone doth trench and ensconce all passes and fords where my lord deputy might get passage."[110]

During the miserably wet autumn of 1600, Lord Mountjoy almost destroyed his army attempting to batter his way through the formidable Irish defences in the Moyry, where one English officer noted: "I vow unto God that I did never see a more villainous piece of work and an impossible thing for an army to pass without intolerable loss."[111] The Irish positions in the Moyry Pass were also cleverly placed on a reverse slope, thereby protecting them from the threat posed by English cannon. Mountjoy's advance into Tyrone the following year also stalled in front of Irish sconces and trenches in the woods between the River Blackwater and Dungannon. Tyrone also raised numerous forts, such as the position at Inisloughan, illustrated by Richard Bartlett in 1602.[112] Six others were raised by the earl to protect the western and southern shores of Lough Neagh.[113]

Not only had the tactics and equipment of the Irish changed, but their military architecture was dramatically altered, as Tyrone eschewed the defensive properties of stout masonry walls. In 1595, the earl ordered his castle and those of his adherents demolished, thus denying the crown any target on which to

---

107 Contrary to the erroneous claims by Klingelhofer that the Irish built few fortifications. See Klingelhofer, *Castles and Colonists*, 43–46.
108 Bagenal's journal, 27 September, 1593 (T.N.A., SP 63/172, f. 114).
109 Sir Theobald Dillon to Cecil, 2 September, 1600 (*Cal. S. P. Ire., 1600*, 409–410); Sir Oliver Lambert to Mountjoy, 22 April, 1600 (T.N.A., SP 63/207 pt. 2, f. 321).
110 Advertisements received by Sir Geffrey Fenton out of MacMahon's country, 7 October, 1600 (T.N.A., SP 63/207 pt. 5, f. 270).
111 Sir Francis Stafford to Sir George Carey, 19 October, 1600 (*Cal. S. P. Ire., 1600*, 489–490).
112 Inisloughan Fort, 1602 (National Library of Ireland, MS 2656, vi).
113 Lough Neagh, 1601–1602 (T.N.A., MPF 1/133).

concentrate their forces.[114] It is unclear to what extent the angular defences of the *trace italienne* were incorporated into the Irish works as ambiguous terms such as sconce, bulwark or flanker were often used by English officers in their descriptions of Irish fortifications. Only two detailed illustrations of Tyrone's fortifications exist, the Irish Blackwater Fort and Inisloughan Fort, neither of which were constructed along modern designs.[115] This does not prove that the Irish never built modern fortifications. In 1602, the castle at Dunboy was heavily modified with a bastioned defence surrounding the masonry tower house.[116] With reference to the formidable defence found in the Moyry Pass in 1601, Mountjoy noted that Tyrone had "time therein to do whatsoever industry could add to the natural strength of the place. Wherein these barbarous people have far exceeded their custom and out expectation."[117] Patently Mountjoy encountered something beyond his experience in Ireland, but there remains no definitive evidence to settle the issue.

## The English Response

Some English officers were quick to notice that the Irish facing them and the manner in which they fought was not the same as before. However, the crown's army had serious difficulties adapting to the new Irish methods.[118] The heavier arms and equipment of the English soldiers, and the baggage associated with their marching columns, meant they were incapable of matching Irish tactical and operational mobility.[119] Furthermore, their unwillingness to leave the perceived safety of the

---

114 Russell and council to the privy council, 20 July, 1595 (*Cal. S. P. Ire., 1592–1596*, 343–344). A letter to the *sugán* earl of Desmond advised him that the war was not to be one of defending castles but "standing and upholding the field." See "Articles of advice to be preferred to the right honourable James [Fitzthomas], earl of Desmond," 13 August, 1601 (T.N.A., SP 63/209 pt. 1, f. 78).
115 Lord Burgh assaults the Blackwater fort, 14 July, 1597 (T.C.D., MS 1209/34); Inisloughan Fort, 1602 (N.L.I., MS 265, vi).
116 Klinglehofer, "The Renaissance fortifications at Dunboy castle," 85–96.
117 Mountjoy and council to the privy council, 28 October, 1600 (*Cal. S. P. Ire., 1600*, 522–524).
118 Sir John Norreys was one, but Captain Tom Lee also recognised the need to change English methods in dealing with the Irish threat. See Thomas Lee, *The discovery and recovery of Ireland with the author's apology*, (B.L., Add. MSS. 33743, ff 18–19, 64, 85–86), transc. John McGurk, viewed at http://www.ucc.ie/celt/published/E590001-005/index.html on 2 February, 2013.
119 Essex and the council to the privy council, 15 July, 1599 (*Cal. S. P. Ire., 1599–1600*, pp. 91–5).

roads or fight in woods or bogs severely hampered their ability to counter Tyrone's men.[120] Traditional English strengths in ordered infantry and artillery proved weaknesses as close-formed infantry were easily outmanoeuvred and heavy guns proved more of a liability given the poor roads in Ireland.

Limitations in equipment and tactics were exacerbated by unimaginative military campaigns, which were essentially blunt applications of military force. The disdain for Irish abilities was evident in 1595 as the army set forth to relieve Monaghan. When a veteran officer questioned why so little ammunition was issued, Lord Deputy Thomas Russell replied: "[C]aptain you are deceived; you are not now in France or the Low Countries, for you shall not be put here to fight as there'; the large number of troops was only to give countenance to the service."[121] This conceit continued to 1600 as English expeditions proved ineffective as the Irish avoided contact, attacked when the occasion suited or they had the advantage, or made the English waste their resources during false truces or facile negotiations.

The arrival of Charles Blount, Lord Mountjoy, as lord deputy in February 1600 marked a change in English policy and fortunes. The new viceroy rearmed and retrained the army similar to the Irish with an emphasis on mobility and firepower.[122] While never matching the speed and manoeuvrability of Tyrone's men, Sir George Carew would later comment that the new lighter infantry deployed by Mountjoy were faster and more flexible than anything fielded in the Spanish *tercios* as, "[w]ith such as light army the commander may go where he lists, and lodge as near the Spaniard (without harm) as he thinks good; for we have the same advantage upon them that the Irish in lightness have of us . . . we shall be able to go into every part of the province and retreat at our pleasure."[123]

Mountjoy also recognized that Tyrone's power was not just in his armies but in the network of alliances and lines of communications. The landing of 4,000 English troops on the Foyle in May 1600, forced Tyrone to withdraw much of his military support from his allies outside Ulster to contain the threat. The lord deputy then campaigned in the midlands cutting Irish lines of reinforcement and supply (especially gunpowder) to the Irish confederates outside Ulster. Tyrone's allies in the midlands were suppressed and Sir George Carew's efforts in Munster quickly pacified the province. Consequently, with Tyrone contained in the north,

---

**120** *The Dialogue of Silvynne and Peregrynne* (T.N.A., SP 63/203, ff 316v –17); The project of service by Sir Ralph Lane, 23 Dec. 1598 (*Cal. S. P. Ire., 1598–1599*, 419–421).
**121** Perrot, *Chronicle*, 94.
**122** For a full examination of Mountjoy's reforms see O'Neill, "A kingdom near lost," 26–47.
**123** Stafford, *Pacata Hibernia*, 255–256.

Mountjoy focused his effort on Ulster. A highly visible part of the renewed English campaigns were the *trace italienne* forts.

The earliest use of polygonal defences recorded in Ireland was the fort at Corkbeg (County Cork) erected at the start of the 1550s.[124] Though modern in design, it did not herald a transformation of English military architecture in Ireland. Subsequent forts such as those built at Philipstown, Maryborough and Essex's Blackwater Fort all followed more traditional designs with circular bastions and square towers.[125] Some towns included angle bastions in their defences, but in some cases they were only one part of a simple rampart and ditch defence. Newry had one angle bastion in 1570 and Carrickfergus was illustrated with a single-angled bastion with orillions at the northern end on an enclosing earth rampart.[126] Clearly *trace italienne* defences were not vital for the overall defence of these positions. When war came in 1593, medieval masonry walls and turrets were the norm for towns fortunate to have any defenses.

For the first seven years of the conflict the English authorities in Dublin were so hard-pressed that they constructed only three new forts; Armagh, Rathdrum and the Blackwater. When Mountjoy reversed the tide of English defeat from 1600 onwards, large numbers of modern style fortifications appeared. Sir Henry Docwra raised numerous modern fortifications after he landed on the Foyle, and after penetrating into southern Ulster in October 1600 Mountjoy started the large campaign fort at Mountnorris. This continued in 1601 to 1603, with large modern fortifications at Charlemont, Augher, the Mullin (the third English Blackwater Fort) and Mountjoy Forts. Huge sums of money were spent to strengthen the defenses of harbors and towns all along the south coast, such as the impressive pentagonal fort at Castle Park, Kinsale. However, did the proliferation of modern fortification have any bearing on the outcome of the war?

Sieges were not a major part of the war, and by the time the crown was raising modern works Tyrone's power was in terminal decline. When sieges did occur it was not modern fortifications holding the attacking armies at bay. Tyrone launched two assaults on the beleaguered garrison of the Blackwater fort during 1597 to 1598 and both were beaten back with heavy losses. The fort was only yielded to Tyrone after the defeat of Bagenal relief army in August 1598. However, the fort did not have modern bastions, but was roughly square in plan with concave ramparts. Light cannon were mounted on platforms at the corners of the structure and the entire site was surrounded by an outer palisade.[127]

---

124 Kerrigan, *Castles and Fortifications in Ireland*, 35.
125 Plot of Maryborough (T.C.D., MS 1209/10); The Blackwater Fort, 1587 (T.N.A., MPF 1/99).
126 A plan of the town of "Craggfargus" [Carrickfergus] (T.C.D., MS 1209/26).
127 A plot of the Blackwater Fort, 1598 (T.N.A., MPF 1/311).

Mountjoy's siege of the Spanish landing force at Kinsale in 1601 was the pivotal moment in the war, as the ensuing battle with Tyrone's broke the military power of the Irish confederates and forced the surrender of the Spanish led by Don Juan del Águila. The siege lasted from October 1601 to January 1602 and forced the crown to commit masses of men, arms and equipment to contain the Spaniards within the town. Carew estimated that the English lost around 6,000 men in the miserable trenches around Kinsale, as the winter weather took a harrowing toll, with men found frozen at their positions. But the English were not checked by modern defenses, but the remains of the medieval town walls dating from possibly the mid-fourteenth century.[128] Águila's men had reinforced the walls with earth and mounted cannon to provide defensive fire, but the town was entirely unsuited to resist a siege. Moreover, Kinsale is overlooked by surrounding hills, which led Águila to describe his position as an *"hoyo"*; a pit or hole.[129] Nevertheless, despite capturing two outlying castles at Rincurran and Castle Park, and surrounding the town with camps, trenches and gun batteries, at no stage in the siege did Mountjoy consider a general assault. Águila only surrendered on generous terms after Tyrone was routed two miles west of the town on Christmas Eve, 1601.

Firearms were thought sufficient to defend many of the forts built late in the war. Charlemont Fort was raised by Mountjoy in 1602 to hold the vital crossing of the Blackwater River, but its ramparts and demi-bastions were not fitted to accept artillery. A firing step for shot lined the inner wall, but no ramps were built to allow the mounting of heavy guns.[130] The same was true of the forts at Mountnorris, Monaghan, Augher and the Moyry Pass.[131] In some cases artillery was available but was not mounted on the modern works. There were two guns in Mountnorris Fort, but they were mounted within a circular earthwork. This was a re-edified bivallate rath; a high-status farmstead dating to the early-Christian period in Ireland (c. 400–1100 AD). The English garrison pushed the inner bank into its ditch to create more room and reinforced the outer rampart with "cadgehouses set upon poles and fastened to the ramparts, which make it both proportionable and defensible."[132] Clearly, while the architecture of modern warfare appeared in

---

[128] Thomas, *The Walled Towns of Ireland*, vol. 2, 137–141.
[129] Silke, *Kinsale*, 112.
[130] Illustrated by Richard Bartlett. See N.L.I., MS 2656, iv.
[131] Also illustrated by Bartlett. See Mountnorris Fort, 1602 (N.L.I., MS 2656, ii); The Carlingford peninsula and Moyry Castle and fort, 1602 (N.L.I., MS 2656, i); Monaghan Fort, 1602 (N.L.I., MS 2656, ix); Augher Island and fort, 1602 (N.L.I., MS 2656, x).
[132] There were five shown in Bartlett's illustration, whereas Markham referred to eight. See Sir Griffin Markham to Cecil, 8 Nov. 1600 (*Cal. S. P. Ire., 1600–1601*, 20–22).

ever-increasing numbers as the war progressed, the impact of polygonal defences was marginal.

A predisposition to modernization and openness to change is often viewed in a wholly positive light. Moreover, in relation to the historiography of early modern Ireland, attempts to show progressivism in native Irish culture has often been used as to ameliorate the long-standing myth of Irish backwardness.[133] The dramatic and enthusiastic adoption of firearms and firepower dominated warfare by Tyrone and his allies is stark evidence for the Irish proclivity for change, at least with regard to military technology and tactics. However, their rapid change also planted the seeds for Irish defeat. As noted earlier, the Irish had come to depend almost completely on gunpowder, without which the thousands of Irish guns became just so many expensive clubs. Tyrone monopolized the supply of gunpowder and kept his major stockpiles in Ulster, possibly as a means to maintaining control of his southern allies.[134] However, keeping key stores of munitions in the north created lengthy and vulnerable lines of supply to his southern allies.

When Mountjoy brought his army into the field in 1600 it was not into Munster or Ulster but against the Irish in Westmeath, Offaly and Laois. His summer campaign forced many of Tyrone's supporters to submit, effectively cutting the earl's lines of communication and supply into Munster, the midlands, and the south-west. The Munster lord's supply of gunpowder was already precarious in 1599. Therefore, with their access to Tyrone's stores of gunpowder severed, the new style Irish infantry found their combat power fatally compromised.[135] Unlike earlier engagements, the Irish could not maintain long skirmishes to wear down and eventually break the English columns.

They Irish were forced to deploy pike in *melee* combat, for which they were neither trained nor equipped.[136] Pike and swordsmen provided security and support for the shot but were only committed to the attack when Irish gunfire had disrupted the enemy formations. Shortages of gunpowder would have had no effect on the old-style galloglass and kerne as their primary weapons did not require it. But by 1600, the traditional Irish means of close hand-to-hand combat

---

133 Bradshaw, "Native reaction to the Westward Enterprise." 66–80; Simms, *From High Kings to Warlords*; Patterson, "Gaelic law and the Tudor conquest of Ireland," 193–215.
134 Jeremy Black noted this as a feature of early-modern state building; See Black, *European Warfare*, 24.
135 "A brief declaration of the state wherein Ireland now standeth" [1599] (*Cal. S. P. Ire., 1599–1600*, 365–370).
136 Pádraig Lenihan demonstrated that in the wars of the 1640s, the Irish only resorted to charging into *melee* when supplies of gunpowder were short. See Lenihan, "Celtic warfare in the 1640s" 120–121.

had disappeared in Tyrone's modernized army. As Smythe predicted, light pike had no chance against close-order heavy pike.[137] Consequently, from 1600 onwards, Irish battlefield success was replaced with failure and defeats. Only in Ulster, where the supply of gunpowder was sufficient, could the crown forces be stopped, as happened at the Moyry Pass in October 1600 and at the Blackwater in 1601. Yet without the island-wide confederation to dissipate English strength, Tyrone could not hope to resist the focused economic and military power of the Tudor state.

In summation, artillery and heavily armed ships had little part to play during the Nine Years War. The landscape and infrastructure or lack thereof precluded the use of anything but the lightest guns. Though construction of angular military architecture associated with military developments on the continent accelerated, their impact on the overall outcome of the war was limited. Naval warfare was more a matter of transportation and logistics than combat, as the preeminence of the English navy was unchallenged by Tyrone's confederation. But these issues must not obscure the fact that three key elements of the military revolution could be associated with the Irish war effort. The rapid expansion of forces available to Tyrone was unmistakable and unmatched by previous Gaelic lords. The Irish campaigns from 1593 to 1600, the Irish displayed a hitherto unrecognised degree of sophistication and cooperation between separate armies, which was essential to achieving success against the crown.

Nevertheless, it was the wholesale transformation of the Irish infantry that was the most striking. As demanded by the military revolution paradigm, firearms and firepower became the foundation of Irish combat power, replacing the melee actions often associated with the Irish. Defensive fieldworks, rarely a feature of Gaelic warfare became commonplace. However, this was no slavish duplication of continental methods. Tyrone created a hybrid force which maximized the effect of firearms while maintaining the Irish advantage of mobility and flexibility. The change was so total that whenever a situation arose when it would be expedient to have troops capable of shock action, Tyrone could not act as there were none to be found. Moreover, when the supplies of gunpowder failed them, Irish tactical advantage turned into a crippling weakness: Mountjoy exposed this failing. He smashed Tyrone's confederation in three years, bringing the end of Gaelic Irish laws and institutions, and completing the Tudor conquest of Ireland.

---

**137** Smythe, *Instructions*, 25–26.

# Bibliography

## Primary Sources – Manuscripts

British Library, Additional MSS. 33743, transcribed by John McGurk, viewed at http://www.ucc.ie/celt/published/E590001-005/index.html on 2 February, 2013.
British Library, Cottonian MS Augustus I/ii, f. 38, the Battle of the Erne Fords, 10 October, 1593.
Hatfield House Archives, Cecil Papers 100/59.
Hatfield House Archives, Cecil Papers 139/54.
Hatfield House Archives, Cecil Papers 177/131.
Lambeth Palace Library, MS 621, f. 136, "A journal of the Lord Lieutenant's proceedings from the 22nd of June to the 1st of July 1599."
The National Archives, Kew, State Papers Ireland (S.P. 63). volumes 136, 143, 151, 172, 178, 180, 182, 195, 197, 199, 203, 207, 209, 210.
The National Archives, Kew, MPF 1/99, The Blackwater Fort, 1587.
The National Archives, Kew, MPF 1/133, Lough Neagh, 1601–1602.
The National Archives, Kew, MPF 1/311, A plot of the Blackwater Fort, 1598.
The National Library of Ireland, MS 2656, vi, Inisloughan Fort, 1602.
The National Library of Ireland, MS 2656, iv, Illustrated by Richard Bartlett, Charlemont Fort, 1602.
The National Library of Ireland, MS 2656, ii, illustrated by Richard Bartlett. Mountnorris Fort, 1602.
The National Library of Ireland, MS 2656, I, The Carlingford peninsula and Moyry Castle and fort, 1602.
The National Library of Ireland, MS 2656, ix, Monaghan Fort, 1602.
The National Library of Ireland, MS 2656, x, Augher Island and fort, 1602.
The National Library of Ireland, MS 2656, vi, Inishloughan Fort.
Trinity College Dublin, MS 1209/35, "The description of the army which was defeated by the earl of Tyrone," 14 August, 1598.
Trinity College Dublin, MS 1209/35, "The description of the army which was defeated by the earl of Tyrone," 14 August, 1598.
Trinity College Dublin, MS 1209/10, Plot of Maryborough.
Trinity College Dublin, MS 1209/26, A plan of the town of 'Craggfargus' [Carrickfergus].
Trinity College Dublin, MS 1209/34, Lord Burgh assaults the Blackwater fort', 14 July, 1597.
University Library, Cambridge, Ms. Kk.1. 15, no. 147, f. 503 "The main strengths of the rebels in Ireland," 1600.

## Primary Sources – Contemporary Publications

Barret, Robert. *The theorike and Practike of Moderne warre*. London, 1598.
Smythe, Sir John. *Instructions, observations, and orders mylitarie*. London, 1594.
Sutcliffe, Matthew. *The practice and proceedings and laws of armes*. London, 1593.

## Primary Sources – Printed and Edited Primary Sources

*Acts of the Privy Council of England.* XXIX (1598–1599), J.R. Dasent, ed. London, 1905.
A.F.M., vi, 2033–2037.
*Calendar of the Carew papers in the Lambeth Library, volume three 1589–1600.* J.S. Brewer and W. Bullen, eds. London: HMSO, 1869.
*Calendar of State Papers relating to Ireland, volume one, 1509–73.* H.C. Hamilton, ed. London: HMSO, 1860.
*Calendar of State Papers relating to Ireland, volume four, August 1588-September 1592.* H.C. Hamilton, ed. London: HMSO, 1885.
*Calendar of State Papers relating to Ireland, volume five, October 1592-June 1596.* H.C. Hamilton, ed. London: HMSO, 1890.
*Calendar of State Papers relating to Ireland, volume six, July 1596-December 1597.* E.G. Atkinson, ed. London: HMSO, 1893.
*Calendar of State Papers relating to Ireland, volume seven, 1598- March 1599.* E.G. Atkinson, ed. London: HMSO, 1895.
*Calendar of State Papers relating to Ireland, volume eight, April 1599- February 1600.* E.G. Atkinson, ed. London: HMSO, 1900.
*Calendar of State Papers relating to Ireland, volume nine, March-October 1600.* E.G. Atkinson, ed. London: HMSO, 1903.
*Calendar of State Papers relating to Ireland, volume ten, November 1600- July 1601.* E.G. Atkinson, ed. London: HMSO, 1905.
Dawtry, Nicholas. "A booke of questions and answers concerning the wars or rebellions of the kingdome of Irelande." In *Analecta Hibernica*, xxxvi (1995), edited by Hiram Morgan, 79–132 (cited as *Q & A*).
Evans, John X., ed. *The works of Sir Roger Williams.* Oxford, 1972.
*Letters and State Papers relating to English Affairs, preserved in the Archives of Simancas (Calendar of State Papers, Spain), volume four, 1587–1603.* Edited by M.A.S. Hume. London: HMSO, 1899.
Lombard, Peter. *The Irish war of defence 1598–1600: extracts from De Hibernia Insula commentaries*, edited and translated by M.J. Byrne. Cork, 1930.
*Cal. Carew MSS, 1589–1600.* 109–110.
Moryson, Fynes. *The Irish sections of Fynes Moryson's unpublished Itinerary*, edited by Graham Kew. Dublin, 1998.
Ó Clérigh, Lughaidh. *The life of Aodh Ruadh O' Domhnaill*, Part 1, edited and translated by Paul Walsh. London, 1948, 159–161.
O'Donovan, John, ed. "Proclamation, in the Irish language, issued by Hugh O'Neill, earl of Tyrone, in 1601." *Ulster Journal of Archaeology*, first series, 11 (1858): 63.
O'Donovan, John, ed. and trans. *Annals of the kingdom of Ireland by the Four Masters*, iv. 3rd ed., 7 vols. Dublin, 1990, 1145–1147 (henceforth cited as *A.F.M.*).
O'Sullivan Beare, Philip. *Ireland under Elizabeth: Chapters Towards a History of Ireland under Elizabeth*, translated by M.J. Byrne. Dublin, 1903.
Perrot, Sir James. *The Chronicle of Ireland 1584–1608*, edited by Herbert Wood. Dublin, 1933.
Stafford, Thomas. *Pacata Hibernia: Ireland appeased and reduced, or a history of the late wares of Ireland, especially in the province of Munster under the command of Sir George Carew*, edited by Standish O'Grady, ii. 2 vols. London, 1896, 255–256.

## Secondary Sources

Andrade, Tonio, Hyeok Hweon Kang, and Kirsten Cooper. "A Korean Military Revolution? Parallel military innovations in East Asia and Europe." *Journal of World History* 25.1 (2014): 51–84.
Andrade, Tonio. *The Gunpowder Age: China, Military Innovation, and the Rise of the West in World History*. Princeton, 2016.
Bartlett, Thomas. *The Academy of Warre: Military Affairs in Ireland, 1600 to 1800, 30th O'Donnell lecture 2002*. Dublin, 2002.
Black, Jeremy. *A Military Revolution? Military Change and European Society, 1550–1800*. London, 1991.
Black, Jeremy. *Beyond the Military Revolution: War in the Seventeenth-century World*. London, 2011.
Black, Jeremy. *European Warfare 1494–1660*. London, 2002.
Bradshaw, Brendan. "Native reaction to the Westward Enterprise: A case study in Gaelic ideology." In *The Westward Enterprise: English Activities in Ireland, the Atlantic and America, 1480–1650*, edited by K.R. Andrews, N.P. Canny, and P.E.H. Hair, 66–80. Liverpool, 1978.
Brewer, John. *The Sinews of Power: War, Money, and the English State, 1688–1783*. New York, 1989.
Connolly, Sean. *Contested Island: Ireland 1460–1630*. Oxford, 2007.
Cruickshank, C. G. *Elizabeth's Army*. 2nd ed. Oxford, 1966.
De Yturriga, José Antonio. "Attitudes in Ireland towards the survivors of the Spanish Armada." *Irish Sword* 17.67 (1990): 244–254.
de hÓir, Siobhán. "Guns in medieval and Tudor Ireland." *Irish Sword* 15 (1982/83): 76–88.
de la Vega, Luis Guntierras. *A compendious treatise entitled, De re militari*, translated by Nicholas Lichefild. London, 1582.
Eltis, David. *The Military Revolution in Sixteenth-century Europe*. London, 1998.
Falls, Cyril. *Elizabeth's Irish Wars*. London, 1950.
Fissel, Mark Charles. *English Warfare, 1511–1642*. London, 2001.
Fissel, Mark Charles. "Military Revolutions." In *Oxford Bibliographies in Military History*, edited by Kaushik Roy. New York: Oxford University Press, https://www.oxfordbibliographies.com/view/document/obo-9780199791279/obo-9780199791279-0212.xml.
Glasgow, Tom. "The Elizabethan Navy in Ireland, 1558–1603." *Irish Sword* 7 (1966): 291–307.
Hall, Bert S. *Weapons and Warfare in Renaissance Europe*. Baltimore, 1996.
Hammer, Paul. *Elizabeth's Wars: War, Government and Society in Tudor England, 1544–1604*. New York, 2003.
Hayes-McCoy, G.A. "The army of Ulster." *Irish Sword* 1.2 (1949–1953): 105–117.
Hayes-McCoy, G.A. "Strategy and tactics in Irish warfare, 1593–1601." *Irish Historical Studies* 2.7 (1941): 255–279.
Hayes McCoy, G.A. "The early history of guns in Ireland." *Journal of the Galway Archaeological and Historical Society* 17.1/2 (1938): 43–65.
Hayes-McCoy, G.A. *Scots Mercenary Forces in Ireland in Ireland, 1565–1603*. 2nd ed. Dublin, 1996.
Hayes-McCoy, G.A. *Irish Battles: A Military History of Ireland*. 2nd ed. Dublin, 1980.
Hill, James Michael. *Celtic Warfare, 1595–1763*. Edinburgh, 2003.

Hill, James Michael. "Shane O'Neill and the campaign against the MacDonalds of Antrim, 1564–5." *Irish Sword* 18.71 (1991): 129–138.
Kerrigan, Paul. *Castles and Fortifications in Ireland, 1485–1945*. Cork, 1995.
Klingelhofer, Eric. *Castles and Colonists: An Archaeology of Elizabethan Ireland*. Manchester, 2010.
Klinglehofer, Eric. "The Renaissance fortifications at Dunboy castle, 1602: a report of the 1989 excavations." *Journal of the Cork Historical and Archaeological Society* 97 (1992): 85–96.
Lenihan, Pádraig. "Celtic warfare in the 1640s." In *Celtic Dimensions of the British Civil Wars*, edited by John R. Young, 116–40. Edinburgh, 1997.
Loeber, Rolf, and Geoffrey Parker. "The military revolution in seventeenth-century Ireland." In *Ireland from Independence to Occupation, 1641–1660*, edited by Jane H. Ohlmeyer. Cambridge, 1995.
McGleenon, C.F. "The battle of Mullabrack 5 September 1595." *Seanchas Ard mhacha* 13.2 (1989): 90–101.
McGurk, John. "The Battle of the Yellow Ford, August 1598." *Dúiche Néill* 11 (1997): 34–55.
Morgan, Hiram. *Tyrone's Rebellion: The Outbreak of the Nine Years War in Ireland*. Dublin, 1993.
O'Báille, M. "The Buannadha; Irish professional soldiery in the sixteenth century." *Journal of the Galway Archaeological and Historical Society* 22.1/2 (1946): 49–94.
O'Byrne, Emmett. "The Battle of Glenmalure, 15 August 1580: cause and course." In *Feagh MacHugh O'Byrne: The Wicklow Firebrand*, edited by Conor O'Brien, 173–179. Rathdrum, 1998.
Ó Domhnaill, Seán. "Warfare in sixteenth-century Ireland." *Irish Historical Studies* 5.17 (1946): 29–54.
O'Neill, James. "A kingdom near lost: English military recovery in Ireland, 1600–3." *British Journal for Military History* 3.1 (2016): 26–47.
O'Neill, James. "Maguire's revolt but Tyrone's war: proxy war in Fermanagh, 1593–4." *Seanchas Ard Mhacha* 26.1 (2016): 43–68.
O'Neill, James. *The Nine Years War, 1593–1603: O'Neill, Mountjoy and the Military Revolution*. Dublin, 2017.
O'Neill, James, and Paul Logue. "The battle of the Ford of the Biscuits, 7 August 1594." In *An Archaeological Survey of County Fermanagh, vol. 1, pt. 2*, edited by Claire Foley, and Ronan McHugh, 913–922. Belfast, 2014.
Parker, Geoffrey. "The 'military revolution,' 1955–2005: from Belfast to Barcelona and the Hague." *Journal of Military History* 69.1 (2005): 205–209.
Parker, Geoffrey. "The 'Military Revolution, 1560–1660' – A myth?" In *The Military Revolution Debate: Readings on the Military Transformation of Early-modern Europe*, edited by Clifford A. Rogers, 37–54. Oxford, 1995; reprinted from *The Journal of Modern History* 48.2 (1976): 196–214.
Parker, Geoffrey. *The Military Revolution: Military Innovation and the Rise of the West, 1500–1800*. 2nd ed. Cambridge, 1996.
Patterson, Nerys. "Gaelic law and the Tudor conquest of Ireland: the social background of the sixteenth century recensions of the pseudo-historical prologue the Senchas Mar." *Irish Historical Studies* xxvii.107 (May 1991): 193–215.
Raymond, James. *Henry VIII's Military Revolution: The Armies of Sixteenth-century Britain and Europe*. London, 2007.

Roberts, Michael. "The Military Revolution, 1560–1660." In *The Military Revolution Debate: Readings on the Military Transformation of Early Modern Europe*, edited by C.J. Rogers, 13–35. Oxford, 1995.

Rogers, Clifford J. "The Idea of Military Revolutions in Eighteenth and Nineteenth-century Texts." *Revista de História das Ideas* 30 (2009): 395–415.

Rogers, Clifford J. "The Military Revolution in History and Historiography." In *The Military Revolution Debate: Readings on the Military Transformation of Early Modern Europe*, edited by C.J. Rogers,1–10. Oxford, 1995.

Rogers, Clifford J. "The Military Revolutions of the Hundred Years War." In *The Military Revolution Debate: Readings on the Military Transformation of Early Modern Europe*, edited by C.J. Rogers, 55–94. Oxford, 1995.

Rowse, A.L. *The Expansion of Elizabethan England*. London, 1973.

Sheehan, A.J. "The overthrow of the Plantation of Munster in October 1598." *Irish Sword* 15. 58 (1982): 11–22.

Silke, John J. *Kinsale: The Spanish Intervention in Ireland at the End of the Elizabethan Wars*. Dublin 1970; reprinted 2000.

Simms, Katharine. *From High Kings to Warlords: The Changing Political Structures of Gaelic Ireland in the Later Middle Ages*. Woodbridge, 1987.

Thomas, Avril. *The Walled Towns of Ireland, vol. 2*. Dublin, 2006.

Turnbull, Stephen. *The Art of Renaissance Warfare*. London, 2006.

Webb, Henry J. *Elizabethan Military Science: The Books and Practice*. London, 1965.

Wheeler, James Scott. "The Logistics of Conquest." In *Conquest and Resistance: War in Seventeenth-century Ireland*, edited by Pádraig Lenihan, 177–209. Leiden, 2001.

White, Dean Gunter. "Henry VIII's Irish Kerne in France and Scotland, 1544–1545." *Irish Sword* 3.13 (1958): 213–225.

Vladimir Shirogorov
# A True Beast of Land and Water: The Gunpowder Mutation of Amphibious Warfare

The early modern upheaval in the practice of warfare, the so-called military revolution, has become a cliché in historical studies. The phenomenon's origins might be traced to either the seminal lecture of Michael Roberts in 1955, in which he proposed the concept of military revolution, or to the *Anti-Dühring* of Frederick Engels, the first-ever proponent of Marxism. As long ago as 1878, Engels asserted a social revolution due to the massive introduction of firearms. Since the 1960s, the concept of the military revolution has gained momentum. It successfully plowed new ground in historiographical "warfare." The bastion fortress and sailing gunship have been identified as Geoffrey Parker's military revolution's major technical hallmarks; infantry salvo fire, the global struggle for sea-domination, the establishment of European world-hegemony, the rise of the fiscal-military state, and utility in nation-building are cited as its accomplishments in the tactical, operational, strategical, social, and political spheres.

Despite its spectacular historiographical "landslide," the concept is counter-charged and debated relentlessly. Unable to deny the existence of a "revolution" completely, skeptics challenge the military revolution's genuinely revolutionary radicalism. Criticism from various quarters points to the concept's principal deficiencies, suggesting that the concept is apocryphal. It is the absence of a fact, or event, which could be undoubtedly interpreted as the cause (or result) of the military revolution that prevents unanimous acceptance among scholars. There are always precedents of military change in history. Their exclusive European pedigree is both overestimated and incorrect. The extra-effectiveness of the "revolutionized" military in comparison with its "stagnating" counterpart is hyperbolic, as well as its causal role in fomenting social and political changes. The proponents of the military revolution search the early modern era for radically new military practices to explain the apparently unparalleled degree of progress in warfare and its ensuing impact on the society. It is the concept's "line of assault."

Comparatively recently historians who had focused on the military revolutions in land warfare and naval expansion have recognized the significance of the amphibious dimension. Amphibious warfare is analyzed in the anthology entitled *Amphibious Warfare, 1000–1700*, with the subtitle *Commerce, State Formation and European Expansion*. The latter themes predominate in military revolution historiography. Amphibious warfare, relatively unexplored historiographically in relation to the military revolution (and often dismissed as merely the practice of special operations involving land and sea), is advancing now as the third spearhead of military revolution studies. Transformations in amphibious warfare occurred roughly simultaneously with those developments that have been associated with the "gunpowder revolution" in general.

## Assault Ships in Amphibious Operations

In antiquity and the Middle Ages, the fighting potential of amphibious warfare was, comparatively speaking, restricted. The relatively modest effectiveness of the weapons available for waterborne assault bound both marine and riverine amphibious operations, compounded by inherent limitations in operational structure and resources. Attack from a sea or river required troops capable of landing in reasonably intact condition to deliver force. In other words, the assault capacity of amphibious attack consisted entirely of its landing party. The naval component of the amphibious force lacked immediate assault capability. Rare were those opportunities where a sailing ship could navigate up to stone walls built upon the shoreline and use ladders and hand-weapons. Heavy bombardment from the deck of a ship upon a "hardened" shore target simply was not practicable because shipborne weapons capable of such destruction did not exist. Likewise, marine troops engaged in landing could not carry massive weaponry. A stone-throwing machine such as a catapult had to be brought disassembled; missile weapons such as the bow and crossbow were ineffective against enemies sheltering in coastal fortifications, stationary or provisional. Marines were on their own: "Ships were little suited for direct engagement against soldiers protected by any substantial fortification."[1] Naval vessels transporting an amphibious force might blockade waterways leading to a besieged coastal fortress or enemy in shoreline defenses but could not attack fortifications aggressively from the water by their own means. The function of ships in medieval amphibious operations was the transportation of troops (and in some cases, horses); launching marines was the only available action against a coastal enemy that could be mounted from aboard ship. Troops wading ashore did not yet enjoy a combined assault partnership with their vessels. They fought alone due to naval incapacity against a shore-based adversary. This relative weakness limited strategic, operational, and tactical abilities of classical

---

[1] Bill, "Scandinavian Warships and Naval Power in the Thirteenth and Fourteenth centuries," 46.

and medieval amphibious attacks compared to what could be practiced with gunpowder weapons in later eras, obviously.

The introduction of shipboard ordnance in the last decades of the fifteenth century genuinely revolutionized naval vessels' functions in amphibious operations. The seaborne component transcended its limited role in providing transportation and disgorgement of fighters that had been the role of sailing vessels since antediluvian times. Ships now pummelled onshore enemy positions. Vessels became platforms for deck-to-shore gunfire concentrated upon shore-based targets. When a ship fired gunpowder weaponry from its deck in preparation for and in tandem with marine landings, the ensuing tactical revolution overturned traditional amphibious warfare. Vessels of conveyance in the gunpowder age enhanced onshore operations directly. Naval force projection gained equal importance with the landing troops in the suppression of shoreline resistance, precisely during this late-medieval epoch. Amphibious operations came of age because both components, ships and landing troops, became potent and aggressive against coastal objectives. Amphibious warfare became truly a formidable beast upon both land and water.

An abundant historiography typically considers the last decades of the fifteenth century as the time of the fundamental transformation of firearms. The introduction of corned powder, a shift to a longer barrel of higher caliber, and proliferation of bronze guns to supplement those of wrought iron all evidence of this ascendancy.[2] The period 1470–1570 is heralded as the dawn of the military revolution because infantry tactics utilizing pike and shot, fortifications incorporating bastions, and artillery-armed sailing ships, likewise matured and dominated warfare.[3] As the late Jan Glete put it, largely state-owned battle fleets were deployed "to control the sea lines of communication around the world".[4]

The topic of the earliest innovative gunpowder-empowered amphibious techniques remains rather unexplored. Yet it was in this era that "deck-to-shore assault ships", specialized or borrowed from naval uses, became prerequisite components for amphibious operations. Although most pillars of the military revolution (pike-and-shot formations, musket salvo fire, ships of the line, and line-of-battle tactics) have vanished, the assault ship of amphibious forces survives if in more industrialized and computerized incarnations. Longevity and immanence of the innovation are criteria for military revolution. Twenty-first century assault ships perform the same basic function as those operating during their

---

[2] Hall, *Weapons and Warfare in Renaissance Europe*, 89–90, 92–93, 95–96.
[3] Parker, *The Military Revolution*, 10–12, 18.
[4] Glete, *Warfare at Sea, 1500–1650*, 2–3.

years of emergence. Deck-to-shore gunfire continues as amphibious warfare's inherent component up to the present time, now inclusive of missiles, aircraft, and drones. Deck-to-shore gunfire, and specialized amphibious assault vessels capable of delivering it are unique and ground-breaking military phenomena without precedent or prototype in the eras prior to the gunpowder revolution. The present essay asks in what ways, where and when exactly, were these pivotal and lasting transformations born? How dramatic was the effect of these weapons and tactics upon the emergence of modern amphibious warfare? How do developments in land-based gunpowder warfare described in the essays in this volume relate to the unique requirements of amphibious operations? Was there an amphibious incarnation of the military revolution?

## Pre-gunpowder Limitations of Amphibious Warfare

David J.B. Trim and Mark Charles Fissel, the editors of *Amphibious Warfare, 1000–1700* defined amphibious warfare as "a form of warfare in which land-based and waterborne forces cooperate" in conjunction for fighting.[5] The definition in the US Armed Forces' *Amphibious Operations Joint Publication 3–02*, issued in 2019, describes the launching of a landing within the littorals.[6] *The American Doctrine for Amphibious Operations* issued in 1962, more straightforward, delineates an amphibious operation as "the concerted military effort against a hostile shore" incorporating integrated naval and landing forces, "involving a landing on a hostile shore."[7] The *Doctrine* concentrates upon three issues: (1) aggression projected from water to shore, (2) integration of naval support and landing troops, and (3) actions taken against onshore opposition upon landing. This narrow definition, devoid of strategical dimensions and divorced from macrocosmic linkages associated with commerce and state formation, eschews expeditionary warfare in general. Nevertheless, this concise characterization of amphibious operations emphasizes the distinctiveness of this type of fighting and the structure of its forces.

The descent, or landing, comprises the crescendo of an amphibious operation, its primary leverage to achieve its objective, accompanied (hopefully) with the secondary tool of naval support gunfire. Disembarked troops were the sole

---

5 Trim and Fissel, "Amphibious Warfare, 1000–1700," 27.
6 *Amphibious Operations*, xi.
7 *Doctrine for amphibious operations*, 1–3:100a, 101a.

aggressive contingent until navies adopted the weapons and tactics of direct water-to-shore firepower made possible by the gunpowder revolution. Qualitatively "marines" were not simply land troops embarked on watercraft. In short, the mode of transport, the physical environment's unusual parameters, and the relative isolation of an amphibious force, meant that "average" soldiers would not suffice. Since amphibious warfare's inception in antiquity, soldiers crossing from boat to shoreline or riverbank possessed specialized skills.[8] While the distinctiveness of amphibious troops is universal, the vessels' cargo capacity determines their destructive potential in personnel, equipment and victuals. The gunpowder revolution transformed their strategical and tactical situation, altered their weaponry, and provided them with greater resources. Medieval landing troops suffered a comparative disadvantage against land-based enemies due to restricted transportation capacity. Given the limited payloads of medieval vessels, the number of shipboard soldiers was as limited as the extent and nature of their weaponry. How much equipment could be transferred to a beach or even a port, particularly when met with hostile resistance?

Medieval hosts were most often equestrian armies; horses were their assault vehicles, their means of maneuver, their symbol of social status and basis of professionalism. The Byzantines were sufficiently proficient to improve their aquatic horse transports from a dozen steeds per boat to as many as 40 horses by the 1100s. Certain Norman craft could sustain a score of mounts. But these successes depended heavily upon the availability of specialized transports and trained personnel. The latter requirement, the primacy of skill and what the late Jan Glete saw as "capabilities" could prove fatal. Consider the example of the crusade of French king Louis IX, who commissioned a hybrid of a galley and a roundship, the *taridae*, in Genoa. This vessel amounted to a vast floating stable accessed via gates at the vessel's stern. When Louis's crusade descended upon Egypt in 1249, the disembarkation was costly in lives because the landing of the horse was not properly planned and managed.[9] Ultimately the crusaders were annihilated by the superior horsemen of the Mamluks led by Baibars al-Bunduqdari. Medieval successes there were, but these made formidable demands on would-be conquerors.[10] Obviously, depending on the type of watercraft used, the logistical realities faced by "Western" knights also bedeviled "Eastern" armies of mounted archers and the "medium" cavalries of Eastern Europe and the Near East.

---

8 Fissel, "Out of Africa."
9 Bennett, "Amphibious Operations from the Norman Conquest to the Crusades of St. Louis c. 1050–c. 1250," 51–67; Vagts, *Landing Operations*, 154–158.
10 Bennett, "Amphibious Operations from the Norman Conquest," 53–54, 67.

Equine combat was thus not well served in early amphibious operations. "Clearly an amphibious assault, whether part of an invasion or part of the siege of a town, would hardly be likely to succeed if the attackers were deprived of this weapon," and sea transport was not "quite as trouble free and routine as is sometimes suggested."[11] Cavalry was rarely wafted across water in numbers adequate to achieve grand strategies, and risks and costs were enormous. John Pryor and Elizabeth Jeffreys assert that "evidence for the capability of any maritime power to transport horses over anything more than short distances before the twelfth century is very meager."[12] These impediments of course meant that equestrian arms delivered amphibiously could be a tremendous advantage, an edge over enemies surprised by a cavalry charge in the situations where their deployment of horsemen was unexpected. Georgios Theotokis attributes the defeat of the Sicilian Muslims (near Messina) in May 1061 to the Norman tactic of feigned cavalry retreat after a properly executed amphibious landing. There were subsequent amphibious equine Norman victories at Hastings (1066) and at Dyrrachium (Durres, present-day Albania, 1081),[13] but the Norman proficiency in amphibious equestrian logistics was exceptional.

Examples of medieval incompetence and ill-luck in transporting cavalry leading to battlefield defeats are more abundant than victories. In 1223, the princes of Rus advanced into the North Black Sea steppes to counter a Mongol invasion, with forces traveling overland as well as on the rivers Dniester, Dnieper, and the Black Sea. Near the island of Khortitsa, the Rus launched a landing and pursued the invaders to the river Kalka, where battle ensued.[14] The Russian army collapsed when its foot array did not receive cavalry support against the Mongolian mounted archers. Even the comparatively impressive Byzantine logistics with their institutional and technical achievements ultimately met failure, as the hurdles involved in transporting horses contributed to the rise of the Ottoman emirate. The Byzantines were unable to ship from Constantinople the necessary mounts for their mercenary Alan horse in the expedition of 1301 against emir Osman, who had besieged Nicaea. The Nicaean militia was deprived of scarce horses available locally in order to mount the Alans. The combined army was crushed by the predominantly equestrian Turks in the battle of Bapheus.[15]

---

11 Rose, *Medieval Naval Warfare 1000–1500*, 44.
12 Pryor, and Jeffreys, *The Age of the ΔΡΟΜΩΝ*, 325.
13 Theotokis, "The Norman Invasion of Sicily, 1061–1072," 392–393.
14 Nicolle, and McBride, *Armies of Medieval Russia 750–1250*, 37.
15 Inalcik, "Osman Ghazi's Siege of Nicaea and the Battle of Bapheus," 94–96.

The English king Edward III mobilized 720 to 747 vessels against France in 1346, 100 to 200 of them "utilized solely for the freighting of horseflesh."[16] Despite these Herculean efforts, the English heavy cavalry remained numerically inferior to mounted French knights. Ironically, the English infantry achieved the victory at Crecy (1346). Medieval amphibious forces' difficulties in transplanting cavalry-based warfare via water were inherently flawed and resolved only with precise execution and good fortune. Adversaries understood this limitation and exploited it.

Moreover, vessels of the Middle Ages were unable to meet the transportation needs for infantry as well. Since the second half of the fifteenth century the development of tactical foot combat, especially by large cohesive squares of partially armored pikemen, dominated the battlefields of Western Europe. In Eastern Europe and the Near East, tacticians obsessed over the disposition of "fighting wagons," like the *oboz* in Poland and the *tabur* in Turkey. However, ships engaged for amphibious operations lacked the capacity to convey optimal amounts of heavy arms, protective gear and field devices (pikes, corselets and other accoutrements of armor such as *paveza* shields, and fighting wagons, etc.). Denied an abundance of cavalry support coupled with a lack of weaponry and equipment, marines mastered (as they had since antiquity) edge weapons for hand-to-hand fighting and carried light projectile weapons. These parameters remained as the increasing commercial and political expansion occurred globally in the late medieval era. Many alluring objectives presented themselves accessible either exclusively aquatically or more advantageously approached by water. The value of amphibious force projection was enhanced as maritime activities increased.

Besides the geographical conditions, there were novel military considerations that encouraged amphibious undertakings. Throughout Europe, medieval armies consisted of horsemen raised by feudal levy. Complementing cavalry were infantry from communal and territorial militias. Varieties of mercenaries also added to the complexity of late medieval – early modern martial enterprises. Foot soldiers, recruited from the lower orders or urban populations, were as requisite as knights because arrayed infantry anchored and stabilized battlefield formations, often shielding horsemen so that the latter could execute tactical maneuvers. Infantry manned siegeworks, the most prominent type of medieval warfare with its expansive outerworks and labor-intensive earthworks. Foot soldiers comprised expeditionary, occupational, and garrison forces.

---

**16** Lambert, *Shipping the Medieval Military*, 96, 140.

The main shortcoming of infantry was its unwieldiness; aquatic transportation overcame the exhaustion and expense of long marches, at least in theory and if transport vessels were equipped to do so. Another incentive for transwater conveyance came with the necessity of transporting guns, ponderous loads for existing overland vehicles and roads. Whether mobile field pieces or cumbersome siege ordnance, moving an artillery train by land posed immense logistical challenges, especially as cannon became more ubiquitous among various strategic cultures. Escort operations coordinating land armies and waterborne force became a feasible solution. Cavalry sought out overland routes, and followed them when possible, whereas the foot and artillery preferably moved by water. The widespread arrangement gave birth to the composite "escort" design of amphibious operations.

## Amphibious Operations of the Escort Design

Table 1, "The design of major amphibious operations circa 1450–1500,"[17] illustrates the surge in amphibious operations commencing in the second half of the fifteenth century in Europe and its adjacent regions. The figures reveal the developing "escort" design. Twenty-two of 32 significant amphibious operations of the epoch can be classified as escort combinations of land armies and amphibious forces. The arrangement proved fruitful, with 14 successful actions; the strategically pivotal Ottoman capture of Constantinople in 1453 and Moscow's first taking of Kazan in 1487 lie between this cross-section of operations.

The amphibious wave arrived first in the Eastern Mediterranean and the Black Sea; the Ottomans, those masters of expansionism, developed an "escort" design of near invincibility. The Ottoman appropriation of Trabzon and Sinope in 1461, Negroponte in 1470, Kiliya and Akkerman in 1484, and Lepanto and Modon in 1499 to 1500 were amphibious victories for the "escort" design, which accelerated Turkey's eclipse of Venice as the superpower of the Eastern Mediterranean and converted the Black Sea into an Ottoman lake.

The stumblings of amphibious ventures utilizing the "escort" design were equally as significant as its successes. The Teutonic Order, Poland's rival in the Eastern Baltic, crumbled when its amphibious force failed to conjunct with its land army. The result was defeat at the battle of the Vistula Lagoon in 1463. The Republic of Novgorod, Moscow's substantial rival, collapsed when its amphibious and land forces were separately destroyed in battle at the river Shelon and

---

17 See the references to the sources of data on the operations in the Tables.

Table 1: Design of major amphibious operations circa 1450–1500.

| Data | Name of operation | Attacker | Defender | Environment | Design; Application | Leader | Outcome; reason | Note |
|---|---|---|---|---|---|---|---|---|
| 1 | 1453 Taking of Constantinople | Turkey | Byzantine empire | Marine | Escort; landing[18] | Mehmed II | Success | 1 |
| 2 | 1456 Relief of Belgrade | Hungary | Turkey | Riverine | Escort; landing, blockade[19] | John Hunyadi; Giovanni da Capistrano | Success | 2 |
| 3 | 1461 Taking of Sinope | Turkey | Emirate of Isfendiyar | Marine | Escort; blockade | Mehmed II; Mehmed Pasha Angelovic | Success | 3 |
| 4 | 1461 Taking of Trabzon | Turkey | Empire of Comnins | Marine | Escort; blockade | Mehmed II; Mehmed Pasha Angelovic | Success | 4 |
| 5 | 1463 Battle of Vistula lagoon | Teutonic Order | Poland | Marine | Escort; transport,[20] landing | Ludwig von Erlichshausen | Failure; miscoordination | 5 |
| 6 | 1463 Taking of Neuhausen | Moscow, Pskov | Livonian Order | Riverine | Escort; transport | Ivan Zvenigorodsky | Success | 6 |

(continued)

[18] Landing is the arrival on a hostile beach;.
[19] Blockade is the breaking of an enemy's communications;.
[20] Transportation is the arrival on a friendly-controlled beach;.

Table 1 (continued)

| | Data | Name of operation | Attacker | Defender | Environment | Design; Application | Leader | Outcome; reason | Note |
|---|---|---|---|---|---|---|---|---|---|
| 7 | 1463 | Taking of Argos and Corinth | Venice | Turkey | Marine | Aside build-up | Alvise Loredan; Bertoldo d'Este | Failure; superior enemy | 7 |
| 8 | 1467 | Advance on Kazan | Moscow | Kazan Khanate | Riverine | Escort; landing | Ivan Obolensky | Failure; miscoordination | 8 |
| 9 | 1468 | Advance on Kazan | Moscow | Kazan Khanate | Riverine | Escort; landing | Constantine Bezzubtsev | Failure; miscoordination | 9 |
| 10 | 1468 | Storm of Asilah | Portugal | Morocco | Marine | Aside build-up | Ferdinand duke of Beja | Success | 10 |
| 11 | 1469 | Advance on Kazan | Moscow | Kazan Khanate | Riverine | Escort; landing | Yury prince of Dmitrov | Failure; miscoordination | 11 |
| 12 | 1470 | Conquest of Negroponte | Turkey | Venice | Marine | Escort; transport and blockade | Mehmed II; Mahmud Pasha Angelovic | Success | 12 |
| 13 | 1471 | Battle of river Shelon | Novgorod | Moscow | Riverine | Escort; landing | Vasily Casimir; Dmitry Boretsky | Failure, miscoordination | 13 |
| 14 | 1471 | Siege of Stockholm | Denmark | Sweden | Marine | Aside build-up | Christian I | Failure; defeat on landing | 14 |
| 15 | 1475 | Taking of Kaffa | Turkey | Genoa | Marine | Aside build-up | Gedik Ahmed Pasha | Success | 15 |
| 16 | 1480 | Taking of Otranto | Turkey | Naples | Marine | Aside build-up | Gedik Ahmed Pasha | Success | 16 |

| | Year | | | | Marine | Aside build-up | | | |
|---|---|---|---|---|---|---|---|---|---|
| 17 | 1480 | Siege of Rhodes | Turkey | Order of Hospitallers | Marine | | Mesih Pasha Palaeolog | Failure; lack of gunfire | 17 |
| 18 | 1480 | Storm of Pskov | Livonian Order | Pskov | Riverine | Escort; landing | Bernhard von der Borch | Failure; defeat on landing | 18 |
| 19 | 1481 | Siege of Otranto | Naples, Spain, Portugal | Turkey | Marine | Escort; landing, gunfire[21] | Ferdinand I of Naples; Garsia Meneses | Success | 19 |
| 20 | 1482 | Advance on Ferrara | Venice | Milan | Riverine | Escort; landing, gunfire | Roberto Sanseverino d'Aragona | Success | 20 |
| 21 | 1484 | Taking of Kiliya | Turkey | Moldavia | Marine | Escort; transport | Bayezid II; Mesih Pasha Palaeolog | Success | 21 |
| 22 | 1484 | Taking of Akkerman | Turkey | Moldavia | Marine | Escort; transport | Bayezid II; Mesih Pasha Palaeolog | Success | 22 |
| 23 | 1487 | Taking of Kazan | Moscow | Kazan Khanate | Riverine | Escort; landing | Daniel Kholmsky | Success | 23 |
| 24 | 1488 | Deny of the coastal pass Bab al-Malik | Turkey | Egypt | Marine | Escort; gunfire | Hersekzade Ahmed Pasha; Hadim Ali Pasha | Failure; superior enemy, weather | 24 |
| 25 | 1494 | Advance on Genoa | Naples, Milan | Genoa, France | Marine | Escort; landing | Federigo prince of Altamura | Failure; defeat on landing | 25 |

(continued)

---

21 Gunfire is the deck-to-shore artillery assault.

Table 1 (continued)

| Data | Name of operation | Attacker | Defender | Environment | Design; Application | Leader | Outcome; reason | Note |
|---|---|---|---|---|---|---|---|---|
| 26 | 1497 Taking of Melilla | Castile | Kingdom of Fez | Marine | Aside build-up | Pedro de Estopiñán y Virués | Success | 26 |
| 27 | 1499 Taking of Lepanto | Turkey | Venice | Marine | Escort; blockade, transport | Kara Nisanci Davud Pasha; Kemal Reis; Mustafa Pasha | Success | 27 |
| 28 | 1500 Taking of Modon | Turkey | Venice, Spain | Marine | Escort; blockade, transport | Bayezid II; Kara Nisanci Davud Pasha; Kemal Reis; | Success | 28 |
| 29 | 1500 Taking of St. George, Cephalonia | Spain, Venice | Turkey | Marine | Aside build up | Gonzalo Fernandes de Cordoba; Benedetto Pisaro | Success | 29 |
| 30 | 1501 Taking of Navarino | Turkey | Venice | Marine | Escort; transport, blockade | Hadim Ali Pasha; Kemal Reis | Success | 30 |
| 31 | 1501 Siege of Mytilene, Lesbos | France, Venice | Turkey | Marine | Aside build-up | Philippe de Clèves et la Marck | Failure; superior enemy | 31 |
| 32 | 1502 Taking of St. Maura, Lefkas | Venice | Turkey | Marine | Aside build-up | Benedetto Pisaro | Success | 32 |

Notes:

1. David Nicolle and Christa Hook, *Constantinople 1453. The End of Byzantium* (Oxford: Osprey Publishing, 2000), 61,71; Marios Philippides and Walter K. Hanak, *The Siege and the Fall of Constantinople in 1453. Historiography, Topography and Military Studies* (Farnham, Burlington: Ashgate, 2011), 438–443, 447.
2. Kenneth M. Setton, *The Papacy and the Levant, 1204–1571* (Philadelphia: The American Philosophical Society, 1978), 2:179–181.

3. Harry J. Doukas Magoulias, *Decline and Fall of Byzantium to the Ottoman Turks* (Detroit: Wayne State University Press, 1975), 20–21.
4. Magoulias, *Decline and Fall of Byzantium*, 20–21.
5. Marian Biskup, *Wojna Trzynastoletnia*, (Krakow: Krajowa Agencja Wydawnicza, 1990), 67; Bernard Nowaczyk, *Chojnice 1454, Swiecino 1462* (Warszawa: Bellona, 2012), 176–178.
6. Shirogorov, *War on the Eve of Nations. Conflicts and Militaries in Eastern Europe, 1450-1500* (New York/London: Lexington Books, 2021), 193.
7. Michael Edward Mallett, "Part I. C. 1400 to 1508," in *The Military Organisation of a Renaissance State: Venice C.1400 to 1617*, ed. M. E. Mallett and J. R. Hale (Cambridge, New York and Melbourne: Cambridge University Press, 2006), 45–47.
8. Shirogorov, *War on the Eve of Nations*, 199.
9. Shirogorov, *War on the Eve of Nations*, 201–202.
10. Peter Purton, *A History of the Late Medieval Siege, 1200–1500* (Woodbridge: Boydell Press, 2010), 346.
11. Shirogorov, *War on the Eve of Nations*, 202–203.
12. Setton, *The Papacy and the Levant*, 2: 300–303.
13. Shirogorov, *War on the Eve of Nations*, 206–210.
14. Kelly DeVries, et al, *Battles of the Medieval World, 1000-1500. From Hastings to Constantinople* (London: Metro Books, 2006), 208–215.
15. Shirogorov, *War on the Eve of Nations*, 136–137.
16. K. Giakoumis, "The Ottoman Campaign to Apulia (1480–81)," in *The Turks*, Vol. 3 Ottomans, eds. Hasan Celal Güzel, C. Cem Oğuz, and Osman Karatay (Ankara: Yeni Türkiye Yayınları, 2002), 191–192; Mesut Uyar and Edward J. Erickson, *A Military History of the Ottomans. From Osman to Ataturk* (Santa Barbara and Oxford: Praeger Security International, 2009), 67.
17. Setton, *The Papacy and the Levant*, 2: 350–63; Purton, *A History of the Late Medieval Siege*, 383.
18. Shirogorov, *War on the Eve of Nations*, 234–235.
19. Vincenzo Scarpello, *Aspetti di Storia Militaria nella Guerra d'Otranto* (2010), 69–70, https://culturasalentina.files.wordpress.com/2010/09/aspettidistoriamilitarenellaguerradotranto.pdf; Setton, *The Papacy and the Levant*, 2: 372; Purton, *A History of the Late Medieval Siege*, 385.
20. Mallett, "Part I. C. 1400 to 1508," 99; Federico Moro, "Venetia Fules the Rivers. La Geo-Strategia Fluviale Veneziana (1431–1509)." *Nuova Antologia Militare* Fascicolo 2.7 (June 2021): 7–62, https://www.nam-sism.org/3.2%20%20fascicoli.html, 29–40.
21. Liviu Pilat and Ovidiu Cristea, *The Ottoman Threat and Crusading on the Eastern Border of Christendom during the 15th Century* (Leiden, Boston and Kohl: Brill, 2018), 208–211.
22. Pilat and Cristea, *The Ottoman Threat and Crusading*, 211–213.
23. Shirogorov, *War on the Eve of Nations*, 268–269.
24. Caroline Finkel, *Osman's Dream: The Story of the Ottoman Empire 1300–1923* (New York: Basic Books, 2006), 92.

25. Cecil H. Clough, "The Romagna Campaign of 1494: A Significant Military Encounter," in *The French Descent into Renaissance Italy, 1494–1495. Antecedents and Effects*, ed. David Abulafia (Abingdon and New York: Routledge, 1995), 196–198.
26. Andrew C. Hess, *The Forgotten Frontier. A History of the Sixteenth-Century Ibero-African Frontier* (Chicago and London: University of Chicago Press, 1978), 37.
27. Daniel Goffman, *The Ottoman Empire and Early Modern Europe* (Cambridge, New York and Melbourne: Cambridge University press, 2009), 143.
28. Setton, *The Papacy and the Levant*, 2: 522.
29. Setton, *The Papacy and the Levant*, 2: 523.
30. N. Bées and A. Savvides, "Navarino" in *The Encyclopedia of Islam, New Edition*, Volume VII: *Mif–Naz. 1037–1039* (Leiden and New York: Brill, 1993), 1038; Palmira Brummett, *Ottoman Seapower and Levantine Diplomacy in the Age of Discovery* (Albany: State University of New York Press, 1994), 105.
31. Halil Inalcik, "The Ottoman Turks and the Crusaders, 1451–1522," in *A History of the Crusades*, Vol. VI *The Impact of the Crusades on Europe*, eds. Kenneth M Setton, Harry W. Hazard and Norman P. Zacour (Madison: The University of Wisconsin Press, 1990), 351; Setton, *The Papacy and the Levant*, 2: 538.
32. Goffman, *The Ottoman Empire*, 144; Setton, *The Papacy and the Levant*, 2: 523.

combats fought at Lake Ilmen in 1471. The impotence of the Livonian Order in attempting a waterborne assault at the city of Pskov in 1480 fractured its coalition with the Polish king Casimir IV and the Grand Horde's khan Ahmed. The latter's defeat in a stand-off against Muscovite forces on the river Ugra, unfolding simultaneously with the attack on Pskov, made Moscow Eastern Europe's great power. In 1494 a Neapolitan amphibious expedition, deprived of the Milanese land army's expected allied conjunction, lost the struggle over Genoa to a French amphibious force, escorted in timely fashion by Genoese troops on the coastline. Neapolitan defeat opened Italy for repeated French invasions.

Miscoordination of the land forces and the amphibious arm of the operation was the main reason of failure. Precise timing was elusive because different types of military forces possessed widely varying composition. To be precise, their mobility was fundamentally different. The forces encountered different types of enemy resistance. The chain of command in the various "service branches" differed, often becoming competitive and quarrelsome. However, the accomplishments and future potential of the "escort" design of amphibious operations were undeniable. The "escort" martial configuration was being exploited widely whenever opportunities of conjunction between force projection upon water and overland marches presented themselves. However, the number of objectives for this kind of operation was decreasing. Contending powers had devoured the most easily accessed territorial routes, making both unimpeded shipping and tramping across unconsolidated territories increasingly problematic. Targets inaccessible via foreign soil required alternative operational designs. To cite a sixteenth- and seventeenth-century example, the "Spanish Road" used by the Habsburgs and monarchs of Spain in suppressing the revolt of the Netherlands could be maritime or a gruelling hike across terra firma. Unsurprisingly, the respective logistics and mechanics differed greatly between the two options. The problem of distance, as described by Fernand Braudel, dominated, especially in the fifteenth century; it was the necessity of administrative and institutional "build-ups" for divergent forms of operational design that posed contrasting logistical obstacles.[22]

Self-sufficient amphibious operations proceeded as a landing followed by an organizational regrouping outside of the objective, then advancing on said objective. But this model, as noted above, was unreliable due to the limited fighting capacity of disembarked troops operating independently, without the immediate assault support of naval force. Looking to improve the odds for success in amphibious warfare, tacticians enhanced their capabilities with firearms. Three decades before the experiments matured, however, the tactic of a supplementary

---

[22] See, for example, Meyer, "States, Roads, Armies, and the Organization of Space," 99–127.

**Figure 1:** The conjunction of Muscovy's foot and horse regiments before Kazan, the capital of the Kazan Khanate, from July to August 1524. This joint action exemplifies the "escort" design in the taxonomy of amphibious operations. The infantry, bearing firearms, wafted downstream

naval component blazed into amphibious warfare, targeting naval artillery against land-based objectives.

## Deck-to-Shore Gunfire Assault

Kelly DeVries and Nicholas A. M. Rodger, in judging the effectiveness of onboard naval artillery, date the emergence of genuinely efficacious ordnance worthy of use against onshore targets to the middle of the fifteenth century onwards.[23] It is unclear whether the examples they consider promoted efficient landings or not. Table 1 documents three deck-to-shore gunfire experiments that promoted amphibious landing. Specifically, in 1482 a Venetian riverine flotilla of 400 varied watercraft navigated the river Po to Ferrara escorting their land army. The flotilla fell into a trap set by the Ferrarese, who unleashed a crossfire from both riverbanks; the Venetian riverboats' ordnance was unable on its own to protect the force. Ultimately, the Venetians dislodged the enemy by executing a landing in conjunction with land forces commanded by condottiere Roberto Sanseverino.

In 1488, a Turkish fleet under Hersekzade Ahmed Pasha cannonaded the Egyptian Mamluks' coastal descent from the mountain pass of Bab al-Malik. The vessels were seconding the Ottoman land forces of Hadim Ali Pasha, struggling to deny passage to the Mamluk cavalry. The Mamluks, apparently undaunted and unimpressed, ultimately destroyed the Ottomans in a battle at the port of Adana. In 1494, a Neapolitan expedition under Federigo, prince of Altamura, landed in Rapallo, erected an earthwork fort onshore, and awaited their Milanese allies, anticipating the arrival of an enemy Genoese land force. Allies of the latter, the French deployed ships and bombarded Federigo's provisional fortifications, then disembarked Swiss pikemen, who acting in the capacity of marines finished off the Neapolitans.

---

**Figure 1** (continued)
on the river Volga to rendezvous with cavalry that had ridden overland. Their joint operation attacked Kazan but ultimately failed due to the loss of the Muscovite heavy artillery in the chaotic disembarkation. The image is reproduced by courtesy of the Department of Manuscripts, the National Library of Russia, Saint-Petersburg.[24]

---

**23** DeVries, "The Effectiveness of Fifteenth-Century Shipboard Artillery," 393; Rodger, "The Development of Broadside Gunnery, 1450–1650," 302.
**24** *The Russian Illustrated Anthological Chronicle*, M. P. Shumilov's Volume, F.IV.232, Л. 858 06.

Did the actions at Ferrara, Bab al-Malik and Rapallo inaugurate a military revolutionary type of amphibious warfare? Contemporary descriptions of these amphibious operations suggest that the 1480s and 1490s witnessed the tactic of deck-to-shore gunfire gain its singular niche. But the Venetians, Turks and French were not such pioneers because they were thinking within the "escort" pattern, not a systematic use of water-borne gunfire to succor land forces and combine them to destroy a land-based adversary.

Both Venetians and Genoese practiced deck-to-shore gunfire as prelude to amphibious assault as early as their confrontation at Chioggia in 1379 to 1380, fought proximate to the littorals of the Venetian lagoon.[25] In the early 1400s, the amphibious forces of Venice and Milan waged war along the rivers Adige and Po, on Lake Garda, and upon multiple canals, in all cases utilizing the technology and tactics of deck-to-shore gunfire as the requisite auxiliary to amphibious landing.[26] Nowhere was it decisive. An epiphany caused by the fall of a "big apple" would have to serve as the catalyst for the proper recognition of the gunpowder revolution's utility in amphibious warfare. The "eureka" moment occurred neither in Italy, enriched with the imagination of the Renaissance, nor in Ottoman Turkey, fed by the interbreeding of Byzantine and Arabian martial legacies. The discovery came in a faraway fringe of the collision of Latin and Islamic arts of war, on the North African coast along the Portuguese-Moroccan frontier.

In 1437, a Portuguese amphibious force descended upon Tangier but was overcome by a Moroccan relief force. The chief of the expedition, Prince Henry, pledged as hostage his brother Fernando to guarantee a safe evacuation of the Christian forces under his command. Henry's vow to release the port of Ceuta, however, was disingenuous; Fernando perished in captivity.[27] The ghost of his captive sibling tormented the prince. Henry then projected Portuguese bellicosity along the Western coast of Africa, searching for conquests to enhance Portugal's maritime status (and salve his defeat by the Moroccans). Henry endeavored to increase the freightage and above all fighting capacity of Portuguese vessels, specifically the ability to land troops efficaciously on foreign shores. The gunpowder revolution offered the most viable option to gain advantage on coastal Africa. Portugal, a relatively poor country with a small population, might use amphibious

---

**25** Dotson, "Venice, Genoa and control of the Seas in the Thirteens and Fourteens Centuries," 133; John Francis Guilmartin, "The Earliest Shipboard Gunpowder Ordnance," 658.
**26** Mallett, "Part I. C. 1400 to 1508," 98–99; Moro, "Venetia Rules the Rivers," 27–29.
**27** Diffie and Winius, *Foundations of the Portuguese Empire, 1415–1580*, 72; Purton, *A History of the Late Medieval Siege, 1200–1500*, 345.

warfare to achieve superiority over the Moors, despite unwieldy contemporary cannon and albeit clumsy firearms.[28]

The institutional and technical requirements to incorporate boarding and bombardment into European shipbuilding design of the final third of the fifteenth century are well known. However, the prerequisites for harnessing amphibious operations are less familiar. For starters, naval vessels garnished with artillery and assault craft facilitating amphibious attacks differed profoundly in form and functions. Even at this early stage in development one must recognize that amphibious warfare was not an offshoot of naval warfare. For naval scholars, amphibious warfare comprises a mere tangential distraction of seagoing fleets, although ironically landing became their prime combat mission in this epoch's historical reality. In the second half of the fifteenth century, naval warfare had a "predominantly amphibious character . . . with battles fought in close conjunction with the taking of port towns and their hinterland."[29] Until the last quarter of the sixteenth century "the basic character of naval warfare – as essentially an amphibious extension of land warfare – remained unchanged."[30] Naturally, therefore, amphibious requirements influenced deeply early modern battleship construction.

The fragility of the native Portuguese vessel, the *barcha*, made it unsuitable for heavy guns; the later-adopted Genoese *carrack* was too cumbersome for littoral action. The *carrack*, fitted with 30 to 40 guns and antipersonnel pieces,[31] presented formidable firepower by contemporary standards. Each broadside gun of the *carrack* "was regarded as fixed,"[32] to be brought into action when the ship's motion aimed the ordnance at a target. The *carrack* possessed one gunner for every two pieces because only cannon on one side of the ship was in action usually.[33] Particularly debilitating for amphibious support was frequently that the vessel's draught was too deep to close on an onshore target. The *carrack* was not a specialized assault vessel for amphibious actions, it was an amalgamated cargo and naval ship. Like later race-built galleons and line-of-battle ships, the *carrack* was designed for marine combat but sometimes utilized against onshore targets *ad hoc*.

---

28 For Portugal's military revolution(s) subsequent to the events discussed above, see Carvalhal, Murteira, and de Jesus, ed. *The First World Empire: Portugal, War and Military Revolution*.
29 Rose, *Medieval Naval Warfare*, 101.
30 Morillo, Black, and Lococo, *War in World History*, 278; Fissel, "Out of Africa."
31 Diffie and Winius, *Foundations of the Portuguese Empire*, 218–219.
32 Rodger, "The Development of Broadside Gunnery," 312.
33 Salgado, "Portuguese Galleon's Armament at the End of the Sixteenth Century," 281.

To further establish Portugal as a maritime power Prince Henry promoted shipyard innovation and produced the *caravel*. Not merely a modification of merchant vessels, like *carracks*, the *caravel* was a purpose-built battleship of "the 150–180 average tonnage." As such, it addressed exclusively "crew needs and artillery, powder and ammunition."[34] The *caravel* possessed medium draft and was flat-bottomed initially. At the end of the 1400s, the *caravel*'s Mediterranean rig of triangular Latin design was supplemented with the Northern rig of rectangular sails, making it quick and agile against wind and when navigating in leeward areas. The *caravel* jettisoned the traditional forecastle, and its smaller aftercastle served "to make handling the sails easier."[35] *Caravels* had no need for "castles" to facilitate boarding actions because their primary purpose was not so much naval as amphibious, and their fighting capacity rested squarely upon gunfire.

In 1481, a Portuguese squadron of twenty *caravels* and one cargo-bearing *carrack* (under command of Garsia Meneses, bishop of Evora) assisted King Ferdinand I of Naples in ousting the Turks from Otranto. It was the *caravel*'s first amphibious campaign against an enemy adept with firearms and ensconced in stone-reinforced coastal fortifications. "The attack was set up by land and sea" but the Ottomans sealed the entrance to the harbor of Otranto, "a likely route" for assault, with sunken boats and coastal batteries.[36] The Crusaders were thus unable to apply amphibiously their naval superiority that had been secured after their destruction of the Turkish fleet at the battle of Saseno on 25 February 1481.[37] On 24 August, the Crusaders hazarded a storm of Otranto, but suffered heavy casualties. However, Mehmed II died, potential Ottoman reinforcements were defeated in Albania,[38] and Otranto's garrison surrendered 6 September, 1481.

Otranto's lesson pushed the Portuguese to equip the *caravel* with heavier guns. Heeding the gunpowder revolution, firepower's integration with inshore amphibious warfare would be increased. The race to improve the *caravel* started in the 1480s, continuing to the close of the 1400s. *Caravels* typically carried around 15 guns and could carry from 30 to 40 pieces.[39] There were "four heavy guns below, and above six falconnets, and ten swivel-guns placed on the quarter deck and in the bows, and two of the falconnets fired astern."[40] The heavy guns

---

**34** Domingues, "The State of Portuguese Naval Forces in the Sixteenth Century," 195.
**35** Unger, *The Ship in the Medieval Economy, 600–1600*, 212.
**36** Purton, *A History of the Late Medieval Siege*, 385.
**37** Scarpello, *Aspetti di Storia Militaria nella Guerra d'Otranto*, 62–65.
**38** Giakoumis, "The Ottoman Campaign to Apulia (1480–1481)," 192–194.
**39** Cipolla, *Guns, Sails and Empires*, 80–81.
**40** Correa, *The Three Voyages of Vasco Da Gama and His Viceroyalty from the Lendas Da India of Gaspar Correa*, 366.

of the *caravel* were of the *cameo* class, long-barrel bronze or wrought-iron muzzle-loaders. Portuguese heavy naval ordnance was now concentrated aboard *caravels*.[41]

The heavy and medium guns of the *caravel* exploited their two-wheel carriages and could be aimed independently from the movement of a ship.[42] The *galleon*, which succeeded the *caravel* as the main Portuguese battleship in the mid-1500s, assigned specialist gunners to each of its larger cannon; their rate of fire was one shot per 30 minutes to one hour,[43] the highest standard of the time. It is reasonable to presume that the *caravel* employed similar teams, approximating the galleon's rate of fire. The heavy pieces were muzzle-loaders. The smaller breech-loaders were fitted with changeable powder chambers for much faster shooting. In an amphibious assault, the heavy guns' initial work was continued by the medium guns that became increasingly effective when the ship moored near the landing site. This schedule created a useful continuance of gunfire, underscoring the role of gunpowder revolution in the development of amphibious operations.

Spanish, Portuguese, Italian, and Maltese forces descended upon the fort at La Goletta in 1535. Their land batteries and naval craft fired off 4,000 rounds in several hours. Compared with 100,000 rounds with which the English batteries pounded Boulogne during the siege of July to September, 1544, and considering the bombardment of Metz by the Imperial army in 1552 at the tempo of 1,000 rounds per day,[44] La Goletta's punishment was extensive for a modest stronghold. The ordnance of the 74 galleys and 30 *galliots* involved in the La Goletta operation, one per vessel, could not have delivered such a hurricane of fire. The Portuguese squadron brought to bear an astounding 600 cannon altogether, including the 366 guns of the *galleon* São João Baptista. The Maltese *carrack* Sant'Anna contributed her 50 pieces of substantial ordnance[45] in providing the bulk of barrage. The *caravels*' contribution to this display of firepower was especially prominent.[46]

*Caravels* accomplished the amphibious breakthrough. Their exemplary design met the requirements of amphibious warfare, especially the tactic of the deck-to-shore gunfire. The *caravel*'s naval service, including the geographical discoveries, was the successful development of its amphibious design. The *caravel* was the

---

41 Guilmartin, "The Earliest Shipboard Gunpowder Ordnance," 665.
42 Salgado, "Portuguese galleon's armament," 279.
43 Salgado, "Portuguese galleon's armament," 281–282.
44 Tallett, *War and Society in Early Modern Europe, 1495–1715*, 166.
45 Atauz, "Trade, Piracy, and Naval Warfare in the Central Mediterranean," 387, 409.
46 Carvalhal, and de Jesus, "The Portuguese Participation in the Conquest of Tunis (1535)," 176.

first example of the ship designed for amphibious practice in the early gunpowder epoch. In the second half of the fifteenth century, this type of gunship was less necessary for Portuguese slave-trafficking on Africa's western coast.[47] Incapable of penetrating the African continent with massive land armies, amphibious warfare was Portugal's sole means of projecting force for commercial advantage. There, on the littoral, indigenous peoples lacked sufficient stone fortifications and masses of garrisoned troops capable of withstanding swift Portuguese interlopers. In particular, the artillery-laden *caravel* was the nemesis of the Moors, designed to extinguish Moorish power upon sea and land. Thus, Portuguese amphibious ways of war were conditioned by African experiences. As in the case of many military revolutions, confrontation with a unique enemy encouraged innovation.

A subsequent Portuguese king, Afonso V, eagerly applied the advantages inherent in the *caravel*. Collaborating with Prince Henry, now aged (but wise in amphibious ways of war), Afonso V landed, stormed, and captured the Moorish fortress of Alcazarseguir with a fleet of 220 ships and 25,000 troops in a mere two days (22–24 October, 1458). Henry assisted by unleashing fire from dozens of wrought-iron monsters waterborne on three sides of the protruding cape fortress and town.[48] Notably, the taking of Alcazarseguir was the first amphibious operation where assault ships themselves employed the tactic of deck-to-shore gunfire, demonstrating techniques of enhanced capabilities. Further improvements in Portugal's gunpowder revolution ensued.

## Common Platforms

Deck-to-shore gunfire gained momentum through the 1500s. Table 2, "Major amphibious operations based on deck-to-shore gunfire circa 1450–1600,"[49] configures data on the tactic's stunning results (when competently practiced). Nineteen of 24 major amphibious operations applying deck-to-shore barrages succeeded impressively.

The storm of Ivangorod in 1496 was arguably the first conquest of Swedish Baltic expansionism. The seizure of the fort of Benasterim in 1511 established Goa as the capital of the Portuguese East Asian empire. The already discussed Spanish-Imperial storming of La Goletta in 1535 widened the clash between

---

47  See the early chapters in Thornton, *Warfare in Atlantic Africa, 1500–1800*.
48  Diffie and Winius, *Foundations of the Portuguese Empire*, 110; Purton, *A History of the Late Medieval Siege*, 346.
49  See the references to the sources of data on the operations in the Tables.

Table 2: Major amphibious operations based on deck-to-shore gunfire circa 1450–1600.

| | Data | Name of operation | Attacker | Defender | Environment | Design | Leader | Outcome | Note |
|---|---|---|---|---|---|---|---|---|---|
| 1 | 1458 | Storm of Alcazarseguir | Portugal | Morocco | Marine | Aside build-up | King Afonso V; prince Henry | Success | 1 |
| 2 | 1472 | Storm of Asilah | Portugal | Morocco | Marine | Aside build-up | King Afonso V | Success | 2 |
| 3 | 1496 | Storm of Ivangorod | Sweden | Moscow | Marine | Direct assault | Knut Jönsson Posse | Success | 3 |
| 4 | 1508 | Taking of Penon de Velez | Spain | Kingdom of Fez | Marine | Direct assault | Pedro Navarro | Success | 4 |
| 5 | 1511 | Taking of Malacca | Portugal | Malacca Sultanate | Marine | Direct assault | Afonso de Albuquerque | Success | 5 |
| 6 | 1511 | Storm of Benasterim | Portugal | Bijapur sultanate | Marine | Aside build-up | Afonso de Albuquerque | Success | 6 |
| 7 | 1513 | Storm of Aden | Portugal | Sultanate of Aden | Marine | Direct assault | Afonso de Albuquerque | Failure | 7 |
| 8 | 1516 | Storm of Jidda | Portugal | Turkey | Marine | Direct assault | Lopo Soares de Albergaria | Failure | 8 |
| 9 | 1530 | Storm of Cherchel | Spain; Genoa | Turkey | Marine | Direct assault | Andrea Doria | Success | 9 |

(continued)

Table 2 (continued)

| | Data | Name of operation | Attacker | Defender | Environment | Design | Leader | Outcome | Note |
|---|---|---|---|---|---|---|---|---|---|
| 10 | 1534 | Relief of Dublin | England | Irish insurgents | Marine | Aside build-up | William Skeffington | Success | 10 |
| 11 | 1535 | Taking of La Goletta and Tunis | Spain, Portugal, Empire | Turkey | Marine | Aside Build-up | Emperor Charles V, Prince Louis | Success | 11 |
| 12 | 1538 | Storm of Preveza | Spain, Empire | Turkey | Marine | Direct assault | Andrea Doria | Failure | 12 |
| 13 | 1543 | Relief of Nice | Spain, Empire | Turkey, France | Marine | Direct assault | Andrea Doria, Alfonso d'Avalos | Success | 13 |
| 14 | 1546 | Relief of Diu | Portugal | Gudjarat Sultanate | Marine | Direct assault | Joao de Castro | Success | 14 |
| 15 | 1550 | Taking of Mahdia | Spain | Turkey | Marine | Aside build-up | Andrea Doria, Juan de la Vega | Success | 15 |
| 16 | 1553 | Landing at Saint-Florent | Spain, Genoa | France, Corsican insurgents | Marine | Aside build-up | Andrea Doria | Success | 16 |
| 17 | 1557 | Storm of Kirkwall | England | Scotland | Marine | Aside build-up | John Clere | Failure | 17 |
| 18 | 1556 | Taking of Astrakhan | Moscow | Astrakhan and Crimean Khanates | Riverine | Direct assault | Ivan Cheryemisinov | Success | 18 |

| | | | | | | | |
|---|---|---|---|---|---|---|---|
| 19 | 1564 | Taking of Peñon de Velez | Spain | Muslim corsairs | Marine | Direct assault | Garcia de Toledo | Success | 19 |
| 20 | 1574 | Relief of Leiden | Netherlandish rebels | Spain, Netherlandish loyalists | Riverine | Direct assault | Louis de Boisot | Success | 20 |
| 21 | 1582 | Taking of Isker | Moscow | Siberian Kh'anate | Riverine | Direct assault | Yermak | Success | 21 |
| 22 | 1583 | Taking of Terceira, Azores | Spain and Portuguese loyalists | France and Portuguese separatists | Marine | Direct assault | Alvaro de Bazan Jr. | Success | 22 |
| 23 | 1585 | Relief of Antwerp | Netherlandish rebels | Spain, Netherlandish loyalists | Riverine | Direct assault | Philip of Hohenlohe-Neuenstein | Failure | 23 |
| 24 | 1587 | Taking of Johor Lama | Portugal | Johor Sultanate | Riverine | Direct assault | Antao de Naronho; Paolo de Lima | Success | 24 |

Notes:

1. Bailey W. Diffie and George Winius, *Foundations of the Portuguese Empire, 1415–1580* (Minneapolis: University of Minnesota Press, 1977), 110; Peter Purton, *A History of the Late Medieval Siege, 1200–1500* (Woodbridge: Boydell Press, 2010), 346.
2. Diffie and Winius, *Foundations of the Portuguese Empire*, 145; Peter Purton, *A History of the Late Medieval Siege*, 346.
3. V. Shirogorov, *War on the Eve of Nations. Conflicts and Militaries in Eastern Europe, 1450–1500* (New York and London: Lexington Books, 2021), 312–313.
4. Cesáreo Fernández Duro, *Armada Española (Desde la Unión de los Reinos de Castilla y Aragón)* (Madrid: Sucesores de Rivadeneyra, 1895–1903), 68–69, https://armada.defensa.gob.es/html/historiaarmada/tomo1/tomo_01_21.pdf.
5. R.W. McRoberts, "An Examination of the Fall of Malacca in 1511," *The Journal of Malaysian Branch of the Royal Asiatic Society (JMBRAS)*, Vol. 57, Pt. 1 (1984): 36–38.
6. Afonso de Albuquerque, *The Commentaries of the Great Afonso Dalboquerque, Second Viceroy of India* (Cambridge/New York/Melbourne: Cambridge University Press, 2010) 3: 215–233.
7. Diffie and Winius, *Foundations of the Portuguese Empire*, 264.

8. John Francis Guilmartin, *Gunpowder and Galleys. Changing Technology and Mediterranean Warfare at Sea in the 16th Century* (London: Conway Maritime Press, 2003), 26–27.
9. Emrah Safa Gürkan, "Ottoman Corsairs in The Western Mediterranean and their Place in The Ottoman – Habsburg Rivalry (1505–1535)," Master's thesis (Bilkent University, Ankara, 2006), 96.
10. James Raymond, *Henry VIII's Military Revolution. The Armies of Sixteenth-century Britain and Europe* (London and New York: Tauris Academic Studies, 2007), 102–106.
11. Gürkan, "Ottoman Corsairs in The Western Mediterranean", 117–133; Helder Carvalhal, and Roger Lee de Jesus, "The Portuguese Participation in the Conquest of Tunis (1535): a Social and Military Reassessment," in *Estudios Sobre Guerra y Sociedad en la Monarquía Hispánicaguerra Marítima, Estrategia, Organización y Cultura Militar (1500–1700)*, eds. Enrique García Hernán and Davide Maffi (Valencia: Albatros, 2017), 170–177.
12. John Francis Guilmartin, *Gunpowder and Galleys. Changing Technology and Mediterranean Warfare at Sea in the 16th Century* (London: Conway Maritime Press, 2003), 62–5; Halil Inalcik, "The Socio-Political Effects of the Diffusion of Fire-Arms in the Middle-East," in *War, Technology and Society in the Middle East*, eds. V.J. Parry and M.E. Yapp (London: Oxford University Press, 1975), 198.
13. Christine Isom-Verhaaren, "Barbarossa and His Army Who Came to Succor All of Us": Ottoman and French Views of Their Joint Campaign of 1543–1544," *French Historical Studies* 30.3 (2007): 411–412; Kenneth M. Setton, *The Papacy and the Levant, 1204–1571* (Philadelphia: The American Philosophical Society, 1978) 3: 523.
14. Roger Lee Pessoa de Jesus, *O Segundo Cerco de Diu (1546): Estudo de História Política e Militar* (Coimbra: Universidade de Coimbra, 2012), 127–131.
15. Bruce Ware Allen, *The Great Siege of Malta. The Epic Battle between the Ottoman Empire and the Knights of St. John* (Lebanon: ForeEdge 2015) 50–51; Cesáreo Fernández Duro, *Armada Española (Desde la Unión de los Reinos de Castilla y Aragón)* (Madrid: Sucesores de Rivadeneyra, 1895–1903) 1: 281–284.
16. J.M. Jacobi, *Histoire Générale de la Corse, depuis les Premiers Temps* (Paris: Aime Andre, Libraire, 1835), 1:329–33, https://play.google.com/books/reader?id=N7sBAAAAYAAJ&hl=ru&pg=GBS.PP9.
17. Steve Murdoch, *The Terror of the Seas? Scottish Maritime Warfare 1513–1713* (Leiden, Boston and Kohl: Brill Academic Publishers, 2010), 66.
18. V. Shirogorov, *Ukrainian War. The Armed Conflict for Eastern Europe in 16th–17th Centuries. Vol. 3 Head-to-head Offensive: Baltics – Lithuania – Steppes (In the Second Half of the Sixteenth Century)* (Moscow: Molodaya Gvardiya: 2019), 125–126.
19. Fernand Braudel, *The Mediterranean and the Mediterranean World in the Age of Philip II* (Oakland: University of California Press, 1996), 2:1000

20. James D. Tracy, *The Founding of the Dutch Republic. War, Finance, and Politics in Holland, 1572–1588* (Oxford and New York: Oxford University Press, 2008), 96–97
21. A.T. Shashkov, "The Beginning of the Takeover of Siberia," in *Studies on Russian History*, Iss. 4 *Eurasian Frontier* (Ekaterinburg: Volot, 2001), 34–35; R. G. Skrynnikov, *Yermak's Siberian Expedition* (Novosibirsk: Nauka, 1982), 155–157.
22. John Francis Guilmartin, *Galleons and Galleys* (London: Cassell, 2002), 155.
23. Tracy, *The Founding of the Dutch Republic*, 219–220.
24. R.O. Winstedt, "History of Malaya," *The Journal of Malaysian Branch of the Royal Asiatic Society (JMBRAS)*, Vol. 13, N 1(121) (1935): 81–82.

Christian and Islamic civilizations in the Mediterranean. Moscow's capture of Astrakhan in 1556 marked its ascendance over the remnants of the Golden Horde. The Netherlandish relief of Leiden in 1574 broke the Hapsburg stranglehold on the Protestant revolution. These amphibious actions constitute major geopolitical events for North and East Europe as well as the Near East and East Asia. The global upheavals of early-modern civilization carried the scent of deck-to-shore gunfire and the military revolution in amphibious warfare.

The geography of operations, displayed in Table 2, demonstrates the rapid dissemination of deck-to-shore gunfire tactics as an innovative amphibious leverage against hostile shores. In the early 1500s assault gunships were adopted in all four principal regions of amphibious operations: the North Atlantic and Baltic, the Mediterranean and the Black Sea, the European and Eurasian interior, and complex waterways connected with the Indian Ocean. The vessel-platforms of deck-to-shore gunfire differed, but its efficacy was proven globally.

In August 1496, 70 Swedish vessels carrying 2,500 to 3,000 men as landing troops, entered the river Narova and penetrated roughly 20 kilometers upstream. Amphibious warfare in the gunpowder age had now arrived at Moscow's frontier fortress of Ivangorod, founded in 1492 on the Livonian border opposite Narva. Moscow's Grand Prince Ivan III planned Ivangorod as the entrepot for Baltic trade in bulk agricultural and forest goods, and luxury furs. The prosperous town presented an excellent strategic and commercial objective for a Swedish counterstrike in retaliation for Moscow's siege of Vyborg in 1495.

The Swedish ships carried guns designed for deck-to-shore fire and for reducing opposing land batteries. The disembarking infantry toted firearms. Knut Jönsson Posse, a trusted servant of the Swedish regent Sten Sture the Elder, led the expedition. Ivangorod relied upon unfinished wooden bastions; its fort lacked sufficient guns; the garrison consisted of its settlers' militia, who generally lacked access to handguns. Posse's crack force easily overwhelmed Ivangorod's defenders. In a few hours, the attackers bombarded the riverbanks and the fortress, landed, transferred the guns to hastily erected batteries ashore, and breached Ivangorod. In an eight-hour assault, the garrison was overrun. The Muscovite light cavalry that undertook observation, screening the shore and border nearby, dared not interfere. The Swedes sacked Ivangorod thoroughly, massacring 3,000 inhabitants, civilians along with the garrison. In a couple of days, the Swedes had loaded the spoil, including precious furs, and departed.

The significance of Posse's amphibious operation against Ivangorod greatly exceeds that of the naval raids which occurred frequently in the Baltic. It was the exemplary direct assault demonstrating how dynamic were amphibious operations carried by troops equipped with handguns and field guns, trained in their application of this style of warfare and supported by the deck-to-shore gunfire.

The *Russian Chronicles* identify most of the ships of Posse's expedition as medium size sailing busses, and some of them as oared vessels.⁵⁰ The next substantive Swedish amphibious expedition was launched in 1555 against the Novgorodian fortress of Oreshek (Noteborg), where the river Neva, flowing into the Finnish Gulf, exited Lake Ladoga. The late Jan Glete supposed that the expedition was sustained by the Swedish galley fleet, purposely built for this kind of endeavor since 1540.⁵¹ The *Russian Chronicles* do indeed depict the type of Swedish ships as the buss again, the latter equipped with a pair of ordnance resting upon two-wheeled carriages,⁵² vessels strikingly different from the oared *snekke* and galley familiar to Russians.⁵³ Russians captured one of the Swedish busses, and it was found to carry a 150 man crew and four guns.⁵⁴ The Swedish fleet included galleys, but the *busses* were its backbone. Posse's campaign confirms the steadiness of the northern *buss* as a gun platform, on par with the *caravel*.

Sweden launched four (in 1574, 1577, 1579, and 1581) substantial amphibious ventures against the Moscow-controlled fortress of Narva. The Swedes abandoned gradually the galley either in the 1560s, concentrating their efforts upon open sea combat against Denmark,⁵⁵ or after the annihilation of their galley fleet in 1574, dashed upon a shoreline by a storm and finished off by Russian coastal troops.⁵⁶ The Swedes wafted later expeditions on various sailing ships of the *carrack*, *caravel*, and *buss* classifications. In 1577, the Swedish fleet razed Russian fortifications in the Narova's estuary by naval bombardment. However, Russian bank-based gunfire prevented the Swedes from sailing upriver beyond Narova. The 1581 campaign saw greater success; it was an escort action conducted in tandem with a Swedish land army under Pontus de la Gardie, which had traversed the frozen Finnish Gulf on foot, compass-navigating over the ice. That force swept the Narova's banks of Russian ordnance and enabled the Swedish fleet to proceed upstream and bombard the fortress of Narva.

The *caravel* was embraced in Northern Europe as superior for amphibious campaigns. The Hansa customized its *caravel*-inspired fleet ships after 1509 to safeguard its commercial ventures and literally to market violence by "renting"

---

50 *The Russian Illustrated Anthological Chronicle, M. P. Shumilov's Volume*, Лл. 541 ОБ., 544.
51 Glete, "Naval Power and Control of the Sea in the Baltic in the Sixteenth Century," 226; Glete, "Amphibious Warfare in the Baltic, 1550–1700," 128.
52 *The Russian Illustrated Anthological Chronicle, the Synod Volume*, СИН-962, Л. 209.
53 Zagoskin, *Russian Waterborne Communications*, 395–398; Sorokin, *The Waterways and Shipbuilding in North-Western Rus' in the Middle Ages*, 128.
54 Zagoskin, *Russian Waterborne Communications*, 398.
55 Glete, "Amphibious Warfare in the Baltic, 1550–1700," 130.
56 Petrov, *The City of Narva*, 102.

**Figure 2:** Swedish *buss*-type craft support the siege of Muscovy's fortress of Oreshek (Nöteborg) at the river Neva's egress from Lake Ladoga, Karelia, in 1555. The *buss*-type ships transported and supported the besiegers by their deck artillery bombarding the fortress. The

it to third parties. Denmark launched two *kraffells* in the same year. The Swedes adopted *caravels*, initially from the Hansa, and then reverse-engineered them in their own shipyards.[57] In the Low Countries, the *boyer* and *flieboot* (flyboat in English) were introduced at the beginning of the 1500s. These swift, nimble, and seaworthy craft utilized shallow draughts, low centers of gravity, light aft upper-works, plus fore and aft rigs.[58] The flyboat of a 40 to 140 tonnage deadweight carried six to 20 guns, including a pair or more bronze muzzle-loaders on two-wheeled carriages, the balance being wrought-iron breech-loaders. It carried a crew of 40 to 140 marines. "The majority of guns were too light to seriously damage proper warships,"[59] however, flyboat were perfect assault gunships for amphibious operations. The amphibious capture of Den Brielle by the rebel "Sea Beggars" under Lumey de La Marck in 1572 was accomplished using flyboats. The Dutch used various flat-bottomed rowing barges for estuarine and riverine operations that could range deeper inland, over rivers and inundated plains. These barges boasted from two to eight heavy and medium bronze guns mounted upon two-wheeled carriages in a shielded bow and from four to six wrought-iron swivel guns along the sides.[60] In October 1574, a rebel flotilla of 70 to 80 vessels, with 3,000 to 4,000 men aboard under Louis de Boisot, burst through the Spanish siege lines to relieve Leiden. The unsuccessful relief of Antwerp in May 1585 witnessed another form of amphibious attack, mine-boats employed as a kind of surface torpedoes. Dutch amphibious tactics (and technologies) empowered their resistance in the Eighty Years' War. These examples also further illustrate the vitality and diversity of amphibious warfare during the military revolutionary era.

In the riverine topography of Russia, the *strug* and *nasad*, traditional types of flat-bottomed rowing vessels, proved adequate for amphibious warfare. These vessels generally had a length of 10 to 25 meters, a cargo capacity 5 to 50 tons, and carried 20 to 60 crewmen.[61] *Strugs* and *nasads* were fitted with guns

**Figure 2** (continued)
siege collapsed due to the Swedes' inability to safeguard their vallations against Russian relief forces.[62] The image is reproduced by courtesy of the Department of Manuscripts, the State Historical Museum of Russia, Moscow. © «Исторический музей».

57 Zwick, "Bayonese Cogs, Genoese Carracks, English Dromons and Iberian Carvels," 669.
58 Unger, *The Ship in the Medieval Economy*, 262–263.
59 de Groot, *Dutch Navies of the 80 Years' War 1568–1648*, 10, 13.
60 de Groot, *Dutch Navies*, 10, 12.
61 Shubin, *The Volga and its Navigation*, 76, 81.
62 *The Russian Illustrated Anthological Chronicle of the Sixteenth Century, the Synod Volume*, СИН-962, Л. 209.

**Figure 3:** Cannon mounted upon two-wheeled carriages were well-designed to facilitate amphibious gunnery in Eastern and Western Europe. Such artillery was adapted for deck-to-shore gunfire, especially in instances of rapid deployment on landing. Here, Muscovite

in the late 1400s, a timeline suggesting innovatory resonance with Western Europe. The *Chronicle*'s miniature of Moscow's 1552 assault on Kazan depicts these craft and their waterborne artillery (utilizing the earlier-mentioned two-wheeled gun carriages).[63] By 1616, Moscow's purpose-built flotilla in Astrakhan consisted of 110 *strugs* for transportation, and 150 "fighting" *strugs* boasting from two to four guns, served by crews of 60 to 80 men each.[64] The amphibious landing at Kazan, under prince Alexander Gorbaty, culminated in combat upon the fords across the River Bulak. This amphibious clash commenced Moscow's final siege of Kazan, and ultimately the conquest of the Kazan Khanate in 1552. Four years later, Ivan Cheryemisinov's amphibious force stormed the town of Astrakhan on the river Volga estuary, defeated Tatar troops, and forced the submission of the Astrakhan Khanate to Moscow. In 1582, an amphibious expedition of Yermak overcame fierce resistance on the Chuvash cape of the river Irtysh. Yermak's *strugs* pummelled opposition on the riverbank with effective firepower. Muscovite handgunners, upon landing, massed their gunfire and annihilated the troops of the Siberian Khanate. Yermak then overran the Khanate's capital of Isker thus opening Siberia for Russian expansion to the Pacific Ocean. These actions prove Russian expertise in estuarine warfare and riverine warfare. By adapting the gunpowder revolution to traditional technologies (and ways of war), such as the adaptations of *strugs* and *nasads*, Moscow mastered diverse variations of inland amphibious warfare in its expansionary strategic culture.

Clearly, though, success in using amphibious assault ships was greatly determined by the degree of resistance they met, and the strength of the objective's defenses. Stone fortifications along the Mediterranean coastlines, improved by bastion-type *trace italienne* design, called for substantial deck-to-shore fire from heavy guns before amphibious attack might be attempted. The amalgam of earthen and wooden fortifications found in the North Atlantic regions, the Baltic, and continental European interior required medium artillery to destroy them.

---

**Figure 3** (continued)
amphibious troops in *strug*-type ships assail the Volga's riverbank at Kazan in 1552. Prefabricated wooden field fortifications helped to secure a bridgehead upon disembarkation.[65] The prominence with which that equipment is depicted in the foreground suggests that the artist wished to draw attention to these ingenious portable redoubts. The image is reproduced by courtesy of the Department of Manuscripts, the State Historical Museum of Russia, Moscow. © «Исторический музей».

---

63 *The Russian Illustrated Anthological Chronicle, the Tsardom Book*, СИН-149, Л. 525.
64 Tushin, *The Russian Navigation in the Caspian*, 37–38.
65 *The Russian Illustrated Anthological Chronicle, the Tsardom Book*, СИН-149, Л. 525.

Smaller guns and heavy firearms worked against shoreline-crowded defenses of many Asian targets. In other words, the amphibious gunpowder revolution must be measured against the defensive capabilities of regional defenses of widely varying styles.

Sixteenth-century naval gunnery, then and by the estimation of many historians today, functioned as "ship-killing." Due to their relatively poor buoyancy, vessels could be sunk by heavy piercing shot aimed at their waterline.[66] However, a few rounds never destroyed coastal fortifications or dislodged fortified shoreline positions. First, defenses along a littoral (or lake or river) required fracturing to disable them as much as possible; second, defensive installations capable of hitting landing craft had to be neutralized; third, marines being most vulnerable during landing operations had to be protected by offshore barrages that suppressed enemy firepower. An amphibious assault vessel itself, as well larger support vessels in deeper water, might maintain prolonged and intensive bombardment of enemy positions whether the adversaries were on the littoral itself or ensconced in coastal fortifications farther inland. Ships' gunfire had to synchronize with the troops' precarious landing, as they consolidated themselves, traversed the shore, and moved forward. To accomplish these tasks assault craft carried weaponry of different calibers and purpose, from larger cannon with high velocity shot to lesser-sized ordnance firing round shot, grapeshot, and scattering shot. Abundant artillery also compensated for slow recharge. That said, too many cannons might compromise a boat's shallow draft that accommodated delivering firepower as close as possible to enemy targets ashore. The closer to shore, the more accessible were important physical features, including shoal littorals, narrow harbors and roads.

Navigating the vessel coincided with the aiming of the fixed ordnance in accordance with the naval tactics of the era, e.g., either in the line-abreast or line of battle formations. While the stationary nature of shoreline defenses was a positive factor, deck-to-shore gunfire required aiming one's ordnance independently from the vessel, the latter possibly jammed in coastal waters or moored to a shore. It is for that reason the two-wheel carriage became universal for heavy guns onboard of the assault ships of amphibious forces whilst the purpose-built ships of marine combat utilized the slide platform or four-wheel carriage.[67] In sum, assault ships of the 1500s designed for deck-to-shore gunfire emerged from complex technical and tactical lessons learned in battle. The amphibious art of

---

66 Rodger, "The Development of Broadside Gunnery," 303.
67 Parker, The Military Revolution, 95; Salgado, "Portuguese Galleon's Armament," 279.

**Figure 4:** A Dutch fleet under Ernest Casimir, count de Nassau-Dietz, masses riverine-based firepower in bombarding Spanish defenders of the Scheldt protective dam near Antwerp on 16 May 1605. The tactic offers evidence of the destructive potential of deck-to-shore gunfire during the early modern era.[68] The engraving is from Willem Baudart's *Les guerres de Nassau, descriptes par Guillaume Baudart* published by M. Colin (Amsterdam 1616), editor's collection.

war that exploited gunpowder must be distinguished from sea combat; different species of navies came into existence as amphibious tactics evolved.

## Amphibious Professionalism

At the turn of the century, the new amphibious "gunpowder tactics" made "marines" highly effective against land objectives, thus pioneering a new way of war. This transformation was evident in landing parties themselves, which

---

[68] Baudart, Les guerres de Nassau, T.2, 386–388.

mustered substantial firepower. Gunpowder weaponry was found in the ranks of landing troops commencing at end of the 1400s. Swedish troops, attacking Ivangorod in 1496, unladed siege guns and harquebuses from their ships. Gonzalo Fernandes de Cordoba pioneered the use of firearms when storming the island of Cephalonia and its capital, St. George, in 1500. Cordoba experimented with gunpowder tactics well before his victories at Cerignola in April 1503 and at the river Garigliano the following December. The former battle is oftentimes heralded as the first major victory attributable to infantrymen using handguns in novel fashion and presaging future techniques. It was upon Cephalonia's beach that Cordoba's talented subordinate, Pedro Navarro, honed his expertise in deploying firearms. The French, renowned for their skill with artillery during their descent into Italy in 1494, practiced the new amphibious gunpowder tactics breaching the fortifications on the island of Lesbos in August 1501. The besiegement of its capital of Mytilene, led by Philippe de Clèves, was aborted only when other adept experts in the use of firearms, the Ottomans, landed relief troops under Hersekzade Ahmed Pasha.

Gunpowder-related innovations, which enhanced firepower through the 1500s, are familiar: harquebuses (long-barrelled handguns fitted with matchlocks), carriage-mounted guns of large caliber replete with aiming devices, pre-measured charges of corned powder, etc. Technical innovations required new skills to deploy guns in battery formation, as well as to drill and practice volley fire from massed infantry. Firearms coupled with the tactical potential of troops made possible not just the military revolution on land, as Michael Roberts argued, but on the sea and the littoral as well. Changes wrought by gunpowder on amphibious operations initially had limited effect on land warfare. Heavy cavalry couching lances while supported by pikemen and halberdiers, operating in close formations, remained ubiquitous in Western Europe. Longbows and crossbows had their place in Western armies. Bows were a staple of mailed horse and foot in the East. Fighting wagons multiplied in Eastern Europe and the Near East as well. Comparatively speaking, from the perspective of a military revolutionary conceptual framework, land warfare exhibited striking inertia. Handgunners seconded by artillery became dominant components of amphibious landing parties almost overnight. The reason behind the precipitous shift is clear. Due to the gunpowder revolution, disembarking troops could now reasonably challenge a land army ashore. They attained comparatively equal defensive and offensive capabilities.

Also emblematic of transformations in personnel that complemented the utility of assault ships, the Swedish attack upon Ivangorod in 1496 was executed by professional troops. Although small in number, these marine warriors were well-equipped, and trained, in contrast with the medieval *ledung* and *uppbad* militias. Pole-weapon armed peasant *uppbad* did not seize Ivangorod, nor did *ledung* fishing boats bring the amphibious party to its walls. Mixed mercenary regulars and hired experienced sailors embarked upon sailing ships that were Swedish state-owned and rented from the Hansa. For the tradition-bound Swedish military, a professional amphibious force was a phenomenon of the most revolutionary kind. The historiographical legend of the militia's basis as the foundations of Sweden's regular army and fleet during the sixteenth century deserves rethinking when amphibious warfare is considered within the equation.

Spanish amphibious operations during the early 1500s saw initiation of regular foot "columns" at the instigation of king Ferdinand II and professional soldiers such as Pedro Navarro. The Portuguese general Afonso de Albuquerque "endeavored to create organized and professionally-led bodies of military"[69] when attacking Malacca and Benasterim in 1511. Professional harquebusiers spearheaded Edward Seymour, the duke of Somerset's, advance upon Leith in 1544.[70] Moscow's amphibious onslaught against the khanates of Kazan, Astrakhan, and the Crimea in the 1550s introduced Russian professional foot, *streltsy*, by command of tsar Ivan IV. Skilled, trained troops were now the order of the day. The Swedish amphibious capture of Narva in 1581, after multiple failures, brought different consequences when the professionalized landing troops collaborated with veteran mercenaries. Many of the latter were Scots[71] or continental professionals, including the leader of the expedition, Pontus de la Gardie, a French turncoat captured from the Danes and naturalized in Sweden.

Amphibious tactics based on the gunpowder revolution embraced what were in effect professional soldiers because higher levels of firearms training and "marine" disembarkation skills were essential for success. This marriage of gunpowder and new tactics matches the Roberts military revolution formula. Escalating amphibious warfare, with its now greater capacity for decisive victory, created careers for regulars. Unlike land warfare, it was the kind of war in which professionals were never interchangeable with feudal levies and communal militias.

## The "Aside Build-up"

The "aside build-up" can be defined as an amphibious landing at a staging point distant from the objective. It is therefore an indirect approach to attack, unlike

---

69 Thomaz, "Factions, Interests and Messianism," 104.
70 Fissel, *English Warfare*, 27; Potter, *Henry VIII and Francis I*, 109, 222–223.
71 Grosjean, *An Unofficial Alliance*, 16–18.

the "direct assault," discussed below. Coordinating tactically sufficient firepower from vessels and marines was the primary challenge to the novel fighting techniques of the gunpowder age. Firearms' slow rate of fire, their limited striking distance, inaccuracy, and unwieldiness, were best compensated for by assembling substantial critical masses of gun-toting troops upon the littoral. Hails of lead could alleviate these shortcomings in the 1500s. Engineers' defensive preparation of terrain, such as the erection of barriers against cavalry and pike formations likewise facilitated. Both solutions involved accumulation of troops and their proper deployment, requiring time and room to maneuver.

**Figure 5:** Flemish, Huguenot, English, and Scottish "marines" commanded by Genlis de Poyet stormed Geertruidenberg by executing an "aside build-up" deployment on 29 August 1573. They captured the town gates and overwhelmed the Spanish garrison. In this case the two-wheeled guns on the decks of the attackers' ships were muted in order to surprise the defenders.[72] The engraving is from Willem Baudart's *Les guerres de Nassau, descriptes par Guillaume Baudart* published by M. Colin (Amsterdam 1616), editor's collection.

---

72 Baudart, *Les guerres de Nassau*, T.1, 131–32.

These prerequisites shaped operational patterns of amphibious warfare, resulting in what the present author has dubbed the "aside build-up." The core of the plan was to select landing zones comparatively remote from the objective and insulated from swift counterattack. Defenders most frequently tried to stymie disembarking troops at the perilous juncture when they traversed water to terra firma; the "aside build-up" used distance and surprise for safe deployment. This tactic enhanced amphibious warfare's prospects for victory (and fulfilment of more aggressive strategies). An abundance of objectives was relatively protected from overland offensives but vulnerable to amphibious attack's power projection. After 1500 the focus of landing operations of amphibious warfare shifted from "escort" to "aside build-up" operational design.

## Transforming the "Escort Design"

In the sixteenth century, the conjunctive amphibious "escort" model developed along with the new pattern of interaction between navies and land armies. Table 3, "The evolution of amphibious escort in the 1500s,"[73] demonstrates the increase in application, and rate of success of the "escort" innovation, referencing too instances of deck-to-shore gunfire as part of the "escort" function. The results were striking.

Table 3 displays 53 amphibious operations of the "escort design" carried out in the sixteenth century; 13 of them involved prominent deck-to-shore gunfire, eight of which succeeded. Three failed due to the inability of riverine flotillas (the Venetians' in 1509; Hapsburgs' in 1537 and Danzig's in 1577) to bring to bear firepower upon the enemy (the Ferrarese, Turks, and Poles) who in turn decimated friendly land forces at the battles of, respectively, Polesella; Valpovo; and Lubschau. The English victories of Pinkie Cleugh in 1547 over the Scots, and Gravelines in 1558 over the French (the latter achieved in alliance with the Netherlands), are "classic" gunpowder revolution escort operations, as is the Portuguese triumph in their sortie from Tangier and defeat of the Moroccans in 1574. During all three events barrages from assault gunships disorganized the enemy's land forces and shifted the advantage to the escorting troops. Additionally, the Swedish naval escort of 1581, assisting its army at Narva was a variation on the trio of actions mentioned above, wherein the pattern of deck-to-shore gunfire escort accompanied successful besiegement. Simultaneously, amphibious escorts proved well suited to coalitional warfare, sixteen operations, including nine

---

[73] See the references to the sources of data on the operations in the Tables.

Table 3: Evolution of amphibious escort in the 16th century.

| | Data | Name of operation | Attacker | Defender | Environment | Use of escort | Leader | Outcome; reason | Note |
|---|---|---|---|---|---|---|---|---|---|
| 1 | 1509 | Advance on Ferrara | Venice | Ferrara | Riverine | Gunfire,[74] Transport[75] | Màr Angelo Trevisàn | Failure; lack of gunfire; superior enemy | 1 |
| 2 | 1511 | Siege of Venlo | England; Netherlands | Duke of Guelders | Marine | Transport, coalition[76] | Edward Poynings | Failure; miscoordination | 2 |
| 3 | 1512 | Invasion of Guyenne | England; Spain | France | Marine | Transport, coalition | Thomas Grey Marquis of Dorset | Failure; miscoordination | 3 |
| 4 | 1513 | Siege of Therouanne and Tournai | England; Holy Roman Empire | France | Marine | Transport, coalition | Henry VIII; Maximilian I | Success | 4 |
| 5 | 1514 | Storm of Bougie | Turkey | Spain | Marine | Aside build-up | Oruc Barbarossa | Failure; lack of gunfire | 5 |

---

74 Gunfire is the deck-to-shore artillery assault.
75 Transportation is the arrival on a friendly-controlled beach;
76 Coalition is the origin of an allied land army and amphibious force from different polities;

| | | | | | | | |
|---|---|---|---|---|---|---|---|
| 6 | 1515 | Siege of Bougie | Turkey; Hafsid Tunisia | Spain | Marine | Landing[77] | Oruc Barbarossa; Ahnad bin al-Kadi | Failure; lack of gunfire |
| 7 | 1516 | Storm of Penon de Alger | Turkey; Algerian locals | Spain | Marine | Landing | Oruc Barbarossa | Failure; lack of gunfire |
| 8 | 1520 | Taking of Stockholm | Denmark; Swedish loyalists | Swedish separatists | Marine | Landing,[78] revolution[79] | Christian II | Success |
| 9 | 1522 | Taking of Belgrade | Turkey | Hungary | Riverine | Blockade | Piri Mehmed Pasha | Success |
| 10 | 1522 | Invasion of Artois and Picardy | England; H.R. Empire | France | Marine | Transport, coalition | Thomas Howard earl of Surrey | Success |
| 11 | 1523 | Advance on Paris, storm of Montdider | England; H.R. Empire | France | Marine | Transport, coalition | Charles Brandon duke of Suffolk | Success |

(continued)

77 Landing is the arrival on a hostile beach.
78 Blockade is the breaking of an enemy's communications.
79 Revolution is the change of a regime inspired by a landing.

**Table 3** (continued)

| | Data | Name of operation | Attacker | Defender | Environment | Use of escort | Leader | Outcome; reason | Note |
|---|---|---|---|---|---|---|---|---|---|
| 12 | 1523 | Foundation of Vasilsursk | Moscow | Kazan Khanate | Riverine | Landing | Shigalyey khan of Kasimov | Success | 12 |
| 13 | 1524 | Advance on Kazan | Moscow | Kazan Khanate | Riverine | Landing | Ivan Byelsky; Ivan Simsky aka Habar | Failure, superior enemy | 13 |
| 14 | 1526 | Advance on Naples | France; Neapolitan pretender | Spain; Neapolitan loyalists | Marine | Gunfire, blockade, coalition | Andrea Doria; Louis de Lorraine comte de Vaudemont | Success | 14 |
| 15 | 1527 | Taking of Genoa | France; Genoese Doria's party | Genoa | Marine | Gunfire, blockade, revolution | Pedro Navarro; Andrea Doria | Success | 15 |
| 16 | 1528 | Advance on Naples | France; Neapolitan pretender; Venice | Spain; Neapolitan loyalists | Marine | Gunfire, blockade, coalition | Odet de Foix vicomte de Lautrec; Pedro Navarro; Andrea Doria | Failure; miscoordination | 16 |
| 17 | 1529 | Taking of Penon de Alger | Turkey | Spain | Marine | Landing | Hayreddin Barbarossa | Success | 17 |
| 18 | 1530 | Advance on Kazan | Moscow | Kazan Khanate | Riverine | Landing | Ivan Byelsky; Mikhael Glinsky | Failure; superior enemy | 18 |

| | | | | | | | |
|---|---|---|---|---|---|---|---|
| 19 | 1535 | Taking of Coppenhagen and Malmo | Sweden; Denmark | Danish pretender; Hansa | Marine | Gunfire, blockade, revolution | Christian III; Johann Rantzau | Success |
| 20 | 1537 | Siege of Essek | Holy Roman Empire | Turkey | Riverine | Gunfire | Hans Katzianer | Failure; lack of gunfire |
| 21 | 1538 | Siege of Diu | Turkey; Gujarat Sultanate | Portugal | Marine | Landing, blockade, coalition | Hadim Suleyman Pasha | Failure; lack of gunfire |
| 22 | 1539 | Siege of Castelnuovo | Turkey | Spain | Marine | Landing, blockade | Hayreddin Barbarossa | Success |
| 23 | 1543 | Siege of Nice | France; Turkey | Savoy; Spain | Marine | Landing, blockade, coalition | Francois de Bourbon count of Enghien; Hayreddin Barbarossa | Failure; battle defeat |
| 24 | 1543 | Siege of Mostargenem | Spain | Turkey | Marine | Blockade | Martin Alonso Fernandez de Cordoba | Failure; lack of gunfire |
| 25 | 1544 | Siege of Montreuil | England; Netherlands; H.R. Empire | France | Marine | Transport, coalition | Thomas Howard duke of Norfolk | Failure; miscoordination |
| 26 | 1544 | Siege of Boulogne | England; Netherlands; H.R. Empire | France | Marine | Transport, coalition | Henry VIII; Charles Brandon Duke of Suffolk | Success |

(continued)

Table 3 (continued)

| Data | | Name of operation | Attacker | Defender | Environment | Use of escort | Leader | Outcome; reason | Note |
|---|---|---|---|---|---|---|---|---|---|
| 27 | 1545 | Landing at Dumbarton | France; Scotland | England | Marine | Transport, coalition | Jacques Montgommery de Lorges | Success | 27 |
| 28 | 1547 | Siege of Mostargenem | Spain | Turkey | Marine | Blockade | Martin Alonso Fernandez de Cordoba | Failure; lack of gunfire | 28 |
| 29 | 1547 | battle of Pinkie Cleugh | England | Scotland | Marine | Gunfire | Edward Seymour; Edward Fiennes de Clinton | Success | 29 |
| 30 | 1551 | Foundation of Sviyazhsk | Moscow | Kazan Khanate | Riverine | Landing | Shigalyey khan of Kasimov | Success | 30 |
| 31 | 1552 | Taking of Kazan | Moscow | Kazan Khanate | Riverine | Landing | Ivan IV; Alexander Gorbaty | Success | 31 |
| 32 | 1555 | Storm of Calvi | France; Turkey | Genoa, Spain | Marine | Landing, coalition | Paul de Thermes; Piali Pasha | Failure; lack of gunfire | 32 |
| 33 | 1555 | Storm of Bastia | France; Turkey | Genoa; Spain | Marine | Landing, coalition | Piali Pasha | Failure; lack of gunfire | 33 |
| 34 | 1555 | Siege of Oryeshek and Korela | Sweden | Moscow | Marine | Landing | Henrik Klasson Horn; Jakob Tordsson Bagge | Failure; lack of gunfire | 34 |

| | | | | | | | | |
|---|---|---|---|---|---|---|---|---|
| 35 | 1558 | Siege of Mostargenem | Spain | Turkey | Marine | Blockade | Martin Alonso Fernandez de Cordoba | Failure; lack of gunfire | 35 |
| 36 | 1558 | Battle of Graveline | England; Spain; Netherlands | France | Marine | Gunfire, coalition | Edward Fiennes de Clinton | Success | 36 |
| 37 | 1560 | Storm of Leith | England | France; Scotland | Marine | Landing | Thomas Howard duke of Norfolk; William Winter | Failure; battle defeat | 37 |
| 38 | 1561 | Taking of Reval | Sweden | Poland-Lithuania | Marine | Landing; revolution | Klas Kristersson Horn | Success | 38 |
| 39 | 1562 | Taking of Le Havre, relief of Rouen | England; French Huguenots | French Catholics | Marine | Landing, revolution | Ambrose Dudley earl of Warwick | Failure; superior enemy | 39 |
| 40 | 1567 | Struggle for Mesopotamian marshes | Turkey | sheyh Ibn' Ulyan' rebels | Riverine | Landing; gunfire | Iskender Pasha | Success | 40 |
| 41 | 1569 | Advance on Astrakhan | Turkey; Crimea | Moscow | Riverine | Landing | Kasym Bey; Khan Devlet Geray | Failure; lack of gunfire | 41 |
| 42 | 1571 | Taking of Bar, Sopot, Ulcinj, Budva | Turkey | Venice | Marine | Blockade, landing | Ahmed Pasha; Muedzinzade Ali Pasha; Pertev Pasha | Success | 42 |

(continued)

Table 3 (continued)

| Data | | Name of operation | Attacker | Defender | Environment | Use of escort | Leader | Outcome; reason | Note |
|---|---|---|---|---|---|---|---|---|---|
| 43 | 1572 | Taking of Den Brielle, riot in Zeeland | Sea Beggars; Netherlandish rebels | Netherlandish loyalists; Spain | Marine | Landing, revolution | Lumey de La Marck | Success | 43 |
| 44 | 1574 | Combat of Tangier | Portugal | Morocco | Marine | Gunfire | Sebastian I | Success | 44 |
| 45 | 1575 | Struggle for Genoa | Party of 'old nobility' | Party of 'new nobility' | Marine | Landing, revolution | Giovanni Andrea Doria | Success | 45 |
| 46 | 1577 | Battle of Lubschau | Danzig | Poland | Riverine | Gunfire | Hans Winckelbruch von Köln | Failure; lack of gunfire | 46 |
| 47 | 1580 | Taking of Smerwick | England | Irish rebels; Papal expedition | Marine | Gunfire, blockade | Willian Winter; Artur Grey de Wilton | Success | 47 |
| 48 | 1581 | Attack on Narva | Sweden | Moscow | Marine | Landing, gunfire | Pontus de la Gardie; Clas Eriksson Fleming | Success | 48 |
| 49 | 1583 | Relief of Kaffa, Struggle for Crimea | Turkey; Crimean Pro-Turkish party | Crimean independence party | Marine | Landing, revolution | Ozdemiroglu Osman Pasha; Kilic-Uluc Ali Pasha | Success | 49 |

| | | | | | | | | |
|---|---|---|---|---|---|---|---|---|
| 50 | 1590 | Taking of Honfleur | French royalists | French opposition | Marine | Gunfire, revolution | Henry IV | Success | 50 |
| 51 | 1594 | Storm of El Lion | England; France | Spain | Marine | Landing, blockade, coalition | Henry Norreys | Success | 51 |
| 52 | 1594 | Storm of Enniskillen | England; Irish loyalists | Irish rebels | Riverine | Landing | John Dowdall | Success | 52 |
| 53 | 1601 | Battle of Kinsale | Spain; Irish insurgents | England; Irish loyalists | Marine | Landing, revolution | Juan del Aguila y Arellano | Failure; superior enemy | 53 |

Notes:

1. Federico Moro, "Venetia Rules the Rivers. La Geo-Strategia Fluviale Veneziana (1431–1509)." *Nuova Antologia Militare* Fascicolo 2.7 (June 2021): 43–53, https://www.nam-sism.org/3.2%20%20fascicoli.html.
2. James Raymond, *Henry VIII's Military Revolution. The Armies of Sixteenth-century Britain and Europe* (London and New York: Tauris Academic Studies, 2007), 15–16.
3. Raymond, *Henry VIII's Military Revolution*, 16–17; David Starkey, *Six Wives. The Queens of Henry VIII* (New York: Harper Perennial, 2004), 128.
4. Raymond, *Henry VIII's Military Revolution*, 17–18.
5. Emrah Safa Gürkan, "Ottoman Corsairs in The Western Mediterranean and their Place in The Ottoman – Habsburg Rivalry (1505–1535)," Master's Thesis (Bilkent University, Ankara, 2006), 47–48.
6. Gürkan, "Ottoman Corsairs in The Western Mediterranean," 48–49.
7. Gürkan, "Ottoman Corsairs in The Western Mediterranean," 52–53.
8. Jan Glete, "Naval Power and Control of the Sea in the Baltic in the Sixteenth Century," in *War at Sea in the Middle Ages and the Renaissance*, ed. John B. Hattendorf and Richard W. Unger (Woodbridge: Boydell Press, 2003), 222–23; Paul Douglas Lockhart, *Denmark, 1513–1660. The Rise and Decline of a Renaissance Monarchy* (Oxford and New York: Oxford University Press, 2007), 15.
9. Mesut Uyar and Edward J. Erickson, *A Military History of the Ottomans. From Osman to Ataturk* (Santa Barbara and Oxford: Praeger Security International, 2009), 73.
10. Raymond, *Henry VIII's Military Revolution*, 20.
11. Raymond, *Henry VIII's Military Revolution*, 21–22.

12. V. Shirogorov, Ukrainian War. The Armed Conflict for Eastern Europe in the 16th–17th Centuries. Vol. 1 Melee of Rus (To the Middle of the 16th Century) (Moscow: Molodaya Gvardiya, 2017), 754.
13. Shirogorov, Ukrainian War. Vol. 1 Melee of Rus, 755.
14. Michael Edward Mallett and Christine Shaw, The Italian Wars 1494–1559. War, State and Society in Early Modern Europe (London and New York: Routledge, 2012), 158.
15. Mallett and Shaw, The Italian Wars 1494–1559, 165.
16. Mallett and Shaw, The Italian Wars 1494–1559, 166–168.
17. N. A. Ivanov, Ottoman Conquest of the Arabian Nations (Moscow: RAN Eastern Literature, 2001), 86.
18. Shirogorov, Ukrainian War. Vol. 1 Melee of Rus, 758–759.
19. Jan Glete, Warfare at Sea, 1500–1650. Maritime Conflicts and the Transformation of Europe (Abingdon and New York: Routledge, 1999), 118–119; Lockhart, Denmark, 1513–1660, 28.
20. James D. Tracy, Balkan Wars. Habsburg Croatia, Ottoman Bosnia, and Venetian Dalmatia, 1499–1617 (New York and London: Rowman & Littlefield Publishers, 2016), 121–22
21. Mansel Longworth Dames, "The Portuguese and Turks in the Indian Ocean in the Sixteenth Century," The Journal of the Royal Asiatic Society of Great Britain and Ireland No. 1 (Jan. 1921): 18–19.
22. Paolo Giovio, La Seconda Parte dell'Istoria del suo Tempo (Venetia, 1560), 485; John H. Pryor, Geography Technology and War Studies in the Maritime History of the Mediterranean 649–1571 (Cambridge, New York and Melbourne: Cambridge University Press, 1988), 80; Idan Sherer, Warriors for a Living. The Experience of the Spanish Infantry in the Italian Wars, 1494–1559 (Leiden; Boston and Kohl: Brill Academic Publishers, 2017), 252–253.
23. Christine Isom-Verhaaren, "Barbarossa and His Army Who Came to Succor All of Us": Ottoman and French Views of Their Joint Campaign of 1543–1544," French Historical Studies No.30 (3) (2007): 411–412.
24. Paul Russ, La Domination Espagnole a Oran sous le Gouvernement du Comte d'Alcaudete (1534–1558) (Paris: Editions Bouchêne, 1998), 119–120
25. David Potter, Henry VIII and Francis I. The Final Conflict, 1540–1547 (Leiden, Boston and Kohl: Brill Academic Publishers, 2011), 172–180
26. Potter, Henry VIII and Francis I, 184–189,197–204.
27. Potter, Henry VIII and Francis I, 113–114.
28. Russ, La Domination Espagnole a Oran, 135–136.
29. Mark Charles Fissel, English Warfare 1511–1642 (London and New York: Routledge, 2001) 32–33; Gervase Phillips, "Anglo-Scottish Wars: Battle of Pinkie Cleugh," Military History Magazine 14.3 (1997): 42–49.

30. V. Shirogorov, *Ukrainian War. The Armed Conflict for Eastern Europe in 16th–17th Centuries. Vol. 3 Head-to-head Offensive: Baltics – Lithuania – Steppes (In the Second Half of the Sixteenth Century)* (Moscow: Molodaya Gvardiya: 2019), 92.
31. Shirogorov, *Ukrainian War.* Vol. 3 *Head-to-head Offensive*, 98–130.
32. Mallett and Shaw, *The Italian Wars 1494–1559*, 266; Xavier Poli, *Histoire Militaire des Corses au Service de la France* (Ajaccio: Dominique de Peretti, 1898), 79–80, https://gallica.bnf.fr/ark:/12148/bpt6k6564540j.textelmage.
33. Mallett and Shaw, *The Italian Wars 1494–1559*, 266; Poli, *Histoire Militaire des Corses*, 79–80.
34. Shirogorov, *Ukrainian War.* Vol. 3 *Head-to-head Offensive*, 174.
35. Russ, *La Domination Espagnole a Oran*, 169–176.
36. Charles Oman, *A History of the Art of War in the Sixteenth Century* (London: Greenhill Books, 1989), 280.
37. Fissel, *English Warfare*, 117–119.
38. Shirogorov, *Ukrainian War.* Vol. 3 *Head-to-head Offensive*, 209–212.
39. Paul E.J. Hammer, *Elizabeth's Wars. War, Government and Society in Tudor England, 1544–1604* (New York: Palgrave Macmillan, 2004), 64–65.
40. I. Metin Kunt, "An Ottoman Imperial Campaign: Suppressing the Marsh Arabs, Central Power and Peripheral Rebellion in the 1560s," *The Journal of Ottoman Studies* Vol. 43 (June 2014): 6–12.
41. Shirogorov, *Ukrainian War.* Vol. 3 *Head-to-head Offensive*, 366–369.
42. Noel Malcolm, *Agents of Empire. Knights, Corsairs, Jesuits and Spies in the Sixteenth-Century Mediterranean World* (Oxford and New York: Oxford University Press, 2015), 140–148.
43. James D. Tracy, *The Founding of the Dutch Republic. War, Finance, and Politics in Holland, 1572–1588* (Oxford and New York: Oxford University Press, 2008) 80–81; D. J. B. Trim, "Transnational Calvinist Cooperation and 'Mastery of the Sea' in the Late-Sixteenth Century," in *Ideologies of Western Naval Power, c. 1500–1815*, eds. J.D. Davies, Alan James and Gijs Rommelse (London and New York: Routledge, 2020), 169–170.
44. Luis Costa e Sousa, "From Tangier to Alcácer Quibir: The Portuguese Military Revolution (Re)visited," *Portuguese Studies Review (SP)* (2015): 13–19.
45. Thomas Kirk, "Chapter Three. Giovanni Andrea Doria: Citizen of Genoa, Prince of Melfi, Agent of King Philip II of Spain," in *Double Agents: Cultural and Political Brokerage in Early Modern Europe*, eds. Marika Keblusek and Badeloch Vera Noldus (Leiden, Boston and Kohl: Brill, 2011), 6566.
46. Radosław Sikora, *Lubieszów 17 IV 1577* (Zabrze: Wydawnictwo Inforteditions, 2005).
47. Hammer, *Elizabeth's Wars*, 109; N. A. M. Rodger, "The Development of Broadside Gunnery, 1450–1650", *The Mariner's Mirror* 82.6 (August 1996): 307.
48. Shirogorov, *Ukrainian War.* Vol. 3 *Head-to-head Offensive*, 713–714.

49. Carl M. Kortepeter, *Ottoman Imperialism During the Reformation: Europe and the Caucasus* (New York: New York University Press, 1972), 85–86.
50. D.J.B. Trim, "Medieval and Early-Modern Inshore, Estuarine, Riverine and Lacustrine Warfare," in *Amphibious Warfare 1000–1700. Commerce, State Formation and European Expansion*, eds. D.J.B. Trim and Mark Charles Fissel (Leiden; Boston and Kohl: Brill Academic Publishers, 2011), 392.
51. Hammer, *Elizabeth's Wars*, 181–182.
52. Fissel, *English Warfare*, 215; Mark Charles Fissel, "English Amphibious Warfare, 1587–1656," in *Amphibious Warfare 1000–1700*, 236–237.
53. S.J. Connolly, *Contested Island. Ireland 1460–1630* (Oxford and New York: Oxford University Press, 2007), 250–252.

English interventions in France, were part of allied efforts. Eight of the allied amphibious escort operations served primarily as troop transport. Gunfire escort by the navy in support of their land forces, and the transportation function of the navy for operations ashore overseas, cultivated joint operations or expeditionary warfare.[80] Another ten amphibious operations of escort design during the 1500s can be classified as an "amphibious revolution," the overthrow of regimes or socio-political disorder provoked by the armed landing, exemplified by the Genoese, Irish, Netherlandish, and Crimean revolts. They are marginal to strictly amphibious operations, as likewise are coalitional and expeditionary warfare, because they either presumed not so much a military but a political conjunction of forces or were devoid of resistance upon a hostile shore.

Among the 22 failures of the epoch's escort operations, only four betrayed the fatal flaws of the operational design, namely miscoordination of amphibious forces with the land army, which underscores the crucial importance of professionalism. All four represented a malaise of coalitional warfare. The siege of Venlo (1511), the invasion of Guyenne (1512), and the siege of Montreuil (1544) were miscarried by an English alliance with Netherlands, Spain and the Holy Roman Empire; the advance on Naples in 1528 collapsed due to the French fall-out with Venice and, primarily, with the sub-par performance of the Genoese contingent of Andrea Doria, which was to provide the bulk of the naval support.

Another 14 escort failures resulted from the tactical deficiency of insufficient deck-to-shore gunfire support. The remaining six instances suffered battlefield misfortunes and the appearance of a superior enemy, a frequent misfortune in war. The 14 "fire-faults" are intriguing because seven were produced by the Turks (two of those allied with the French) and three by the Spaniards. The Ottomans' escort misfortunes include the storming of Bougie in 1514 and 1515, Penon de Alger in 1516, the siege of Diu in 1538, the advance on Astrakhan in 1569, and attacks against Corsican targets Calvi and Bastia in 1555 jointly with French forces. The trio of Spanish escort disasters (1543, 1547, 1558) transpired under the command of Martin Alonso Fernandez de Cordoba and around the same problematic location, the fortress of Mostagenem (near the Spanish stronghold of Oran). A discernible trend in these cases remains elusive. However, the

---

**80** For the concept, see Liles, "The Development of Amphibious/Expeditionary Warfare"; Cahill, "An Unassailable Advantage."

**Figure 6:** Dutch forces under Adolf von Meurs and Maarten Schenk van Nijdeggen, along with allied English troops under Sir John Norreys, storm the Spanish sconce at Ijsseloord at the bifurcation of the rivers Rhine and Ijssel on 14 October 1585. The attackers employed the "escort" design of simultaneous amphibious and overland assault against fortified troops. This action consisted of land-based artillery fire coordinated with the disembarkation of troops seconded by the landing crafts' deck-to-shore gunfire.[81] The engraving is from Willem Baudart's *Les guerres de Nassau, descriptes par Guillaume Baudart* published by M. Colin (Amsterdam 1616), editor's collection.

Ottomans' misfortunes occurred within varied Mediterranean geographical contexts, and outside – in the Eurasian interior (Astrakhan), and South Asia (Diu). They befell despite the assertion in contemporary historiography that the Ottomans were the amphibious paradigm of the sixteenth century.[82] They were not.

---

**81** Baudart, *Les Guerres de Nassau*, T.2, 57–60.
**82** Goffman, *The Ottoman Empire and Early Modern Europe*, 145–146; Hattendorf, *Naval Strategy and Power in the Mediterranean*, 8.

## Fire-fault

The Ottomans' comparative degradation from their operational pinnacle in the second half of the fifteenth century, achieved mostly with the escort model, was a direct result of the mutation of amphibious warfare, specifically the emergence of the assault ship and its tactic of deck-to-shore gunfire. The dissonance between actual Ottoman performance and recent scholarship requires addressing the amphibious operations of other designs as well. Table 4, "Divergence of designs of amphibious operations circa 1500–1600,"[83] sets 65 major amphibious operations of the 1500s which are unquestionably amphibious because they featured the "aside build-up" and "direct assault" designs.

The Ottomans were the authors of fourteen of the former enterprises. No Turkish operation followed the "direct assault" design. None of the operations based on deck-to-shore gunfire, presented in Table 2, involved Ottoman forces. Among the numerous Turkish amphibious operations of the epoch are grandiose epeditions such as the sieges of Rhodes (1522) and Corfu (1537) as well as protracted struggles for Malta (1565) and Cyprus (1570–1571). The success to failure ratio in them is half-to-half; one learns nothing if the mechanics beneath the operational surface are not scrutinized closely. Ottoman narratives reveal suspicious similarities. After having delivered its troops and established a blockade, the Turkish fleet invariably drifted idly during sieges. No serious attempts were made to suppress the fortifications of Rhodes, Corfu-town, Birgu, St. Elmo, St. Michael, and Famagusta by ship-based gunnery (or by the outflanking of the defenders by some amphibious diversion) even though the objectives featured proximate promising landing beaches and were well within range of Ottoman naval gunnery.

In 1529, Hayreddin Barbarossa successfully combined a frontal attack overland with an amphibious landing against the Spanish Penon de Alger. On 15 July, 1565, Hayreddin Barbarossa's son, Hassan, the governor of Algiers, launched boats against the seaside wall of Birgu, island of Malta, leading Muslim corsairs and North African volunteers at his own initiative. The fleet of Piali Pasha did not support Hassan's assault with firepower. The attack failed miserably in the face of the Hospitallers' counter-amphibious gunnery.[84] Unlike Barbarossa's son, Ottoman naval leaders avoided amphibious risks as much as possible. A half-hearted advance landing of Piali Pasha on Cyprus, near Limassol, on 1 July, 1570 was repulsed by the Venetians despite Turkish advantages. Hesitant Ottoman attempts

---

83 See the references to the sources of data on the operations in the Tables.
84 Guilmartin, "The siege of Malta," 173–174.

Table 4: Divergence of designs of amphibious operations circa 1500–1600.

| | Data | Name of operation | Attacker | Defender | Environment | Design | Leader | Outcome | Note |
|---|---|---|---|---|---|---|---|---|---|
| 1 | 1505 | Taking of Mers el-Kebir | Spain | Kingdom of Tlemcen | Marine | Aside build-up | Diego Fernandes de Cordoba | Success | 1 |
| 2 | 1506 | Siege of Kazan | Moscow | Kazan Khanate | Riverine | Aside build-up | Dmitry prince of Uglich | Failure; superior enemy | 2 |
| 3 | 1507 | Submission of Hormuz | Portugal | Hormuz Sultanate | Marine | Aside build-up | Afonso de Albuquerque | Failure; politics | 3 |
| 4 | 1509 | Taking of Oran | Spain | Kingdom of Tlemsen | Marine | Aside build-up | Pedro Navarro | Success | 4 |
| 5 | 1510 | Taking of Bougie | Spain | Kingdom of Tlemsen | Marine | Aside build-up | Pedro Navarro | Success | 5 |
| 6 | 1510 | Taking of Tripoli | Spain | Hafsid Tunisia | Marine | Aside build-up | Pedro Navarro | Success | 6 |

| | | | | | | | | |
|---|---|---|---|---|---|---|---|---|
| 7 | 1510 | Submission of Djerba | Spain | Djerba sheikh | Marine | Aside build-up | Pedro Navarro | Failure; superior enemy | 7 |
| 8 | 1510 | Storm of Calicut | Portugal | Zamorin kingdom | Marine | Aside build-up | Fernando Coutinho | Failure; lack of gunfire | 8 |
| 9 | 1510, May | First taking of Goa | Portugal | Bijapur Sultanate | Marine | Aside build-up | Afonso de Albuquerque | Success | 9 |
| 10 | 1510, Nov | Second taking of Goa | Portugal | Bijapur Sultanate | Marine | Aside build-up | Afonso de Albuquerque | Success | 10 |
| 11 | 1514 | Storm of La Goletta | Genoa | Turkey | Marine | Aside build-up | Andrea Doria | Success | 11 |
| 12 | 1515 | Submission of Hormuz | Portugal | Hormuz Sultanate | Marine | Aside build-up | Afonso de Albuquerque | Success | 12 |
| 13 | 1516 | Advance on Algiers | Spain | Turkey | Marine | Aside build-up | Diego de Vera | Failure; weather | 13 |
| 14 | 1516 | Storm of Aden | Egypt | Aden Sultanate | Marine | Direct assault | Salman Reis | Failure; lack of gunfire | 14 |

(continued)

**Table 4** (continued)

| Data | Name of operation | Attacker | Defender | Environment | Design | Leader | Outcome | Note |
|---|---|---|---|---|---|---|---|---|
| 15 1517–18 | Struggle for Tlemcen | Turkey | Spain | Marine | Aside build-up | Oruc Barbarossa | Failure, superior enemy | 15 |
| 16 1519 | Advance on Algiers | Spain | Turkey | Marine | Aside build-up | Hugo de Moncada | Failure; weather | 16 |
| 17 1520 | Submission of Djerba | Spain | Djerba sheikh | Marine | Aside build-up | Hugo de Moncada and Diego de Vera, | Success | 17 |
| 18 1520 | Taking of Tenes | Turkey | Spain | Marine | Aside build-up | Hayreddin Barbarossa | Success | 18 |
| 19 1522 | Conquest of Rhodes | Turkey | Order of Hospitallers | Marine | Aside build-up | Suleiman I, Piri Mehmed Pasha | Success | 19 |
| 20 1525 | Submission of Djerba | Turkey | Djerba sheikh | Marine | Aside build-up | Hayreddin Barbarossa | Failure; lack of gunfire | 20 |
| 21 1528 | Taking of Genoa | Spain, Genoese Doria's party | Genoa | Marine | Direct assault | Andrea Doria | Success | 21 |

| | | | | | | | | |
|---|---|---|---|---|---|---|---|---|
| 22 | 1531 | Taking of Modon | Order of Hospitallers | Turkey | Marine | Direct assault | Bernardo Salviati; de Boniface | Success | 22 |
| 23 | 1532 | Taking of Coron | Spain, Genoa | Turkey | Marine | Direct assault | Andrea Doria | Success | 23 |
| 24 | 1532 | Taking of Patras | Spain, Genoa | Turkey | Marine | Direct assault | Andrea Doria | Success | 24 |
| 25 | 1532 | Taking of Lepanto | Spain, Genoa | Turkey | Marine | Direct assault | Andrea Doria | Failure; lack of gunfire | 25 |
| 26 | 1534 | Taking of Tunis, La Goletta, Bizera | Turkey, Tunisian pretender | Hafsid Tunisia | Marine | Aside build-up | Hayreddin Barbarossa | Success | 26 |
| 27 | 1537 | Siege of Corfu | Turkey | Venice | Marine | Aside build-up | Suleiman I, Hayreddin Barbarossa | Failure; lack of gunfire | 27 |
| 28 | 1538 | Taking of Castelnuovo | Spain | Turkey | Marine | Direct assault | Andrea Doria | Success | 28 |
| 29 | 1541 | Attack on Suez | Portugal | Turkey | Marine | Direct assault | Estevao da Gama | Failure; lack of gunfire | 29 |

(continued)

Table 4 (continued)

| | Data | Name of operation | Attacker | Defender | Environment | Design | Leader | Outcome | Note |
|---|---|---|---|---|---|---|---|---|---|
| 30 | 1541 | Siege of Algiers | Spain, H.R. Empire | Turkey | Marine | Aside build-up | Charles V | Failure; weather | 30 |
| 31 | 1544 | Taking of Leith | England | Scotland | Marine | Aside build-up | Edward Seymour earl of Hertford | Success | 31 |
| 32 | 1550 | Taking of Katif | Portugal; Hormuz and Moghistan sultanates | Turkey | Marine | Aside build-up | Antao de Noronha; Turan Shah; Amir Magid; | Success | 32 |
| 33 | 1551 | Taking of Tripoli | Turkey | Spain | Marine | Aside build-up | Sinan Yusuf; Turgut Reis | Success | 33 |
| 34 | 1553 | Taking of Bastia | France, Corsican insurgents | Genoa, Spain | Marine | Aside build-up | Antoine Escalin des Eymars Baron de la Garde; Paul de Thermes | Success | 34 |
| 35 | 1553 | Taking of Bonifacio | France, Turkey | Genoa, Spain | Marine | Aside build-up | Antoine Escalin des Eymars, Baron de la Garde; Paul de Thermes; Turgut Reis | Failure; lack of gunfire | 35 |

| | | | | | | | | |
|---|---|---|---|---|---|---|---|---|
| 36 | 1556 | Attack on Ochakov | Moscow | Turkey, Crimea | Riverine | Aside build-up | Matthew Rzhevsky aka Djak; Dmitry Wisniowiecki | Success | 36 |
| 37 | 1558 | Taking of Brest | England, Netherlands | France | Marine | Aside build-up | Edward Fiennes de Clinton; Adolph of Wakken | Failure; lack of gunfire | 37 |
| 38 | 1559 | Advance on Crimea | Moscow | Turkey, Crimea | Riverine | Aside build-up | Daniel Adashev | Success | 38 |
| 39 | 1559 | Siege of Azov | Moscow | Turkey, Crimea | Riverine | Aside build-up | Dmitry Wisniowiecki | Failure; lack of gunfire | 39 |
| 40 | 1559 | Taking of Bahrein | Turkey | Portugal | Marine | Aside build-up | Mustafa Pasha | Failure; lack of gunfire | 40 |
| 41 | 1560 | Siege of Azov | Moscow | Turkey, Crimea | Riverine | Aside build-up | Dmitry Wisniowiecki | Failure; lack of gunfire | 41 |
| 42 | 1560 | Struggle for Djerba | Turkey | Spain | Marine | Aside build-up | Piali Pasha | Success | 42 |
| 43 | 1563 | Taking of Oran | Turkey | Spain | Marine | Aside build-up | Hasan Pasha Hayreddinzade | Failure; lack of gunfire | 43 |

(continued)

Table 4 (continued)

| | Data | Name of operation | Attacker | Defender | Environment | Design | Leader | Outcome | Note |
|---|---|---|---|---|---|---|---|---|---|
| 44 | 1565 | Struggle for Malta, siege of Birgu. | Turkey | Spain, Order of Hospitallers | Marine | Aside build-up | Kizilahmedli Kara Mustafa; Piali Pasha | Failure; lack of gunfire | 44 |
| 45 | 1565 | Relief of Malta | Spain; Order of Hospitallers | Turkey | Marine | Aside build-up | Garsia de Toledo | Success | 45 |
| 46 | 1568 | Storm of Malacca | Aceh sultanate | Portugal | Marine | Direct assault | Alauddin al-Kahar | Failure; lack of gunfire | 46 |
| 47 | 1570–71 | Taking of Cyprus, siege of Famagusta | Turkey | Venice | Marine | Aside build-up | Lala Mustafa Pasha | Success | 47 |
| 48 | 1571 | Storm of Durrës | Venice | Turkey | Marine | Aside build-up | Sebastian Venier | Failure; lack of gunfire | 48 |
| 49 | 1571 | Taking of Castelnuovo | Venice | Turkey | Marine | Aside build-up | Sebastian Venier | Failure; lack of gunfire | 49 |

| | | | | | | | | |
|---|---|---|---|---|---|---|---|---|
| 50 | 1571 | Taking of Klis | Venice | Turkey | Marine | Aside build-up | Sebastian Venier | Failure; lack of gunfire | 50 |
| 51 | 1572 | Taking of Novigrad | Venice | Turkey | Marine | Aside build-up | Giacomo Foscarini | Success | 51 |
| 52 | 1572 | Taking of Modon, storm of Navarino | Spain; Venice; Papal State | Turkey | Marine | Direct assault | Don Juan of Austria; Giacomo Foscarini; Marcantonio Colonna; Alexander Farnese | Failure; lack of gunfire | 52 |
| 53 | 1573 | Storm of Malacca | Aceh sultanate | Portugal | Marine | Direct assault | Ali Ri'ayat Syah I | Failure; lack of gunfire | 53 |
| 54 | 1573 | Taking of Tripoli | Spain | Turkey | Marine | Aside build-up | Don Juan of Austria | Success | 54 |
| 55 | 1574 | Taking of Tripoli | Turkey | Spain | Marine | Aside build-up | Uluc-Kilic Ali Pasha | Success | 55 |
| 56 | 1574 | Attack on Narva | Sweden | Moscow | Marine | Aside build-up | Jakob Tordsson Bagge | Failure, weather | 56 |
| 57 | 1577 | Attack on Narva | Sweden | Moscow | Marine | Aside build-up | Jakob Tordsson Bagge | Failure, lack of gunfire | 57 |

(continued)

**Table 4** (continued)

| | Data | Name of operation | Attacker | Defender | Environment | Design | Leader | Outcome | Note |
|---|---|---|---|---|---|---|---|---|---|
| 58 | 1578 | Battle of Alcazarquivir | Portugal | Morocco | Marine | Aside build-up | Sebastian I | Failure; superior enemy | 58 |
| 59 | 1579 | Attack on Narva | Sweden | Moscow | Marine | Aside build-up | Henrik Klasson Horn; Bengt Severinsson | Failure; superior enemy | 59 |
| 60 | 1586 | Siege of Malacca | Johor Sultanate | Portugal | Marine | Aside build-up | Ali Jalla Abdul Jalil Shah II | Failure; lack of gunfire | 60 |
| 61 | 1589 | Storm of Corunna | England | Spain | Marine | Aside build-up | John Norris; Francis Drake | Failure; lack of gunfire | 61 |
| 62 | 1595 | Advance on Panama | England | Spain | Marine | Aside build-up | Francis Drake | Failure; superior enemy | 62 |
| 63 | 1596 | Taking of Cadiz | England | Spain | Marine | Aside build-up | Robert Devereux earl of Essex; Charles Howard earl of Nottingham | Success | 63 |

| | | | | | | | |
|---|---|---|---|---|---|---|---|
| 64 | 1598 | Combat of Irmen | Moscow | Siberian Khanate | Riverine | Direct assault | Andrew Voyeikov | Success | 64 |
| 65 | 1600 | Battle of Nieuwpoort | Netherlands | Spain | Marine | Aside build-up | Maurice of Nassau | Success | 65 |

Notes:

1. Andrew C. Hess, *The Forgotten Frontier. A History of the Sixteenth-Century Ibero-African Frontier* (Chicago and London: University of Chicago Press, 1978), 38.
2. V. Shirogorov, *Ukrainian War. The Armed Conflict for Eastern Europe in the 16th–17th Centuries. Vol. 1 Melee of Rus (To the Middle of the 16th Century)* (Moscow: Molodaya Gvardiya, 2017), 687.
3. Mohammed Hameed Salman, "Aspects of Portuguese Rule in the Arabian Gulf, 1521–1622," doctoral thesis, University of Hull, 2004, 79–80.
4. Emrah Safa Gürkan, "Ottoman Corsairs in The Western Mediterranean and their Place in The Ottoman – Habsburg Rivalry (1505–1535)," Master's thesis, Bilkent University, Ankara, 2006, 35; Henry Kamen, *Spain's Road to Empire. The making of a world power, 1492-1763* (London: Penguin Books, 2003), 31.
5. Gürkan, "Ottoman Corsairs in The Western Mediterranean," 41.
6. Gürkan, "Ottoman Corsairs in The Western Mediterranean," 41.
7. Gürkan, "Ottoman Corsairs in The Western Mediterranean," 41–42; Kamen, *Spain's Road to Empire*, 163.
8. Francisco Bethencourt, "The Political Correspondence of Albuquerque and Cortes," in *Correspondence and Cultural Exchange in Europe, 1400–1700*, eds. Francisco Bethencourt and Florike Egmond (Cambridge, New York and Melbourne: Cambridge University Press, 2007), 225; Diffie and Winius, *Foundations of the Portuguese Empire*, 247–48
9. Diffie and Winius, *Foundations of the Portuguese Empire*, 250–252.
10. Bethencourt, "The Political Correspondence," 226; Diffie and Winius, *Foundations of the Portuguese Empire*, 253.
11. Angus Konstam and Gerry Embleton, *The Barbary Pirates 15th–17th Centuries* (Oxford: Osprey Publishing, 2016), 13.
12. Bethencourt, "The Political Correspondence," 238; Diffie and Winius, *Foundations of the Portuguese Empire*, 269–270
13. Gürkan, "Ottoman Corsairs in The Western Mediterranean," 58–61; Hess, *The Forgotten Frontier*, 64.
14. Diffie and Winius, *Foundations of the Portuguese Empire*, 274.
15. Gürkan, "Ottoman Corsairs in The Western Mediterranean," 61–67; Hess, *The Forgotten Frontier*, 65.
16. Gürkan, "Ottoman Corsairs in The Western Mediterranean," 66–67; Hess, *The Forgotten Frontier*, 66.
17. Gürkan, "Ottoman Corsairs in The Western Mediterranean," 72; Hess, *The Forgotten Frontier*, 67.

18. Gürkan, "Ottoman Corsairs in The Western Mediterranean," 84.
19. Zeynep Nevin Yelce, "The Making of Sultan Süleyman: a Study of Process/es of Image-making and Reputation Management," doctoral dissertation, Sabancı University, Tuzla/Istanbul, 2009, 306–307.
20. Gürkan, "Ottoman Corsairs in The Western Mediterranean," 88.
21. Michael Edward Mallett and Christine Shaw, *The Italian Wars 1494–1559. War, State and Society in Early Modern Europe* (London and New York: Routledge, 2012), 170.
22. Dennis Castillo, *The Maltese Cross. A Strategic History of Malta* (Westport and London: Praeger, 2005), 53; Rene-Aubert Verto, *The History of the Knights of Malta* (London: G.Strahan, 1728), 46–50, https://play.google.com/books.
23. Theodore Spandounes, *On the Origins of the Ottoman Emperors. Translated and edited by Donald M. Nicol* (Cambridge: Cambridge University Press, 1997), 72–73
24. Spandounes, *On the Origins of the Ottoman Emperors*, 72–73.
25. Spandounes, *On the Origins of the Ottoman Emperors*, 72–73.
26. Gürkan, "Ottoman Corsairs in The Western Mediterranean," 112; Hess, *The Forgotten Frontier*, 72.
27. Daniel Goffman, *The Ottoman Empire and Early Modern Europe* (Cambridge/New York/Melbourne: Cambridge University Press, 2009), 148; Kenneth M. Setton, *The Papacy and the Levant, 1204–1571* (Philadelphia: The American Philosophical Society, 1978) 3: 431–432.
28. John Francis Guilmartin, *Gunpowder and Galleys. Changing Technology and Mediterranean Warfare at Sea in the 16th Century* (London: Conway Maritime Press, 2003), 69.
29. Giancarlo Casale, *The Ottoman Age of Exploration* (Oxford and New York: Oxford University Press, 2010), 70–71; Diffie and Winius, *Foundations of the Portuguese Empire*, 291.
30. Charles Oman, *A History of the Art of War in the Sixteenth Century* (London: Greenhill Books, 1989), 698.
31. Mark Charles Fissel, *English Warfare 1511–1642* (London and New York: Routledge, 2001), 27–28; J. Balfour Paul, "Edinburgh in 1544 and Hertford's Invasion," *The Scottish Historical Review*, Vol. 8, No. 30 (1911): 120–122.
32. Salih Ozbaran and Dom Manuell de Lyma, "The Ottoman Turks and the Portuguese in the Persian Gulf (1534–1581)," *Journal of Asian History* Vol. 6, No. 1 (1972), 58–59.
33. Fernand Braudel, *The Mediterranean and the Mediterranean World in the Age of Philip II* (Oakland: University of California Press, 1996) 2: 921–922.
34. J. M. Jacobi, *Histoire Générale de la Corse, depuis les Premiers Temps* (Paris: Aime Andre, Libraire, 1835) 1: 329–330, https://play.google.com/books
35. Jacobi, *Histoire Générale de la Corse*, 330.

36. V. Shirogorov, *Ukrainian War. The Armed Conflict for Eastern Europe in 16th–17th Centuries. Vol. 3 Head-to-head Offensive: Baltics – Lithuania – Steppes (In the Second Half of the Sixteenth Century)* (Moscow: Molodaya Gvardiya: 2019), 153-154.
37. John Cassell's *Illustrated History of England* (London, Paris & New York: Cassell Petter & Galpin, 1909), 2:389, https://archive.org/details/cassellsillustra02lond/page/n8/mode/2up.
38. Shirogorov, *Ukrainian War. Vol. 3 Head-to-head Offensive*, 155–156.
39. Shirogorov, *Ukrainian War. Vol. 3 Head-to-head Offensive*, 157–158.
40. Ozbaran and de Lyma, "The Ottoman Turks and the Portuguese," 67–68.
41. Shirogorov, *Ukrainian War. Vol. 3 Head-to-head Offensive*, 159.
42. Guilmartin, *Gunpowder and Galleys*, 142–144.
43. Hess, *The Forgotten Frontier*, 82.
44. Castillo, *The Maltese Cross*, 60–75; John Francis Guilmartin, "The Siege of Malta", in *Amphibious Warfare 1000–1700. Commerce, State Formation and European Expansion*, eds. D.J.B. Trim and Mark Charles Fissel (Leiden/Boston/Kohl: Brill, 2011).
45. Tim Pickles and Christa Hook, *Malta 1565. Last Battle of The Crusade* (Oxford: Osprey Publishing, 1998), 84–85.
46. R.O. Winstedt, "History of Malaya," *The Journal of Malaysian Branch of the Royal Asiatic Society (JMBRAS)* 13.1 (121) (1935): 79–80.
47. Niccolo Capponi, *Victory of the West. The Story of the Battle of Lepanto* (London: Macmillan, 2006), 136–144; Tamás Kiss, "Cyprus in Ottoman and Venetian Political Imagination c.1489–1582," The Doctoral dissertation (Central European University, Budapest, 2016), 106–108.
48. Noel Malcolm, *Agents of Empire. Knights, Corsairs, Jesuits and Spies in the Sixteenth-Century Mediterranean World* (Oxford/New York: Oxford University Press, 2015), 137.
49. James D. Tracy, *Balkan Wars. Habsburg Croatia, Ottoman Bosnia, and Venetian Dalmatia, 1499–1617* (New York/London: Rowman & Littlefield Publishers, 2016), 210.
50. Tracy, *Balkan Wars*, 210.
51. Tracy, *Balkan Wars*, 210–211.
52. Braudel, *The Mediterranean and the Mediterranean World*, 1122–1123.
53. Winstedt, "History of Malaya," 81.
54. John Francis Guilmartin, *Galleons and Galleys* (London: Cassell, 2002), 150; Malcolm, *Agents of Empire*, 190–192.
55. Ivanov, *Ottoman Conquest of the Arabian Nations*, 230–33; N.A. Ivanov, *Ottoman Conquest of the Arabian Nations* (Moscow: RAN Eastern Literature, 2001), 227–229.
56. Ian Glete, "Amphibious Warfare in the Baltic, 1550–1700," in *Amphibious Warfare 1000–1700. Commerce, State Formation and European Expansion*, eds. D.J.B. Trim and Mark Charles Fissel, 129; Shirogorov, *Ukrainian War. Vol. 3 Head-to-head Offensive*, 549.

57. Glete, "Amphibious Warfare in the Baltic, 1550–1700," in *Amphibious Warfare 1000–1700*, 129; Shirogorov, *Ukrainian War*. Vol. 3 *Head-to-head Offensive*, 592.
58. Luís Costa e Sousa, "From Tangier to Alcácer Quibir: The Portuguese Military Revolution (Re)visited," *Portuguese Studies Review (SP)* (2015): 27–28.
59. Shirogorov, *Ukrainian War*. Vol. 3 *Head-to-head Offensive*, 682–683.
60. John Villiers, "Aceh, Melaka and the *Hystoria dos Cercos de Malaca* of Jorge de Lemos," *Portuguese Studies* Vol. 17 (2001): 78–82.
61. Paul E.J. Hammer, *Elizabeth's Wars. War, Government and Society in Tudor England, 1544–1604* (New York: Palgrave Macmillan, 2004), 157–158.
62. Hammer, *Elizabeth's Wars*, 192.
63. Hammer, *Elizabeth's Wars*, 196.
64. V.V. Trepavlov, *Siberian Yurt after Yermak* (Moscow: RAN Eastern Literature, 2012), 57–58.
65. Bouko, de Groot, *Nieuwpoort 1600: The First Modern Battle* (Oxford: Osprey Publishing, 2019); Christer Jorgensen, et al, *Fighting Techniques of the Early Modern World* (New York: Thomas Dunne Books, 2006), 138–139,144; Bas Kist, "A Spanish-Dutch Military Confrontation. The Battle of Nieuwpoort, July 2nd 1600," *Militaria. Revista del Cultura Militar* 7 (1995): 297–302.

to force their galleys into the harbor of Famagusta in 1570 was repulsed by the fortress' guns.[85] The fatal consequences of that Turkish stratagem compounded the costly onslaught fought on the landward of the defenders' fortifications. Ottoman failures at Corfu (1537) and Malta (1565) resulted from the absence of deck-to-shore gunfire in support of the landing troops. The sieges of Rhodes in 1522 and Famagusta in 1570 to 1571 likewise became slaughterhouses by failing to succor disembarkation with gunpowder tactics. Only the Ottomans' senselessness in absorbing such losses pressed the defenders to surrender. Smaller scale operations, such as the attempts on Bahrein in 1559 and the siege of Oran in 1563, collapsed due to Turkish unwillingness to embrace the new tactics. For the Ottomans, who are considered by many historians as the architects of effective amphibious strategies and tactics, these lapses in harnessing the tactical advantages of waterborne bombardment in coordination with amphibious attack are baffling.

Ottoman obliviousness to the impact of the military revolution upon amphibious warfare occurred at a time when deck-to-shore gunfire was an amphibious imperative. Thirty-three of 65 operations shown in Table 4 can be judged as failures, and of these, 22 faltered due to the lack of naval gunfire support. One also ought to consider the Portuguese descent upon Calicut in 1510, the Egyptian storming of Aden in 1516, the Turkish invasion of Djerba in 1525, the Portuguese attack on Suez in 1541, the Franco-Turkish siege of Bonifacio in 1553, the English-Netherlandish attempt on Brest in 1558, Moscow's siege of Azov in 1559, Aceh's efforts to take Malacca in 1568 and 1573, the Holy League's assaults on Navarino in 1572, the Swedish attack on Narva in 1577, the English designs against Corunna in 1589, and so forth. Topped with the failures from Table 2 such as the Portuguese mishap at Aden in 1513, the Spanish defeat of Preveza in 1538, and the Netherlandish rebels' setback at Amsterdam in 1585, the catalogue of amphibious disasters through the sixteenth century is self-revealing. In the age of the sixteenth-century military revolution, the lack of proper naval gunfire raining down upon the onshore objective meant that an amphibious operation had little chance of success.

## Direct Assault

Another striking gap in the Ottomans' performance of amphibious operations is the absence of "direct assaults." "Aside build-up" accentuated the imperative of logistically magnifying amphibious force projection by bringing ashore

---

85 Capponi, *Victory of the West*, 136, 176.

ammunition, field guns, etc. Without such swift and substantial support, the potential for an overwhelming assault exceeding the capacity of the land-based defenders could not be fulfilled. The "direct assault" was born of the "aside build-up" model as a result of tacticians' conviction that rearmed and "professionalized" troops could land, be arrayed, be deployed, and attack the objective with uninterrupted continuity. Tacticians also believed that ships could decisively secure a tactical foothold for the disembarked troops via deck-to-shore barrage. The difference between the "direct assault" and the "aside build-up" is the immediate attack on the objective from water without any tactical pause. Afonso Albuquerque authored such a maneuver in his attack on the Melaka bridge during the storm of Malacca in 1511. A "desperate business"[86] before, the "direct assault" became the accepted tactic of successful landing.

Table 2 demonstrates that 16 of 24 operations in the 1500s relying upon massive naval gunnery were designed as "direct assaults," and a dozen succeeded. Table 4 showcases 12 "direct assaults" dominated by landing troops. These include a pair of victories achieved by Andrea Doria at Cherchel in 1531 and in Morea in 1532. The Portuguese relief of Diu under Joao de Castro in 1546, Spain's capture of Penon de Velez under Garcia de Toledo in 1564, and Moscow's attack on the Siberian camp on the river Irmen under Andrew Voyeikov in 1598 provide the best examples of the paradigm. All four commanders respectively grasped the crucial advantage provided by deck-to-shore gunfire and the landing carried outright on the suppressed beach defenses under the continuous barrage of ships' guns. Successful performance of such operations required not only well-planned and effective naval gunnery and capable landing troops; it also necessitated precise coordination coupled with the tactical imagination of senior command working in tandem with those leading troops ashore.

## Changing Amphibious Doctrines

As argued above, it has been alleged that the Ottomans possessed unsurpassed amphibious expertise, capitalizing on their large and well-equipped fleet with its advanced ordnance and numerous, well-trained and motivated marines, including consummate professionals filling the ranks of the firearms-bearing janissary elite and *topcu* artillery corps. The Turks also enjoyed an ample choice of objectives in the Mediterranean and Near East for their amphibious operations.

---

86 Trim, "Medieval and Early-Modern Inshore, Estuarine, Riverine and Lacustrine Warfare," 384.

**Figure 7:** A Dutch fleet under Philip von Hohenlohe coordinated with Antwerp's flotilla commanded by Philips van Marnix to attack from opposite sides the dam of Kouwendijk, 26 May 1585. Spanish forces held that position in order to deny access to the waterway to any attempt to relieve the city. The violent conjunction was an amphibious "direct assault," the disembarkation performed under a barrage of waterborne gunfire. The Spanish counterattack was directed by Alessandro Farnese, Duke of Parma.[87] The engraving is from Willem Baudart's *Les guerres de Nassau, descriptes par Guillaume Baudart* published by M. Colin (Amsterdam 1616), editor's collection.

However, the Ottomans obstinately avoided adopting the direct assault design. Their adversaries' direct assaults apparently startled them frequently. The logistically conscious Ottomans instead embraced the aside build-up and carefully worked to simplify the operation into transportation of troops and naval escort of the land army. John Guilmartin's scholarship makes explicit this Ottoman strategic and tactical adherence.[88]

---

87 Baudart, *Les guerres de Nassau*, T.2, 29–32.
88 Guilmartin, *Gunpowder and Galleys*, 229.

In 1543, Savoyard Nice was besieged by an Ottoman amphibious force under Hayreddin Barbarossa, in conjunction with a French fleet and army under Francois de Bourbon, Count of Enghien, soon to be the famed victor of the battle of Ceresole on 11 April, 1544. Hayreddin Barbarossa, reputed as the unrivaled naval tactician of his time, was astounded when on one stormy day the Spanish-Genoese fleet under Andrea Doria descended directly on his landing foothold and anchorage at Villafranca, bombarded his facilities ashore, and landed Spanish infantry under Alfonso d'Avalos. Doria lost four galleys but achieved a stunning surprise victory and got the most from his seasoned Spanish *tercios*. The French were broken in street-to-street fighting and retreated. Barbarossa withdrew his fleet a few hours before the attack and many Turks were evacuated with little time to spare.[89] This violent contest demonstrated the newest features of amphibious incorporation of the direct assault, in contrast with the escort design and aside build-up preferred by the French and Turks in the same time and place.

In 1550, Mahdia, on the North African coast, was besieged by Spanish forces under Andrea Doria and Juan de la Vega, the Viceroy of Sicily. The Turkish governor of Algeria, Turgut Reis, the foremost pupil of Barbarossa, spied a gap in the sea blockade and stealthily brought in a relief force. He could not imagine the use of direct assault as his tactic. He landed outside the crucial area, and found his attack cut short on the contravallation. In the end, the soldiers of Turgut Reis were decimated. After Turgut's flight, Garcia Alvarez de Toledo engineered a floating battery of galleys, "steepled" and cleared of masts. Upon this innovative gun platform were positioned heavy ordnance aimed broadside. The contraption battered the stubborn fortress to rubble. The stronghold was stormed, its garrison and inhabitants massacred.[90]

The Ottoman amphibious doctrine stagnated virtually between the conquest of Venetian Negroponte in 1470 and Genoese Kaffa in 1475. The former again utilized the escort design. Mahmud Pasha Angelovic was its operational leader. His protégé, Mesih Pasha Palaeolog, guided the Ottoman fleet in an escort role in the taking of Chiliya and Akkerman in 1484. Another promising pupil, Kara Nisancı Davud Pasha, commanded the Ottoman fleet in the battles of Zonchio in 1499 and 1500 (assisted by the talented Kemal Reis, formerly a corsair). These actions culminated in an escort of forces to the capture of Lepanto and Modon. Gedik Ahmed Pasha conquered Kaffa in 1475 as well as Apulian Otranto in 1480 following the aside build-up design. However, in 1482 he was killed in the feud among

---

[89] Isom-Verhaaren, "Barbarossa and His Army Who Came to Succor All of Us," 411–412; Setton, *The Papacy and the Levant, 1204–1571*, 523.
[90] Allen, *The Great Siege of Malta*, 50–51; Duro, *Armada Española*, 281–284.

Mahmed II's heirs. Turkish expertise in aside build-up perished with him and the escort design became a legacy of Ottoman amphibious doctrine. The latter was fixed by the unique origin of Ottoman naval commanders as the pupils of the slave schools and "graduates of palace service," with Hayreddin Barbarossa as the single exclusion,[91] and "the linear promotion system" in the Ottoman forces; automatic appointment of commanders of the operational level to the vacancy next upper in hierarchy.[92]

While Ottoman amphibious doctrine atrophied, the increased potential of amphibious operations changed other nations' military strategies. The amphibious arm became one of the principal methods of war in the sixteenth century. After the amphibious successes of Afonso de Albuquerque in the 1510s, Portugal changed its strategy in East Asia from sea domination to the conquest of the territorial empire, adapting its uses of gunpowder weaponry as well as its vessels. England turned to amphibious invasion as a strategy against Scotland after the success of William Skeffington's expedition to Ireland in 1534, where he landed against the Geraldine rebels and pacified them. The attack on Leith in 1544, cited above, expressed the new English doctrine. Moscow harnessed the amphibious onslaught on the Crimea, instead of costly overland ventures, after the taking of Astrakhan by amphibious landing in 1556. In 1559, Crimea was devastated, releasing Moscow's land army for deployment to the Livonian War, an excellent example of the strategic benefits of the new amphibious warfare. Both France, after the struggle for Nice in 1543, as well as Spain, after the taking of the Azores in 1582, sought amphibious solutions in their grand strategy against England (in 1545 and 1588, respectively). England, by virtue of its geography, explored naval counter-amphibious operations when diverting invading continental amphibious armadas, like the aborted French invasion of 1445,[93] and prevented the Spanish landing of 1588.[94] Sweden, the nation that Michael Roberts credited with developing the military revolution, exhibited mastery of amphibious conquest by building her Baltic empire after successes at Reval in 1561 and Narva in 1581. The visceral reality of the impact of the amphibious spearhead of the military revolution is not a paper exercise by contemporary scholars. The strategists of the 1500s wrestled with a gritty reality that smelled of both saltwater and coastal sands.

---

**91** Imber, *The Ottoman Empire, 1300–1650*, 297–299.
**92** Uyar and Erickson, *A Military History of the Ottomans*, 40.
**93** Potter, *Henry VIII and Francis I*, Ch. 8.
**94** Martin and Parker, *The Spanish Armada*, P.II.6 and IV.14.

**Figure 8:** Another instance of how Muscovy married amphibious riverine warfare with poliorcetics occurred when its troops invaded the fortress of Dorpat (Tartu) on the river Embach (Emajõgi) in Eastern Livonia (Estonia) in 1558. Again, an "aside build-up" stratagem

## The Worst of Amphibious Craft

An additional reason accounts for Ottoman conservatism besides solely doctrinal considerations. It was their fleet. Contrary to the current dominant historiography,[95] the Ottoman fleet was simply unfit for the prosecution of amphibious warfare during the gunpowder epoch. The Ottoman navy's main vessel, the galley, has the historiographical reputation of "the *sine qua non* of offensive strategic operations in Mediterranean waters,"[96] "the only viable warships" of a great "usefulness in amphibious warfare and sieges"[97] and "the perfect craft for amphibious raids and assaults on enemy coasts and ports."[98] At the very least, it has been argued that the galley was a "better choice at the time."[99] However, for amphibious operations it was the worst choice given the nature of the gunpowder revolution of the 1500s.

The galley barely had room for troops besides some cockpits on the deck, in the prow and aft, where soldiers travelled in extremely cramped conditions. Galleys had strictly limited space for heavy armor, pikes, powder, ladders, assorted fighting gear, food, and water. Troops transported by galleys arrived at the point of disembarkation tired, thirsty, hungry, disorganized, deprived of the tools of fighting and in need of rest. The battle crew of the galley consisted of sailors and land troops; in particular they carried the weight of combat in the event of a boarding action. Should circumstances mandate debarkation of auxiliary fighting crew members in support of an amphibious assault, a galley became defenseless against an adversary's fully manned vessel. Sea combat might commence at any time, with the marine troops for the amphibious action ashore. The galley's sole prow heavy gun was another assault tool of the galley. Due to its limited cargo capacity, the galley was unable to bring extra guns

---

**Figure 8** (continued)
carried the day. Handgunners spearheaded the landing. Their foothold was shielded by the erection of earth-filled baskets that enabled the positioning of artillery. The siege perimeter was then defended by army contingents, predominantly cavalry, countering the relief forces of the Livonian Order.[100] The image is reproduced by courtesy of the Department of Manuscripts, the State Historical Museum of Russia, Moscow. © «Исторический музей».

---

95 Imber, *The Ottoman Empire*, 311.
96 Guilmartin, "The Earliest Shipboard Gunpowder Ordnance," 653.
97 Glete, *Warfare at Sea*, 111, 144.
98 Morillo, Black and Lococo, *War in World History*, 280.
99 Phillips, "Navies and the Mediterranean in the Early Modern Period," 14.
100 *The Russian Illustrated Anthological Chronicle, the Synod's Volume*, СИН-962, Л. 372 Об.

besides its service piece, and it was routinely dismounted for the land batteries. Quite easily, the standard galley became fatally inferior to the well-armed galley of an opponent.

Galley fleets could not wage naval battle and perform amphibious operations simultaneously. The shortcomings of the galley virtually doomed amphibious assaults when the enemy's seaborne counter-amphibious operation endangered the attacking force. For example, the Spanish fleet at Djerba in 1560 and the Ottoman navy at Malta in 1565 were unable to intercept at sea and divert the enemy's relief armadas because, in both cases, they had neither the extra-crew troops nor heavy artillery in readiness onboard. They landed their troops onshore, the Spaniards to construct fortifications, the Ottomans to besiege Birgu. The Spaniards unladed artillery from their galleys to equip the fort. The Ottomans debarked ordnance to arm their siegeworks. Both the Spaniards and the Ottomans strove to complete their business (at Djerba and Malta respectively) in anticipation of a relief fleet's arrival. When the sails of the awaited force appeared, both armies hurried to rescue their ships, with inevitable casualties due to forced evacuation. The Ottomans were lucky to reembark and save their vessels; the Spaniards did not have time and lost theirs.

The galley's fit as a landing craft is likewise far from perfect. Its agility in leeward waters and a shallow draft were advantages; its comparatively feeble structure could put an entire enterprise at risk, however. The galley was also a perilous craft, notoriously unseaworthy in a storm. The operations of the aside build-up design carried normally outside of the objective's harbor were naturally endangered by rough seas, which could swamp a galley. The Mediterranean was more or less open sea on the North African coast. Ports, harbors, inlets, channels were the objectives of amphibious operations there, but they did not provide abundant venues for "aside" landing and build-up. Such amphibious operations required shelter of some sort for proper execution. This geographical prerequisite, one to which galleys were highly vulnerable, is why many amphibious ventures in the Mediterranean were cut short by the onset of violent weather. Algiers occupied an especially unfortunate location; Spanish operations there were doomed by storms in 1516, 1519, and 1541 with the significant losses of men, ships, and guns. The Swedish fleet was destroyed by a tempest, carrying a landing at Narva in 1574,[101] a pitiful mishap for the Swedish *dominium maris Baltici*.

The galley's ordnance was insufficient for bombardment of coastal defenses according to the new tactics because it consisted of one heavy piece that

---

**101** Petrov, *The City of Narva*, 102.

was reloaded slowly. Galleys had neither the firepower nor structural strength to heap firepower upon coastal fortifications, which in the 1500s were prepared to return fire vigorously. These considerations help explain why the Ottomans evaded "direct assaults."

The galley did well comparatively, briefly enjoying the status of the assault ship in the second and third decades of the sixteenth century. The Spaniards commenced their amphibious climb with the direct assault on Penon de Velez provided by the special floating battery constructed by Pedro Navarro in 1508. However, they diverted from the direct assault as soon as the initial reaction of the coastal defenders, shocked by the deck-to-shore gunfire, changed to the boom of the coastal fortification and counter-naval artillery since the 1530s. In 1538 at Preveza, Andrea Doria dared not contest the mighty ordnance of Preveza's castle, strengthened with ship guns hauled there by Barbarossa. In the coming inclement weather, his galleys were not seaworthy. Doria abandoned his littoral position, and Hayreddin Barbarossa preyed on Doria's ships that lagged behind. The fleet of Barbarossa was soon, in October 1538, wrecked by tempestuous weather; Doria immediately attacked the objective nearest at hand, the smaller fort of Castelnuovo (Herceg Novi). The direct assault by deck-to-shore gunnery and landing was carried out in exemplary fashion. Barbarossa emulated Doria in 1539, when he brought four artillery ships at Castelnuovo. The Spanish garrison obstinately engaged in a fire fight with the attacking ships and crushed a Turkish spearhead contingent. The stalemate was broken by the intervention of a Turkish land army, escorted by the navy, ending the bloody siege.

Since the 1540s, the function of the assault ship of amphibious operations had rendered the galley obsolete. The refitting of Spanish galleys, rearmed in the 1560s to 1570s, was unimpressive; their attack on problematic Penon de Velez in 1563, under Garcia de Toledo, reminiscent of Tunis in 1535, was supplemented by a battering from Portuguese *caravels*. Nor did Venice invest in substantial improvement of the galley's ordnance. The Republic's fleet under the formidable Sebastian Venier defaulted hopelessly at Durrës, Klis, and Castelnuovo in 1571. Amphibious assault by galley fleet, aside from raiding, required either escort by a land army (the customary Turkish practice) or ancillary gunships and carriers (the "Christian" way). The Turkish practice declined due to a lack of objectives achieved in both ways. John Guilmartin concurred with Jan Glete's calculation that sailing warships sailed into oblivion in the Mediterranean during the 1540s.[102] The effectiveness of amphibious operations of Christian states in the region

---

[102] Glete, *Warfare at Sea*, 102; Guilmartin, "The Earliest Shipboard Gunpowder Ordnance," 653.

disappeared with them. The balance of power between Ottomans and Hapsburgs, allegedly achieved in the Mediterranean in the last third of the sixteenth century, was a self-imposed stupor caused by chronic amphibious impotence. The absence of amphibious leverage to accomplish a decisive assault heavily influenced the waning of traditional Mediterranean strategy to close the 1500s.

## Control of the Littorals

The interplay between offensive tools and the means of resistance in armed struggles is a universal dialectic. Innovations in counter-amphibious measures reflect the revolutionary changes in amphibious warfare. Counter-amphibious methods in the early 1500s maintained their general consensus. The four types of sea control, consisting of domination over (1) the lines of communication, (2) the immediate area of operation, (3) the embarkation point of the invasion army, and (4) access to landing areas, remained classic naval counter-amphibious precautions. They were apparent in marine and riverine counter-amphibious struggle. While such military science is part and parcel of naval history, one must note that the developments that upturned amphibious warfare in the 1500s were not "naval" but shore based. Unlike the circumstances of land forces, water-borne attackers have a completely different type of opponent, the shore-based defenders. The Turks became acquainted with the "negative fire-control of the littorals" as early as June 1480 when their two consecutive amphibious attacks on fort St. Nicholas during the siege of Rhodes were repulsed by fierce Hospitallers' gunnery. In 1511, Afonso de Albuquerque wrestled the "fire-control of the littorals" from the fortress of Benasterim in the Goa's suburb and compelled the defenders to surrender.

Andrea Doria's failure to force an entrance into the Gulf of Preveza in 1538, an example of sixteenth-century counter-amphibious defense accomplished by Hayreddin Barbarossa, had precedent in amphibious operations fought in the region of the Red Sea. In 1516, Lopo Soares de Albergaria, the Portuguese governor of India, had sailed into the Red Sea, pursuing the Egyptian-Turkish fleet under the Egyptian admiral Amir Hussein and Turkish mercenary Salman Reis. The Turks and Egyptians found shelter in the harbor of Jidda (Jeddah). Amir Hussein had not spent in vain those years after his defeat in the harbor of Diu in 1509. He understood that weak coastal batteries had allowed Francisco de Almeida's incursion inside the harbor where the latter annihilated the Egyptian fleet at closequarters with waterborne artillery. Hussein rearranged the coastal defenses of the harbor at Jidda to prevent a repeat of that massacre of 1509. Albergaria tried to penetrate the harbor, but Hussein's heavy guns that guarded the port's entrance prevented him.[103] Albergaria was unable to elevate the barrels of his ships' guns to sight onshore targets in the bending and sloped entrance of the harbor. Stymied, Albergaria abandoned both the landing and the enterprise.

The setback of the Holy League's fleet under Don Juan in 1572 at Modon followed a similar pattern. The Portuguese attempt to sail from the Persian Gulf up the Euphrates to Basra in 1550 was stymied by Ottoman artillery positioned on the riverbanks. Similarly, Russian banks-based gunfire repulsed the Swedish naval invasion originating in the Gulf of Finland and proceeding upon the river Narova to the fortress of Narva in 1577. In 1592 a Dutch flotilla supporting French Protestants in the fortress of Caudebec "suffered badly . . . and eventually withdrew further downstream" from the bank-based gunnery of the Spanish army under Duke of Parma.[104] At the end of 1601, the Spanish onshore battery drove out the English naval squadron of Richard Leveson from the harbor of Castlehaven, Ireland. The battery was improvised with the guns of the Spanish ship forced "to run aground" by "Leveson's cannonades."[105] Defensive amphibious warfare had risen to the occasion and countered the offensive technical and tactical advances of the early 1500s. The amphibious gunpowder revolution had come full circle.

The events cited above demonstrate the need for re-evaluation of the sea littoral as the area of amphibious warfare. When coastal defenses boasted long-range guns, the expansive marine and riverine spaces, which were open waters from an operational point of view, became narrow waters tactically.[106] Harbor areas, minor gulfs, and straits between islands, many lakes, and almost all rivers fell under control from their shores by the positioning of land-based gunpowder weapons.

However, though disputed by offense and defense, those prime areas of amphibious operations, the waters nearby and the respective landing zones remained the stage upon which the amphibious Gunpowder Revolution proceeded and evolved. The conversion of open sea areas into narrow sea lanes had changed the rules of amphibious warfare dramatically. Now the intruder had to wrestle from the defender the control over the narrow waters, dictated by the weapons deployed ashore. It was a sharp twist in the evolution of amphibious warfare. The revolution continued into the twenty-first century as the

---

103 Guilmartin, *Gunpowder and Galleys*, 26–27.
104 Trim, "Medieval and Early-Modern Inshore, Estuarine, Riverine and Lacustrine Warfare," 392.
105 Fissel, "English Amphibious Warfare, 1587–1656," 239.
106 See terminology in Vego, *Naval Strategy and Operations in Narrow Seas*, 7.

improving range and effectiveness of land-based weapons converted all seas and oceans of the globe into narrow waters.

The new counter-amphibious defenses of threatened nations mirrored the seriousness with which they confronted the aggressive potential of amphibious warfare. England sensed herself one of the vulnerable geographies and fenced her beaches and harbors with coastal forts which "emphasized height and firepower, not tactical defensibility and siege-worthiness."[107] Gunfire control of the littorals was at the core of their design. The model created by the Portuguese empire in East Asia was strikingly counter-amphibious. The Portuguese not only maintained sea domination but also never ceded gunfire control of the littorals in defending their holdings. At Malacca their vigilance was expressed in the mighty coastal gun-tower. Aceh's attacks on Malacca in 1551, 1568, and 1573 failed due to the invader's inability to wrestle gunfire control of the littorals from the defenders. "Following the collapse of Spain's military position in 1576," the rebels of Netherlands hastily erected the network of the earth-timber bastions of the "Netherlands fortification style" for the firepower control of inner waterways, especially in the Rhine-Meuse-Scheldt delta.[108] Former port towns and even fishing villages were transformed into forts bristling with gunpowder armaments along the coasts of the Mediterranean, the Atlantic, the Baltic, and South Asia, as well as within the riverine interior of Europe and Northern Eurasia. Seemingly everywhere, strongholds embraced the advanced firepower that enabled them to control their littorals as a counter-amphibious safeguard. The arms race between ship-based and onshore weapons had commenced.

## Amphibious Mastery

The mutual relationship in the conjunction of shipboard gunnery with the tactical initiative embodied in landing troops' assaults was illustrated by the interactions between them. Ships blazed away at coastal defenses with their indispensable gunfire, and then staged the descent of the marines, watching over them as the disembarked warriors secured their foothold on terra firma. The marines were not on their own, as in the pre-gunpowder era, but enjoyed a symbiotic relationship with the vessels that had brought them thus far. Landing parties repaid their comrades aboard ship by neutralizing the enemy's gunnery from the coastal area, putting to the sword anyone and anything that endangered their support vessels

---

**107** Fissel, *English Warfare*, 37.
**108** Tracy, *The Founding of the Dutch Republic*, 148–150.

riding at anchor; landing parties might provide the ships with information on hostile gunfire positions via signaling. New operational complexities required expert management of the ships and landing troops, which were not strictly navy and army in the amphibious context, but components of a combined amphibious force. A new art of war was born, amphibious mastery, that can be characterized fairly as a military revolution. It is the specific military art of armed struggle in littorals, consisting of the combined application of the assault potential of the navy and landing troops against coastal (or shoreline) objectives. Amphibious mastery brought tactical superiority for forces which might be inferior to the enemy in general terms.

## Summarizing Mutation

Diagram 1 displays 174 operations from 1453 to 1601, registered for this essay in Tables 1 to 4. It includes 75 operations of the escort design, 72 operations of the aside build-up, and 28 direct assaults. Although it is saw-like in format, due to relatively small numbers, Diagram 1, "Major amphibious operations circa 1450–1600. Dynamic of designs," conveys the essential conclusions.

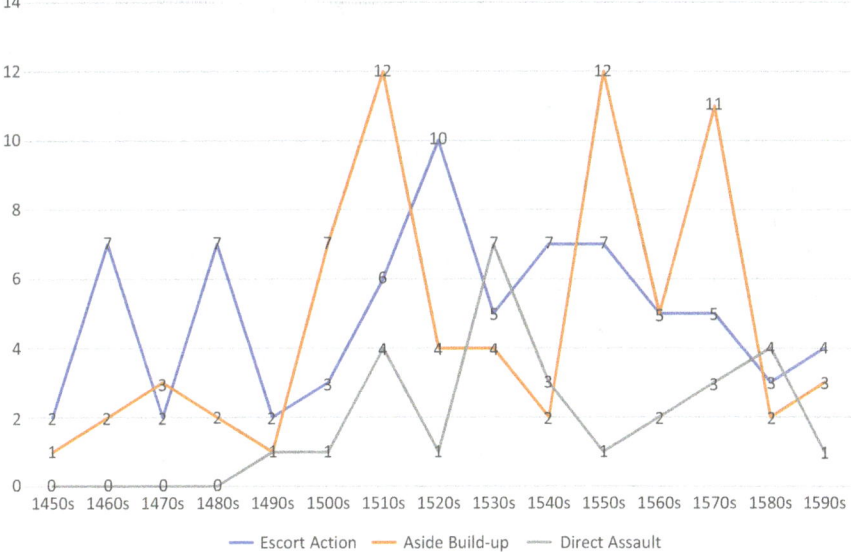

**Diagram 1:** Amphibious operations, dynamic of designs.

The intensity of amphibious operations grew in the second half of the fifteenth century; the "escort design" dominated the epoch when amphibious warfare entered its military revolution. In the sixteenth century, the volume of expeditions adhering to the "escort" paradigm slowly decreased; operations shifted to waterborne gunfire and transportation support by vessels to aid land (and landing) armies and converted themselves into expeditionary warfare or joint operations.

The rearmament of landing troops with handguns and field guns boomed after 1500. The fighting capacity of disembarked amphibious forces sharply improved. They became increasingly self-sufficient, though yoked with the support available from their ships. The "aside build-up" design was the tactical expression of the marines' new efficiency. The offensive performance of such troops rose sharply through the sixteenth century. Circa the 1510s, the adoption of the assault ship for amphibious operations, an unprecedented innovation of the early gunpowder era, perfected deck-to-shore gunfire. The advantage to precede and accompany the landing with direct deck-to-shore firepower was grasped by tacticians of the first half of the sixteenth century quickly and imaginatively. The utilization of a direct assault, incepted in the 1510s, may only have been favored by some of the powers conducting amphibious warfare, but these pioneers proved that tactic to be the most effective operational design. The consequent swelling wave of amphibious operations exhibiting all three designs circa 1490 to 1539 encapsulates the era of the gunpowder "mutation" of amphibious warfare. It also expresses how belligerent nations dared develop amphibious doctrines that helped them establish regional hegemonies. The new effectiveness of amphibious warfare promised victorious campaigns and strategically enlarged wars in regions once relatively inaccessible.

From the 1550s through the 1570s, intensive amphibious force projection clashed with counter-amphibious obstacles and offensives. Over the century, the previously mentioned defensive amphibious art of war accumulated the experience and equipment needed to repulse the novel amphibious attacks that gunpowder had made possible. Coastal fire-control of the littorals, a tactic no less innovative than the deck-to-shore gunfire assault, somewhat stalled the amphibious wave. From the 1580s onwards, amphibious warfare gravitated to a level of reciprocal normality until the next technological or organizational breakthrough, powered by the challenge-response mechanism discussed in military revolution historiography.

George Raudzens, profoundly skeptical of the notion of miracle weapons,[109] concluded that the initial colonial successes of Europeans overseas were achieved not by superior armament but by their transportation efficiency. It brought the West a superiority of numbers and logistical capabilities into areas of confrontation.[110] Raudzens may well be right if, after their long voyage, the Europeans did not land against the outward armed resistance of indigenous rulers, population, and troops. Precipitous landings upon arrival, tenacious contests for onshore footholds, the initial collisions of armed men penetrating unsecured ground, and subsequent (frequently violent) exploration of the environs, all these factors mattered more than Raudzens acknowledges. The transportation advantage, besides bringing onsite the intruders' numbers and material force, also had to force a point of entry. History chronicles how superior forces failed miserably due to their inability to implement their advantages and consolidate them. The transportation advantage requires amphibious superiority. The gunpowder mutation of amphibious warfare brought such leverage to Europeans. That appears to be true, at least until further global studies of comparative amphibious warfare are proffered, and this less known facet of the military revolution analyzed.

When historians estimate the profundity of the military revolution they ought to appreciate that initially the phenomenon did not achieve the impressive scale of the triumphal deeds of Maurice of Nassau's House of Orange and Gustavus Adolphus' spectacular Swedish adventures. The military revolution in amphibious warfare instead occurred within cramped perimeters of a few thousand square meters of land and water, precariously struggling in the littorals of a hostile shore in the midst of a hazardous landing. Such risky endeavors in a confined space remains an undeniable dimension of the military revolution and its decisiveness. For example, the taking of Malacca by the Portuguese in 1511 and Moscow's seizure of Isker, the capital of the Siberian khanate in 1582, were accomplished by direct assault of professional landing troops augmented by the deck-to-shore gunfire of the ships' ordnance. These emblematic and pivotal episodes of the European advances in East Asia and Northern Eurasia confirm the "Rise of the West" theory[111] by providing solid evidence of the military revolution in amphibious warfare.

---

109 Raudzens, "War-Winning Weapons," 403–434.
110 Raudzens, "Military Revolution or Maritime Evolution?," 631–641; for an insightful variation on that theme, see Sharman, *Empires of the Weak*.
111 Parker, *The Military Revolution*, Ch. 4.

## "Ideal Types" of Military Revolution on Warfare's "Periphery"

The gunpowder-armed assault ships of amphibious operations, especially their tactic of deck-to-shore gunfire, are phenomena that did not exist prior to the era of the military revolution. Although these operations belong not to mainstream arenas of warfare but to its environmental periphery, the specific instances of amphibious military revolution we have cited above exemplify the military revolution and its variations. Elaboration upon the general concept of military revolution and its accompanying debates generated Weberian-style "ideal types" or "orientational models" serving as researchers' "yardsticks"[112] to assess "how people would behave if they were guided by only specified motives under unrealistically simplified conditions."[113] One need not be misguided that the principal "ideal types" or "orientational models" of the military revolution (and the concept itself as their generalization) are reportage-style descriptions of actual military events and even processes. They are rather the instruments for gaining "insight into the real causal forces at work."[114] The amphibious operational designs of escort, aside build-up, and direct assault are "orientational models;" the "assault ship of amphibious operations," "tactics of deck-to-shore gunfire," and "shore-based fire-control of the littorals" are similarly "ideal types" created to analyze how major "ideal types" of the early-modern military revolution "worked" in the specific peripheral conditions of amphibious warfare.

The gunpowder revolution is undoubtedly the primary military historical yardstick linking the "linear tactic" of Michael Roberts, "bastion fortification" and "artillery sailing ship" of Geoffrey Parker, and "open sea domination" of Jan Glete. Similar benchmarks date from as long ago as the writings of Machiavelli, who proposed the "orientational model" of the "permanent military organization throughout the territory based on part-time soldiers,"[115] i.e., conscripted regular army. They are accompanied by sociological yardsticks like "state formation" and geopolitical ones such as the "Rise of the West." Creating the subordinating "ideal types" for early modern amphibious warfare not only helps outline patterns among the military changes of the epoch within this specialized field, but

---

112 Kalberg, *Max Weber's Comparative Historical Sociology*, 90, 93.
113 Derman, *Max Weber in Politics and Social Thought*, 144–145.
114 Derman, *Max Weber in Politics and Social Thought*, 145.
115 Hornqvist, "Perché non si usa allegare i Romani: Machiavelli and the Florentine Militia of 1506," 154.

also defines how the major "orientational models" reflect in a dimension of warfare that differs greatly from continental land warfare and naval warfare which were analyzed to generate them. The application of the major "ideal types" of the concept of military revolution to the "periphery of warfare" reveals how the real historical phenomena behind them emerged and developed and in what way military transformation, described by the concept of military revolution, consolidated.

Early modern warfare was practiced in idiosyncratic and unique peripheries; amphibious warfare engaged a natural environmental periphery totally different from the peculiarities of Eurasian steppe warfare. The elaboration of the main "orientational models" of the military revolution concept upon the mass of Western Europe formed not only its geopolitical overseas periphery, in East Asia, South Asia, and the Middle East, but also its semi-periphery within adjacent regions (wherein the military changes if measured by the mainstream "yardsticks" look distorted and underdeveloped). First of all, these are the regions of Eastern Europe and the Near East, with a trio of principal participants often regarded as paradigmatic imitators (or simulators) of the military revolution: the Ottoman Empire, Poland, and Russia (Muscovy). Until recently no sound theory of the emergence of their revolutionized military practices has been proposed besides their presumed adoption from the center of emission, Western Europe, which itself wasn't "revolutionary homogenous" inside, and where the epicenter of the military revolution moved, closing the sixteenth century, from South to Northwest (allegedly excluding Britain, which until comparatively recently was considered a backwater).

The idea of a "central emission – peripheral adoption" dynamic, integral within the military revolution, resides in the theory of international relations of "uneven and combined development." The theory is attributed to the Russian Marxist and founder of the Red Army, Leon Trotsky, and referred to the two miniscule and rather vague paragraphs among his 1,000-page *History of the Russian Revolution*, written in the 1920s.[116] Justin Rosenberg, who introduced intimations of Trotsky's thinking into the sociology of international relations as a full-scale theory, finds its core in the "spreading outward" of advanced technical and organizational forms from the international epicenter of development to its periphery, and their "fusion" with the outdated national forms in there. Rosenberg illustrates his thesis with the transfer of the economic and social forms of the "industrial capitalism" born in Northwestern Europe, which enormously boosted its might, to "tsarist" Russia, striving to catch with it, and mixed there in the "amalgama" with her

---

116 Trotsky, *History of the Russian Revolution*, 4–5.

intrinsic "elements."[117] Trotsky and Rosenberg propose the "whip of external necessity" as the force driving the adoption and "amalgama"-making adaptation of the domestic structures of less developed nations in relation to the West European spearhead of modernity.[118] They insist on "the centrality of international relations to understanding any national path of development."[119] Followers of the views of Trotsky and Rosenberg extrapolate this rule to the general "examining world history over the *longue durée*," including epochs preceding (and following) the advent of "industrial capitalism."[120] Accordingly, "the notion of core/periphery relations is a fundamental concept in this theoretical approach."[121] The actual theory of the diffusion of advanced military innovation from Western Europe's center to its geographical periphery and semi-periphery is the affiliate of this approach. In other words, from continental land warfare and naval warfare as impetuses, to amphibious warfare and Eurasian steppe warfare in their guises as environmental peripheries.

However, the development of the amphibious way of war of the fifteenth and sixteenth centuries, as outlined above, clashes with the core-periphery rule. The assault ship of amphibious operations was not an adoption of the naval ship designed for marine combat. As soon as technical progress made firearms available for onboard installation, both kinds of gunpowder warfare utilizing maritime vessels, naval ("central") and amphibious ("peripheral"), make use of this military revolution, developing different species of it and applying novel tactics in its use. When the handgun became "hit-effective" and "user-friendly," overland infantry and marines capitalized upon it as well. The latter forces developed various tactics for its use; the land-based version became known as "pike and shot," while the amphibious incarnation is not yet thoroughly researched. This is what above is described as the model labeled the aside build-up. It is conceivable that the advanced practices of warfare emerging in the epoch of the military revolution did so randomly. It is possible that the hierarchy of the environmental center and periphery of warfare did not truly exist and that transitions in military techniques were multi-directional.

The same could be argued about the interrelations of the seemingly geographical center, periphery, and semi-periphery of the military revolution. We

---

117 Rosenberg, "Isaac Deutscher," 6–8.
118 Rosenberg, "Uneven and Combined Development," 23; Trotsky, *History of the Russian Revolution*, 5.
119 Rosenberg, "Isaac Deutscher," 8.
120 Anievas and Matin, "Introduction," 3.
121 Chase-Dunn and Grell-Brisk, "Uneven and Combined Development in the Sociocultural Evolution of World-Systems," 205.

have proposed that the "semi-peripheral" Ottomans tried the tactics of deck-to-shore gunfire first at Bab al-Malik (1488). Their taking of Kaffa (1475) occurred between first amphibious operations of the aside build-up design, reinforced with firearms, and they were pioneers of the counter-amphibious "negative" control of the littorals through firearms in 1538 at Preveza. The escort design of amphibious operations, which as we have argued the Ottomans preferred, was not outdated when it was introduced in the middle of the fifteenth century, and it integrated advanced use of artillery by both the escorting navy and the escorted land forces. On the opposite fringe of Europe, Sweden lay distant from the supposed geopolitical "center" of the military revolution until the blossoming of Michael Roberts' military revolution of 1560 to 1660. Despite their semi-peripheral remoteness, in the last decade of the fifteenth century the Swedes pioneered a direct assault mode of amphibious operations. They did so via the employment of assault ships, deck-to-shore gunfire, and outright storming of objectives, e.g., Ivangorod, with professional marines equipped and trained to use handguns and artillery. The ways of war within the "semi-peripheries" were not so much the recipients of innovation; rather, they synthesized arts of war and diffused them further.

Amphibious warfare was not something unique in this process. "Pike and shot" tactics exemplify the military revolution concept's "ideal types." That art of war's origin is attributed to the supposed geopolitical center of the military transformations occurring during the 1500s, South Western Europe. The paradigms hold only if the military revolution is proven to be the monolithic constant of a period lasting one to two centuries. However, implacable and inconvenient factual details from sixteenth-century historical reality grind the monolith to rubble. The janissaries of Sultan Selim I, after lessons learned at the hands of Mamluk horsemen at Marj-Dabiq (1516), were rearmed with pikes allotted to half of the Ottoman infantry (whilst the other half consisted of archers and handgunners).[122] Although Gábor Ágoston ascribes Ottoman success in the battle of Raydaniyah (1517) to the adept use of "matchlockmen and artillery,"[123] the counterattack undertaken by the "pike and shot" janissaries, charging out of a wagon-camp, *tabur*, their customary tactical configuration,[124] looks more plausible. This instance may not conform with the commonly held image of "pike and shot" tactics practiced in the Eighty Years' War or Thirty Years' War;[125] but do the practices in actual combat contradict the paradigmatic formula? The battle of Raydaniyah, esteemed

---

122 Irwin, "Gunpowder and Firearms in the Mamluk Sultanate Reconsidered," 136.
123 Ágoston, "War-Winning Weapons?," 139.
124 Uyar and Erickson, *A Military History of the Ottomans*, 50–51.
125 See Roberts, *Pike and Shot Tactics*.

in Turkish collective historical memory, was arguably one of the finest applications of "pike and shot" tactics. Subsequently, "pike and shot" tactics were abandoned apparently. And Sultan Selim I is now commemorated as an advocate and practitioner of the *tabur* array.

The Polish infantry led into the battle at Obertyn in 1531, under *hetman* Jan Tarnowski, boasted a significant percentage of pikemen, perhaps as much as one third of his foot soldiers. The latter were native Polish troops, not German or Czech mercenaries.[126] Tarnowski's host operated differently from the "ideal model" of "pike and shot" tactics. Tarnowski's horse and foot were confined inside the wagon-camp, *oboz*, not independently on a field as their West-European counterparts, and operated from the *oboz* in the manner of repulse and sorties. When Tarnowski sent his best infantry to counter-charge out of the *oboz*, the assault array boasted a composition of 300 pikemen and 600 arquebusiers. The "shot" detachment was deployed not in front of the pike "square" or on its "wings" but behind the ranks of pikemen. The arquebusiers fired over the heads of the pikemen by exploiting the sloped trajectory without direct aiming.[127] The operational conduct of Tarnowski's foot was recognizably and fundamentally "pike and shot" tactics. The battle of Obertyn was so pivotal in the transformation of the Polish state that its consequences attest to the influence of "pike and shot" tactics on state-development. Then, "pike and shot" tactics lost favor among native Polish infantry. *Hetman* Tarnowski is now heralded as an exemplary practitioner of the *oboz* tactical array.

Tsar Ivan the Terrible's Muscovite infantry, transformed by his radical military reforms of 1550 to 1551, consisted of *streltsy* musketeers and dismounted retainers of aristocratic cavalrymen, the latter skilled in the use of the sabre. When Kazan was stormed (1552) after a breach in the city walls, the *streltsy*, along with the above-mentioned retainers under *voyevoda* prince Michael Vorotynsky, demonstrated a form of urban warfare consisting of charging through the streets with heavy long spears (*rogatina*) whilst handgunners heaped fire from the rooftops.[128] Although this instance was clearly not the "ideal type" of "pike and shot" tactics it raises questions about how standardized was the "classic" Western European model. The taking of Kazan was so instrumental in the consolidation of the Russian empire that its centrality could be viewed as a spectacular variant of "pike and shot" tactics that assumed geopolitical significance in contested "peripheral" lands. Subsequently, however, pike and shot tactics

---

126 Boldyrew, "The Armament of Polish Mercenary Infantry," 81.
127 Plewczynski, *Obertyn 1531*, 203.
128 Trofimov, *Campaign at Kazan*, 92–93.

were eclipsed among Muscovite foot soldiers for nearly a century. Prince Vorotynsky is now acclaimed as an exalted practitioner of *gulyaj-gorod* array,[129] an adaptation of the wagon-camp, resembling the Turkish *tabur* and Polish *oboz*.

Contrasting amphibious operational practices with traditional continental warfare, summarized above, shifts discourse regarding the diffusion of the military revolution. Discussion moves from "center-periphery" relations to analysis as to how (and why) similar practices emerged simultaneously, grew somewhere to a degree of revolutionary change (e.g., Northwestern Europe) while elsewhere they lapsed or deteriorated (e.g., Poland, Muscovy, and Turkey). Then there is the third option, which co-existed with "traditional" practices (England is a well-documented example[130]), sometimes remaining stillborn rather than stimulating revolutionary change in a nation's military practice and strategic culture. Answering these questions involves "disclosure of the enigma." In other words, how does one explain the unique emergence of military innovations which are enshrined as "ideal types" within the military revolution concept?

## Revealing the Enigma

Our analysis of amphibious warfare in the epoch of the military revolution reveals militarily based springs of innovation that bore revolutionary consequences. Technological (and tactical) innovations are not confined to drawing boards and military manuals. The crucible of combat itself generates new incarnations, frequently by improvisation and adaptation. Assault ships' tactics of deck-to-shore gunfire, innovative operational models of escort, the aside build-up, and direct assault amphibious operational designs were constructed and made manifest via action taken by commanders. In the "theory of cultural and social selection" of Walter Runciman, inventions such as these receive the accurate label of "mutant practices."[131] These invented "mutant practices" of fighting derived from the best practices of armies which, as Jan Glete established, were the complex professional and administrative organizations in a constant interaction with the economies, societies and political regimes.[132] The crossroads of implementation and development, refusal or isolation of "mutant practices" are navigated within the interaction of armies and nations.

---

**129** Davies, "Guliai-gorod, Wagenburg, and Tabor Tactics," 95.
**130** See Fissel, *English Warfare*.
**131** Runciman, *The Theory of Cultural and Social Selection*, 144.
**132** Glete, *War and the State*, 42–66.

**Figure 9:** This pair of miniatures displays the panoply of weapons used in Muscovy's wars: matchlock-firing infantry, bowmen, cavalry and pikemen. Practiced routinely in Eastern Europe, pike and shot tactics are regarded as emblematic of Western Europe's military. revolution. The illustrations depict Muscovite forces storming the city of Kazan in 1552. Probably based upon eyewitness evidence, these depictions appeared in the

**Figure 9** (continued)
*Moscow Illustrated Anthological Chronicle* of the sixteenth century.[133] The images are reproduced here by courtesy of the Department of Manuscripts, the State Historical Museum of Russia, Moscow. © «Исторический музей».

---

133  *The Russian Illustrated Anthological Chronicle, the Tsardom Book*, СИН-149, Л. 547 Об., 614 Об.

Four theories explain the outcomes. The first was expressed by Frederick Engels in his preparatory chapter to *Anti-Dühring*, written in 1877 but published only in 1935, with the self-explanatory title "The tactics of infantry and its material bases." The social composition of armies was presented as the channel transferring technical innovations into tactical models.[134] This social channel filtered as well and sorted the tactical models that were potentially militarily successful but socially unacceptable as advanced tactics for less developed societies. Following Engels' thinking, it is probable that "pike and shot" tactics were refuted in Poland, the Ottoman Empire, and Muscovy, due to what Engels regarded as their lagging behind in social-economic development. Innovative forms perished because they failed to conform to the social and material culture of the army as a whole, or its soldiers and officers.

The second theory was hinted at by Leon Trotsky, whose ideas of "external necessity" are referred to above. The "mutant practices" of Walter Runciman are particularly urgent in war because "if outdated military practices are not replaced in time, the army will go down to defeat and the country be conquered."[135] Trotsky's idea is founded upon the well-known Marxist maxim of the relative autonomy of the state as an institution.[136] The state, when endangered from outside, adopted tactical models considered superior, either from domestic practices or via importation, arranged the necessary supply of weaponry, human resources, trained commanders, and insisted on their application. Trotsky's well-honed ideas of "uneven and combined development" permeated contemporary political practice (ironically, for example, Josef Stalin's charge to the "socialism in a single country"[137]) and remain valuable insights. Considering Trotsky's views on state formation and warfare, we might infer that the "pike and shot" tactics in Poland, Turkey, and Muscovy were first imported by expansionist states from Western Europe, then utilized and ultimately rejected due to poor capabilities in organizing their military via state bureaucracy. In other words, political failure associated with state formation was paramount in the implementation of military revolutionary transformation, as suggested by Aliaksandr Kazakou in the current volume.

A third theory is that proposed by sociologist Charles Tilly, who cites the example of Poland as illustrative of how state formation may be reversed due to the

---

134 Engels, "Infantry Tactics and its Material Bases 1700–1870," 20, 655–662.
135 Runciman, *The Theory of Cultural and Social Selection*, 145.
136 For more detail, see Haldon, *The State and the Tributary Mode of Production*, 33–39.
137 Day, *Leon Trotsky and the Politics of Economic Isolation*, 161; Trotsky, *The Stalin School of Falsification (1937)*.

decline of "capital concentration" and "concentration of coercion means."[138] Tilly combines the ideas of Engels and Trotsky positing that the Polish state lost its integrating momentum due to the decline of its social base demanding such reforms. In the same way, it could be affirmed that the reverse of the military revolution is possible for some nations' armies, as the peril of retrogression cited in Campbell Dalrymple's cyclical view circa 1761.[139] Some powers experienced a steadily advancing dynamic characterized as military revolution. Other polities oscillated, advancing only to reverse progress consciously or unawares. And then there were nations that suffered repeated retrogression when attempting to implement substantive change in their way of war. In the case of "pike and shot," tactics serving as the "orientational model," Poland and Muscovy experienced alternating advancing-reversing trends; while the Ottomans succumbed to singular reversal despite their spectacular advance previously.

The fourth theory is that of Michael Mann, who sees societies as "multiple overlapping and intersecting sociospatial networks of power," which includes four sources of "ideological, economic, military, and political (IEMP) relationships."[140] Mann's "sources of power" are not static pools. They are the streams of conflicts, "their stable formation in a particular historical period are called 'crystallizations.'"[141] Walter Runciman insists that "the prerequisite for change is that mutant practices must appear, no matter from where, which either modify existing roles or create new ones."[142] This suggests that the military revolution as an agent of radical change "crystallizes" only if the fighting innovations of the military "network of power" alters the social roles in other "networks." If they are not able to produce the changes, they die out. The political, economic, and ideological "networks of social power" in Muscovy, Poland, and the Ottoman Empire occurred were insufficiently pliable to be transformed by the "mutant practices." In the end, these transformations were rejected despite their fighting efficiency. The "networks of social power" in some countries of Western Europe allowed the changes (Glete [2002], generalized them as the "fiscal-military state"), and the "mutant practices" of warfare "crystallized" as military revolution.

Combining Mann's theory with Glete's insights suggests two distinctive stages of the military revolution that sometimes coincided. Elsewhere, as in the semi-periphery of Eastern Europe and the Near East, these phases did not align closely. The first is the "military" stage when the fighting innovations are

---

138 Tilly, *Coercion, Capital and European States*, 59–60.
139 Dalrymple, *A Military Essay*.
140 Mann, *The Sources of Social Power*, 1–2.
141 Collins, "Mann's Transformation of the Classic Sociological Traditions," 20.
142 Runciman, *The Theory of Cultural and Social Selection*, 144.

**Figure 10:** Makeshift field fortifications were East-European and Near-Eastern battlefield countermeasures against the offensive capabilities of cavalry similar to the West-European massive pike and shot infantry formations. Muscovy's foot toting handguns, ensconced in a

generated and assimilated. The second is a "state-building" phase wherein the consequences of military change are either absorbed by the political, social, and economic structures (and hence transformed) or the results are rebuffed. Perceiving military revolution as an integral continuity of its two stages makes that concept a useful yardstick for assessing military development in different regions and nations.

The four theories described above are not mutually contradictory. Viewed via the military revolution as it was actually practiced within the geographical semi-periphery these theories appear as processes involving the emergence of revolutionary "mutant practices"; the latter in turn were rejected or isolated by impotent state systems and/or inflexible societies due to unfavorable political conditions, backward class structure, and rather comparatively underdeveloped economies. Military revolution in Muscovy, Poland, and the Ottoman Empire in its earlier incarnation possessed powerful and innovatory dynamics similar to those in Western Europe. However, some states were unable to accommodate the economic, social and political changes requisite for the paradigmatic military revolution. Of the "peripheral" states, only Russia managed to harness the military-charged economic, social and political dynamic in the second half of the seventeenth century. Social dynamics and geography created conditions conducive for the birth of a Russian "fiscal-military state" that ultimately made Russia a hegemonic empire.

The proper understanding of the nature of military revolution on the geopolitical semi-periphery of Eastern Europe and the Near East is problematic because its specific "ideal types" are neither adequately developed nor universally accepted. The military revolutions in Russia and Poland are measured mainly by Western European benchmarks. Only the Ottoman version of military revolution has its own "orientational models," thanks due largely to numerous studies that

---

**Figure 10** (continued)
wooden field fortification, an *ostrog*, repels Crimean and Turkish soldiers at Sudbishchi, near Tula, Russia. On 4 July 1555, the second day of battle, the Crimean-Turkish army deployed their field artillery against the Muscovite *ostrog*-anchored array as well as handgunners.[143] Note the presence of the traditional composite bow archers on both sides, a fixture of steppe warfare since at least the first millennium BCE. These mounted and dismounted archers are a reminder that elements of continuity existed even in the midst of military revolutions. The image is reproduced here by courtesy of the Department of Manuscripts, the State Historical Museum of Russia, Moscow. © «Исторический музей».

---

143 *The Russian Illustrated Anthological Chronicle, the Synod Volume*, СИН-962, Л. 195.

**Figure 11:** The Baltic Crusading Orders were among the most innovative military forces in Eastern Europe. The army of the Livonian Order (under the Master's coadjutor Gotthard von Kettler) moved against the Muscovy-held castle of Ringen (Rõngu, Estonia) using a

have placed Ottoman developments in a broad geographical and chronological perspective (such as that of Wayne E. Lee in the present volume). One must approach the subject comparatively and regionally. Neither Brian Davies, venturing an extensive study of the wagon-camp array in Eastern Europe,[144] nor Gábor Ágoston, accentuating the tactical importance of the wagon-camp for the Ottoman gunpowder innovations,[145] propose *oboz*, *gulyaj gorod*, and *tabur* as orientational models for the military revolution in their respective regions. Both authors do explain how the tactics of the wagon-camp were applied by the Muscovite, Polish, and Turkish armies in field array by infantry equipped with firearms. Undoubtedly the wagon-camp array was "more defensive and static" than Western European pike formations (which were surprisingly mobile) but its tactical function was identical. It was the "the protection and defensive role . . . for the arquebus/musket carrying infantry."[146] Pike and spear formations among foot soldiers existed along with wagon-camp arrays. What remains significant is how the phenomenon of the military revolution prompted their integration with the "shot," the detachments of handgunners. Both practices exemplify how the gunpowder revolution incorporated traditional tactical forms into innovative practice of fighting.

Taking an even broader perspective, note that key victories in East Europe and the Middle East during the 1500s were achieved via wagon-camp types of deployment of infantry bearing firearms: Kletsk (1506), Lopuszno (1512), Caldiran (1514), Marj Dabiq (1516), Raydaniyah (1517), Mohács (1526), Jam (1528), Obertyn (1531), Sudbishchi (1555), and Molodi (1572) and so forth. Maneuvering gun-toting infantry in a wagon-camp array differs from the "pike and shot" tactics, but must revolutions in military affairs be perfectly identical? The euphemism of *tabur*, *oboz*, *gulyaj gorod*, Turkish, Polish, Muscovite versions of wagon-camp as the field infantry array, substituting in the description of some military events in Eastern Europe and the Near East regionally modified "pike and shot" tactic, is

---

**Figure 11** (continued)
wagon-camp array strengthened with an impressive amount of artillery and handguns. One should note the wide variety of battlefield fortifications depicted in this chapter's illustrations. Here the attacking Livonian Order holds out defensively against the numerically superior troops fielded by Moscow.[147] Again, these figures include bowmen among the horse and foot. The image is reproduced here by courtesy of the Department of Manuscripts, the State Historical Museum of Russia, Moscow. © «Исторический музей».

---

144 Davies, "Guliai-gorod, Wagenburg, and Tabor Tactics."
145 Ágoston, "Firearms and Military Adaptation," 85–124.
146 Uyar and Erickson, *A Military History of the Ottomans*, 50.
147 *The Russian Illustrated Anthological Chronicle, the Synod Volume*, СИН-962, Л. 406.

an example of the widely mentioned but not properly abstracted "orientational model" of the military revolution.

Similar euphemisms, inadequately abstracted to fit within the parameters of "ideal types" of military revolution in semi-peripheries and peripheries both environmental and geopolitical, exist in the discourses on these regions' military technology, organization, recruitment patterns, command structure, tactics, and strategies. Reconciling ideal types with reality encompasses explaining economics, social affairs, and state-formation. Until these variations on a concept are properly analyzed and defined, proposing "the" military revolution as a major phenomenon that has dictated the course of history will be a dubious exercise. Studying military revolutions occurring on peripheries, geopolitical and environmental, requires similar but unique yardsticks that comprehend the role of "ideal types" and "orientational models" as tools for understanding historical reality. Amphibious warfare demonstrates how insightful the application of this Weberian instrument is when assessing military revolutions on the "peripheries" of warfare.

# Bibliography

Ágoston, Gábor. "War-Winning Weapons? On the Decisiveness of Ottoman Firearms from the Siege of Constantinople (1453) to the Battle of Mohács (1526)." *Journal of Turkish Studies* 39 (December 2013): 129–143.

Ágoston, Gábor. "Firearms and Military Adaptation: The Ottomans and the European Military Revolution, 1450–1800." *Journal of World History* 25.1 (2014): 85–124.

Albuquerque, Afonso de. *The Commentaries of the Great Afonso Dalboquerque, Second Viceroy of India, Volume 3*. Cambridge/New York/Melbourne: Cambridge University Press, 2010.

Allen, Bruce Ware. *The Great Siege of Malta. The Epic Battle between the Ottoman Empire and the Knights of St. John*. Lebanon: ForeEdge, 2015.

*Amphibious Operations. Joint Publication 3-02*. The Joint Chiefs of Staff of the US Armed Forces. (Washington, DC, 4 January 2019).

Anievas, Alexander, and Kamran Matin. "Introduction: Historical Sociology, World History and the 'Problematic of the International.'" In *Historical Sociology and World History. Uneven and Combined Development over the Longue Durée*, edited by Alexander Anievas, and Kamran Matin. London/New York: Rowman & Littlefield International, 2016.

Atauz, Ayse Devrim. "Trade, Piracy, and Naval Warfare in the Central Mediterranean: The Maritime History and Archaeology of Malta." Doctoral dissertation. Texas A&M University, 2004.

Baudart, Guillaume. *Les Guerres de Nassau*. Amsterdam, 1616. https://play.google.com/books

Bées, N., and Savvides, A. "Navarino." In *The Encyclopedia of Islam, New Edition, Volume VII: Mif–Naz*, 1037–1039. Leiden/New York: Brill, 1993.

Bethencourt, Francisco. "The Political Correspondence of Albuquerque and Cortes". In *Correspondence and Cultural Exchange in Europe, 1400–1700*, edited by Francisco Bethencourt and Florike Egmond, 219–74. Cambridge/New York/Melbourne: Cambridge University Press, 2007.
Bill, Jan. "Scandinavian Warships and Naval Power in the Thirteenth and Fourteenth centuries." In *War at Sea in the Middle Ages and the Renaissance*, edited by John B. Hattendorf, and Richard W. Unger. Woodbridge: Boydell Press, 2003.
Biskup, Marian. *Wojna trzynastoletnia*. [Thirty Years' War; in Polish] Krakow: Krajowa Agencja Wydawnicza, 1990.
Boldyrew, Aleksander. "The Armament of Polish Mercenary Infantry in the First Part of the 16th Century." *Fasciculi Archaeologiae Historicae* 27 (2014): 79–85, https://journals.iaepan.pl/fah/article/view/1475.
Braudel, Fernand. *The Mediterranean and the Mediterranean World in the Age of Philip II. V. 2*. Oakland: University of California Press, 1996.
Brummett, Palmira. *Ottoman Seapower and Levantine Diplomacy in the Age of Discovery*. Albany: State University of New York Press, 1994.
Cahill Matthew J. *An Unassailable Advantage: The British Use of Principles of Joint Operations from 1758–1762*. Fort Leavenworth: United States Army Command and General Staff College, 2017.
Capponi, Niccolo. *Victory of the West. The Story of the Battle of Lepanto*. London: Macmillan, 2006.
Carvalhal, Helder, and Roger Lee de Jesus. "The Portuguese participation in the conquest of Tunis (1535): a social and military reassessment." In *Estudios sobre guerra y sociedad en la Monarquía Hispánicaguerra marítima, estrategia, organización y cultura militar (1500–1700)*, edited by Enrique García Hernán and Davide Maffi, 169–87. Valencia: Albatros, 2017.
Castillo, Dennis. *The Maltese Cross. A Strategic History of Malta*. Westport and London: Praeger, 2005.
Casale, Giancarlo. *The Ottoman Age of Exploration*. Oxford/New York: Oxford University Press, 2010.
Chase-Dunn, Christopher, and Marilyn Grell-Brisk. "Uneven and Combined Development in the Sociocultural Evolution of World-Systems." In *Historical Sociology and World History. Uneven and Combined Development over the Longue Durée*, edited by Alexander Anievas, and Kamran Matin. London/New York: Rowman & Littlefield International, 2016.
Cippola, Carlo M. *Guns, Sails and Empires: Technological Innovation and European Expansion 1400–1700*. New York: Pantheon Books, 1965.
Clough, Cecil H. "The Romagna campaign of 1494: a significant military encounter." In *The French Descent into Renaissance Italy, 1494–1495. Antecedents and Effects*, edited by David Abulafia. 191–216. Abingdon/New York: Routledge, 1995.
Collins, Randall. "Mann's Transformation of the Classic Sociological Traditions." In *The Anatomy of Power. The Social Theory of Michael Mann*, edited by John A. Hall, and Rolf Schroeder. Cambridge/New York/Melbourne: Cambridge University Press, 2006.
Connolly S. J. *Contested Island. Ireland 1460–1630*. Oxford/New York: Oxford University Press, 2007.
Correa, Gaspar. *The Three Voyages of Vasco Da Gama and His Viceroyalty from the Lendas Da India of Gaspar Correa. Accompanied by Original Documents. c. 1583*. Translated by Henry E.J. Stanley. New York: Burt Franklin Publisher, 1849.

Costa e Sousa, Luis. "From Tangier to Alcácer Quibir: The Portuguese Military Revolution (Re) visited." *Portuguese Studies Review (SP)* (2015): 1–29.

Crowley, Roger. *Empires of the Sea. The Final Battle for the Mediterranean, 1521–1580.* London: Faber and Faber, 2008.

Dalrymple, Campbell. *A Military Essay: Containing Reflections on the Raising, Arming, Cloathing, and Discipline of the British Infantry and Cavalry, with Proposals for the Improvement of the Same.* London: D. Wilson, 1761, https://play.google.com/store/books.

Davies, Brian L. "Guliai-gorod, Wagenburg, and Tabor Tactics in 16th–17th Century Muscovy and Eastern Europe." In *Warfare in Eastern Europe, 1500–1800*, edited by Brian L. Davies. Leiden/Boston: Brill, 2012.

Day, Richard B. *Leon Trotsky and the Politics of Economic Isolation.* Cambridge: At the University Press, 1973.

de Groot, Bouko. *Dutch Navies of the 80 Years' War 1568–1648.* Oxford: Osprey Publishing, 2018.

de Groot, Bouko. *Nieuwpoort 1600: The First Modern Battle.* Oxford: Osprey Publishing, 2019.

de Jesus, Roger Lee Pessoa. *O segundo cerco de Diu (1546): estudo de história política e militar.* [The second siege of Diu (1546): study of political and military history; in Portuguese] Coimbra: Universidade de Coimbra, 2012.

Derman, Joshua. *Max Weber in Politics and Social Thought. From Charisma to Canonization.* Cambridge/New York/Melbourne: Cambridge University Press, 2013.

DeVries, Kelly. "The Effectiveness of Fifteenth-Century Shipboard Artillery." *The Mariner's Mirror* 84 (1998): 389–399.

DeVries, Kelly, et al. *Battles of the Medieval World, 1000–1500. From Hastings to Constantinople.* London: Metro Books, 2006.

Diffie, Bailey W., and George Winius. *Foundations of the Portuguese Empire, 1415–1580.* Minneapolis: University of Minnesota Press, 1977.

Doctrine for amphibious operations. US Departments of the Army and the Navy. Washington DC (July 1962).

Domingues, Francisco Contente. "The State of Portuguese Naval Forces in the Sixteenth Century." In *War at sea in the Middle Ages and the Renaissance*, edited by John B. Hattendorf, and Richard W. Unger, 187–198. Woodbridge: Boydell Press, 2003.

Dotson, John. "Venice, Genoa and control of the Seas in the Thirteens and Fourteens Centuries." In *War at Sea in the Middle Ages and the Renaissance*, edited by John B. Hattendorf, and Richard W. Unger, 119–136. Woodbridge: Boydell Press, 2003.

Duro, Cesáreo Fernández. 1895. Armada Española (desde la unión de los reinos de Castilla y Aragón). Tomo I. [Spanish Navy (from the union of the kingdoms of Castile and Aragon). Volume I; in Spanish], https://armada.defensa.gob.es/html/historiaarmada/tomo1/tomo_01_21.pdf.

Engels, Frederick. "Infantry Tactics and its Material Bases 1700–1870." *Anthology.* K. Marx, and F. Engels. Moscow: Politizdat, 1967.

Finkel, Caroline. *Osman's Dream: The Story of the Ottoman Empire 1300–1923.* New York: Basic Books, 2006.

Fissel, Mark Charles. "English Amphibious Warfare, 1587–1656." In *Amphibious Warfare 1000–1700. Commerce, State Formation and European Expansion*, edited by D.J.B. Trim, and Mark Charles Fissel, 217–261. Leiden/Boston/Kohl: Brill, 2011.

Fissel, Mark Charles. *English Warfare 1511–1642.* London and New York: Routledge, 2001.

Fissel, Mark Charles. "Out of Africa: The Egyptian Origins of Amphibious Warfare." In *The Routledge Handbook of the Global History of Warfare*, edited by Kaushik Roy and Michael Charney. London/New York: Routledge, forthcoming.

Giakoumis, Konstantinos. "The Ottoman campaign to Apulia (1480–81)." In *The Turks, Vol. 3. Ottomans*, edited by Hasan Celal Güzel, C. Cem Oğuz, and Osman Karatay, 189–197. Ankara: Yeni Türkiye Yayınları, 2002.

Giovio, Paolo. *La seconda parte dell'istoria del suo tempo*. Venetia, 1560.

Glete, Jan. *Warfare at Sea, 1500–1650. Maritime Conflicts and the Transformation of Europe*. Abingdon/New York: Routledge, 1999.

Glete, Jan. *War and the State in Early Modern Europe. Spain, the Dutch Republic and Sweden as Fiscal-Military States, 1500–1660*. London/New York: Routledge, 2002.

Glete, Jan. "Naval Power and Control of the Sea in the Baltic in the Sixteenth Century." In *War at Sea in the Middle Ages and the Renaissance*, edited by John B. Hattendorf, and Richard W. Unger, 217–32. Woodbridge: Boydell Press, 2003.

Glete, Jan. "Amphibious Warfare in the Baltic, 1550–1700." In *Amphibious Warfare 1000–1700. Commerce, State Formation and European Expansion*, edited by D.J.B. Trim and Mark Charles Fissel. Leiden/Boston/Kohl: Brill, 2011.

Goffman, Daniel. *The Ottoman Empire and Early Modern Europe*. Cambridge/New York/Melbourne: Cambridge University Press, 2009.

Grosjean, Alexia. *An Unofficial Alliance, Scotland and Sweden 1569–1654*. Leiden/Boston: Brill, 2003.

Gürkan, Emrah Safa. "Ottoman Corsairs in The Western Mediterranean and their Place in The Ottoman – Habsburg Rivalry (1505–1535)." Master's Thesis. Bilkent University, Ankara, 2006.

Guilmartin, John Francis. "The Earliest Shipboard Gunpowder Ordnance: An Analysis of Its Technical Parameters and Tactical Capabilities." *The Journal of Military History* 71 (2007): 649–669.

Guilmartin, John Francis. *Galleons and Galleys*. London: Cassell, 2002.

Guilmartin, John Francis. *Gunpowder and Galleys. Changing Technology and Mediterranean Warfare at Sea in the 16th Century*. Revised Edition: Conway Maritime Press, 2003.

Guilmartin, John Francis. "The Siege of Malta, 1565." In *Amphibious Warfare 1000–1700. Commerce, State Formation and European Expansion*, edited by D.J.B. Trim, and Mark Charles Fissel. Leiden/Boston/Kohl: Brill, 2011.

Haldon, John. *The State and the Tributary Mode of Production*. London and New York: Verso, 1993.

Hall, Bert S. *Weapons and Warfare in Renaissance Europe. Gunpowder, Technology, and Tactics*. Baltimore and London: The John Hopkins University Press, 1997.

Hammer, Paul E.J. *Elizabeth's Wars. War, Government and Society in Tudor England, 1544–1604*. New York: Palgrave Macmillan, 2004.

Hattendorf, John B. *Naval Strategy and Power in the Mediterranean. Past Present and Future*. London/Portland: Frank Cass, 2000.

Hess, Andrew C. *The Forgotten Frontier. A History of the Sixteenth-Century Ibero-African Frontier*. Chicago/London: University of Chicago Press, 1978.

Hornqvist, Mikael. "Perché non si usa allegare i Romani: Machiavelli and the Florentine Militia of 1506." *Renaissance Quarterly* 55.1 (2002): 148–191.

Imber, Colin. *The Ottoman Empire, 1300–1650. The Structure of Power*. New York: Palgrave Macmillan, 2002.

Inalcik, Halil. 'The Ottoman Turks and the Crusaders, 1451–1522.' In *A History of the Crusades, Volume VI. The Impact of the Crusades on Europe*, edited by Kenneth M Setton, Harry W. Hazard, and Norman P. Zacour, 311–353. Madison: The University of Wisconsin Press, 1990.

Inalcik, Halil. "Osman Ghazi's Siege of Nicaea and the Battle of Bapheus." In *The Ottoman Emirate (1300–1389). Halcyon Days in Crete I. A Symposium Held in Rethymnon 11–13 January 1991*, edited by Elizabeth Zachariadou, 77–98. Rethymnon: Crete University Press, 1993.

Inalcik, Halil. 'The Socio-Political Effects of the Diffusion of Fire-Arms in the Middle-East' In *War, Technology and Society in the Middle East*, edited by V.J. Parry, and M.E. Yapp, 195–217. London: Oxford University Press, 1975.

Irwin, Robert. "Gunpowder and Firearms in the Mamluk Sultanate Reconsidered." In *The Mamluks in Egyptian and Syrian Politics and Society*, edited by Michael Winter, and Amalia Levanoni. Leiden/Boston: Brill, 2004.

Isom-Verhaaren, Christine. "Barbarossa and His Army Who Came to Succor All of Us": Ottoman and French Views of Their Joint Campaign of 1543–1544." *French Historical Studies* 30.3 (2007): 395–425.

Ivanov, N.A [Иванов Н. А.] *Османское завоевание арабских стран. 1516–1574* [Ottoman conquest of the Arabian nations; in Russian] Москва: Изд. фирма «Восточная литература» РАН, 2001.

Jacobi J.M. *Histoire générale de la Corse, depuis les premiers temps: Volume 1*. Paris: Aime Andre, Libraire, 1835, https://play.google.com/books/reader?id=N7sBAAAAYAAJ&hl=ru&pg=GBS.PP9.

*John Cassell's Illustrated History of England. Volume 2*. London/Paris/New York: Cassell Petter & Galpin, 1909, https://archive.org/details/cassellsillustra02lond/page/n8/mode/2up.

Jorgensen, Christer, et al. *Fighting Techniques of the Early Modern World*. New York: Thomas Dunne Books, 2006.

Kalberg, Stephen. *Max Weber's Comparative Historical Sociology*. Cambridge: Polity Press, 1994.

Kamen, Henry. *Spain's Road to Empire. The Making of a World Power, 1492–1763*. London: Penguin Books, 2003.

Kiss, Tamás. "Cyprus in Ottoman and Venetian Political Imagination c.1489–1582." Doctoral dissertation. Central European University, Budapest, 2016.

Konstam, Angus, and Gerry Embleton, Gerry. *The Barbary Pirates 15th-17th Centuries*. Oxford: Osprey Publishing, 2016.

Kortepeter Carl M. *Ottoman Imperialism During the Reformation: Europe and the Caucasus*. New York: New York University Press, 1972.

Kirk, Thomas. "Chapter Three. Giovanni Andrea Doria: Citizen of Genoa, Prince of Melfi, Agent of King Philip II of Spain." In *Double Agents: Culturaland Political Brokerage in Early Modern Europe*, edited by Marika Keblusek, and Badeloch Vera Noldus. Leiden/Boston/Kohl: Brill, 2011.

Kist, Bas. "A Spanish-Dutch military confrontation. The battle of Nieuwpoort, July 2nd 1600." *Militaria. Revista del cultura military* 7 (1995): 297–302.

Lambert, Craig L. *Shipping the Medieval Military. English Maritime Logistics in the Fourteenth Century*. Woodbridge: Boydell Press, 2011.

Liles, Christian Frederick Michael. "The development of amphibious/expeditionary warfare in the United States and the United Kingdom, 1945–1968: a study in comparison, contrast and compromise." Doctoral thesis. King's College, London, 2010.

Lockhart, Paul Douglas. *Denmark, 1513–1660. The Rise and Decline of a Renaissance Monarchy*. Oxford/New York: Oxford University Press, 2007.

Longworth Dames, Mansel. "The Portuguese and Turks in the Indian Ocean in the Sixteenth Century." *The Journal of the Royal Asiatic Society of Great Britain and Ireland*, 1 (1921): 1–28.

Magoulias, Harry J. Doukas. *Decline and Fall of Byzantium to the Ottoman Turks*. Detroit: Wayne State University Press, 1975.

Malcolm, Noel. *Agents of Empire. Knights, Corsairs, Jesuits and Spies in the Sixteenth-Century Mediterranean World*. Oxford/New York: Oxford University Press, 2015.

Mallett, Michael Edward. "Part I. C. 1400 to 1508." In *The Military Organisation of a Renaissance State: Venice C.1400 to 1617*, edited by M.E. Mallett and J.R. Hale, 1–210. Cambridge/New York/Melbourne: Cambridge University Press, 2006.

Mallett, Michael Edward, and Christine Shaw. *The Italian Wars 1494–1559. War, State and Society in Early Modern Europe*. London and New York: Routledge, 2012.

Mann, Michael. *The Sources of Social Power. Vol. 1. A History of Power from the Beginning to AD 1760*. Cambridge/New York/Melbourne: Cambridge University Press, 2012.

Martin, Colin, and Geoffrey Parker. *The Spanish Armada: Revised Edition*. Manchester: Mandolin, 1999.

McRoberts R. W. "An Examination of the Fall of Malacca in 1511." *The Journal of Malaysian Branch of the Royal Asiatic Society (JMBRAS)* 57.1 (1984): 26–39.

Meyer, Jean. "States, Roads, Armies, and the Organization of Space." In *War and Competition between States*, edited by Philippe Contamine. Oxford/New York: Oxford University Press, 2000.

Morillo, Stephen, Jeremy Black, and Paul Lococo. *War in World History. Society, Technology, and War from Ancient Times to the Present, V.1. To 1500*. Boston/New York: McGraw Hill, 2009.

Moro, Federico. "Venetia Rules the Rivers. La Geo-Strategia Fluviale Veneziana (1431–1509)," *Nuova Antologia Militare* Fascicolo 2.7 (Giugno 2021): 27–29, https://www.nam-sism.org/3.2%20%20fascicoli.html.

Murdoch, Steve. *The Terror of the Seas? Scottish Maritime Warfare 1513–1713*. Leiden/Boston/Kohl: Brill, 2010.

Nicolle, David, and Angus McBride. *Armies of Medieval Russia 750–1250*. Oxford: Osprey Publishing, 1999.

Nicolle, David, and Christa Hook. *Constantinople 1453. The End of Byzantium*. Oxford: Osprey Publishing, 2000.

Nowaczyk, Bernard. *Chojnice 1454, Swiecino 1462*. [In Polish] Warszawa: Bellona, 2012.

Oman, Charles. *A History of the Art of War in the Sixteenth Century*. London: Greenhill Books, 1989.

Ozbaran, Salih, and Dom Manuell de Lyma. "The Ottoman Turks and the Portuguese in the Persian Gulf (1534–1581)." *Journal of Asian History* 6.1 (1972): 45–87.

Parker, Geoffrey. *The Military Revolution. Military Innovation and the Rise of the West 1500–1800*. Cambridge/New York/Melbourne: Cambridge University Press, 1996.

Paul, J. Balfour. "Edinburgh in 1544 and Hertford's Invasion." *The Scottish Historical Review* 8. 30 (1911): 113–131.

Pedani, Maria Pia. *The Ottoman-Venetian Border (15th–18th Centuries)*. Venezia: Edizioni Ca' Foscari, 2017.

Petrov, A.V. [Петров А.В.] *Город Нарва. Его прошлое и достопримечательности в связи с историей упрочения русского господства на Балтийском побережье, 1223–1900* [The city of Narva. Its past and sights linked to the history of the consolidation of the Russian domination on the Baltic coast; in Russian] СПб.: Типография Министерства внутренних дел, 1901.

Philippides, Marios, and Walter K Hanak. *The Siege and the Fall of Constantinople in 1453. Historiography, Topography and Military Studies*. Farnham/Burlington: Ashgate, 2011.

Phillips, Carla Rahn. "Navies and the Mediterranean in the Early Modern Period." In *Naval Policy and Strategy in the Mediterranean: Past, Present and Future*, edited by John B. Hattendorf, 3–29. London: Frank Cass Publishers, 2000.

Phillips, Gervase. "Anglo-Scottish Wars: Battle of Pinkie Cleugh." *Military History Magazine* 14.3 (1997): 42–49.

Pickles, Tim, and Christa Hook. *Malta 1565. Last Battle of The Crusade*. Oxford: Osprey Publishing, 1998.

Pilat, Liviu, and Ovidiu Cristea. *The Ottoman Threat and Crusading on the Eastern Border of Christendom during the 15th Century*. Leiden/Boston/Kohl: Brill, 2018.

Plewczynski, Marek. *Obertyn 1531*. [in Polish] Warszawa: Bellona, 2008.

Poli, Xavier. *Histoire militaire des Corses au service de la France*. T. I. Ajaccio: Dominique de Peretti, 1898, https://gallica.bnf.fr/ark:/12148/bpt6k6564540j.texteImage.

Potter, David. *Henry VIII and Francis I. The Final conflict, 1540–1547*. Leiden/Boston/Kohl: Brill, 2011.

Pryor, John H. *Geography Technology and War Studies in the Maritime History of the Mediterranean 649–1571*. Cambridge/New York/Melbourne: Cambridge University Press, 1988.

Pryor, John H., and Elizabeth M. Jeffreys. *The Age of the ΔΡΟΜΩΝ. The Byzantine Navy ca 500–1204*. Leiden/Boston/Kohl: Brill, 2006.

Purton, Peter. *A History of the Late Medieval Siege, 1200–1500*. Woodbridge: Boydell Press, 2010.

Raudzens, George. "Military Revolution or Maritime Evolution? Military Superiorities or Transportation Advantages as Main Causes of European Colonial Conquests to 1788." *The Journal of Military History* 63.3 (1999): 631–641.

Raymond, James. *Henry VIII's Military Revolution. The Armies of Sixteenth-Century Britain and Europe*. London/New York: Tauris Academic Studies, 2007.

Roberts, Keith. *Pike and Shot Tactics 1590–1660*. Oxford: Osprey Publishing, 2010.

Rodger, N.A.M. 1996. "The Development of Broadside Gunnery, 1450–1650." *The Mariner's Mirror* 82.3 (1996): 301–324.

Rose, Rose. *Medieval Naval Warfare 1000–1500*. London/New York: Routledge, 2002.

Rosenberg, Justin. "Isaac Deutscher and the lost history of international relations." *New Left Review* (Jan.–Feb. 1996): 3–15.

Rosenberg, Justin. "Uneven and Combined Development. 'The International' in Theory and History." In *Historical Sociology and World History. Uneven and Combined Development over the Longue Durée*, edited by Alexander Anievas, and Kamran Matin. London/New York: Rowman & Littlefield International, 2016.

Runciman, Walter Garrison. *The Theory of Cultural and Social Selection*. Cambridge/New York/Melbourne: Cambridge University Press, 2009.

Russ, Paul. *La domination espagnole a Oran sous le gouvernement du comte d'Alcaudete (1534–1558)*. Paris: Editions Bouchène, 1998.

Salgado, Augusto António Alves. "Portuguese galleon's armament at the end of the sixteenth century." In *Fernando Oliveira and his Era. Humanism and the Art of Navigaion in Renaissance Europe (1450–1650). Proceedings of the IX International Reunion for the History of Nautical Science and Hydrography*, 277–91. Cascais: Patrimonia, 1999.

Salman, Mohammed Hameed. "Aspects of Portuguese Rule in the Arabian Gulf, 1521–1622." Doctoral thesis. University of Hull, Kingston upon Hull, 2004.

Scarpello, Vincenzo. "Aspetti di Storia Militaria nella Guerra d'Otranto" (2010), https://culturasalentina.files.wordpress.com/2010/09/aspettidistoriamilitarenellaguerradotranto.pdf.

Servantie, Alain. "Venetian assistance to Turkish Oceanic shipping (1517–1541)," 2017, www.academia.edu/33430270/Venetian_assistance_to_Turkish_Oceanic_shipping?auto=download.

Setton, Kenneth M. *The Papacy and the Levant, 1204–1571*. Philadelphia: The American Philosophical Society, 1978.

Sharman, J.C. *Empires of the Weak. The Real Story of European Expansion and the Creation of the New World Order*. Princeton: Princeton University Press, 2019.

Sherer, Idan. *Warriors for a Living. The Experience of the Spanish Infantry in the Italian* Wars, *1494–1559*. Leiden/ Boston/Kohl: Brill, 2017.

Shirogorov, Vladimir. *War on the Eve of Nations. Conflicts and Militaries in Eastern Europe, 1450–1500*. Lanham / Boulder / New York / London: Lexington Books, 2021.

Shirogorov, Vladimir. *Украинская война: Вооруженная борьба за Восточную Европу в XVI-XVII вв. Кн. I: Схватка за Русь (До середины XVI в.)* [Ukrainian War. The armed conflict for Eastern Europe in XVI–XVII centuries. Volume I. Melee of Rus (to the Middle of XVI century); in Russian] Москва: Молодая гвардия, 2017.

Shirogorov, Vladimir. *Украинская война: Вооруженная борьба за Восточную Европу в XVI–XVII вв. Кн.3. Встречное наступление Балтика – Литва – Поле (вторая половина XVI в.)* [Ukrainian War. The armed conflict for Eastern Europe in XVI–XVII centuries. Volume III. Head-to-head Offensive: Baltics – Lithuania – Steppes (In the Second Half of Sixteen century); in Russian] Москва: Молодая Гвардия, 2019.

Shubin, I.A. *Волга и волжское судоходство.* [The Volga and its navigation; in Russian] Москва: Транспечать, НКПС, 1927.

Sikora, Radosław. *Lubieszów 17 IV 1577.* [In Polish] Zabrze: Wydawnictwo Inforteditions, 2005.

Sorokin, P.E. [Сорокин П.Е]. *Водные пути и судостроение на Северо-Западе Руси в средневековье.* [The waterways and shipbuilding in North-Western Rus' in the Middle Ages; in Russian] Санкт-Петербург: Издательство Санкт-Петербургского универсистета, 1997.

Spandounes, Theodore. *On the Origins of the Ottoman Emperors*. Translated and edited by Donald M. Nicol. Cambridge: Cambridge University Press, 1997.

Starkey, David. *Six Wives. The Queens of Henry VIII*. New York: Harper Perennial, 2004.

Tallett, Frank. *War and Society in Early Modern Europe, 1495–1715*. London/New York: Routledge, 1992.

The Russian Illustrated Anthological Chronicle of the Sixteenth Century, M. P. Shumilov's Volume, [Русский лицевой летописный свод XVI века, Шумиловский том] Moscow, Russia, the Sixteenth Century, the National Library of Russia, OP F.IV.232. http://nlr.ru/

manuscripts/RA1527/elek-tronnyiy-katalog?ab=9765898F-CEB8-41E6-BC07-E6E7CC33226D

The Russian Illustrated Anthological Chronicle of the Sixteenth Century, the Synod Volume, [Русский лицевой летописный свод XVI века, Синодальный том] Moscow, Russia, the Sixteenth Century, the State Historical Museum of Russia, СИН-962.

The Russian Illustrated Anthological Chronicle of the Sixteenth Century, the Tsardom Book, [Русский лицевой летописный свод XVI века, Царственная Книга] (Moscow, Russia, the Sixteenth Century, the State Historical Museum of Russia), СИН-149.

Theotokis, Georgios. "The Norman Invasion of Sicily, 1061–1072: Numbers and Military Tactics" *War in History* 17.4 (2010): 392–393, https://doi.org/10.1177/0968344510376463.

Thomaz, L.F.F.R. "Factions, interests and messianism. The politics of Portuguese expansion in the East, 1500–1521." *Indian Economic & Social History Review* 28.1 (1991): 97–109.

Thornton, John. *Warfare in Atlantic Africa, 1500–1800*. London: Routledge, 2021.

Tilly, Charles. *Coercion, Capital and European States, AD 990–1990*. Oxford: Basil Blackwell, 1990.

Tracy, James D. *Balkan Wars. Habsburg Croatia, Ottoman Bosnia, and Venetian Dalmatia, 1499–1617*. Lanham/Boulder/New York/London: Rowman & Littlefield Publishers, 2016.

Tracy, James D. *The Founding of the Dutch Republic. War, Finance, and Politics in Holland, 1572–1588*. Oxford/New York: Oxford University Press, 2008.

Trepavlov, V.V. [Трепавлов В.В.] *Сибирский юрт после Ермака* [Siberian Yurt after Yermak; in Russian]. Москва: Издательская фирма «Восточная литература» РАН, 2012.

Trim, D. J. B. "Transnational Calvinist Cooperation and 'Mastery of the Sea' in the Late-Sixteenth Century." In *Ideologies of Western Naval Power, c. 1500–1815*, edited by J.D. Davies, Alan James, and Gijs Rommelse, 153–187. London/New York: Routledge, 2020.

Trim, D. J. B. "Medieval and Early-Modern Inshore, Estuarine, Riverine and Lacustrine Warfare." In *Amphibious Warfare 1000–1700. Commerce, State Formation and European Expansion*, edited by D.J.B. Trim, and Mark Charles Fissel, 357–419. Leiden/Boston/Kohl: Brill, 2011.

Trim, D. J. B., and Fissel, Mark Charles. 2011. "Amphibious warfare, 1000–1700: concepts and contexts." In *Amphibious Warfare 1000–1700. Commerce, State Formation and European Expansion*, edited by D.J.B. Trim, and Mark Charles Fissel, 1–50. Leiden/Boston/Kohl: Brill, 2011.

Trofimov, V.O. *Campaign at Kazan, its Siege and Taking in 1552*. Kazan: Military Headquarter, 1890.

Trotsky, Leon. *History of the Russian Revolution*. Chicago: Haymarket Books, 2008.

Trotsky, Leon. *The Stalin School of Falsification (1937). Two Speeches at the Session of the Central Control Commission*. Trotsky Internet Archive, 2013, https://www.marxists.org/archive/trotsky/1937/ssf/sf09.htm.

Tushin, Yu. P. [Тушин Ю.П.]. *Русское мореплавание на Каспийском, Азовском и Черном морях (XVII век)*. [The Russian navigation in the Caspian, Asov and Black Seas (the XVII century); in Russian] Москва: Издательство «Наука», 1978.

Uyar, Mesut, and Edward J. Erickson. *A Military History of the Ottomans. From Osman to Ataturk*. Santa Barbara, Denver; Oxford: Praeger security international, 2009.

Unger, Richard W. *The Ship in the Medieval Economy, 600–1600*. London/Montreal: Groom Helm; McGill-Queen's University Press, 1980.

Vagts, Alfred. *Landing Operations: Strategy, Psychology, Tactics, Politics, from Antiquity to 1945*. Harrisburg: Military Service Publishing Co., 1946.

Vego, Milan N. *Naval Strategy and Operations in Narrow Seas*. London/Portland: Frank Cass, 2003.

Verto, Rene-Aubert. *The History of the Knights of Malta: Volume 2*. London: printed for G. Strahan, 1728.

Villiers, John. "Aceh, Melaka and the *Hystoria dos Cercos de Malaca* of Jorge de Lemos." *Portuguese Studies* 17 (2001): 75–85.

Winstedt, R.O. "History of Malaya." *The Journal of Malaysian Branch of the Royal Asiatic Society (JMBRAS)* 13.1(121) (1935): iii–270.

Yelce, Zeynep Nevin. "The making of Sultan Süleyman: a study of process/es of image-making and reputation management," Doctoral dissertation. Sabancı University, Tuzla/ Istanbul, 2009.

Zagoskin, N.P. *Russian Waterborne Communications and Shipping in pre-Petrine Russia*. Kazan: Kharitonov, 1910.

Zwick, Daniel. "Bayonese cogs, Genoese carracks, English dromons and Iberian carvels: Tracing technology transfer in medieval Atlantic shipbuilding." *Itsas Memoria. Revista de Estudios Marítimos del País Vasco* 8 (2016): 647–680.

Mark Charles Fissel
# From the Gunpowder Age Military Revolution to a Revolution in Military Affairs

What conclusions emerge as we transition from gunpowder age military revolutions to our pair of chapters on revolutions in military affairs? "Military revolutions" occur but differ and diverge according to the characteristics of their respective strategic cultures. Michael Roberts' military revolution, the initial template, broadened spatially as Geoffrey Parker inherited the concept. The subsequent historiography veered literally and figuratively in other directions (for example Jeremy Black's insistence on global validation of the theory). The preceding chapters have shown the influence of regionality in military revolutions and patterns of diffusion through peripheries and borders.

We may also conclude that geographically based "mutations" (to use Vladimir Shirogorov's term) indicate that using binary conceptualization to silhouette military revolutions does not categorize well. As Alex Roland points out, one need not bifurcate military revolutions exhibiting characteristics suggesting "determinism" from those showing developments exemplifying "indeterminism."[1] Military revolutions and RMAs share common assumptions about determinism and lean heavily on technology as a causal factor. Inexorably technology remains the focal point of military revolution historiography. Technological determinism permeates subtly, and the present chapter is emblematic of how (frustratingly) inescapable is gadgetry when analyzing changes in warfare. Nevertheless, "systems" incorporating societally imposed realities have stripped technology of presumed autonomy and its inherent determinism (although, again, in the chapter below discussion of technological aspects of military revolutions feels implicitly determinist despite attempts to the contrary).

The corresponding binary of "revolutionary" versus "evolutionary" has been shown to be ahistorical, by Clifford J. Rogers, primarily through adaptation of the explanation of punctuated equilibrium from the biological sciences to the study of history. In terms of process, the directionality of military revolutions can be cyclical in the sense of looking back (and presuming usually incorrectly there are universals in military history) yet clearly progressive, hence linear. Participants in military revolutions temper application of advanced weaponry with notions

---

[1] Roland, "Is Military Technology Deterministic?," 31.

that the art of war is universal. This assumption persists into the era of the RMAs. Cyclical patterns and esteem for the military achievements of their forebearers endow a sense of historical pedigree to those who have taken up soldiering. Convergent evolution, wherein contrasting strategic cultures confront common problems associated with hand-to-hand combat for example, generates similar armor, edge weapons and headgear. Cyclical yet linearly progressive configurations that appear congruent yet still incongruent are reconciled by punctuated equilibrium.[2]

Revisiting Michael Roberts' original theory (proposed before punctuated equilibrium had been postulated and applied to disciplines outside the biological sciences) has been seized upon to explain the RMA. However, the (once) notorious "dominant paradigm" has evolved into overlapping scholarship from "Geoffrey Roberts" and gravitated eastward. Like the initial template, Asian military revolutions wrestled with fundamentally technological-centric dynamics. Globalization of variant species of military revolutions forced the field to recognize the primacy of social context. The latter entails interactions between warfare practices and revolutions, such as industrialization, radical transformation of political systems, class struggle, and cultural revolutions (the latter example including the Athenian Republic and the People's Republic of China). Geographic breadth is also complemented by the realization that macrohistory and microhistory tend to tip military revolutions in different directions, microhistory being less susceptible to determinist optical illusions. The relativity that confounds tidy definitions of military revolutions, and the fluidity of the field's historiography, have goaded scholars to appropriate methodologies from the social sciences and elsewhere. Both Jeremy Black and Alex Roland have encouraged military historians to theorize more, though not to the extent that theory undermines historicism. Daring to theorize, then, military revolutions appear exogenous, the catalyst for military revolution originating via external influences that can range from technological innovation to disaster on the battlefield to new strategic threats posed by competing states. Revolutions in military affairs in contrast are endogenous, or so the strategists and generals tell us. The genesis of the American RMA of the 1980s to the 1990s came about through the maturation of domestic resources: digital technology, stealth engineering, robotics, the evolution of computers, innovative missile manufacture, satellite-based intelligence-gathering, funding for aerospace technology, creativity analysis within "think tanks," etc. Therefore, while military revolutions and RMAs seem to have rather different DNA the punctuated equilibrium paradigm, adapted from biological science to social science, explains the patterns of causation in both phenomena. We accept that there are various

---

[2] For context, see Gould, *Punctuated Equilibrium*, 265–269, 298–300, 359–361.

species of military revolutions, and that RMAs (which appear to mutate faster than military revolutions) being endogenous will bear the hallmarks of the societies within which they spawn.

Nevertheless, exogenous or endogenous, both exhibit "[q]ualitative jumps in the strategic character of the international system and its units that occur in addition to more gradual incremental changes in weapons and warfare."[3] That definition fits perfectly what Wayne E. Lee, James O'Neill, Aliaksandr Kazakou, Vladimir Shirogorov and Hyeok Hweon Kang have described above, in their narratives of strategic situations and their description of adaptations in weapons and tactics. Strategic culture is at the core of the process. The products of military revolutions and RMAs are designed to be used against a foe dictated by the strategic situation (and conditioned by strategic culture). Those actions, perceived as offensive or defensive does not matter, compel states to devise weapons systems and individual tools of war. The latter, especially once used in combat and applied tactically, take on a life of their own. From thence comes the optical illusion of technological determinism.

The anointed heirs of technological-centered military revolutions are of course the RMAs. Technology still maintains its hegemony in the creation of RMAs. In the following essays by Vicente and Mandeles, technology occupies a central place in ongoing RMAs, but it is placed squarely within the sweeping definition that Jacques Sapir (and Laurent Henninger) propose:

> A revolution in military affairs is characterized, from a military point of view, as the need for radical changes in the structure and posture of forces, in the nature of economic and social system providing the bedrock of those forces, finally in the relationship between offensive and defensive, and the nature of the general correlation of forces.[4]

Those "radical changes" are most profoundly manifested due to three developments that changed the trajectory of military revolutions into what has become the RMA. Those are naval military revolutions, nationalism, and (most significantly) industrialization. Aquatic warfare, as Shirogorov argues in the preceding essay, created its own evolutionary pathway, punctuated with technological innovation as commercial expansion transformed the strategic cultures that pursued warfare upon the seas. Nationalism, then and now, stamped its imprimatur upon the process, perhaps most evidently in the various incarnations of imperialism. Driving the expansion was capital, empowered by the technologies of industrialization.

---

[3] Gould, 3–4.
[4] Henninger, "La 'révolution militaire,'" 92–93 citing Sapir 456, translated from the French, https://journals.openedition.org/mots/16312.

Naval warfare therefore is the starting point for a lengthy treatment of how military revolutions became RMAs.

## Naval Military Revolutions

Military revolution origins are terrestrial, as civilization (and thus human conflict) developed from the sedentary life of working the soil. However, as urban settlements fashioned river craft, and boats capable of traversing coastlines, the tools of land warfare were laded upon aquatic vessels. With the dawning of the gunpowder era in Asia, stretching from the 1200s and beyond, maritime endeavor accelerated. Naval warfare flourished. Medieval Asian ships plied the Eastern Indies and South China Seas wielding gunpowder weapons with which they scattered would-be predators preying upon commerce.[5] Europe's hunger and relative poverty (compared to Asia) provided incentives to brave the oceans and invest resources in voyages of exploration.[6] Christian Europe's impulse toward territorial and commercial expansion stemmed from covetousness and dependence upon the acquisition of raw materials beyond one's borders.[7] The latter truism has ranged across times and cultures, from pharaonic campaigns to secure Lebanese cedar to keep Egypt's barges afloat to near-contemporary US interventions in the Middle East to access petroleum. However, Asian polities' comparative affluence conditioned how their respective strategic cultures approached competition. "India was so wealthy and rich in natural resources and manpower that the Empire [the Mughals] did not have to look abroad for many vital commodities."[8] Therefore the sheer variety of strategic cultures and degrees of prosperity around the globe made the motivation and hence conduct of warfare diverse and fluid.

The comparative facility of water transportation endowed naval military revolutions[9] with a geographical aggressiveness that could outreach strictly land-based expansion. In most physical environments, military revolutions meshed

---

5 This assertion is based on marine archaeological evidence, maps, quantitative analysis and calculations by Brian Fahy presented in Hong Kong in December 2017. Dr. Fahy's methodology is expounded upon in https://www.academia.edu/9669623/A_Seat_at_the_Table_Addressing_Artefact_biases_in_Asian_Shipwreck_Assemblages.

6 Roy, *Military Transition in Early Modern Asia, 1400–1750*, 1–2.

7 Latham, "Warfare Transformed," 231.

8 De la Garza, *The Mughal Empire at War*, 185.

9 For purposes of this chapter, we employ the loose term "naval military revolutions" in recognition of their tangled provenance and conflicting historiography. Elaboration follows.

with expansionary activities due to the capabilities of firearms in projecting force dramatically and oftentimes spectacularly. Strategic cultures were especially vibrant when hosting nautical traffic that sparked technological transfer via imported materials. Intercultural exchanges led to seagoing vessels implementing early forms of technology and accompanying technical innovations, such as during antiquity the proliferation of brailed sails and their effect upon hull design. When the gunpowder revolution was securely aboard ship in the 1400s, commercial interactions over more than two millennia had achieved impressive seaworthiness, improved navigation, and cannonades in support of deck-borne combat.

Furthermore, technical improvement integrated seagoing activity into the societal fabric. The long seventeenth century saw in places such as the Indian Ocean sophisticated maritime systems that "witnessed profound shifts in power dynamics, commercial trading practices and sociopolitical alliances. These transitions were often effected at sea, the structureless zone of contact and interaction that spawned new and often unexpected allegiances as well as hostilities."[10] A strategic culture approached seagoing commerce according to its unique view of the surrounding world, meaning that naval military revolutions were tailored to fit a society's perceived needs. Again drawing an example from the Mughals, the latter derived their practices from their riverine and coastal experiences which "did not conform to the Mahanian ideal of the mature Western military revolution . . . . The Mughals focused on points instead of lines, controlling ports and other strategic locations" having "no plans to acquire overseas colonies."[11]

Incipient European colonialism became progressively sea-orientated through the further augmentation of maritime proficiencies, especially after the mid-1700s, following the eclipse of land-based Eurasian steppe empires. Expansion was not yet informed by a coherent state-sponsored strategy, and the volume of trade and its inherent value, especially after 1650, was more transforming than the germination of seaborne technologies. Regarding the latter, the perennial problem of distance was addressed by increasing ship speed, organizing collective action on shipboard and in ports, expanding capacity, strengthening vessel endurance through use of new materials such as metal, and ultimately marrying masted vessels to steam power. The aggregate effects further promoted the union of mercantile endeavor and naval warfare. Oceans opened gateways for acquisition and coercion. This had not always been the rule in empire-building. Polities engaged in incursions in Central Asia (e.g., the cultures occupying parts of what today is Russia and China in reference to inner Asian empires) spread out via

---

10 Reddy, "Disrupting Mughal Imperialism," 138.
11 De la Garza, *The Mughal Empire at War*, 184–185.

land warfare and through coalitions sometimes involving mass migrations of peoples. Asian contact zones possessed varying degrees of permeability: "The Muslim states in Central Asia and the Middle East were not as impermeable a barrier for the Mughals as they were for Christian countries."[12] Again, the nuances of characteristics of respective strategic cultures qualify, sometimes defy, generalizations that attempt to define military revolution upon the waters.

Thus, in reference to Geoffrey Parker's invocation of the "rise of the West," there was indeed something dissonant (though not necessarily uniformly superior) about the arrival of aquatic harbingers of Christian European power on the coasts of Africa, the shores bordering the Indian Ocean, and the seas of East Asia. Lacking the interconnectivity and regional technological affinities of, for example again, the Mughals, in contrast the Portuguese, the Dutch and their fellow Europeans chose to inflict their own ways of war upon the peripheries of these strange new lands. Thus, as their cannon blasted against aquatic, littoral, and land-based interfaces interconnecting the continents, rarely did the Europeans prior to the 1700s penetrate contact zones substantively and subjugate the hinterlands. Such strategic achievements exceeded the means of Western agents of economic proto-imperialism, exemplified by the Portuguese *nau*, the famed carrack boasting immense carrying capacity that made it simultaneously mercantile and military. Such vessels never encroached deeply into estuaries and riverine systems.

Colonial enterprise challenged Western concepts of three-dimensional space. European efforts to juxtapose their territorial conventions overseas lay beyond reach. Srinivas Reddy notes the European "goal of territorializing the unwieldiness of the ocean, to bring order to an otherwise undifferentiated realm" was elusive because Westerners "viewed the oceans just as they saw the land, as a space to be parsed, regulated and ultimately dominated."[13] The extensive manufacture of maps and nautical charts via copper-plate printing, particularly in the 1700s, was both cause and effect of this *mentalité*. In contrast, the Mughals "were almost flippant toward the great ocean that surrounded them. By nature, convenience or arrogant lethargy, the Mughal emperors never invested in building their maritime power or even securing their long coastlines."[14]

Maritime endeavors continued to exploit gunpowder to propel a juggernaut of competition. Societies increasingly traded and fought ferociously amongst themselves for profits as their tools became more versatile and efficacious.[15]

---

12 De la Garza, *The Mughal Empire at*, 185.
13 Reddy, "Disrupting Mughal Imperialism," 139.
14 Reddy, "Disrupting Mughal Imperialism," 128–142, quotation from 139.
15 Appel, "How competition drove social complexity," 728–745, especially, 739–743.

Initially the ideology of mercantilism guided the contesting powers. Genuinely capitalist endeavor would not blossom fully until the era of industrialization. The distinction, of course, is that mercantilism perceived wealth as finite and palpable, for example the stereotypical Spanish passion for silver bullion. Capitalism, building on Adam Smith's concept of markets, is more complex. Both ideologies shared a common denominator: the application of force, or at least the implied threat of violence, to open and dominate markets. Both systems, ultimately, depended upon force projection, frequently across the waters. Fernand Braudel opined that "warfare, which was becoming more and more expensive, contributed to the development of mercantilism," the craving for wealth spurred on by the "double imperfection of the [territorial state's] fiscal system and . . . the administrative organization of the state."[16] The symbiotic relationship between state-sponsored overseas endeavor and burgeoning commerce underlay the earliest naval military revolutions. The late John Guilmartin remarked that "the military revolution at sea is inseparable from changes in the economics of maritime commerce."[17] Given the evolutionary nature of the adaptation of gunpowder technology to seagoing enterprise, and the process' commercial context, the military revolution afloat was more than an exercise in the imposition of a weapons system.

What were the naval military revolutions? From the 1400s onwards it appears that the "military revolution on land was paralleled by a simultaneous revolution at sea."[18] Michael Duffy's observation does not necessarily mean these were identical or predetermined. Given gunpowder's utility on land and sea, one might assume that naval military revolutions were fundamentally the maturation of terrestrial military revolution embarked upon watercraft. However, the current generation of naval historians warns of one-dimensional historical templates. Nicholas Rodger, Gijs Rommelse, Louis Sicking, and others, caution against defining naval revolutions as approximations of celebrated land-based military revolutions.[19] Naval military revolutions were not aquatic mirror images of land-based military revolution; they resembled each other only partially, primarily through (1) exploitation of explosives, (2) tactical adaptations to technology, and (3) macrocosmic changes wrought upon societies. This trio aligns with Michael Roberts' template in terms of categories if not in specific ways of war.

---

16 Braudel, *The Wheels of Commerce*, 532 and 544.
17 Guilmartin, "The military revolution in warfare at sea," 130.
18 Duffy, "The Foundations of British Naval Power," 49–85, especially 49, though how "simultaneous" might be questioned in a global context.
19 Rodger, "The military revolution at sea," 59–76; Rommelse, "An early modern naval revolution?," 138–150; Sicking, "Naval warfare in Europe, c. 1330–c. 1680," 236–263.

An attempt at gauging the technical and tactical complexity of European naval military revolutions was proffered by John Guilmartin, who arranged in quadripartite and chronological sequence the military revolutions at sea, emphasizing technology transfer in the maritime world. Proliferation of technologies is evident in his first early-modern European military revolution at sea, the "Iberian fusion of Atlantic and Mediterranean shipbuilding technology," citing the example of malleable iron-refining processes of the so-called Catalan forge.[20] Ironworking's Catalunyan pedigree dated back to Phoenician settlements. Coastal urban communities frequently pioneered successive technical improvements in a commodity that proved useful for application on land and sea. Yet despite the centuries-old ironworking industry (and that is not an anachronistic application of the latter term), when the Spanish Crown needed iron artillery to suppress the Dutch rebels in 1574, Don Luis Requesens looked askance at Catalunya and instead coveted upstart English guns for use aboard his ships. Frustratingly, he could not obtain these for political reasons. No region enjoyed a lasting monopoly on the manufacture of iron guns, just as chariot technology was competitive in the ancient world.

Improvements in furnace design, metallurgy, "fining" (refining molten metal in a separate hearth), "hammering" (e.g., the drome-beam hammer to "forge, fine, and draw out" iron,[21] achieved piecemeal and experimentally, defy national or regional categorization. An entrepreneurial prince or an abundance of ore could shift the location of foundries and their metalworkers. The geographic range, and the chronological progression, of pre-modern technology transfers meant that while innovative cannon foundries might inspire imitators in distant lands it is difficult to "freeze frame" seminal developments and dub the solitary achievement a military revolution. Furthermore, hybrid techniques such as Styrian Walloon forging methods make "national" categorization of military revolutions dubious.[22] As the competition in artillery-founding demonstrates (see for example Aliaksandr Kazakou's comments above), the "lead" in technological innovation could be claimed for decades, then lost for extended periods of time or permanently.[23]

---

[20] Guilmartin, Jr., "The military revolution in warfare at sea," 129–137, especially 130; Steven Walton questions the centrality of the Catalan forge in sorting out naval military revolutions (personal communication). See Estanislau, "The Catalan process," 225–232; also Cipolla, *Guns, Sails and Empires*, 45–48.

[21] Awty, "The Development and Dissemination of the Walloon Method of Ironworking," 791.

[22] Awty, "The Development and Dissemination of the Walloon Method of Ironworking," 784–785, 787, 792–793, 799–803.

[23] Theobald, "European Weapons in China."

So, while indeed military revolutionary episodes in technical improvement occurred, they did so in discontinuous streams of the development of material culture sometimes lasting centuries.[24] That intermittent evolutionary dynamic provides a backdrop, especially considering nautical technological innovation, given the frequency and intensity of technical knowledge transfers that occurred in port cities (such as Barcelona, for example).[25] The obsession with maintaining an edge and avoiding potential enemies from achieving "parity" (as H.H. Kang might put it) is universal. Guilmartin may have overemphasized the Catalan forge. Nevertheless, the competition between Mediterranean cleverness in fashioning tools of war versus the brash Atlantic interlopers represented the age-old competition in technology transfer that has risen to its greatest heights today, partially on the shoulders of the RMA.

Guilmartin traces a second revolution to the discovery in the 1510s "of a means to fuse the tactical mobility under oars of an ordinary Mediterranean war galley with the firepower of heavy gunpowder ordnance, a fusion first achieved by Venetian shipwrights towards the end of the Ottoman War of 1499–1503."[26] The mutability, resiliency and longevity of galley warfare in the Mediterranean, and merchant men-of-war in northern European waters, ultimately make a case for naval transformation, not so much a naval revolution or military revolution afloat.[27] The interplay between technical problem-solving and tactical practices remains inherently dynamic, recommending the model of punctuated equilibrium as an explanatory causal framework, as discussed below.[28]

Guilmartin's third naval revolution centered upon galleons. Galleons overcame hostile environments inaccessible to land-based campaigns. Capacity again factored in naval design. What mattered was not solely galleons' martial cargo but specifically the judicious deployment of shipboard artillery which made them formidable. Innovatory fluidity makes assessment of galleons a process, not simply recognition of a static change in the way war was conducted in an aquatic environment. As has been argued above, the evolutionary processes of land and sea warfare differed qualitatively, as well as in their respective environmental applications of technology. Galleons constitute a major component in

---

[24] Rogers "Gunpowder Artillery in Europe, 1326–1500," 39–71.
[25] Trim and Fissel, "Conclusion," 429–32.
[26] Guilmartin Jr., "The military revolution in warfare at sea," 132.
[27] Palmer, "The 'Military Revolution' Afloat," 123–149.
[28] The greatest proponent of said model is Clifford J. Rogers, who describes punctuated equilibrium evolution as "a mix of incremental, Darwinian evolution and more radical 'mutations,'" in "The Artillery and Artillery Fortress Revolutions Revisited," 75.

Geoffrey Parker's declaration of the West's "victory at sea."[29] Parker (influenced by Braudel more than is recognized) notes how differing environmental conditions were addressed by customizing design, which in turn realized strategic potential. Guilmartin's galleons fit the bill as the next phase in the amalgamation of maritime entrepreneurship and naval warfare.

Guilmartin's fourth military revolution at sea arrived with the hegemony of ships of the line dispensing even greater firepower, exhibiting remarkable swiftness and maneuverability. A centuries-long cadence of technical innovation coupled with hands-on tactical experiences produced this sleek naval military revolution, again with gunpowder as the binding agent.[30] Each technical innovation was applied in unique environmental settings which in turn precipitated a tortuous evolution in tactics, complications that had to be factored in when commanding a sailing vessel. Human skills had to develop in tandem with technical innovation. These synthetic processes are evident in Vladimir Shirogorov's "transit warfare" (or transport warfare, in others' parlance), which captures accurately the variety of adaptations of gunpowder to aquatic traffic. His essay explores the amalgam of commerce and war as practiced upon seas, estuaries, and riverine systems. Shirogorov links the primary function of vessels, namely transport, to military revolution, a dimension sometimes underemphasized in military revolution historiography (though not underestimated by Geoffrey Parker or Jeremy Black).

George Raudzens' Braudelian perspective encompasses incarnations of seapower within nautical environments as venues for force projection and cultural encroachment. Raudzens proposes a "sea transport revolution," or more accurately a "maritime evolution" overshadowing "military revolution."[31] The evolution of maritime transportation systems enabled gunpowder revolutions to have effects beyond regional phenomena via naval military revolutions' continuous reinvention of seapower socio-economically and technologically. What of aquatic-based firepower in relation to land warfare? Shirogorov argues that airborne explosives (in the form of ordnance) changed ship-to-shore operations forever. Gunpowder introduced a new dimension to amphibious warfare once the appropriate vessel design had been achieved and tactics developed to maximize its fightingpotential. Gunpowder revolution shaped conflict on the littoral, on *terrafirma*, and upon the seas comprehensively. The chemical compound's ubiquity stemmed

---

**29** Parker's "Victory at Sea" chapter in the various editions of his *The Military Revolution. Military Innovation and the Rise of the West*. Doubtless Parker's forthcoming comprehensive study will revise and showcase the significance of the galleon(s).
**30** Guilmartin, Jr. "Technology and Strategy," 11–46.
**31** Raudzens, George. "Military Revolution or Maritime Evolution?," 631–641.

from its sustainability and the portability of managed violence. The gestation of the gunpowder age, which occurred over centuries (like the industrial revolution being in chronological unfolding evolutionary in nature) conjoined numerous military revolutions. However, the Devil's distillate had to be utilized in tandem with factors such as human choice, organizational culture, twists of fate, and ultimately the aquatic environment that dominates most of the globe, oceans. The necessity of adapting to geographical space prompted technical innovation, oftentimes practical rather than theoretical.

The mutability of naval warfare and the precariousness of vessels fighting upon heaving expanses amid unpredictable winds made for a uniquely challenging art of war and more than one "tactical conundrum"; which leads M.A.J. Palmer to proclaim, "[n]owhere were the pressures of the military revolution greater than at sea."[32] Resultant line ahead tactics, however, developed in the pattern of punctuated equilibrium over a period of nearly two centuries. Continuity should be stressed considering the evolutionary nature of adaptive innovations in naval warfare.[33] Technological innovations and structural modifications meant that commanders faced greater choices when fighting upon the seas thus spawning tactical innovations (e.g., line ahead formations). In a Robertisian-like formula, technology (which affected capacity as well as speed) and tactics engaged in the lengthy dialectic between the ancient art of sailing and the gradual progression set in motion by the scientific revolution. The cautionary observations of George Raudzens and Jeremy Black characterize the process of naval military revolution as evolutionary even if its results were revolutionary, particularly when industrial mechanization increasingly became available to men under sail.

"Evolutionary" and "revolutionary" as noted above are regarded as contradictory terms and thus words shape our conceptualization in a fashion that suggests that these conditions are mutually exclusive. In doing so we hamstring our understanding of the complexity of naval military revolutions. Indeed, in some historical phenomena evolutionary and revolutionary characteristics co-exist.[34] Semantics aside, the confluence of factors that produced military revolutions included gradual improvement with more abrupt technical innovations. Naval military revolutions also occurred through an aggregation of elements (gun carriages, lidded ports, consequent tactical innovations, etc.). Fixation upon naval armaments and design is not antiquarian. Missile weapons powered by

---

[32] Palmer "The 'Military Revolution' Afloat," 123–124, 146; Parker, "The Dreadnought Revolution of Tudor England," 357–388.
[33] Lambert, "Naval Warfare," 172–192.
[34] Rogers, "The Artillery and Artillery Fortress Revolutions Revisited," 78–79; see also 3–4, 314–319 above, and below, 328–335, 358.

gunpowder, even more so than shock, spearheaded naval military revolutions. Looking back, shipborne gun carriages (coupled with vessel mobility) exemplified aquatic adaptations of the gunpowder revolution. It has been argued that the adoption of gunports and heavy guns "did not necessarily entail the development of specialized or purpose-built warships"; however, "improvements in sailing-ship technology made possible the rise of the sailing ship . . . ."[35] The ingenious incarnations of sailing ships right through to the industrial era illustrate accurately punctuated equilibrium in naval military revolutions. Regarding purpose-built vessels, Louis Sicking's assertion quoted above is generally true with the important exception of the sturdy bomb ketch and its mortars. Strategically and tactically, lobbing explosive shells into ports might equalize contests of bombardment between stationary coastal fortifications and naval vessels firing from heaving seas. One recalls the old maxim that "a ship's a fool to fight a fort." The shoreline and riverbank emplacements that had contended with waterborne attack as described by Vladimir Shirogorov (above) soon adapted.[36]

Coastal and estuarine operational effectiveness was integral in the evolution of seapower over two centuries. In the case of Britain, at the vanguard of these developments, the goal was attained when its "battleships dominated most oceanic sea routes by concentrating on the terminal points of these routes."[37] During the period 1650 to 1830, the battleship was the catalyst for domination (even if "domination" of the seas was still inchoate in the minds of those who managed the state). Battleship hegemony allowed frigates and lesser vessels to attack enemy merchantmen along shorelines and elsewhere while forestalling the enemy's ability to concentrate their battleships. Richard Harding remarks, "[by] the end of the Napoleonic Wars in 1815, [battleships] denied the advantages of the sea to their enemies while reaping maximum advantages for themselves. This was the classic exercise of seapower – sea control and sea denial."[38] Thus the commercial context of naval military revolutions enabled the mercantile expansion that served as a prerequisite for industrialization (the premier example occurring in Britain). Naval military revolutions are, even more so than their terrestrial cousins, inseparable from their broader socio-economic contexts. They ultimately developed robustly because their fortunes were enmeshed with capitalist endeavors such as the commercial revolution and subsequent industrial revolution.

---

**35** Sicking, "Naval warfare in Europe, c. 1330–c. 1680," 238.
**36** Stapleton, Jr., "The Bluewater Dimension of King William's War," 320, 333, 349, and Rowlands, "The King's Two Arms," especially 287–288.
**37** Harding, *Seapower*, 37–38.
**38** Harding, *Seapower*, 38.

This chapter posits the industrial revolution as the pivotal point in explaining the sequence of military revolutions. The impact and range of industrialization's influence upon technological development and its scale, as well as upon the speed and frequency of technological transformation, add up to a military revolution that exceeds Roberts' of 1560 to 1660 (or Parker's or Black's for that matter). To hazard becoming repetitious, the industrial revolution is the great turning point in human history. Naval military revolutions were swept up and became part of the industrial tide. Following in the wake of the post-1650 aquatically based commercial revolution the oceanic environment ultimately experienced this second military revolution. Navies integrated components of industrialization (c.1800–c.1918) not unlike how the commercial revolution (c.1650–c.1800) melded with the "Geoffrey Roberts" military revolution.[39]

This second "industrial" military revolution coincided with the flooding of overseas markets with manufactured goods, steady demand that had been cemented by colonization. The scale of mercantile expansion and its flotillas had made overseas expansion sustainable long before the mills and factories were erected. England and the United Provinces "turned to the sea" largely through private enterprise which in turn matured synthesized financial institutions, e.g., the foundation of the Bank of England in 1697, thus the purview of the state in partnerships with a new capitalist ruling class (championed by England's Whigs). Financiers needed fledgling but vigorous government as well as vice versa in order to prosper in the new commercial world. Representative governmental bodies such as parliaments and states general pondered overseas matters as prototypical industrialization expanded markets in ways and on a scale unforeseen. Institutions (private, civil, governmental, etc.) grew and adapted in response to burgeoning trade and attendant maritime conflict. "To the seventeenth- and eighteenth-century mind the role of seapower and the advantages of battleships were less obvious. The dominance of the battleship was based on its ability to appear in almost all waters of the world in sufficient numbers for a long enough period to overwhelm local naval resistance."[40] However, this hegemony required 200 years of great exertion, in addition to lots of money allocated for such purposes. Nicholas Rodger assesses this macrocosmic transmogrification: "In a world that was steadily moving away from land and population as the sole basis of wealth and towards commerce and industry, naval powers and open societies were much better placed to adapt than

---

**39** The "Geoffrey Roberts" allusion has been appropriated from Laurent Henninger; see also Headrick, *The Tools of Empire*, a reference owed to Wayne E. Lee, and to Headrick's *Power over Peoples*, 151–154, 186–217, 257–726.
**40** Harding, *Seapower*, 38.

military powers and autocratic monarchies."[41] States, and thus the future of their respective societies, were transformed when seapower was embraced (or not embraced).

A comparison with the Mughal Empire is again instructive. To reprise Srinivas Reddy,

> Mughal naval weakness was the root cause of the empire's downfall, and in turn, the rise of European colonial dominance. In this light, there appears to be an arc, a curve that bends from imperialism to colonialism. In that sense, is the ocean perhaps the defining difference between these two modalities? In other words, does colonialism as a practice and as a principle arise from controlling the sea?[42]

Imposing any degree of sovereignty upon immense expanses posed organizational demands on those states who would presume to do so. According to Richard Harding, "The major changes in the maritime world between 1650 and 1830 were not so much technological but in the scale and diversity of operations."[43] Considerable socio-economic bureaucratic structure (public and private) was prerequisite for colonization. Colonial organization rested upon substantial tax bases. Later imperialist incarnations by which an industrialized naval military revolutionary system was unleashed upon the world rested on bureaucracy and wealth generated by decades of voluminous overseas trade. The commercial contributions to early-modern state formation were a fundamental building block, as acknowledged in the Parkerian military revolution formula. Classic Robertsian military revolution theory also linked the above-mentioned administrative imperatives to ways of war.

Hefty military budgets reflect a state formation model reflective of what Roberts proposed for 1560 to 1660, namely deliberate investment in infrastructure to enable competitive force projection. Naval military revolutions encompassed what Gijs Rommelse terms "economic reason of state."[44] The military revolutionary implications of wafting gunpowder upon the waves ensnared sources of economic power within the grasp of territorial states, for example in symbioses of private capital and the public policies of mercantilist governmental institutions. As Nicholas Kyriazis's revisionist work on military revolutions and fiscal-military states declares,

---

[41] Rodger, "The military revolution at sea," 70–71; Rommelse, "Introduction: The military revolution at sea," 117–118.
[42] Reddy, "Disrupting Mughal Imperialism," 128–142, quotation from 139.
[43] Harding, *Seapower*, 13.
[44] Rommelse, "An early modern naval revolution?," 138–150.

the choice of seapower by a state leads to a different regime than the choice of land military power, because sustainable seapower necessitates a wide alliance of interests, which brings with it more democratic regimes, develops new more efficient and complex forms of organizations, requires the acquisition and diffusion of new knowledge and expertise, which brings with it institutional change and economic growth.[45]

Discerning the intentions of "the state" in adjusting the nuts and bolts of naval military revolution is not straightforward. The technologies involved, as discussed below, are regarded as an inscrutable "black box" by scholars. In other words, in terms of adaptation, do acquisitions of naval technologies "drive" policy when states spy rivals exploiting an innovation which must be copied or countered? Dovetailing with this obscure sequence of causation is the fact that "the state" itself is regarded as a "black box" by neorealists and others, citing the state's fluid and evolutionary nature, the inner workings of which can be inscrutable.[46] What is the process by which the will of a strategic culture is made manifest, given that (despite neorealist assertions of near-universal political responses) every society possesses "different constitutional arrangements"? One can examine the material evidence of naval vessels and institutional records. However, the social process by which the Mughals (to revisit an example used in this essay) were prompted to deal with issues of seapower are sufficiently murky that comparisons with, say, Western Europe are tentative at best. State-sponsored commitment to a policy of seapower should be coupled with other factors such as agricultural advances, fiscalism, and population growth in the confluence of causes of industrial revolution. The latter process frequently depended upon a maritime flow of raw materials. Desirous of external natural resources on ever greater scales, post-commercial revolution imperialist states scoured the world with "iron" navies, manifesting their aspirations via physical embodiments of the mechanical features of industrialization.[47] To achieve these ends the incestuous relationship between capitalism and violence continued unabated into the industrial age.[48] Explaining and "measuring" the impact of a series of military revolutions includes consideration of the more violent features of commerce, specific global technology transfers, and contrasts between pre- and post-industrial imperialism. Imperialism and industrialization, the former promoted by private capital as well as state-based

---

**45** Kyriazis,. "Seapower and socioeconomic change," 71–108 using a formal quantitative model, especially 71.
**46** Frank, "Military Revolutions," 23.
**47** On the Industrial Revolution as turning point, see Walton, "Technological Determinism(s) and the Study of War," 10.
**48** Satia, *Empire of Guns*, who observes that in the 1770s the "state did not have a monopoly on violence, because it was not yet institutionally coherent enough to have one" (245–246).

mercantile policy, and the latter requiring specialized ingredients (geography, climate, population, capital, and infrastructure) mutated the strategic culture that determined a polity's ability to wage war. Allan Dafoe's "military-economic competition" though focused upon the modern world bridges military revolutions across time and vindicates Marxian interpretation of the multi-faceted material influences of forces of production, discussed below.[49]

Those forces of production intertwined economics with technological science. The maritime contentions of the commercial revolution fomented (and absorbed) piecemeal advances in mechanization that might not have been maritime in their genesis but affected ship design, e.g., steampower. Iron-hulled sailing ships and masted cruisers harnessed natural forces such as wind in tandem with steam locomotion, liberating them from dependence upon a solitary energy source. By possessing artificial power that complemented sails they were more independent in maneuver and could persevere longer distances by alternating and coupling propellants. Vessels reliant upon early engines were not entirely trustworthy. Endurance resulted from uniting wind and industrial power when that latter source broke down mechanically. Combining wind and steam power more than enhanced propulsion. Traversing greater distance and doing so independently from the vagaries of weather inspired grander strategies. Seakeeping abilities were expanded and honed, a blessing for both commerce and naval warfare. Early iron-cladding had imposed design problems such as vulnerability to capsizing. "Seakeeping" ensured stability and maneuverability in rough seas, while seaworthiness paid dividends strategically and tactically in naval warfare.

The heyday of masted cruisers reprised the amalgam of commercial and naval action we have cited, for example as carried out by Federal commerce raiders during the American Civil War. While there may have been few if any major fleet actions or splendid naval contests fought with these "synthesized" craft, the versatility of what became the storied steam-powered ships-of-the-line made possible imperial expansion. To repeat, the consequential yet protracted processes resemble the punctuated equilibrium model of military revolution.[50] Halting yet progressive advances certainly appear more evolutionary than stunningly "revolutionary" one must concede. Improving upon shipboard muzzle-loading guns had taken centuries, for example. These long-term innovatory confluences lead to "punctuations", such as when the "introduction of the ironclad coincided with a revolution in naval ordnance during which rifled guns replaced

---

[49] Dafoe, "On Technological Determinism," 1047–1076, cited in Zimmerman, "Neither Catapults nor Atomic Bombs," 49.
[50] Discussed below, 330–332, 334–335.

smooth bores and metal carriages replaced wooden truck mountings."[51] Taking the wider chronological perspective, the scope of the industrial impact of the second military revolution upon naval warfare brought about through the products of industrialization is prodigious: steel, electricity, screw propulsion, explosive shells, hydraulic power (such as used in lifts), self-propelled torpedoes, wireless communications, refrigeration and air conditioning, rangefinders, searchlights, etc.

Further illustration of how industrialization sequentially transformed naval warfare can be seen by contrasting how even prior to the industrial revolution the speed of naval warfare was qualitatively different from force projection dependent upon land transport. Pre-industrial carriage's maximum pace was equine. Wind and oar-driven vessels, unlike horses, derived their energy from the environment in which they fought, whereas steeds needed food and water. That said, seapower's advantages were often negated abruptly since weather could disrupt aquatic activities, including implementation of tactics at critical moments. Nor should it be forgotten that indigenous peoples, for example in the Americas, forged coalitions that could set into motion military action briskly and over great distances. The French in North America, the Incas in South America, and imperial powers of the Eurasian steppes, used swift couriers to induce allies to strike unexpectedly. More numerous examples could be cited from Eurasian military history. The latter arrangements, though, involved complicated diplomacy and the presence of obliging physical features. Navies, despite the vagaries of weather, were more consistent, autonomous and self-reliant when properly supplied. Naval warfare offered range that enabled strategic action that was not as feasible in land campaigns. Despite caveats, appending gunpowder weapons to water transport opened new vistas for strategic and tactical operations (as is apparent in Shirogorov's essay). Subsequent industrialization enhanced yet further the qualitative distinctions of naval campaigning. Industrial revolution constitutes Clifford Rogers's "tipping point" broadly applied, as discussed immediately below.

Convergent evolution, where related entities develop similarly as they encounter roughly equivalent types of challenges, applies to military organizations. This evolutionary pattern's contours fit the multifarious yet simultaneously analogous shapes of military revolutions. Geographically distant strategic cultures facing the challenge of human combat quite understandably fashioned similar though not identical ways of war (e.g., how to balance missile and shock in that particular context, fighting mobility versus the need for steadfast formation, etc.). The origins and paths of convergent evolution are in the case of military

---

51 Chesneau, and Kolesnik, ed, *Conway's All the World's Fighting Ships*, 4.

revolutions compatible with the stuttered and oscillating nature of punctuated equilibrium, where abrupt and short-lived advances provide an unpredictable dynamic. Convergent forces contribute to the momentum of punctuated equilibrium. This Darwinian and evolutionary theorem of punctuated equilibrium was applied to the artillery and artillery fortress revolutions of the late Middle Ages. That template conforms with the interactions between industrial innovation and the consequent naval military revolution of, for example, 1850 to 1905. Punctuated equilibrium's flexible conceptual framework reconciles evolutionary and revolutionary features. Clifford Rogers remarks that "evolution can reach a tipping point that's revolutionary"[52] and the result is substantive "punctuation" deemed revolutionary. Language helps explain rather than confound historical reality in this case.[53] Rogers reconciles "evolutionary" with "revolutionary" by noting that all revolutions are spawned from impellent conditions over time. Drawing from the biological sciences, Rogers points out that in the artillery-related military revolutions of the 1400s and early 1500s "sudden change . . . occurs when a basically linear advance reaches what scientists who study non-linear phenomena call a 'tipping point.'"[54] The challenge-response dynamic Rogers identifies, when medieval fortification was overcome by more effective siege guns only to see military architecture regain the initiative through the defensive qualities of the bastion, is not historically unique. There were earlier punctuations in poliorcetics. The impotent siegecraft of Athens' combined forces in the campaign against Syracuse circa 415 to 413 B.C. was emblematic of the hegemony of military architecture in the late archaic era. When in 409 B.C. a Carthaginian force assaulted the coastal *polis* of Selinus in southwestern Sicily, the invaders brought to bear novel (for Sicily) projectile-throwing devices, rams, and associated siege engines that pummeled the traditional mud brick fortification walls. Acragas, Himera, Gela, and Camarina fell in succession to the Carthaginian "revolution in siege warfare." In the decades that followed, innovations in Sicilian fortifications were hurriedly implemented.[55] The process resonates what Clifford Rogers has described as occurring in the wars of late-fifteenth and early sixteenth century Europe. Likewise, the formula combining convergent evolution with punctuated equilibrium shares resonance with "continuous reinvention," introduced in our introductory chapter with reference to Marshall Poe's "three halting stages" of development.[56]

---

52 Personal correspondence with Cliff Rogers initiated by Mark Danley.
53 Peers, "Revolution, Evolution, or Devolution?," 81–106.
54 Rogers, "The Artillery and Artillery Fortress Revolutions Revisited," 79.
55 Holloway, *The Archaeology of Ancient Sicily*, 141–147.
56 Above, 12.

Convergent evolution also accommodates the sundry geographical and cultural variations that bedevil application of a paradigmatic template. Convergent evolution, when conjoined with punctuated equilibrium, explains how regionally proximate states, such as those described by Aliaksandr Kazakou above, could possess similar weapons and tactics, but evolve differently in their respective ways of war. We have seen analogous processes, too, in the naval military revolutions that occurred during industrialization. Punctuated equilibrium, then, characterized martial technology on both sides of that great divide, the industrial revolution. After 1800, steam technology combined with burgeoning infrastructure, such as dockyards nourished by state financing, and "tipped" the scales. The tentative yet sometimes frantic tinkering with technological applications resembles these evolutionary models. Iron-cladding confounded centuries-old traditions in muzzle-loading cannon. Tactical responses included recourse to ramming. The challenge-response mechanism moved vessels away from the old formula by which wooden ships were rated by number of guns carried (e.g., 74 cannon as standard for the basic ship of the line). Combat power as measured by ship rating was altered permanently. By the time of the Greek war of independence in the 1820s, the sequencing of technical innovation at sea accelerated. The introduction and experimentation with central batteries, iron-cladding, turrets, chain hoists for ammunition, etc. punctuated the accumulative tempo and substance of the industrial transformation of naval military revolutions.

Furthermore, convergent evolution explains how second and third tier powers, lacking abundant resources such as a substantial arms industry and oftentimes obliged to incorporate obsolescent naval technologies into their defensive systems, kept pace with "great powers." In the case of the above-mentioned Greek conflicts of the 1820s to 1830s, necessity was indeed the mother of invention. Arguably the war for independence saw the first use of a steam warship in combat.[57] The Greek revolutionaries employed steam corvettes, grappling (fruitfully) with attendant mechanical problems and thus advancing what would be a component of naval military revolutions. In the 1880s the Greeks would experiment with submarine design. The nation's socio-political situation and economic resources inhibited full-blown development of these technologies, but their predicament instilled within the Greek navy resourcefulness and a willingness to take risks.[58]

Another intensive phase of punctuated equilibrium occurred circa 1850–1870. A dialectic between substantive changes in naval gunnery and defensive design

---

[57] Mark Danley provided commentary. Any misinterpretations are entirely my own.
[58] Chesneau, and Kolesnik, ed *Conway's All the World's Fighting Ships*, 387–388.

propelled striking advances. Around 1860 an innovative spike can be identified among sequences of punctuated equilibria within the period 1850 to 1906. The mid-century impetus culminated in the 1906 *Dreadnought* launch, a showcase of improved propulsion, imaginative application of hydraulics, steel construction, and guns utilizing slower-burning gunpowder. Episodes of stasis are evident as well, as exemplified by relative stagnation from 1890 to 1914 (though punctuated by the 1906 Dreadnought design revolution that owed so much to advances earlier in the century). Irreversible progress there was, even if uneven in its emergence. By 1900 the cascade of innovations constituting punctuated equilibrium of the second military revolution made shipborne guns themselves altogether different. From the smokeless powder that propelled them to the hydraulic systems that replaced chain hoists and chambered large projectiles, the principle may have remained intact but the mechanics had undergone revolutionary changes when measured by the quantitative and qualitative results of bombardment. The battleships of 1905 more resembled the design norm of 1935 than they did fin-de-siècle vessels. So, while not identical, force projection on land, across the littoral, upon the waves, and ultimately in the air, shared genuinely military revolutionary characteristics while sharing a common denominator, explosives. To reiterate, the cadence of naval military revolutions reveals evolutionary processes that can produce revolutionary results.[59]

The aggregate effect of technological innovations (most small, some grand) by the turn of the century made battleships uncontested masters of seapower. There may have been no abrupt and decisive turning point in the ascent of the battleship, but its hegemony was made by the inclusion of technological improvements within the construction of a "system" replete with institutional reinforcement. To be precise, between 1650 and 1830 global maritime expansion was lionized not so much by the evolution of technologies but the fortifying of systems, some commercial some military and naval. Maturing industrialization then realigned naval organizations and their application of increasingly mechanized aquatic warfare. As Harding put it, there "was no clearly defined 'military revolution' or technological revolution that ousted one form of war and replaced it by the battlefleet."[60] The dovetailing of the "battlefleet revolution" with the First World War saw experimentation with seaborne aviation, such as the 1913 trials associated with the HMS *Hermes*, a cruiser. Commercial ships were redesigned and fitted with hangars, as various adapted incarnations of seaplane carriers.[61]

---

[59] The chronological "phases" above may be gleaned from Chesneau and Kolesnik, ed, *Conway's All the World's Fighting Ships*.
[60] Harding, *Seapower*, 287; Herwig, "The battlefleet revolution, 1885–1914," 114–131.
[61] Chesneau and Kolesnik, ed, *Conway's All the World's Fighting Ships*, 64–71.

This series of punctuations that produced three-dimensional naval combat in the 1940s was compounded by an "exponential" increase in naval power projection in the late nineteenth and early twentieth centuries.

The process of punctuated equilibrium that incrementally industrialized navies acuminated in the ultimately revolutionary maturation of aircraft carriers, confirming that advances in seaborne warfare were on one level an "evolution" in military affairs with military revolutionary consequences.[62] The intersection of all three dimensions of warfare, literally, developing over time and culminating in a "Carrier Revolution" achieved legitimate military revolution status, specifically as to how the dimension of air power was grafted onto formidable seapower.[63] The three-dimensional battlespaces brought into existence in the twentieth century (then amplified to a fourth dimension via cyberwarfare) demonstrate that previously unimagined and unprecedented types of warfare must be judged revolutionary no matter how deeply rooted in seemingly timeless, cyclical, centuries-old arts of war. The process and results of the second military revolution created via industrialization resemble Michael Roberts' formula proposed in the 1950s. In the modern era, new weapons brought new tactics, bolder strategies and profound social change. Interposed between Roberts' early modern military revolution and the RMA, the industrial revolution generated a plethora of innovative technologies that were produced in varying lengths of gestation. The schoolboy's definition of the industrial revolution as an engulfing "wave of gadgets" bears a kernel of truth. The abundance of industrial innovations created a new state of affairs for military science. Coupled with the politics of nationalism and imperialism, "industrialized" warfare possessed (and coveted) its complex pedigree. The celebrated revolution(s) in military affairs acknowledged linkages with earlier forms warfare that were in turn reconciled with now-extant revolutionary new incarnations of combat.

Aircraft opened a new battle space and metamorphosed the conditions under which navies and armies operated, arguably revolutionary consequences. The synthesis entailed widening strategical and tactical dimensions in multiple environmental settings. Aircraft carriers working in tandem with fleets consisting of vessels with highly "industrialized" ordnance, sophisticated navigation systems, and mechanically enhanced capabilities that improved speed and maneuverability required infrastructure for sustenance. The "Carrier Revolution" burnished its credentials as a military revolution by producing a new organizational culture as

---

62 Herwig, "The battlefleet revolution, 1885–1914," 114–131, especially 115.
63 Stulberg, "Managing Military Transformations," 489–528; the land-based dimension of the Carrier Revoution would include the dockyards and naval bases that sprung up from 1900 to 1950, which also provided continuity.

well as a novel art of war.⁶⁴ The appearance in 1982 of the study by Hone and Mandeles constituted one of the earliest OSD/NA studies on what was later viewed as an RMA stemming from joining aviation and carrier vessels.⁶⁵ Ocean-going warfare had been retrofitted into the RMA strategic framework in response to the advent of aircraft carriers. Aircraft carriers, like later missile-bearing nuclear submarines, were harbingers of a new era and lent credence to the validity of the RMA concept. A model proposed by Andrew Liaropoulos, "The Radical Transformation Paradigm," aligns closely with Mark Mandeles' conclusions regarding aviation, carriers, and future war. Visionaries, specialists with the imagination to conceive radically new approaches, are catalysts for revolutions in military affairs. Liaropoulos' view, too, resonates the traditional Robertsian model. In both cases insightful organizational innovation or rigorous administrative reform complemented, frequently enabled, the hardware.⁶⁶

The "Carrier Revolution" example should be tempered, though. Carriers had little effect on land warfare, according to Jeremy Black.⁶⁷ Airpower itself was overhyped. The marriage of high explosives and swift airborne attacks, touted by Guilio Douhet and Billy Mitchell, was not independently "decisive."⁶⁸ Although air power captivated the public as well as strategists, the conjuncture between navies and aviation hung upon the coattails of centuries-long dominant seapower; likewise, the above-mentioned nuclear submarines owed much to the earlier naval military revolutions. Waging war below the surface, like flying above the waves, added physical dimensions to naval warfare, however. The roots lay in the evolutionary process from torpedo-carrying surface craft to submarines, yet another example of punctuated equilibrium. The crucial developmental process commenced around the 1860s, leading to various incarnations of vessels (designed for defense primarily) e.g., torpedo boats, torpedo gunboats, and destroyers (such as French and British torpedo boat destroyers of the 1890s). Deck-mounted guns were often at the center of the challenge-response mechanism. Refinements in torpedo technology ("range, speed, reliability and explosive power") were countered by surface ship ordnance's "improvement in range,

---

64 Stulberg, "Managing Military Transformations," 489–528.
65 Hone, and Mandeles, "Innovation in Naval Aviation," This seminal study compared the introduction of carrier-based aviation into the interwar USN, Royal Navy, and Imperial Japanese Navy; Hone, Friedman, and Mandeles, *The Introduction of Carrier Aviation into the U.S. Navy and Royal Navy*. This study, updated, was later published as Thomas C. Hone, Norman Friedman, and Mark D. Mandeles, *American and British Aircraft Carrier Development, 1919–1941* (Annapolis: Naval Institute Press, 1999).
66 Liaropoulos, "The Radical Transformation Paradigm," 363–384, especially 369–370.
67 Black, *Combined Operations*, 128.
68 Buckley, "Air Power," 159; Roland, "Is Military Technology Deterministic?" 19–33.

accuracy and rate of fire of guns, made possible by improvements in metallurgy, explosives and hydraulics."[69] Interaction between design and mechanical improvements that originated in industrial science exemplify the stuttering yet progressive characteristics of punctuated equilibrium. Michael O'Hanlon, channeling Clifford Rogers to a degree, makes explicit the evolutionary nature of technology's application in how the industrial catalyst of the second military revolution birthed the revolution in military affairs.[70] So, too, Mark Mandeles' analysis of avionics and innovation in warfare. Both these scholars' work resonate Andrew Liaropoulos' "Continuity and Evolution Paradigm" wherein an evolutionary process, innovation inspiring and sustaining further innovation (usually through bureaucracies designed for such purposes), create an environment fertile for military revolutions and RMAs.[71]

Gradual change can still shock the beholder. The realization that war upon the seas had changed forever was a direct result of the "unnatural" appearance of ironclads and their successors.[72] This was apparent in the spectacle of proto-industrial skylines of dockyards at Portsmouth, Woolwich, Plymouth and elsewhere, where monstrous agglomerations of equipment for vessel maintenance and construction created a landscape deserving of William Blake's derision. Radical changes in appearance indicated substantive transformation. Revolutionary consequences were readily visible in many cases. Making the naval military revolution comprehensible to the popular imagination as progressive rather than cyclical, the *Kansas City Star* in 1898 contrasted the physical appearance of warships on the eve of the twentieth century with vessels employed in the American Civil War (that had occurred a mere thirty-three years earlier). The latter more resembled Noah's Ark than contemporary flotillas.[73] The *Kansas City Star* editorial contained a grain of truth: pre-dreadnaught battleships of 1890 resembled battleships on the eve of the First World War more so than a wooden steam screw ship-of-the-line appeared to be a forerunner of 1880s ironclads. Profundity of change could also be veiled. The 1906 *Dreadnought* design was, indisputably, revolutionary. But its physical appearance was not as unnervingly futuristic to the observer as had been the early ironclads, and the latter by no means made masted vessels obsolescent. Popular, especially journalistic, evidence is reminiscent of a controversy of the gunpowder age, that of the "moderns" versus the "ancients," the latter

---

69 Chesneau, and Kolesnik, ed. *Conway's All the World's Fighting Ships 1860–1905*, 86–87, 323.
70 O'Hanlon, *Technological Change and the Future of Warfare*.
71 Liaropoulos, "Revolutions in Warfare," 363–384.
72 Deogracias, *Battle of Hampton Roads*.
73 Rogers, "The Idea of Military Revolution in Eighteenth and Nineteenth Century Texts," 413.

asserting that the more war changed the more it remained the same.[74] Despite unprecedented technological changes, some contemporaries preferred dressing up naval military revolutions in a garb of timelessness. Steven A. Walton wonders if developments such as the reprise of ramming were not regarded as universals in naval warfare. Were torpedo launches mechanical emulations of Sir Walter Scott's accounts of jousts? Had the designers of ironclads been reading Thucydides?[75] Nothing was new under the sun given a cyclical view of military practices.

Strategic cultures of antiquity recognized that the sands shifted continually while the littoral itself remained. Demosthenes' assertion that "we are living today in a very different world from the old one, I consider that nothing has been more revolutionized and improved than the art of war" sounds like something uttered by RMA proponents in 1991, or by strategists of the late 1960s.[76] The Athenian was reacting to what has been described as the Macedonian revolution in military affairs.[77] Like the Spanish sword-and-buckler men of the 1500s who cut and stabbed like Roman legionaries, Philip II of Macedon drew inspiration from contemplating the past, in his case the *Iliad*. "[A]ccording to Diodorus (16.3.2) Philip II got the idea of an infantry force armed with long pikes from Homer."[78] Warriors often "advanced" via the past. Demosthenes' oft-cited declaration if not contextualized can be referenced as an intimation of linear progress in the conduct of warfare in classical antiquity. In fact, Demosthenes was a proponent of moving forward by harkening backward, not unlike what Machiavelli did when penning *L'Arte della Guerra*.[79] Linking the High Renaissance's taste for classical aesthetic models (for example, in art and architecture circa 1470–1520) with the military revolution of the same period reveals a shared and deliberate connection with the ancients. Bert Hall's merging of the resurrection of weapons such as the pike and the "rediscovery of the crossbow" with the gunpowder revolution suggests that inspiration from the past can be synthesized with scientific progression.[80]

While the process of synthesizing with the past to create something new is textbook military revolution, warriors of classical antiquity saw war differently.

---

74 For new light on old controversies, see Neill, "Ancestral Voices," 487–520.
75 Personal correspondence with Professor Walton, who graciously shared his thoughts on this topic.
76 See Possony and Pournelle's effusive *The Strategy of Technology. Winning the Decisive War*.
77 Brice, "Philip II, Alexander the Great, and the Question of a Macedonian 'Revolution in Military Affairs,'" 137–147. Brice's model, intended to be applied universally, concedes that a military revolution or RMA need not fulfill every criterion for the term to be applied usefully.
78 Rey, "Weapons, Technological Determinism," 32, especially note 28.
79 Machiavelli, *The Art of War*.
80 See the analysis in Heuser "Denial of Change," 13; Hall, *Weapons and Warfare*.

The past was paradigmatic (and thus deserving of study), not something to be improved upon either by technology or virtue. The ancients, including Demosthenes, neither subscribed to our "idea of progress" nor did they fixate on weaponology as a potential determining factor in the unfolding of human events. Where the universal permeates the "forward into the past" mentality is in drawing upon history to assist one's present predicament. The Greeks, the practitioners of sixteenth-century infantry tactics, and the RMA think-tankers, all would agree on the value of looking backwards to face the perils of the day.[81] Such was, and is, the authority of history. New arts of war often drape themselves in the image of past glories (e.g., Special Forces teams as bands of Spartans), embracing conscious anachronism. Such views suggest a cyclical view of the art of war can co-exist with the reality of linear progression. Universalist tendencies in military science, whilst exhibiting mutability, were regarded in the timeline of martial advancement as applying a "technological logic that rhymes."[82]

If certitude in universals persisted in some circles, particularly cyclical views of coordinating weapons and tactics, then coming to terms with authentic permutation (such as a newly found "decisiveness" in a weapons systems) is requisite, especially for practitioners. Accurate analysis of warfare within the linear evolution of technological innovation requires understanding macrocosmic and microcosmic aspects. How wide and inclusive is the framework in which we examine military historical development? Were its permutations inevitable? David Zimmerman's answer, inspired by Allan Dafoe's observations regarding interpretations of "macro" and "micro" timeframes, indicate that military revolutions can be genuinely revolutionary even when qualitatively different in nature. Chronological framework influences how we see "inevitability," the latter implicit in any chain of development. Long-term (macro) studies are indeed prone to embrace inevitability and determinism when assessing material culture's role in historical causation. It is not the historical development's true nature that is in dispute; rather, it is the framework and language used to isolate and define it. Categorization, classification, and comparison clarify the complexities of warfare's chaos. Planting milestones can show the way even if the process is artificial. In short, military revolution studies aim to conciliate what appears to be cyclical with the stark reality of a linear progression which cannot be denied.

---

[81] One gets the impression that Asian cultures shared this reverence for the achievements of ancestors; see also Heuser "Denial of Change," 3–27.
[82] The musings and brief quotation are taken from correspondence with Steven Walton, 2021. Any misinterpretation is my own, however.

Zimmerman's binary view is that industrialization divided military revolutions between two realms, pre-modern and modern.[83] Nicholas Rodger (writing both as macro-historian and micro-historian) described post-1600 navies as manifestations of modernity, harbingers of imperialism.[84] Pursing further useful lines of demarcation between pre-modern and modern, one can reckon that the "modernity" of the British achievement, adorned in the neo-classicism of the 1700s, sought its provenance in the exemplary rise of ancient Athenian imperial naval power. That image was of course inconsistent with the reality of Athens's defensive alliance system that consisted of "shaking down" reluctant city-states.[85] Nevertheless, British sailors (and soldiers) were fond of considering themselves as heirs to the warriors of classical Greece. Such pretensions are not farfetched, really. If military revolutions abound and (while fanning out chronologically) possess symmetry, then referencing antiquity can be a return to origins accompanied by consciousness of one's place in the developmental cycle. Therefore, we cast a wide net and consider ancient tactics and the amphibious origins of the "hoplite revolution" (keeping in mind the philhellenism of the builders of the British empire).

Military revolutions are exogenous, as the diffusion of weaponry demonstrates and the assimilation of tactics from other armies proves. Appropriation from "outside" is not limited to inspiration from contemporaries. Practitioners of military revolutions also borrowed from other eras, oftentimes deep in the past drawn from their ancestors (or their ancestors' rivals), sometimes as practiced in contrasting physical environments. The latter point can be sketched out by revisiting classical antiquity. The hoplite revolution (the ascendancy of Greek heavy infantry), chronologically speaking one the earliest specimens of military revolution, was planted squarely on *terrafirma*.[86] However, the hoplite revolution, amphibian-like, perhaps emerged from the Mediterranean Sea. Classical amphibious operations may have "inseminated" the hoplite revolution. In transitioning from epic individualized combat and "swarming" tactics to the precision of tight formations and interlocking Argive shields, did the Greeks draw upon lessons learned during archaic coastal raiding?[87] John R. Hale asks

---

[83] Walton, "Technological Determinism(s) and the Study of War," 10; Zimmerman, "Neither Catapults nor Atomic Bombs," 48–49, etc.
[84] Rodger, "The military revolution at sea," 59–76.
[85] This colorful characterization is owed to Mark Danley.
[86] For but one example of the historiography, see Viggiano, "The Hoplite Revolution and the Rise of the Polis," 112–133.
[87] Illustrated by the Amathis bowl, in Hale, "Not Patriots, Not Farmers, Not Amateurs" 183. See also Crouwel, "Fighting on Land and Sea in Late Mycenaean Times," 100–105.

if teamwork, the collective and disciplined regimen necessary to navigate the seas and execute a disembarkation, provided a proven model for the hoplite revolution? A nucleus of trained men that adapted the cross-littoral shield wall to the comparatively more easily navigated level field of battle upon terrain may well have gestated one of the earliest military revolutions.[88]

Classical naval warfare presented predicaments ripe for innovation, particularly tactical experimentation with a weapons "system" (in this case hoplite arms). Where resonances between seagoing warfare and its terrestrial cousins differed was apparent in the contrasting venues in which tactics were initiated and carried out. To quote J.E. Lendon, "At the first sea battle of which a detailed description survives, the battle of Artemisium in 480 BC, the Greeks formed their triremes into a defensive circle . . . [Greek naval warfare] was more consistently tactical than land fighting . . . [I]t has been proposed that tactical fighting on land grows from the example of tactical fighting on the sea."[89] Hale's thesis suggests that the binary of land warfare as distinct from naval military revolutions (which unfortunately has been resorted to in the present work) is artificial and sometimes inaccurate, especially with reference to the formative years of antiquity. Tangentially, if one accepts that amphibious warfare predates true naval warfare, then a pattern was established early on wherein the advancement of military science drew from practices of a sister form of combat.[90] Such a conclusion is significant because it demonstrates that military revolutions, unsurprisingly, assimilate innovations from warfare fought in disparate physical settings as well as from different eras.

Geographical cross-pollination, especially in "peripheries," has catalyzed military revolutions. Adaptation of a given technology to a different physical environment (or, strategic culture) can produce unprecedented results. This was true when improved ordnance was integrated with amphibious warfare, a new generation of missile weapons with tactical consequences that opened up regions for the incursions of the commercial revolution. That geographical diffusion returns us to naval military revolutions and combined operations.

Interactions between weaponry and tactics (a theme dear to the late Michael Roberts) reveal contoured details of what actually constituted naval military revolution. Martial technology had been shipborne for millennia and hand-to-hand combat on decks has been remarkably unrevolutionary, though when it did change in the late 1400s to early 1500s that major punctuation coincided with

---

88 Hale, "Not Patriots, Not Farmers," 186, 189–191.
89 Lendon, *Soldiers and Ghosts*, 161.
90 On naval warfare's debt to amphibious operations, see Fissel, "Out of Africa" (forthcoming).

the artillery-fortification revolution dialectic on land, and with those European infantry battles from 1480 to 1525 cited as the efflorescence of a military revolution.[91] As argued above, proven battle techniques inspired experimentation in a different physical environment.

Why did naval warfare rely so heavily on well-coordinated tactical expertise? Sailors cannot quite roam the battlefield as cavalry and infantry do. Confined to their vessels (and doomed if plunged into the surrounding waters) seamen had to work even more collectively than did infantry upon land. There was little heroic individualized warfare and the sphere of action was circumscribed by the sea. Comparatively speaking, sailors lacked individual mobility and compensated for this deficiency by perfecting innovative tactics. The latter were partly made possible by technical improvements. Weaponry had changed, but cross-pollination among early amphibious, land-based, and naval forces occurred more widely (e.g., adaptation of missile weapons, tactical configurations, drill) than is the case in contemporary specialized forces. British amphibious fighters of the industrial age had reason to consider themselves as reincarnations of the Myrmidons. Operationally their successful amphibious attack upon Havana in 1762 was achieved with deployment of a hodge-podge of grenadiers, sailors, and others.

In some parts of globe, Bengal for example, the binary of land warfare versus naval combat was an artificial and oftentimes inaccurate distinction given the ubiquity of lacustrine, riverine and estuarine battle spaces.[92] The interplay is evident, for example in how amphibious practices may have inspired both naval warfare and land tactics, discussed above. Technical variation in vessels, mutable coastlines, and borrowed tactical approaches inconveniently complicate simplification of naval military revolutions. Nonetheless, the relationship between tactics and technology even though multifaceted appears to be cyclical and universal as historians and practitioners impose their linear progressive framework upon military revolutions. The numerous technological and tactical variants, the multifarious battle spaces, the significance of uneven refinements over time, suggest authentic revolutionary characteristics of war upon littorals, landscapes, and seascapes. The macro-historical framework, however, conspires against seizing striking military revolutionary occurrences as "decisive" due to the challenge-response mechanism. Turning points such as the application of aviation to warfare, discussed briefly above, to the domain of naval warfare

---

**91** Warming, "An Introduction to Hand-to-Hand Combat at Sea," 117–119.
**92** De la Garza *The Mughal Empire at War*, 51; Trim, "Medieval and early-modern inshore, estuarine, riverine and lacustrine warfare," 357–413.

precipitate undeniably revolutionary consequences even if the gestation of the development is prolonged over decades.

## Nationalism and Industrialization Forge the Strategic Cultures of the RMAs

Military revolutions (and RMAs for that matter) range across contrasting physical features. They also, as we have suggested, range across time periods for inspirations that have on occasion brought battlefield success. The initial scholarship on military revolutions focused on early modern European nation-states, usually monarchies. Assumptions were made that proto-national polities provided structures within which military revolutions flourished. As scholarship broadened and military revolutions were identified in strategic cultures outside Europe, military revolutions were clearly far from being limited to a solitary template.[93] Military revolutions are found globally, and in eras before Roberts' allegedly paradigmatic examples. What emerged from early modern Europe was a more ravenous species. The gunpowder age's military revolutionary social wave unfolded by stages and followed different paths, especially outside Europe. The most explicit documentation of the latter is a comparison of military systems of the Ottomans, Safavids, Mughals, Ming, Manchu, and others within the chronological framework of 1400 to 1750.[94] While military revolution in Europe may have been associated with advances in the use of the matchlock musket, for those polities contending with European colonization in a later era, it was (to cite one example) a "breechloader revolution."[95] Kaushik Roy's unusually broad cross-cultural approach to institutionalization underscores once again the significance of diversity in state formation, and like Jeremy Black identifies the late eighteenth century as the crucial phase, at the expense the early modern era.[96] The 1700s witnessed the confluence of imperial military revolutions, national military revolutions, and industrialization, at the interface of unprecedented imperialist expansion.

Where the various forms of nationalism have persisted in military revolution scholarship is in the exportation of Western imperialism in its multifarious

---

[93] On global variations of nationalism and strategic cultures, see the format and mode of analysis used in Sondhaus, *Strategic Cultures*.
[94] Roy, *Military Transition in Early Modern Asia, 1400–1750*.
[95] Peers, "Revolution, Evolution, or Devolution?," 93–94, citing Headrick's *The Tools of Empire*, 96–102; Satia, *Empire of Guns*, 283–288.
[96] Roy, "Military Synthesis in South Asia," 651–690.

incarnations, primarily the "rise of the west" theory. Indisputably, global socioeconomic contours were reshaped by the transoceanic expansion of European nation-states. Nation states fixated on boundaries. However, European colonization and imperialism struggled to delineate global frontiers: "the motivations and goals of a wide range of maritime actors . . . [and] the connection between the use of maritime violence and the bid of maritime polities to 'territorialize' the realm of the ocean . . . [was designed] to extend their territory out to sea."[97] A symbiotic relationship had grown between national states and mercantilism. As Fernand Braudel put it, "it was nation, or embryonic would-be nation which, by inventing itself, invented mercantilism."[98] Post-1650 Europe's military fiscalism, and subsequent interloping into other corners of the globe in collaboration with entrepreneurs, facilitated state formation. Nation-states developed fiscal infrastructures capable of maintaining military establishments tailored for the unique threats they faced. Inherent within a nation-state's strategic culture was a political equation (and value judgment) calculating that society's collective force projection. Whether preindustrial military expenditure spurred or retarded overall economic growth determined choice and abundance of weaponry as well as logistical potency.

Nationalism's effects upon socio-political structures assembled a propitious organizational culture for capitalist economic growth. This synthesis occurred within the parameters of the greatest revolution in history, the industrial revolution. As is well known, industrialization ushered in unprecedented technological innovation coupled with proliferation of that technology via scale in the guise of mass production. The capitalist dynamic that grew symbiotically with industrialization, in tandem with the politics of nationalism, created the militarist juggernaut of the second military revolution and ultimately the RMA. These convergences lie at the heart of the second military revolution of industrialization. As we shall see, nationalism created strategic cultures conducive to technological (and tactical) innovations by binding to the military revolutionary impulse in new ways the human and material resources of powerful states.[99]

Strategic cultures assimilated (1) the unifying ideology of nationalism, (2) increased financial resources available particularly through taxation, and (3) technical expertise from ongoing industrialization. These largely post-1650 developments altered Europe's trajectory not only in relation to other parts of the world but within the European state system, even though the ruling classes of

---

[97] Margariti, "Mercantile Networks, Port Cities, and 'Pirate' States," 546; quoted in Reddy, "Disrupting Mughal Imperialism," 139.
[98] Braudel, *The Wheels of Commerce*, 544.
[99] Knox, "Mass Politics and Nationalism as Military Revolution," 57–73.

that era appeared fairly oblivious to the fundamental changes occurring in the social (and economic) order. The impulse towards hostilities which was (and is) prosecuted vigorously by nationalist strategic culture seized firmly tools that could be adapted for war (e.g., steam engines). Nationalism, being a nebulous yet formidable *geist*, is more conducive of conquest and external force projection than simple defense of the *patria* prior to 1790.[100] Western nationalism, prone to xenophobia and exclusionary ideology, conforms closely to the neorealist view of international relations, namely that "survival mandates aggressive behavior."[101] Nationalism inclined governments toward international conflict, increasing the impetus for war, for example as in the "second wave" of Western imperialism. Furthermore, following Kenneth Waltz's formula, "competitive pressures compelled" European powers to develop "similar military forces," hence what appears to be a fundamental military revolution template that has spawned an historiography of national military revolutions.[102]

Nationalism's palpable industrialized manifestations today shape (and extinguish) polities as occurred in the nineteenth and twentieth centuries. The latter centuries proved subsequent highwater marks of what are commonly described as "military individual nation-states." Military revolution historiography has gravitated toward critical analysis of "great nation states of Europe," giving way to a collective realization that a global context was requisite in order to understand military revolutions.[103]

How an intensified nationalism harnessed technology politically for transnational aggression (explained via narration of diplomacy) is complicated by the methodologies of "thick" and "thin" narratives. The latter pair are complex subjects which threaten to divert us from our focus, the relationship between military revolutions and revolutions in military affairs. However, military history favors a "thin narrative" arranging chronological sequences that trace how wars come about, or the unfolding events of battle. Military revolutions and RMAs require more of a "thick" narrative because of their inherent linkages with strategic culture, economics, politics, etc. A tangential complexity applies to technological determinism, treated below. The infusion of nationalism into military revolutions likewise begs the application of thick narrative, for example how wars occur in the first place.

---

100 Orwell, "Notes on Nationalism." Steven Walton pointed out the relevance.
101 Frank, "Military Revolutions, Evolution, and International Relations Theory," 15.
102 Frank, "Military Revolutions, Evolution, and International Relations Theory," 16.
103 Stradling, "'A Military Revolution,'" 272, an authority on military practice in Spain, reinforcing the point driven home by Dorn, "The 'Military Revolution,'" 656–658.

Medieval rulers had justified their resort to war with formulas such as dynastic claims, frequently campaigning for "personal" reasons such as territorial ambitions which today appear whimsical and unnecessary. Or (as in the Crusades), divine imperative was cited as justification, and religion became even more a motivation in the 1500s. Today, "national security" appears to be a more justifiable motive to fight, but the marriage of industrialization (with its attendant technological dynamic) and nationalism makes that terminology euphemistic, especially when framing the most recent RMAs against the backdrop of petro-politics, anti-globalism, racism, and displaced domestic anxiety about deindustrialization. Western species of nationalism and industrialization, ideologically and materially, permeated the membranes of world cultures and thus compelled countermeasures in global military practices. Those processes, illumined by social constructivism and the social shaping of technology (discussed below), couple the early-modern military revolution's mutation with the twentieth and twenty-first century revolutions in military affairs.

Such is the conundrum linking the RMA backwards to the early modern military revolution. Historical interpretation usually seeks out origins to explain outcomes. Such was the case with the ebullience regarding the RMA in the late 1980s and 1990s, which stemmed from an instinctive sense that a new era had dawned. Proponents of the RMA felt obligated to connect with the past, contextualizing a practitioner's teleological presentism with the perceived glories of the gunpowder age and even the stoic legacy of classical antiquity. Ironically, the second military revolution, involving industrialization's amalgamation with nationalism and imperialism, sought legitimation via a storied provenance. Such tendencies reveal ambivalence regarding military revolutions' linear progression as opposed to practitioners' ambivalence regarding what appear to be cyclical characteristics in military science.

Military history possesses inherent utilitarian concerns in the form of military science as an applied discipline. The new dimensions of the RMA and future war were bridged to earlier military revolutions to make sense of it all and exhibit the scientific logic a practitioner craves. As discussed earlier, a move forward sometimes prompts a look back, even as far as to classical antiquity. This instinct was challenged by the profundity of change wrought by the industrial revolution. Technologies that had been made possible by industrialization raised the stakes. Manufactured commodities, and the systems which produced them, were part of strategic culture. Strategic culture largely determines how technology is applied to warfare. Alex Roland accepts that "technology has driven the evolution of warfare" and transformed it more than any other "variable."[104] In the case of nationalism, technology is placed in that peculiar context of national politics and culture, etc. Political structure is paramount because the invention and application of military technology is channeled by institutions sustaining human collaboration in warfare. The weapons and tactics so often featured in studies of military revolutions are manifestations of society's collective efforts to secure their strategic culture. Political institutions in particular (in the Robertisan paradigm, the statecraft of the Nassaus and then that of Gustavus Adolphus) made possible the utilization of military technology as state-sponsored force projection. The political dimension is even more pronounced in the fate of RMAs because, it has been argued, they are endogenous hence entities created by domestic material culture.[105]

In addition to military historians, international relations theorists have been intrigued with the relationship between politics and military innovation, especially technologies. The latter theorists, particularly the neorealist school that has been subject to revisionism, saw in military revolutions (and RMAs) templates that conformed to their vision of international relations. Drawing again from Waltz it was asserted that "contending states imitate the military innovations contrived by the country of greatest capability . . . so the weapons of major contenders, and even their strategies, begin to look much the same all over the world." Holger Herwig's battlefleet revolution of 1885 to 1914 is highlighted as a paradigmatic example.[106] Neorealists were more attuned to external technological innovation (even determinism) than to the inner workings within the state (in contrast to social constructivists and "ways of war" historians, for example). The question arises, does technology (determinist or not) shape socio-political institutions or vice versa?

Technology became central to human conflict once violence escalated beyond blows from bare fists. At the dawn of the atomic age, the idea of "technological determinism" emerged. Historians of warfare took note as did military professionals. Michael Roberts' military revolution theory likewise appeared in the early years of the Cold War. Born of the same era, these intellectual constructs intertwined. Ultimately, students of military revolutions debated the nature of technological determinism.

Steven Walton defines technological determinism: "[T]echnology has an exogenous life of its own, unresponsive to social concerns or pressures, and that

---

104 Roland, *War and Technology*, 1.
105 Mandeles, *The Future of War*, 35–36, 41–42; Black, *War and Technology*, 173–227.
106 Frank, "Military Revolutions," 16–17, from whom the Kenneth Waltz quotation is taken.

technology is the main determinant of social structure."[107] From the outset, assumptions about the primacy of technological determinism permeated the "Geoffrey Roberts" military revolution arguments.[108] What was implicit in the latter became explicit in the revolution in military affairs of the 1980s to the 1990s. Geoffrey Parker's (still) predominant early modern military revolution model and the RMA remain technology-centered. Technological determinism is frequently cited as a common denominator shared between the RMA and the sixteenth-century gunpowder revolution. Indeed, the official definition of a revolution in military affairs, as pronounced by government think tanks, proclaimed connections between military revolutions and RMAs. Technology was the core of these phenomena. Determinism, progression, and the materiality of political leverage also were ingredients. Joint Vision 2010 expressed this explicitly:

> Technology-driven RMAs have been occurring since the dawn of history, they will continue to occur in the future, and they will continue to bestow a military advantage on the first nation to develop and use them . . . . [The nation must] be aware of technology developments that could revolutionize military operations, and [seek] revolutionary ways in which to employ those technologies in warfare . . . . Regarding past revolutions in military affairs (RMAs), –What lessons can we learn from the historical record?[109]

Joint Vision 2010 confirmed the twentieth-century redefinition of technology's role in human conflict.

While interactions, or reciprocities, flourish between mechanical innovation and the culture that spawns new technology, context is paramount, obviously. Dating technological developments conveniently sequenced military revolutions. In the process, we endowed technology with a unique causal role in history. The allure of the military revolution theory lies partly in the tidiness of its unilateral causation. RMA theorists and practitioners, implicitly and explicitly, expounded "the view that technology is an autonomous historical force."[110] Unquestioningly

---

107 Walton, "Technological Determinism(s) and the Study of War," 5.
108 Alex Roland somewhat dissents, stating that Roberts and Parker "built their arguments around causal mechanisms other than technology" and that Parker "eschewed deterministic claims." All this hinges on how one classifies the gunpowder revolution; see Roland, "Is Military Technology Deterministic?," 23.
109 From Joint Vision 2010. See Warner, *Implementing* Joint Vision 2010, *a Revolution in Military Affairs for Strategic Air Campaigns*. The Defense Advanced Research Projects Agency (DARPA) sponsored studies, often conducted by the Acquisition and Technology Policy Center of RAND's National Defense Research Institute (NDRI), a "federally funded research and development center sponsored by the Office of the Secretary of Defense, the Joint Staff, the defense agencies, and the unified commands"; Hundley, *Past Revolutions, Future Transformations*.
110 Stone, "Technology and War," 29.

and unquestionably technology has been infused with determinism. How and in what circumstances does military technology fate historical outcomes? Impressive capabilities resulting in splendid victories were potent but not necessarily "revolutionary." An assumption was made, nevertheless, that new technologies possessed inherently the potential to achieve a decisive result on the battlefield. Current scholarly consensus is that technological determinism is an oversimplification, requiring qualification at least; if not the latter, then entirely fallacious. Insistence upon technology's perceived deterministic causal role prompted critics to upbraid military revolution theorists for conceptualizing technology as a mysterious black-box that explained military revolutions without thoroughly and genuinely defining technology itself or its processes.[111] Social scientists understood the utilitarian goals of RMA theorists, but rejected the technological determinism inherent within the RMA: "[T]he 'black-box' of technology must be opened, to allow the socio-economic patterns embedded in both the content of technologies and the processes of innovation to be exposed and analysed."[112] The assertion that "technology" created its own destiny posed frightening implications. The Robertisan formula suggested that when humans engaged in warfare their tools, though not sentient, carved out the contours of society, hence human existence itself. Theo Ferrell attempted to clarify the issue of causation. He depicted technology as the "engine of military change," because the noun "driver" mistakenly suggests technology itself is "determining the outcomes of processes of change."[113]

Technological determinism trains attention upon weaponry. However, the place of technology changed after industrialization. David Zimmerman points out in reference to the RMA that technology invades every facet of human life now. Such was not the case in the early modern world. Or, as F. E. Rey points

---

[111] An explicit statement of the latter position (Hall and DeVries 1990) critiqued Parker ([1988] 1996), advocating a more focused definition of "technology" and re-examination of the causal role that technology plays in history. In turn, Hall and DeVries were accused of judging Parker ([1988] 1996) on the basis of microcosmic technical discrepancies while missing the broader historical context of the work, namely the role of military revolutions in the formation of centralized states.

[112] Williams, and Edge, "The Social Shaping of Technology," 866 in printed edition; Langdon Winner expresses it clearly: "The term black-box in both technical and social science parlance is a device or system that, for convenience, is described solely in terms of its inputs and outputs. One need not understand anything about what goes on inside such black boxes. One simply brackets them as instruments that perform certain valuable functions." Winner, "Upon Opening the Black-Box and Finding It Empty," 365; on the "black-box," see also below, paragraphs on social constructivism, 349–353, 356–357.

[113] Ferrell, "The Limits of Security Governance: Technology, Law, and War" 118.

out, in antiquity.¹¹⁴ Social context may well be the line of demarcation between pre-modern military revolutions and what occurred after industrialization (and in the case of the RMA). Rey and Zimmerman both point to the fundamentally different nature of technology (and hence determinism) in pre- and post-industrial societies.

Technological development it is said precipitates degrees of change, and when utilized in warfare, evolvement "drives" military revolutions in tandem with contextualizing factors such as strategic culture. Crossing the industrial watershed in the 1700s, we entered a new world, especially in the relationship between human society and the military demands of the state. Though many students of military revolution insist that state-sponsored conflict transformed society and led humankind toward "modernity," in the Western sense of the word, the historical reality of the 1800s suggests that socio-economic change drives the "real" military revolution. Generally speaking, societies shape states not the other way around, providing the broadest context for military revolutions.¹¹⁵ Socio-economic forces foment political struggles made manifest in the state. For example, the military transformation of France 1790–1815 followed by the nation's "modern warfare" (that emerged in the latter half of the nineteenth century) suggests that warfare and its attendant military establishments do not precipitate military revolutions so much as weather them, especially if military revolutions are exogenous. Mark Grimsley reminds us that all institutions are buffeted by social waves and economic developments.¹¹⁶

We have returned to where we started. Much near-recent military history echoed classical writers. Research undertaken since the 1970s that has been labeled "war and society" predominantly took its cue from the ancients, intending to "focus on warfare and its impact on human life, rather than on the weapons themselves."¹¹⁷ The technological determinist school has been perceived as a materialist and mildly antiquarian competitor to the "war and society" genre. In time the RMAers (if one dares generalize) conceived themselves as having reconciled for their limited purposes these contradictory schools of thought. While confidence in technology was affirmed with the successes of Desert Storm, still the "social dimension" increasingly occupied scholars in forecasting future developments in the

---

114 Rey, "Weapons, Technological Determinism, and Ancient Warfare," 21–56.
115 Rodger, "The military revolution at sea," 59–76; Rodger, "From the 'military revolution' to the 'fiscal-naval state,'" 119–128. Ideologized groups, such as the Khmer Rouge and Taliban, of course engaged in social engineering.
116 Grimsley, "Surviving military revolution," 74–91.
117 Rey, "Weapons, Technological Determinism, and Ancient Warfare," 34.

RMA.[118] Were "socio-economic forces" rather than technology, then, causal and determinist? Challenges to this causal interpretation came forth from the social sciences, emphasizing the "social shaping of technology'" (SST). It was argued that it was more often likely that society shaped technological innovation and application than vice versa. Socio-cultural influences played the causal role, and accordingly technological innovation and utilization resulted. By bringing to bear the scholarship of "social shaping" this revisionism widened the disciplinary and methodological assessment of military revolutions and RMAs. Technology-driven military revolutions were subsumed within larger socio-cultural realities governing innovation, invention, and "progress."[119] RMA scholars now recognize that deciphering paths of causation requires "multiple levels of analysis."[120] While the "socially constructivist approach" to military technologies can itself assume a form of determinism it has the advantage of insisting that "monocausal explanations are exposed to the corrective of multiple causal variables."[121] In other words, grandiose claims of causation must be qualified by considering the strategic cultural context specifically, and the broader societal framework in which military revolutions occur.

The social shaping of technology is plainly evident within strategic culture. The theory of SST rests upon the "concept that there are 'choices' (though not necessarily conscious choices) inherent in both the design of individual artefacts and systems . . .."[122] Tangentially, social constructivists (or constructionists, as the terms are sometimes used interchangeably) offer "models of a dynamic, multi- centered process of social selection"[123] that rendered technology as something crafted by humans, altered by humans, in a complex societal and cultural context.[124]

Science and Technology Studies, applying elements of social constructivism, suggest that "sociotechnical systems" are in play with the RMA, not an amorphous "technology" harnessed by technicians. According to Steven Walton, scholarship now focuses upon "two non-mutually exclusive ways" to test the veracity of the existence of technological determinism: first, "as an *autonomous* entity,

---

[118] Howard, "How much can technology change warfare?."
[119] Hacker, "Military Institutions, Weapons, and Social-Change," 768–834; Raudzens, "War-Winning Weapons," 403–434.
[120] Roland, *War and Technology*, 1, 5.
[121] Roland, "Is Military Technology Deterministic?," 30.
[122] Williams, and Edge, "The Social Shaping of Technology," 866 in print copy, electronic version, 2.
[123] Winner, "Upon Opening the Black-Box and Finding It Empty," 371.
[124] Williams, and Edge, "The Social Shaping of Technology," 1–50, particularly the comments on "The 'social constructivist' analysis of software," 21–24 in the electronic version.

developing by internal logics and without social input or control, and second, as society inexorably *changing in response* to new technologies."[125] By illumining military technology's linkage to the macrocosm of human society (and more pointedly to interaction among social classes), the connection between military revolutions and the RMA is that of resonant military systems within unique strategic cultures that have capitalized on their technological tools. Colonel Vicente, a practitioner of the RMA as well as a theorist, emphasizes choice and recognizes validity in SST's argument that "[d]ifferent routes are available, potentially leading to different technological outcomes."[126] The military revolution orthodoxy from its outset implied that warfare created a reality in which humans were not fully in control. With the emergence of the devices described by Colonel Vicente, military "technology" has gone from shaping and being shaped by human society to the point of eradicating society.

Robertsian military revolution theory posited how warfare wrought changes upon society circa 1560 to 1660. That simple cause-and-effect relationship oversimplifies what social constructivists identify as an interactive and mutually initiative relationship. As a system, following that social constructivist formula, it is clear that crafted arms grew within a strategic culture. Once weaponry was chosen (Roberts might have argued), his theoretical chain of causation followed as consequences of that choice, creating institutions (implicitly) in bondage to their weapons systems (e.g., an Ordnance Office). To reiterate, strategic culture was still the primary (though not solitary) impetus, especially in the pre-industrial world. Over time, more complex organizational (and socio-economic) forces such as industrialization shaped the conduct of war, and society accommodated its military needs. Identifying the nuances of causal social elements is a subtle exercise more commonly attempted by academicians and journalists. SST makes interplay between RMA technology and socio-economic forces more explicit and understandable to a wider audience.

Social constructivism and the social shaping of technology contextualize the essays by Kang, O'Neill, Lee, and Kazakou because they show how regional context, peripheries, led to decisions to embrace certain military technologies. Indeed, the contributions to this volume might be classified as examples of the social construction of technology (SCOT). The latter theorizes that "battlefield dilemmas [have] been neither inevitable nor predictable but rather dependent

---

[125] Walton, "Technological determinism(s) and the Study of War," 9, also 12–13.
[126] Williams, and Edge, "The Social Shaping of Technology," 866 in original print article, in electronic version, 2.

on the specific social choices and contextual circumstances the various interest groups have found themselves in at any one time."[127]

While social constructivism expands and admittedly blurs technology's context, RMAs such as cyberwarfare obscure older definitions of warfare and what constitutes "battlespace."[128] Military historians know that battlespaces are individualized. That contextual factor is a dominant theme in studies of military revolutions within peripheries. Those military revolutions grouped as peripheral in this anthology followed unique courses of development, as described by our contributors. Wayne E. Lee's portrait of the Ottomans reveals how social constructivism explains apparent paradoxes such as those posed by the Ottoman Empire, a broker in artillery technology between East and West. The Ottomans interchanged expertise with Asian powers, oftentimes in the person of the artillerist. During the sixteenth century military revolution the Sultan exceeded Western Europe in institutionalizing, on a semi-permanent but centralized basis, ordnance and its attendant personnel. Ottoman adroitness waned in the following centuries, for example in maintenance of standardized bore and regimentation of artillery train and siege equipment. Despite their geographical advantage (especially proximity to the Balkan cockpit) as a crossroads of military technologies, and their centralized yet flexible imperial infrastructure, Ottoman hegemony was eclipsed in what is referred to above as the second military revolution era of post-1650 world history.[129]

Oftenimes a "lead sector" (as the late W.W. Rostow might say) can be identified, as in our era electronic technology such as cyberwarfare, by distinguishing its many points of contact with diverse sectors of an economy. We conclude that "while advances in technology typically underwrite a military revolution, they alone do not constitute the revolution."[130] Exclusive focus on technology, especially when defined vaguely, undermines global contextualization of military revolutions, as it is far too easy to generalize, inaccurately, about how some cultures cannot readily adopt those technological innovations that provide military advantage. We refer here not to an infantryman's revulsion toward a cartridge lubricated with pork fat. For decades western historians, mostly English-speaking, assumed Asian cultures (Chinese and Japanese especially) only reluctantly seized upon the savage potentiality of firearms.[131] Strategic culture, state infrastructure,

---

127 Ford, "The British Army and the politics of rifle development, 1880 to 1986," 240.
128 Stone, "Cyberwar *will* take place!," 101–108.
129 Agostin, "The Ottoman Empire and the Technological Dialogue Between Europe and Asia," 28–29, 37–39 in Gunergun and Raina, ed., *Science Between Europe and Asia*; Agostin, "Behind the Turkish War Machine," 105–106, 109–111, 119–126.
130 Cuoco, *The Revolution in Military Affairs*, 17; Krepinevich, "Cavalry to Computer," 30–42.
131 Mokyr, *The Lever of Riches*, 209–238, especially 221.

and resources, not cultural aversion to technology, affected the degree to which peoples adapted European weapons systems. Mark Mandeles and Colonel Vicente both raise issues of social constructivism in assessing the RMA, specifically how organizational structures we utilize shape the "artifacts" of weaponry. Colonel Vicente suggests that the RMA assumed an incarnation encompassing more than what in the past has been deemed military. As we move from focus upon transfer of a few technologies that promise decisive results to a more realistic "systems" scenario that derives from cultural context, the historian's task becomes rather more challenging, especially now that genuinely global studies of military revolution have appeared. Likewise the task for RMAers has become more complicated and dependent upon social scientists as well as the ubiquitous technicians.

In addition to simultaneously managing while conceptualizing military revolutions, there is the tangential complication regarding who writes about them, and in the case of the RMA, materializes them. One can be caught between the historicism requisite for accurately identifying a military revolution and the potential ahistoricism of contemporary applied military science as practiced by many RMA researchers. Alex Roland put it well: "As soldiers and scholars contemplated technological changes in warfare in the 1990s two arcs of analysis intersected without really having much impact on each other, passing instead like proverbial ships in the night."[132] Roland distinguishes between apparently partitioned military, and intellectual, undertakings. We have history for history's sake, a manifestation of historicism; the other, a utilitarian approach to past experiences. Some scholars ably straddle both worlds, for example Geoffrey Parker.[133] The architects (and interpreters) of military revolutions and RMAs are frequently at cross purposes, however. Each constituency manifests the mentality of a self-interested professional class, creators of these phenomena as well as the those who try to make sense of what has been created. In Andrew Latham's words,

> [M]ost of the scholarship on the RMA reflects a profoundly superficial and *ahistorical* understanding of the changing nature of warfare. A truly historical approach to the phenomenon of war, of course, would embrace what Robert Cox calls the historical mode of thought – that is to say, analysis would start from the premise that historical structures and institutions (including those related to organized political violence) are socially constructed and therefore vary considerably across time and space.[134]

---

132 Roland, *War and Technology*, 105.
133 For just one example of his ability to engage both audiences, see Parker, "The Limits to Revolutions in Military Affairs," especially 367–372.
134 Latham, "Braudelian," 232; Winner, "Upon Opening the Black-Box and Finding It Empty," 362–378; Steen, "Upon Opening the Black Box and Finding it Full," 389–420; for a

We have seen, above, that mainstream military historians borrow from social constructivism to explain the nature of military revolutions, and that RMAers, too, have availed themselves of the social sciences (even if their goals in studying these phenomena differ). Both groups, especially those engaged in utilitarian pursuits, realize that historical accuracy is imperative. More subversive than judged simply "ahistorical," the military revolution might be condemned as "dangerously close to what David Edgerton has called 'anti-history': the invention of imaginary explanations to account for things which never happened"[135] The latter observation points out that if generating misinterpretation, military revolution studies potentially do a disservice to both the study of history and the implementation of military science. There is an imperative in getting it right. That urgency accounts partially for why scholars avail themselves of the social sciences (as discussed above) but refrain from adopting rigid models (a cautionary note that applies to Marxian interpretations as well).

The complexity of fully explaining technology's interactions with society and culture is all the more daunting because the rapidity and scale of the world's military revolutions appear uneven when measured strictly in terms of technical innovation. RMAs, like military revolutions, do not move in coordinated fashion on a broad front. At risk, therefore, of distorting the veracity of the development of military revolutions, let us consider how a dominant interpretation within the social sciences, Marxism, might reveal the connections between the military revolution and the RMA. Michael Roberts' military revolution template with its larger armies, grander strategies, more organized ways of war, reflects the social dynamic that brought into being competition among states that were developing increasingly capitalist economies and social structures. The "productive force" that Karl Marx identified as a catalyst for the leap into modernity is in the case of military revolution theory a destructive force, the gunpowder revolution.[136] Whether one subscribes to Marxian views or not, it is clear that the amalgamated "Geoffrey Roberts" military revolution posits the erosion of medieval military elites. From pole-armed urban militias to humble artillerymen, the actors of the military revolution contested the nobility's hegemony (globally, not just in Europe) over the application of violence. Clifford J. Rogers' artillery revolution (and subsequent artillery fortress revolution) are grounded in proto-industrial materialist development. Grimy artillery-founding and construction of massive fortifications created small but valuable military-industrial themed townscapes. Artillery, produced in

---

definition of social constructivism, see Kukla, *Social Constructivism and the Philosophy of Science*, 1–6.
**135** Rodger, "From the 'military revolution' to the 'fiscal-naval state,'" 120.
**136** Kurz, "The destructive origins of capitalism."

an industrial fashion created a qualitatively different weapons producing sector from the craftsmen who fashioned armor in an agricultural economy. The foundries did not change society, obviously. Rather metallurgical advances and molten metal were products of an evolutionary process made by the trajectory of strategic culture.[137] Aliaksandr Kazakou's essay on Eastern Europe's peripheries illustrates how the acquisition and use of cannon varied widely within that peripheral region. The ordnance-studded vessels of the commercial revolution fed a dominant bourgeois mercantile class. Ample evidence suggests both military revolution and RMA fit within a Marxian dialectic, and not just in the West.

The literature of the Marxian dialectic accommodates military revolution more so than military revolution theorists have been inclined to come to terms with Marx. In terms of causation, the conviction that economics shape war (rather than the methods of warfare fomenting macroeconomic change) is more conceivable than military revolution as the prime mover. Andrew Liaropoulos writes, "[s]tarting with the very invention of agriculture, every revolution in the system for creating wealth triggered a corresponding revolution in the system for making war."[138] Even before industrialization, patterns of conscription reveal at work omnipresent stresses generated by frequent warfare.[139] The demands of accumulation unleashed by warfare upon societies on the cusp of expanding market economies moved toward the development of a class society in war as in peace.

From antiquity to the present, effective force projection has been based upon "acquisitional" ability, by the state and/or individuals. Using Brett Steele's definition, this capability entailed "the technological transformation of civilian resources into military assets, including weapons, ammunition, and armor, in addition to suitable vehicles, food, and fodder."[140] These processes are material, and Marx would have recognized human resources as well as material resources and capital as essential ingredients. Organizational culture, again, is requisite in how technology, however brilliantly conceived and created, can be deployed effectively. What Marxian class consciousness brings to the picture is both quantitative and qualitative, specifically (1) ever-greater sized armies and (2) more specialized, highly trained soldiers such as artillerists, brought into being by the military revolution. Could the gunpowder age's expanding cadre of infantrymen

---

[137] López-Martin, "Historical and Technological Evolution of Artillery."
[138] Liaropoulos, "Revolutions in Warfare: Theoretical Paradigms and Historical Evidence," 368.
[139] For example, Øystein, "State and society in seventeenth-century Norway," 337–363 and Jespersen, "Social Change and Military Revolution in Early Modern Europe," 1–13.
[140] "Introduction" in Steele and Dorland, ed. *The Heirs of Archimedes*, 3.

be a growing proletariat?[141] The exploitation of ever larger numbers of infantry is compatible with a Marxian view of the social order. Warriors of the military revolution were proto-typical wage earners, a sort of proletariat for the production of war, emblematic of a calcified feudalism and obsolescent medieval arms.[142] Conscripts' experiences resonated a brutal, regimented, and quasi-industrialized life. The technology and methodical brutality foot soldiers dealt with included the speed and quantity of lethal missiles from firearms, a reality persisting to the modern day.

If elements of social exploitation and class conflict stirred in the military revolution, then did the military revolution weaken aristocracies? Or, did the ruling classes (e.g., the nobility) adapt and exploit the transformation to the fiscal-military state allegedly caused by the military revolution? Post-1650 global military revolutions originated and thrived in cooperation between capital and governmental policy. This view incorporates the military revolution into class conflict as a conjoined dynamic that had the potential to undermine old structures of government, and precipitate fundamental socio-economic change.[143] Military revolutions were undertaken and sustained to maintain a society's security against violence from a foreign political entity. Scholars of international state systems encounter difficulties incorporating military revolutions into their models, however. In Aaron Frank's words, "[s]tates seem to be equally fearful of domestic upheaval, rebellion and dissolution as foreign aggression and conquest."[144] That is particularly true of pre-industrial societies where internal security was not so institutionalized as in later centuries. Frank notes that proliferation of certain types of weaponry among the population (and the social promotion of certain military cadres among the elites) retarded the mastery of military revolution in many cases (Japan, Prussia, etc.). Domestic unrest and manifestations of class struggle therefore figured in the strategic calculations made by the state.

The proto-industrialization of war necessitated state formation, or more accurately, state transformation. In later centuries, overwhelming an enemy by mass, be it soldiers or bombs dropped from the skies, prolonged this industrial revolution of warfare. In the West, the application of industrialization to warfare coincided with formulation of an ideology of class struggle. Herein lie the roots of the RMA as "industrialised warfare."[145] Interpreting the RMA as the

---

141 Sherer, *Warriors for a Living*; DeMesa, *The Irish in the Spanish Armies in the Seventeenth Century*.
142 Kurz, "The destructive origins of capitalism."
143 Downing, *The Military Revolution and Political Change*.
144 Frank, "Military Revolutions, Evolution, and International Relations Theory," 22.
145 Blackmore, *War X*, 36–125.

means by which capitalist superpowers gorge on petroleum fits Marxian logic. Capitalist states use novel forms of violence to secure raw materials. Thus, the social shaping of technology is consistent with Marx's dialectical materialism. The problem is that causation is reversed when compared with the Robertsian military revolution. SST suggests military technology is a result, not a source of causation; technological innovations are enmeshed within a socio-economic web of markets, an assertion sustained by the most recent commercial and governmental forays in information science, which enable the kind of warfare Colonel Vicente describes.[146]

Capitalism provides the means, through the marketing of telecommunications and innovative transport systems, to influence and coerce societies (e.g., drones which have military as well as peaceful application) and, most revealing, cyberwarfare rooted in information technology rather than the arsenal of traditional weaponry. The result is Alain Joxe's description of flourishing "predatory empires" that wage semi-permanent global techno-warfare, sowing discord among foreign polities in order to enrich themselves.[147] Exploitative capitalism figures largely in the social shaping of technology. Considering punctuated equilibrium's cadences of innovation, SST might be regarded as at least a variant of the socio-economic dialectic of Marxism. Implicit determinism permeates the orthodox Robertsian military revolution, just as it does Marxist theory, and all ahistorical systems of thought designed to explain the configuration of present reality. Fredrich Engels synthesized the determinism of social construction with technological determinism when he wrote, "[t]he producer of more perfect tools, *vulgo* arms, beats the producer of more imperfect ones."[148]

Regarding present-centered historiography Jeremy Black has shown how the determinist Whig interpretation of history, too, can be enmeshed within case studies of military revolutions.[149] Profound changes in modern society can on occasion be traced back to how technical innovations and practices were integrated into the social fabric. Jacques Sapir identifies the point of convergence at which military revolutions and RMAs meet, namely innovation. "The revolution in military affairs is therefore located basically at the intersection of technical innovation,

---

146 Williams, and Edge, "The Social Shaping of Technology," 877–878, 893.
147 Joxe, *Empire of Disorder*, especially 182–189; Keane, "Epilogue: does democracy have a violent heart?," 403–404.
148 Cited by Steven Walton, citing J.F.C. Fuller, in the former's "Technological Determinsism(s) and the Study of War," 4. Steven Walton supplied advice and references to the present author more generously than can be fully acknowledged. Errors, however, are the responsibility of this editor.
149 Black, "Military History and the Whig Interpretation," 5–23.

social innovation and organizational innovation."¹⁵⁰ The process that foments innovation is dialectical, mirroring Marxian views of "progress." However, where Marxian views of military revolutions and RMAs are weakest is in their inherent determinism. Considering the "Eureka" moments of scientific discovery and the frequently haphazard process of technology transfer, "innovation" as a phenomenon within itself is not deterministic. Like fortunes on the battlefield, it is not predestined so much as subject to the vagaries of free will and accident. Innovation is more like the wheel of fortune emblazoned on Ehrenhold's livery depicted on the cover of this volume.

The determinism of Marxian and Whig interpretations of history inherently fashions a chain of causation which explains (but doesn't necessarily vindicate) modernity's status quo. As noted, critics of social constructivism and SST have challenged those theories' claims to transcend determinism.

> As a program of inquiry, social constructivism is careful to avoid the technological version of the 'Whig theory of history,' in which the past is read as a sequence of steps leading inevitably to the accomplishments of today. But although social constructivism escapes the bind of Whig history, it seems not to have noticed the problem of elitism, the ways in which even a broad, multicentered spectrum of technical possibilities is skewed in ways that favor some social interests while excluding others.¹⁵¹

One facet of that "elitism" is economic elitism, the longstanding subordination of the application of violence to enrich the perpetrator. The RMA's appropriation by globalist capitalism is delineated clearly in the works of Alain Joxe, Neal Balan, Jacques Sapir, Laurent Henninger and others. Although these commentators differ ideologically, they agree that the complexities of socio-economic factors, especially in the wake of the second military revolution, propel military revolutions and RMAs. Interpreting the views of Sapir, and himself, Henninger explains why "determinism and reductionism" are impediments when "discussing military matters with other historical areas." In explaining how and why a strategic culture conducts its warfare in the present, it construes the past to rationalize, justify, and serve a concurrent reality. In other words, an exercise such as that of proposing revolutions in military affairs is fundamentally ahistorical. History is dismantled and reassembled in the image of contemporary military practice. Even the earliest attempts at defining the military revolution were fraught with this danger. The convenience of employing the shorthand of "military revolution" while writing

---

150 Laurent Henninger on Jacques Sapir, translated from the French, online in "La 'révolution militaire'. Quelques éléments historiographiques" in *Mots. Les Langages du Politique* vol. 73 (2003), 93 citing Sapir 456. https://journals.openedition.org/mots/16312.
151 Winner, "Upon Opening the Black-Box and Finding It Empty," 370.

history risks (as Laurent Henninger has put it more than once) determinism and reductionism.[152]

The advent of a striking invention or innovation captures the attention of subsequent generations that in turn reckon how that device or technique created their contemporary world. Historical possibility is narrowed and oversimplified. Henninger has cautioned that military history in particular is susceptible to this linear and present-minded view of how human reality unfolds. Determinism (and in Henninger's analysis, reductionism), inherent in military revolution theory and assumptions regarding RMA development, infects other progressive views of history as well (e.g., Marxian and Whig). All historical interpretation distorts to a degree. Seen through the eyes of an economist, Jacques Sapir, military historians risk focusing on a recognizable theoretical microcosm while underappreciating the highly complex (non-linear) macrocosm (though ironically macro studies tend to favor determinism more than do exercises in microhistory, as was argued above).

An example of the latter, where a chronologically tightly focused study on a specific topic reveals a developmental and evolutionary pattern that is not so jarring as to be characterized as "revolutionary," is the detailed technical analysis of Burgundian ordnance circa 1363 to 1477. Although compatible with punctuated equilibrium, this case study reveals glacially paced experiments and small-scale innovations that, coupled with organizational growth, are clearly "evolutionary."[153] "Revolutionary" establishes that the decisive change was so profound it altered the subsequent course of events. So, determinism and revolution are bedfellows in the chain of causation. High resolution examination tempers the broad generalizations that one is tempted to draw. The problem is not the act of fixating on a single technological innovation; rather it is when we tie that object to a broader historical (and social) context and attempt to forge a causal linkage. Jacques Sapir, for one, forbids gadgets to become totemic.

The RMA "reflects no mechanical application of the technical field on the other because technological innovation is naturally a result, at a time of economic and scientific progress, and transformations that affect companies and

---

[152] Henninger, "The Military Revolution and the Birth of Modernity"; Henninger, "La révolution militaire et la naissance de la modernité"; Henninger, "La 'révolution militaire'. Quelques éléments historiographiques," 87–94; republished as Henninger, "The 'military revolution,'" 81–86; Henninger, "Military Revolutions and Military History," 8–19.

[153] Smith and DeVries, *The Artillery of the Dukes of Burgundy 1363–1477*, for example, 33, 54, 70, 317. The process was evolutionary yet the improvements in artillery were genuinely revolutionary. See above, 9.

organizations."[154] In other words, military historians should consider process before objects. Identifying "the" military revolution, like delineating a textbook definition of the RMA, is akin to taking a snapshot while realizing that focusing exclusively upon a solitary moment in time, an event, an invention, distorts reality by "freeze-framing." This aspect of existence was best encapsulated by a philosopher, Henri Bergson, in *An Introduction to Metaphysics*. Colin Gray, the prolific authority on strategy in the era of the RMA, recognizes what Bergson discerned as existence's highly fluid mutability and applies it to military history. Against the sense of compartmentalization and the coupling of distinct historical developments, Gray emphasizes "the force of continuity in strategic history" (in Carlo Alberto Cuoco's words). Gray draws from a British military historian of an earlier generation, Cyril Falls, who advised,

> the student should not believe everything moves only when he sees the process at a glance, and stands still when he does not see it moving. It is his eyes which are at fault. They see movement of a pattern and in circumstances which are familiar to them; they fail to detect it when those are unfamiliar. The more scholarly the enquirer becomes, the more conscious is he of endless change.[155]

Though the past may be impossible to comprehend authentically, we are still left with the present as our only point of reference. For example, the somewhat flawed concept of "total war" (the inclusive notion that warfare can be the all-consuming Moloch) has shaped our contemporary strategic culture, positively and negatively. Like the nuclear weapons that evolved ultimately out of the global agonies of the twentieth century, "total war's" destructiveness is defined as wholesale and indiscriminate. "Consequences so catastrophic" (in Sir Michael Howard's assessment of 1914–1918) begged for military technology systems less destructive of industrialized society.[156] Naturally, military historians as well as professional soldiers sought a point of origin for "total war." Our reading of the past constructed our strategic culture and led to unrealizable expectations. Initially, hope was expressed that the RMA would "civilize" warfare by eradicating exclusively "hostiles" and their resources, without incurring collateral damage. However, as Colonel Vicente's essay shows, warfare is far from being tamed. Indeed, combat is destined to humble humanity, even if inadvertently. Military technology which birthed "total war" is poised to ravage its creators apparently.

---

154 Henninger, "La 'révolution militaire'. Quelques éléments historiographiques." Subsequently translated into English and republished the following year as Henninger, "The 'military revolution,'" 81–6.
155 Cuoco, *The Revolution in Military Affairs*, 42, citing Colin S. Gray quoting Cyril Falls.
156 Philpott, "Total War," 147.

Ironically, this volume joins the chorus of voices questioning a direct causal role ascribed to technology (or, technological determinism), with the caveat that our final essay may reveal the point at which technology actually does, ultimately unilaterally, impose itself on humankind. Colonel Vicente describes a technology that will direct itself. Strauss and Ober have exclaimed: "It is a supremely dangerous error to suppose that technology is a solution for the problems of war. A strategy devised by technocrats, based solely on superiority in weaponry is no strategy at all. Machines do not win wars."[157] However, is the institutionalized international violence of the twenty-first century (rooted in the RMA) yet another incarnation in civilization's dialectic in military technology? Or are we undergoing something qualitatively different, an unprecedented brutality that transcends morality and human control? When human combatants no longer risk life and limb in battle (though civilian populations will remain vulnerable, as they have been since time immemorial), governments will be more willing to gamble on the application of force projection. Decisions will have to be made quickly, ultimately in nanoseconds, eschewing unavoidably consultation with assemblies or advisors. Indeed, the deployment of "unmanned" weapons such as drones lies beyond the constraining reach of the Congressional War Powers Act.[158]

Current global trends toward autocratic government are abetted by autonomous military technologies. Political costs can be minimized for the ruling classes. Warfare will be less visceral, more abstract. Indeed, a narcotic complacency accompanies RMAs, for if they are indeed endogenous and products of internal initiatives, a sense of manageability infects their designers and managers. Revolutions in military affairs, wherein the technological initiatives and strategies are internal, imply autonomous and guided control from the state's military organizations. This self-absorption is partially to blame for the Soviet and American wars fought in Afghanistan being won via insurgency.

Wars of national and corporate aggression have conveniently distanced themselves from morality whilst lethality magnifies itself.[159] Perhaps Strauss and Ober's observation was accurate in the past, but now fails to comprehend our immediate future. This volume's conclusions concur that social context and environment-specific factors must create circumstances in which technology can perform it magic. However, with the advent of autonomous lethal systems (for example the oft-fantasized killer robots) and airborne weapons networks

---

**157** Rey, "Weapons, Technological Determinism, and Ancient Warfare," 54, footnote 83, quoting Strauss and Ober, *The Anatomy of Error*.
**158** Shaw, "The Future of Killer Robots."
**159** Blackmore, *War X*. On reality and lethality, see 10–35, 164–199.

described by Colonel Vicente, we enter an era where humans must turn to sentient machines simply to survive even when we cannot "win" wars.

That said, we should be mindful of the high resolution definition proffered by Clifford Rogers, wherein RMAs confine themselves largely to the way in which military force wages combat (and thus inhabit exclusively the realm of Mars, god of war, as Rogers puts it).[160] In the final measure we are concerned with the clash of arms itself. Military revolution through the centuries has consistently increased lethality. The final phase of the RMA does not guarantee peace through deterrence. Rather, twenty-first century military technologies take military revolutions to the pinnacle of its development, almost casual and instant annihilation. The present writer would agree that humankind has passed the point of no return due to the inevitability of employing weapons utilizing artificial intelligence. By necessity we have already submitted our fate to the god of war.

In the next chapter Mark Mandeles, who sees both the historical and utilitarian aspects of the revolution in military affairs, addresses the degree to which humankind have been (and will be) masters of our own destiny. Strategic culture shapes military revolution. However, to what degree do governments comprehend and steer the strategic culture that they predominate?

# Bibliography

Ágoston, Gábor. "The Ottoman Empire and the Technological Dialogue Between Europe and Asia: The Case of Military Technology and Know-How in the Gunpowder Age." In *Science Between Europe and Asia*, edited by Feza Gunergun, and Dhruv Raina, 28–29, 37–39. Dordrecht: Springer. 2011.

Ágoston, Gábor. "Behind the Turkish War Machine: Gunpowder Technology and War Industry in the Ottoman Empire, 1450–1700." In *The Heirs of Archimedes. Science and the Art of War through the Age of Enlightenment*, edited by B. Steele, and T. Dorland, 105–106, 109–111, 119–126. Cambridge: The MIT Press, 2005.

Andrade, Tonio. *The Gunpowder Age. China, Military Innovation, and the Rise of the West in World History* Princeton: Princeton UP, 2016.

Appel, Tiago Nasser. "How competition drove social complexity: the role of war in the emergence of States, both ancient and modern." *Brazilian Journal of Political Economy* 40.4 (October–December 2020): 728–745.

Awty, Brian. "The Development and Dissemination of the Walloon Method of Ironworking." *Technology and Culture* 48.4 (October 2007): 783–803.

Black, Jeremy. *Combined Operations. A Global History of Amphibious and Airborne Warfare*. Lanham: Rowman and Littlefield, 2018.

---

[160] Rogers, "'Military Revolutions' and 'Revolutions in Military Affairs,'" 24, 30–34.

Black, Jeremy. "Military History and the Whig Interpretation." In *Military History. Some Introductions Designed to Begin a Debate*, Jeremy Black, 5–23. Rome: Società Italiana di Storia Militare, 2020.

Black, Jeremy. *War and Technology*. Bloomington and Indianapolis: Indiana University Press, 2013.

Blackmore, Tim. *War X: Human Extensions in Battlespace*. Toronto: University of Toronto Press, 2006.

Braudel, Fernand. *The Wheels of Commerce*. New York: Harper and Row, 1986.

Brice, Lee L. "Philip II, Alexander the Great, and the Question of a Macedonian 'Revolution in Military Affairs.'" *The Ancient World* 42.2 (2011): 137–147.

Buckley, John. "Air Power." In *Palgrave Advances in Military History*, edited by M. Hughes, and W. Philpott. Basingstoke: Palgrave Macmillan, 2006.

Chesneau, Roger, and Eugene Kolesnik, ed. *Conway's All the World's Fighting Ships 1860–1905*. New York: Mayflower, 1979.

Cipolla, Carlo M. *Guns, Sails and Empires: Technological Innovation and the Early Phases of European Expansion 1400–1700*. New York: Minerva Press, 1965.

Crouwel, Joost H. "Fighting on Land and Sea in Late Mycenaean Times." In *Selected writings on chariots and other early vehicles, riding and harness*. M.A. Littauer and J.H. Crouwel 100–105. Leiden: Brill 2002.

Cuoco, Carlo Alberto. *The Revolution in Military Affairs: Theoretical Utility and Historical Evidence*. Athens: RIEAS, 2010.

Dafoe, Allan. "On Technological Determinism: A Typology, Scope Conditions, as a Mechanism." *Science, Technology, & Human Values* 40.6 (2015): 1047–1076.

De la Garza, Andrew. *The Mughal Empire at War. Babur, Akbar and the Indian Military Revolution*. London: Routledge, 2016.

DeMesa, Eduardo. *The Irish in the Spanish Armies in the Seventeenth Century*. Woodbridge: Boydell, 2014.

Deogracias, Alan J. *Battle of Hampton Roads: A Revolution in Military Affairs*. New York: E.P. Dutton, 2004.

Dorn, Harold. "The 'Military Revolution': Military History or History of Europe." *Technology and Culture* 32.3 (July 1991): 656–658.

Downing, Brian, *The Military Revolution and Political Change. Origins of Democracy and Autocracy in Early Modern Europe* Princeton: Princeton UP, 1992.

Duffy, Michael. "The Foundations of British Naval Power." In *The Military Revolution and the State 1500–1800*, edited by Michael Duffy, 49–85. Exeter Studies in History. Exeter: University of Exeter Press, 1980.

Estanislau, Tomás. "The Catalan process for the direct production of malleable iron and its spread to Europe and the Americas." *Contributions to Science* 1.2 (1999): 225–232.

Fahy, Brian. "A Seat at the Table: Addressing Artefact Biases in Asian Shipwreck Assemblages", https://www.academia.edu/9669623/A_Seat_at_the_Table_Addressing_Artefact_biases_in_Asian_Shipwreck_Assemblages.

Ferrell, Theo. "The Limits of Security Governance: Technology, Law, and War." In *Technology and Security. Governing Threats in the New Millenium*, edited by B. Rappert, 111–131. Basingstoke: Palgrave Macmillan, 2008.

Fissel, Mark Charles. "Asia's Ancient Mediterranean Littoral as a Military Contact Zone: 2500–498 BCE." In *Handbook of Asian Military History*, edited by Kaushik Roy. Oxford University Press – India, forthcoming.

Fissel, Mark Charles. "Military Revolutions." In *Oxford Bibliographies in Military History*. Ed. Kaushik Roy. New York: Oxford University Press, https://www.oxfordbibliographies.com/view/document/obo-9780199791279/obo-9780199791279-0212.xml.

Fissel, Mark Charles. "Out of Africa: The Egyptian Origins of Amphibious Warfare." In *The Routledge Handbook of the Global History of Warfare*, edited by Kaushik Roy, and Michael Charney, forthcoming.

Ford, Matthew Charles. "The British Army and the politics of rifle development, 1880 to 1986." PhD Thesis. King's College London, 2008.

Frank, Aaron. "Military Revolutions, Evolution, and International Relations Theory," https://www.academia.edu/51313171/Military_Revolutions_Evolution_and_International_Relations_Theory.

Gould, Stephen Jay. *Punctuated Equilibrium*. Cambridge: Belknap Harvard, 2007.

Grimsley, Mark. "Surviving military revolution: The U.S. Civil War." In *The Dynamics of Military Revolution 1300–2050*, edited by Macgregor Knox, and Williamson Murray, 74–91. Cambridge: Cambridge UP, 2009.

Guilmartin, John F, Jr. "Technology and Strategy: What Are the Limits?" In *Two Historians in Technology and War*, edited by Michael Howard and John F. Guilmartin, Jr., 11–46. Carlisle Barracks, 20 July 1994, US Army War College, Strategic Studies Institute.

Guilmartin, John F. Jr. "The military revolution in warfare at sea during the early modern era: technological origins, operational outcomes and strategic consequences." *Journal for Maritime Research* 13.2 (June 2011): 129–137.

Hacker, Barton C., "Military Institutions, Weapons, and Social Change: Toward a New History of Military Technology." *Technology and Culture* 35.4 (October 1994): 768–834.

Hale, John R. "Not Patriots, Not Farmers, Not Amateurs: Greek Soldiers of Fortune and the Origins of Hoplite Warfare." In *Men of Bronze. Hoplite Warfare in Ancient Greece*, edited by D. Kagan, and G. Viggiano, 176–191. Princeton: Princeton UP, 2013.

Hall, Bert S. *Weapons and Warfare in Renaissance Europe: Gunpowder, Technology and Tactics*. Baltimore: The Johns Hopkins UP, 1994.

Hall, Bert S. and K.R. DeVries, "Essay Review: The 'Military Revolution' Revisited." *Technology and Culture* 30 (1990): 147–154.

Harding, Richard. *Seapower and Naval Warfare 1650–1830*. Annapolis: Naval Institute Press, 1999.

Heuser, Beatrice. "Denial of Change: The Military Revolution as seen by Contemporaries." *International Bibliography of Military History* 32 (2012): 3–27 (pagination varies with format and mode).

Headrick, Daniel R. *The Tools of Empire: Technology and European Imperialism in the Nineteenth Century*. New York: Oxford UP, 1981.

Headrick, Daniel R. *Power over Peoples. Technology, Environments, and Western Imperialism, 1400 to the Present*. Princeton: Princeton UP, 2010.

Henninger, Laurent. "The Military Revolution and the Birth of Modernity" (23 October 2014), http://theatrum-belli.com/la-revolution-militaire-et-la-naissance-de-la-modernite-par-laurent-henninger/.

Henninger, Laurent. "La révolution militaire et la naissance de la modernité." (16 September 2017), https://www.youtube.com/watch?v=9azybDTWuR8&t=29s.

Henninger, Laurent. "La 'révolution militaire'. Quelques éléments historiographiques." *Mots. Les langages du politique* 73 (2003): 87–94; https://journals.openedition.org/mots/16312

Henninger, Laurent. "The 'military revolution': some historiographical elements." *Revue Historique des Armées* (1 March 2004): 81–6.
Henninger, Laurent. "Military Revolutions and Military History." In *Palgrave Advances in Modern Military History*, edited by M. Hughes, and W. Philpott, 8–19. Basingstoke: Palgrave Macmillan, 2006.
Herwig, Holger. "The battlefleet revolution, 1885–1914." In *The Dynamics of Military Revolution 1300–2050*, edited by Macgregor Knox, and Williamson Murray, 114–131. Cambridge: Cambridge University Press.
Holloway, R. Ross. *The Archaeology of Ancient Sicily*. London: Routledge, 1991.
Hone, Thomas C., and Mark D. Mandeles. "Innovation in Naval Aviation," Final Technical Report, Delex Systems, Inc., 31 July 1982.
Hone, Thomas C., Norman Friedman, and Mark D. Mandeles. *The Introduction of Carrier Aviation into the U.S. Navy and Royal Navy: Military-Technical Revolutions, Organizations, and the Problem of Decision*. Washington, DC: OSD/Net Assessment, 1994.
Hone, Thomas C., Norman Friedman, and Mark D. Mandeles, *American and British Aircraft Carrier Development, 1919–1941*. Annapolis: Naval Institute Press, 1999.
Howard, Sir Michael. "How much can technology change warfare?" In *Two Historians in Technology and War*, Sir Michael Howard and John F. Guilmartin, Jr., 1–46. Carlisle Barracks 20 July 1994, US Army War College, Strategic Studies Institute.
Hundley, Richard O. *Past Revolutions, Future Transformations: What Can the History of Military Revolutions in Military Affairs Tell Us About Transforming the U.S. Military?* RAND Corporation, May 21, 1999, Santa Monica CA.
Jespersen, Knud. "Social Change and Military Revolution in Early Modern Europe: Some Danish Evidence." *The Historical Journal* 26.1 (March 1983): 1–13.
*Joint Vision* 2010 The Defense Advanced Research Projects Agency DARPA (Washington DC 1999). Acquisition and Technology Policy Center of RAND's National Defense Research Institute (NDRI).
Joxe, Alain. *Empire of Disorder*. Los Angeles: Semiotext(e), 2002.
Keane, John. "Epilogue: does democracy have a violent heart?" In *War, Democracy, and Culture in Classical Athens*, edited by David M. Pritchard, 378–408. Cambridge: Cambridge University Press, 2010.
Khan, Iqtidar. *Gunpowder and Firearms. Warfare in Medieval India*. Oxford: Oxford University Press, 2004.
Knox, Macgregor. "Mass Politics and Nationalism as Military Revolution: The French Revolution and After." In *The Dynamics of Military Revolution, 1300–2050*, edited by Macgregor Knox, and Williamson Murray, 57–73. Cambridge: Cambridge University Press, 2009.
Krepinevich, Andrew. "Cavalry to Computer. The Pattern of Military Revolutions." *The National Interest* (Fall 1994): 30–42.
Kukla, Andre. *Social Constructivism and the Philosophy of Science*. London: Routledge, 2000.
Kurz, Robert. "The destructive origins of capitalism," https://libcom.org/history/destructive-origins-capitalism-robert-kurz, translated by "Alias Recluse" (14 June 2011).
Kyriazis, Nicholas. "Seapower and socioeconomic change." *Theory and Society* 35.1 (February 2006): 71–108.
Lambert, Andrew. "Naval Warfare." In *Palgrave Advances in Modern Military History*, edited by M. Hughes and W. Philpott, 172–192. Basingstoke: Palgrave Macmillan, 2006.

Latham, Andrew. "Warfare Transformed: A Braudelian Perspective on the 'Revolution in Military Affairs'"; *European Journal of International Relations* 8.2 (2002): 231–266.

Lendon, J. E., *Soldiers and Ghosts. A History of Battle in Classical Antiquity*. New Haven: Yale University Press, 2006.

Leonard, Steven. *Inevitable Evolutions: Punctuated Equilibrium and the Revolution in Military Affairs*. School of Advanced Military Studies United States Army Command and General Staff College. Carlisle, 2001.

Liaropoulos, Andrew. "Revolutions in Warfare: Theoretical Paradigms and Historical Evidence – The Napoleonic and First World War Revolutions in Military Affairs." *Journal of Military History* 70.2 (April 2006): 363–384.

López-Martin, Francisco Javier. "Historical and Technological Evolution of Artillery From its Earliest Widespread Use Until the Emergence of Mass-Production Techniques." Ph.D. thesis, London Metropolitan University, June 2007.

Lynn, John A. "Reflections on the History and Theory of Military Innovation and Diffusion." In *Bridges and Boundaries. Historians, Political Scientists, and the Study of International Relations*, edited by C. Elman, and M. Fendius Elman, 359–382. Cambridge: MIT Press, 2001.

Machiavelli, Niccolò. *The Art of War*, edited by Neal Wood. New York: Da Capo, 1963.

Mandeles, Mark D. *The Future of War. Organization as Weapons*. Washington, DC. Potomac Books, 2005.

Margariti, Roxani Eleni. "Mercantile Networks, Port Cities, and 'Pirate' States: Conflict and Competition in the Indian Ocean World of Trade before the Sixteenth Century." *Journal of the Economic and Social History of the Orient* 51 (2008): 543–577.

Mokyr, Joel. *The Lever of Riches. Technological Creativity and Economic Progress*. Oxford: Oxford University Press, 1990.

Neill, Donald A. "Ancestral Voices: The Influence of the Ancients on the Military Thought of the Seventeenth and Eighteenth Centuries." *Journal of Military History* 62.3 (July 1998): 487–520.

O'Hanlon, Michael. *Technological Change and the Future of Warfare*. Washington DC: Brookings Institution Press, 2000.

Orwell, George. "Notes on Nationalism" (1945) reproduced electronically by the Orwell Foundation, https://www.orwellfoundation.com/the-orwell-foundation/orwell/essays-and-other-works/notes-on-nationalism/.

Palmer, M.A.J. "The 'Military Revolution' Afloat: The Era of the Anglo-Dutch Wars and the Transition to Modern Warfare at Sea." *War in History* 4.2 (April 1997): 123–149.

Parker, Geoffrey. *The Military Revolution. Military Innovation and the Rise of the West*. 1988. Cambridge: Cambridge University Press, 1996.

Parker, Geoffrey. "The Dreadnought Revolution of Tudor England." In *Warfare in Early Modern Europe 1450–1650*, edited by Paul Hammer, 357–388. Aldershot: Ashgate, 2007.

Parker, Geoffrey. "The Limits to Revolutions in Military Affairs: Maurice of Nassau, the Battle of Nieuwpoort (1600), and the Legacy." *The Journal of Military History* 71.2 (April 2007): 331–372.

Peers, Douglas. "Revolution, Evolution, or Devolution? The Military Making of Colonial India." In *Empires and Indigenes. Intercultural Alliance, Imperial Expansion and Warfare in the Early Modern World*, edited by Wayne E. Lee, 81–106. New York: NYU Press, 2011.

Philpott, William J. "Total War." In *Palgrave Advances in Modern Military History*, edited by M. Hughes, and W. Philpott, 131–147. Basingstoke: Palgrave Macmillan, 2006.

Possony, S. and J. Pournelle. *The Strategy of Technology. Winning the Decisive War.* Cambridge: Dunellen, 1970.

Raudzens, George. "War-Winning Weapons: The Measurement of Technological Determinism in Military History" *Journal of Military History* 54.4 (October 1990): 403–434.

Raudzens, George. "Military Revolution or Maritime Evolution? Military Superiorities or Transportation Advantages as Main Causes of European Colonial Conquests to 1788." *The Journal of Military History* 63.3 (July 1999): 631–641.

Reddy, Srinivas. "Disrupting Mughal Imperialism: Piracy and Plunder on the Indian Ocean." *Asian Review of World Histories* 8 (2020): 128–142.

Rey, Fernando Echeverría. "Weapons, Technological Determinism, and Ancient Warfare." In *New Perspectives on Ancient Warfare*, edited by Garrett Fagan, and Matthew Trundle, 21–56. Leiden: Brill, 2010.

Rian, Øystein. "State and society in seventeenth-century Norway." *Scandinavian Journal of History* 10.4 (1985): 337–363.

Roberts, Michael. *The Military Revolution, 1560–1660. Inaugural lecture delivered before the Queen's University Belfast.* M. Bond (Belfast 1956), reprinted in *The Military Revolution Debate. Readings on the Military Transformation of Early Modern Europe*, edited by Clifford J. Rogers, 13–35. Boulder: Westview, 1995.

Roberts, Michael. *Essays in Swedish History.* London: Weidenfield and Nicolson, 1967.

Rodger, N.A.M. "From the 'military revolution' to the 'fiscal-naval state.'" *Journal for Maritime Research* 13.2 (June 2011): 119–128.

Rodger, N.A.M. "The military revolution at sea." In *Essays in Naval History, from Medieval to Modern*, edited by N.A.M. Rodger, 59–76. Farnham: Ashgate Variorum, 2009.

Rogers, Clifford J. "'Military Revolutions' and 'Revolutions in Military Affairs': A Historian's Perspective." In *Toward a Revolution in Military Affairs? Defense and Security at the Dawn of the Twenty-First Century*, edited by T. Gongora, and H. von Riekhoff. Westport: Greenwood, 2000.

Rogers, Clifford J. "The Idea of Military Revolutions in Eighteenth and Nineteenth Century Texts." *Revista de História das Ideias* 30 (2009): 395–415.

Rogers, Clifford J. "The Artillery and Artillery Fortress Revolutions Revisited." In *Artillerie et Fortification 1200–1600*, 75–79. Rennes: Presses Universitaires de Rennes, 2011.

Rogers, Clifford J. "Gunpowder Artillery in Europe, 1326–1500: Innovation and Impact." In *Technology, Violence, and War Essays in Honor of Dr. John F. Guilmartin, Jr.*, edited by Robert S. Ehlers Jr., Sarah K. Douglas, and Daniel P.M. Curzon, 39–71. Leiden: Brill, 2019.

Roland, Alex. *War and Technology. A Very Short Introduction.* Oxford: Oxford University Press, 2016.

Roland, Alex. "Is Military Technology Deterministic?" *Vulcan* 7 (2019): 19–33.

Rommelse, Gijs. "Introduction: The military revolution at sea." *Journal for Maritime Research* 13.2 (June 2011): 138–150.

Rowlands, Guy. "The King's Two Arms: French Amphibious Warfare in the Mediterranean under Louis XIV, 1664 to 1697." In *Amphibious Warfare 1000–1700*, edited by D.J.B. Trim and Mark Charles Fissel, 265–311. Leiden: Brill, 2006.

Roy, Kaushik. *Military Transition in Early Modern Asia.* London: Bloomsbury, 2014.

Roy, Kaushik. "Military Synthesis in South Asia: Armies, Warfare, and Indian Society, c. 1740–1849." *The Journal of Military History* 69.3 (July 2005): 651–690.

Satia, Priya. *Empire of Guns. The Violent Making of the Industrial Revolution.* New York: Penguin, 2018.

Shaw, Ian G. R. "The Future of Killer Robots: Are We Really Losing Humanity?" *e-International Relations* (12 November 2012), https://www.academia.edu/2276635/Killer_Robots_Are_we_really_Losing_Humanity.

Sherer, Idan. *Warriors for a Living. The Experience of the Spanish Infantry in the Italian Wars, 1494–1559*. Leiden: Brill, 2017.

Showalter, Dennis. "Caste, Skill, and Training: The Evolution of Cohesion in European Armies from the Middle Ages to the Sixteenth Century." *The Journal of Military History* 57 (July 1993): 407–430.

Sicking, Louis. "Naval warfare in Europe, c. 1330–c. 1680." In *European Warfare 1350–1750*, edited by F. Tallett and D.J.B. Trim, 236–263. Cambridge: Cambridge University Press, 2010.

Smith, R.D., and K. DeVries. *The Artillery of the Dukes of Burgundy 1363–1477*. Woodbridge: Boydell, 2005.

Smythe, Sir John. *Certain Discourses Military*, edited by J.R. Hale. Ithaca: Cornell, 1964.

Sondhaus, Lawrence. *Strategic Cultures*. Abingdon: Routledge, 2006.

Stapleton, John M. Jr. "The Bluewater Dimension of King William's War: Amphibious Operations and Allied Strategy During the Nine Years' War, 1688–1697." In *Amphibious Warfare 1000–1700*, edited by D.J.B. Trim and Mark Charles Fissel, 315–353. Leiden: Brill, 2006.

Steele, B. and T. Dorland, ed. *The Heirs of Archimedes. Science and the Art of War through the Age of Enlightenment*. Cambridge: MIT Press, 2005.

Steen, Marc. "Upon Opening the Black Box and Finding it Full: Exploring the Ethics in Design Practices." *Science, Technology, and Human Values* 40.3 (2015): 389–420.

Stone, John. "Cyberwar *will* take place!" *The Journal of Strategic Studies* 36.1 (2013): 101–108.

Stone, J. "Technology and War: A Trinitarian Analysis." *Defense & Security Analysis* 23.1 (2007): 27–40.

Stradling, R.A. "'A Military Revolution': The Fall-Out from the Fall-In." *European History Quarterly* 24 (1994): 271–278.

Strauss, Barry S., and Josiah Ober. *The Anatomy of Error: Ancient Military Disasters and Their Lessons for Modern Strategists*. New York: St. Martin's Press, 1990.

Storrs, Christopher, and H.M. Scott. "The Military Revolution and the European Nobility 1600–1800." *War in History* 3.1 (January 1996): 1–41.

Stulberg, Adam N. "Managing Military Transformations: Agency, Culture, and the U.S. Carrier Revolution." *Security Studies* 14.3 (July/September 2005): 489–528.

Theobald, Ulrich. "European Weapons in China: Muskets and Cannons in the late Ming (1368–1644) and early Qing (1644–1912) periods," (2013), https://www.academia.edu/43903705/European_Weapons_in_China_Muskets_and_Cannons_in_the_Late_Ming_1368_1644_and_Early_Qing_1644_1912_Periods.

Tilly, Charles. *Coercion, Capital, and European States, A.D. 990–1990*. Oxford: Wiley-Blackwell, 1992.

Trim, D.J.B. "Medieval and early-modern inshore, estuarine, riverine and lacustrine warfare." In *Amphibious Warfare 1000–1700*, edited by D.J.B. Trim and Mark Charles Fissel, 357–413. Leiden: Brill, 2006.

Trim, D.J.B. and Mark Charles Fissel, ed. *Amphibious Warfare 1000–1700*. Leiden: Brill, 2006.

Trim, D.J.B. and Mark Charles Fissel. "Conclusion." In *Amphibious Warfare 1000–1700*, edited by D.J.B. Trim and Mark Charles Fissel, 429–432. Leiden: Brill, 2006.

Viggiano, Gregory F. "The Hoplite Revolution and the Rise of the Polis." In *Men of Bronze. Hoplite Warfare in Ancient Greece*, edited by D. Kagan, and G. Viggiano, 112–133. Princeton: Princeton University Press, 2013.

Vince, J., ed. and trans. *The Speeches of Demosthenes*, vol. 1. London/New York: Loeb Classical Library, 1930.

Walton, Steven A. "Technological Determinism(s) and the Study of War" *Vulcan* 7 (2019).

Warming, Rolf. "An Introduction to Hand-to-Hand Combat at Sea –General Characteristics and Shipborne Technologies from c. 1210 BCE to 1600 CE." In *On War on Board. Archaeological and Historical Perspectives on Early Modern Maritime Violence and Warfare*, edited by Johan Rönnby, 99–124. Stockholm: Södertörns högskola, 2019.

Warner, Christopher G. *Implementing Joint Vision 2010: A Revolution in Military Affairs for Strategic Air Campaigns*. Maxwell Air Force Base, Alabama: Air University Press, 1999, reprinted 2012.

Williams, R. and D. Edge. "The Social Shaping of Technology." *Research Policy* 25.6 (1996): 856–99 [1–50 in electronic version].

Winner, Langdon. "Upon Opening the Black-Box and Finding It Empty – Social Constructivism and the Philosophy of Technology." *Science Technology & Human Values* 18.3 (1993): 362–378.

Zimmerman, David. "Neither Catapults nor Atomic Bombs: Technological Determinism and Military History from a Post-Industrial Revolution Perspective." *Vulcan* 7 (2019): 45–61.

Mark D. Mandeles
# To Dream the Impossible Dream: Feasibility of Deliberate Government Guidance of Revolutions in Military Affairs

> Because history is a debate without end it matters far less who is right or wrong than how we conduct that debate.[1]

## Introduction

Under the banners of "military revolution," "revolution in military affairs" (RMA), and "military transformation," the role and impact of advanced technologies in war are popular topics for discussion, prediction, and hyperbole.[2] This chapter takes the fascination with advanced military technology and its promise of decisive impacts in war as its point of departure, focusing on the obstacles inherent in organizational processes of military innovation and transformation, and premises officials use to design military innovation programs. I argue that United States national security leaders overestimate their ability to manage and control the pace and outcomes of advanced technology development programs, and pose two questions to test this hypothesis: (1) can we identify any general laws of innovation that will help these leaders better guide advances in military technology and generate revolutionary operational capability; and (2) are current management strategies, organizational structures, and processes in place to achieve desired revolutionary changes?

---

[1] Lambert, 244.
[2] In 2018, a search of titles on the term "Military Revolutions" at the online site, academia.edu, revealed 61,462 titles contained the term in a set of 20 million full-text research papers. In 2021, a search on "Military Revolutions" revealed 88,614 papers contained the term in the full text in a set of 22 million research papers. See Fissel, "Military Revolutions."

---

**Note:** The author is responsible for all conceptual, analytical, and factual errors in the text below, and recognizes and thanks the following individuals for criticism and comments: Mark Charles Fissel, Thomas C. Hone, Lt.Col. Frank D. Kistler (USAF, ret.), and Laura Mandeles. This essay is dedicated to Jacob Neufeld, Thomas C. Hone, Frank D. Kistler, and LTC (USA, ret.) John Ricca – four extraordinary people who taught me more than they can imagine.

https://doi.org/10.1515/9783110661415-008

To address these questions, it will be useful first to examine briefly the history of research on military revolutions in this country, including how the Soviet concept of military-technical revolutions influenced the US RMA concept. This material will provide a useful context to evaluate the feasibility of efforts to intentionally deploy and implement revolutionary military advances – and apply a research tactic championed by Andrew W. Marshall, first director of the US Department of Defense's Office of Net Assessment, to ask the "right" questions.

The argument presented here seeks to elucidate what is entailed by decisions to create new military capabilities, and fuses ideas developed within different thought communities, including policymakers, defense intellectuals, senior acquisition managers, military officers, and civilian academics. We begin with discussions about the history of early modern Europe.

# Military Revolutions, the Soviet Concept of Military-Technical Revolutions, and Its Impact on the US RMA Concept

## Military Revolutions

In 1956, Michael Roberts, an historian of early modern Europe, sparked historians' interest in military revolutions by examining tactical reforms initiated by sixteenth-century Swedish military commander Gustavus Adolphus during the Thirty Years' War.[3] These reforms included increased use of linear formations of infantry armed with shoulder-fired gunpowder weapons, adoption of drill and army-wide exercises, regular payments to troops, and reorganization of military forces into standardized units. Roberts argued that such changes in the means of warfare had widespread social and political impacts, including stimulating an increase in the authority of the nation-state. Historians investigating Roberts' seminal ideas have examined military revolutions through research on specific events – such as the Hundred Years' War between 1337 and 1453, the Wars of Italy that raged between 1494 and 1559, Thirty Years' War in central Europe between 1616 and 1648, and the nineteenth-century British conquest of India. Research on political and social phenomena has also been conducted on topics such as the development of administrative bureaucracy in Europe and the formation of nation-states. Roberts' hypothesis of great social consequences attending the use of gunpowder weapons

---

3 Roberts, "The Military Revolution."

concerns broad questions of how social changes emerge through interaction of with technology and organization. Historian Barton Hacker noted Roberts' claims about the emergence of social change has not been challenged as widely as his claims about nature and timing of impacts of gunpowder-fired weapons in battles between states.[4]

Historical assessment of the origins, development, diffusion, military outcomes, and global impacts of military revolutions involve consideration of many questions, including (1) how particular military organizations developed qualitatively higher capabilities than their adversaries; (2) how factors enabling a military revolution interacted with society at large; (3) what role military revolutions played in building global empires; (4) whether technological advances alone have been sufficient to cause a military revolution; (5) whether the militaries of different early modern European nations came to resemble each other because common security problems forced similar solutions; and (6) whether patterns of military advances may be explained by differences in institutions and policies.[5]

Despite the potential value of historical research to inform the design and implementation of military innovation and transformation programs, American and Soviet officials responsible for managing innovation programs probably did not devote time and attention to reading and digesting historical research about military revolutions.[6] As we shall see below, Soviet military theorists employed Marxism-Leninism to set boundaries on the discussion, and senior American national security officials relied on the Department of Defense Office of Net Assessment to inform their choices.

## Soviet Concept of Military-Technical Revolution

Soviet military theorists developed analyses of war to guide their own efforts to field a lethal challenge the West. Soviet theorists believed Marxism-Leninism provided the intellectual tools – the variables, concepts, theory, and levels of analysis – to formulate physical, social, and economic laws of nature – and apply this

---

4 Hacker, "Engineering a New Order," 176, 191.
5 For example, as discussed by Mokyr, "King Kong and Cold Fusion"; Mantzavinos, North, and Shariq, "Learning, Institutions, and Economic Performance," 75–84.
6 A creative effort to assess the problem of academic experts' influence on policymakers is Lindblom and Cohen, *Usable Knowledge*. See also Avey and Desch, "What Do Policymakers Want From Us?"; Ricks, "Put down that broad brush, Steve!"; Farrell, "If policymakers had listened to political scientists, we wouldn't have invaded Iraq."

knowledge to develop national plans to introduce and improve military technology and operational concepts.

The epistemological basis of Marxism-Leninism is different from that of mainstream Western historical and social science-type research,[7] and Marxism-Leninism created at least three practical obstacles in implementing a MTR: Marxism-Leninism (1) set a deductive context for military theory and analysis that could not be tested empirically,[8] (2) restricted intellectual freedom to identify variables and test causal hypotheses,[9] and (3) relied on a centrally planned economy to invent, re-combine, produce, and distribute technologies supporting and enabling new military concepts.[10] The following paragraphs illustrate and describe barriers Marxism-Leninism placed on Soviet military theorists' thinking about military innovation.

---

[7] Karl R. Popper neatly described the relevant difference between the two epistemologies in his definition of science: "science is one of the very few human activities – perhaps the only one – in which errors are systematically criticized and fairly often, in time, corrected. This is why we can say that, in science, we often learn from our mistakes, and why we can speak clearly and sensibly about making progress there"; Popper, *Conjectures and Refutations*, 216. Geoffrey Parker's comparison of ancient and modern military revolutions and the epistemology underlying Western analysis of war-making technology is consistent with Popper's definition; Parker, "The Limits to Revolutions in Military Affairs."

[8] Zvorikine, "The History of Technology as a Science and as a branch of Learning," 1–4; Zvorikine, "Technology and the Laws of Development," 443–458; Joravsky, "The History of Technology in Soviet Russia and Marxist Doctrine," 5–10. See especially intelligence analyst Dziak's *Soviet Perceptions of Military Power* for discussion of how Soviet military doctrine is the Communist party's guide to the future direction of the Soviet military.

[9] Reliance on *ad hominem* argument to refute a factual argument is one type of obstacle Marxist-Leninist theorists placed in the path of research into the phenomenon of military revolutions. For example, according to Adolf Rosengarten, a World War Two Army staff officer, the 1950 *Soviet Encyclopedia* described Jean de Bloch's *The Future of War* as having been written in the "spirit of bourgeois pacifism" (Rosengarten, Jr., "John Bloch – A Neglected Prophet," 39). This is not the place to provide a detailed description and review of Bloch's insightful work. It should suffice to note that the late celebrated military historian Sir Michael E. Howard stated that Bloch's *The Future of War* was the first modern systems analysis and operational analysis of war and warfighting; Howard, "Men Against Fire," 41. See also Mandeles, "Military Technology, Tactics and Operations, and Social Change."

More than 35 years after the 1950 *Soviet* Encyclopedia was published, entries on Bloch in Ogarkov's *Military Encyclopedic Dictionary* and the *Soviet Military Encyclopedia* (416) did not contain the reference to Bloch's "spirit of bourgeois pacifism."

[10] Friedrich von Hayek published extensively on practical obstacles to central planning. His paper, "The Use of Knowledge in Society" is an exemplary empirical analysis of barriers to centralized economic planning.

During the 1950s, Soviet leaders had been reluctant to acknowledge the possibility of a nuclear-weapons-based military revolution because it was not predicted by Marxist-Leninist political and social doctrine, and the new type of warfare stimulated by the advent of nuclear weapons required soldiers "possess a high degree of initiative, self-reliance, and ability to act independently." Soviet military theorists recognized "that such training may cultivate independent minds, one day ready to . . . overthrow the party's dictatorship."[11]

In 1962, official discussion of a nuclear weapons-based revolution started in response to a question about an on-going military-technical revolution (MTR) submitted to the editorial staff of *Krasnaya Zvezda* (*Red Star*) – the prestigious Soviet military journal. Once official discussion of the MTR concept began, the Soviet military press published many articles that repeated the main points of *Voyennaya strategiya* (*Military Strategy*), a book edited by Marshal Vasily Sokolovskiy and first published near the end of 1962.[12] In 1964, Marshal Sokolovskiy (and Maj. Gen. M. Cherednichenko) embraced the view that "the most spectacular revolution in history is now taking place in the methods of armed struggle, in military art."[13]

Some *Krasnaya Zvezda* MTR articles cited Friedrich Engels, a Marxist authority on military science, who argued that the technical basis of military development depended on an economy's stage of production and new productive forces, rather than intellectual creativity.[14] Soviet military theorists believed that military and technological superiority over the forces of imperialism required that theoretical work precede military practice.[15] Soviet theorists avoided long-term tests of doctrine by establishing short intervals between research and weapon systems' production schedules.[16] In parallel to theorizing about the

---

11 Galay, "The Soviet Approach to the Modern Military Revolution," 20–21.
12 Galay, "The Soviet Approach to the Modern Military Revolution," 22–23.
13 Sokolovskiy and Cherednichenko, "The Revolution in Military Affairs, Its Significance and Consequences." A few years later, Cherednichenko would write a chapter, "Conventional Weapons and the Prospects of Their Development," in *Scientific-Technical Progress*.
14 Galay, "The Soviet Approach to the Modern Military Revolution," 24.
15 Adamsky, *The Culture of Military Innovation*; Buchholz and Blakeley, "The Role of the Scientific-Technological Revolution in Marxism-Leninism," 146.
16 In effect, the Soviet military acquisition process exploited advantages of incremental decision-making to build knowledge and minimize decision-making under uncertainty. See the extensive body of work on Soviet design practice produced by Arthur J. Alexander at the RAND Corporation, including the following monographs: Alexander, "R&D in Soviet Aviation"; Alexander, "Weapons Acquisition in the Soviet Union, United States, and France"; Alexander, "The Process of Soviet Weapons Design"; Alexander, "The Linkage between Technology, Doctrine, and Weapons Innovation"; Alexander, "Soviet Science and Weapons Acquisition";

MTR, Soviet theorists also discussed the management of ongoing technological progress in great detail – the "scientific-technological revolution." Class struggle was the primary variable that explained social and economic development; thousands of articles and books were written to examine all conceivable aspects of the topic.[17]

In 1964, Col. General N. A. Lomov proposed that military technology had acquired revolutionary significance with the appearance of many new weapons, and especially nuclear weapons. The revolution occurred with a rapid technological advance and "in the minds of Soviet military personnel, in their views on war and their approach to their duties." For Lomov, the Soviet MTR operated differently from the way military technology operated when employed by American forces.[18]

In the mid-1960s, Soviet writers identified three twentieth century MTRs that resulted from military applications of scientific-technological progress.[19] The first twentieth-century MTR coalesced during World War One from the introduction of motorized transport, chemical weapons, and aviation. The second was enabled by interdependent elements: nuclear weapons, ballistic missiles, nuclear missile submarines, and automated control systems. The assumption of interdependency among elements allowed Soviet theorists to deduce how improvements in one element – e.g., warheads – compelled additional improvements in the warhead platforms and means of support and control.[20]

In 1977, Soviet Marshal Nikolai Ogarkov became chief of the General Staff, and as such was considered the leading Soviet defense intellectual. Ogarkov identified a third MTR arising from Western technological advances in sensors, guidance, communications, computers, and conventional weapons anticipating that these technologies could combine to increase the lethality of conventional high-explosive weapons to the level of nuclear or other weapons of mass destruction.[21] He also argued that higher conventional weapons lethality would be enabled by centralized planning and command and control.[22]

---

Alexander, "Patterns of Organizational Influence in Soviet Military Procurement"; Alexander, "Knowing about Soviet Weapons Acquisition and Strategic Weapons."
**17** Buchholz and Blakeley, "The Role of the Scientific-Technological Revolution in Marxism-Leninism," 145. See also, Lieberstein, review of *Chelovek-nauka-tekhnika*, 691–693.
**18** Galay, "The Soviet Approach to the Modern Military Revolution," 24–25.
**19** Odom, "Soviet Force Posture: Dilemmas and Directions,"; Odom, "Soviet Military Doctrine."
**20** Lomov, ed., *Scientific-Technical Progress and The Revolution in Military Affairs*.
**21** Herspring, "Nikolay Ogarkov and the Scientific-Technical Revolution in Soviet Military Affairs."
**22** Galay, "The Soviet Approach to the Modern Military Revolution," 24.

In the 1970s and 1980s, Soviet military theorists became concerned that integration of new organizational, sensor, and weapons technologies into NATO forces might overcome the existing Soviet conventional military advantage in numbers – and change what Soviet military doctrine asserted could be accomplished in tactical-, operational-, and strategic-level engagements. Soviet theorists valued centralized planning and thought that organizational changes accompanying new sensor and weapons capabilities would enable NATO military planners to exercise a greater degree of centralized control from headquarters – and thus increase the operational effectiveness of Western forces.

Israeli political scientist Dima Adamsky noted that although the "Soviets used the West's scientific and technological superiority as a conceptual starting point for their own doctrinal innovations. . . . [T]he gap between MTR conceptualizations and the actual capabilities of the Soviet military forces was never bridged."[23] Philosopher of science Karl R. Popper and Nobel laureate economist Friedrich von Hayek (among others) explained the source of the gap. Flaws intrinsic to the Marxist-Leninist theory of history – applied to the organization and structure of the Soviet economy – made impossible realization of the goal to direct the growth of human knowledge.[24] Soviet failure to apply market mechanisms limited the number and variety of possible tools and products that could be applied to military problems and added to the difficulties Soviet managers encountered as they tried to design and implement plans for far-reaching and qualitatively higher military capability.[25]

Soviet military organizations did not permit effective experimentation on and implementation of theoretical concepts, and fealty to Marxism-Leninist doctrine increased the difficulties faced by military theorists to diagnose adversaries accurately and design strategies to defeat them. One indicator of Soviet organizational propensity for mis-diagnosis of military problems was discovered during the Cold War by the US Department of Defense's Office of Net Assessment (an office created to diagnose and assess national security issues for the Secretary of Defense): when Soviet military theorists discovered battles in which the side with superior correlation of forces lost, they faulted the indices,

---

23 Adamsky, *The Culture of Military Innovation*, 4.
24 See, for example, Popper, *The Poverty of Historicism*; Popper, *The Open Society and Its Enemies* Vol. 2; Popper, "Chapter 10: Truth, Rationality, and the Growth of Scientific Knowledge," and "Chapter 16: Prediction and Prophesy in the Social Sciences" in *Conjectures and Refutations*; Hayek, *The Road to Serfdom*; Hayek, "The Use of Knowledge in Society"; Hayek, "The Pretence of Knowledge"; Hayek, *The Counter-Revolution of Science*.
25 Berliner, *Factory and Manager in the USSR*.

not the theory – and repeatedly re-wrote the historical record to adjust the correlation of forces.[26]

## The Influence of the Soviet MTR Concept on Office of Net Assessment's RMA Concept

The "conventional account," as Stephen P. Rosen put it, regarding the impact of the RMA concept on the American military is that in the mid-1970s, Andrew W. Marshall, then-director of the Office of Net Assessment, read translations of Soviet military journals that referred to a military-technical revolution associated with the use of non-nuclear weapons – guided munitions, sensors, and communications technologies. Marshall did not accept *in toto* the Soviet view of the MTR. Instead, the Soviet MTR hypothesis inspired Marshall to advance his own hypothesis that a military revolution composed of guided munitions, sensors, and communications technologies may be underway that would enable the US to defeat Soviet tank formations in Europe.[27] Marshall investigated this hypothesis by assigning Office of Net Assessment staff to conduct research, contracting for operational analyses, historical analyses, and war games and simulations, and conducting discussions with military service thought leaders.

Thus, American RMA initiatives were profoundly influenced by interactions within and among (1) congressional authorization and appropriations processes, (2) military services, government and academic research laboratories, (3) government and non-government policy analysts, and (4) business corporations developing and producing military technologies. Rosen's account of the Office of Net Assessment's impact on US Defense Department thinking about the RMA reveals complementary activity. First, the fact of Soviet MTR theorizing inspired Marshall to investigate whether a military revolution phenomenon existed, to search for examples of revolutionary military advances, and to ask questions about how to initiate a military revolution. Second, the substance of US initiatives and research were more influenced by internal organizational and institutional incentives than by reports about Soviet military doctrine.

The genesis of the Office of Net Assessment's revolution in military affairs hypothesis began in 1973 when Secretary of Defense James R. Schlesinger appointed Marshall to direct the Office of Net Assessment. Marshall was a former

---

[26] Krepinevich, Jr. and Watts, *The Last Warrior*, 116.
[27] Rosen, "The Impact of the Office of Net Assessment on the American Military in the Matter of the Revolution in Military Affairs," 470.

RAND colleague of Schlesinger. The Office's mission was to identify military, technological, political, economic, and other factors that contribute to military capability, develop comparative studies of military balances over a long planning horizon, and alert senior defense officials to problems and opportunities.[28]

Prompted by lethal precision-guided munitions (PGMs) used by Egyptian, Syrian, and Israeli forces during the 1973 Yom Kippur War, Marshall commissioned studies and workshops that suggested improvements in munition accuracy would generate significant changes in military operations.[29] In the mid-1970s, Marshall and his military assistants examined Soviet military theoretical assessments of long-term military consequences of the use of aircraft to deliver PGMs that destroyed the Thanh Hoa and Paul Doumier bridges in Vietnam. As RAND analyst James Digby noted in 1976,

> For a good many years now, engineers could foresee a new generation of weapons that would force major changes in the posture and tactics of nearly every military power. But, as is so often the case, it took practical battle demonstrations – in Southeast Asia and in the October War of 1973 – to *convince* decision-makers that a new generation of weapons was at hand.
>
> Until recently most of the things shot by military men at their enemies would miss the target. Not so the new generation. These weapons have a high probability of hitting the target with their first shot. And the weapons being relatively cheap, a force may outfit itself with large numbers of them.[30]

Leaders of the military services responded to evidence and analysis of US and Israeli combat experience. The US Army (enabled by congressional appropriations) initiated a precision-guided weapons-development and replacement program that by the 1980s had led to the most complete rearming in the Army's history. The US Air Force and US Navy followed suit.[31] Lieutenant General William

---

[28] Krepinevich and Watts, *The Last Warrior*, 93–100; Mandeles, "Interdisciplinary Inquiry and Preparation for Net Assessment."

[29] Rosen, "The Impact of the Office of Net Assessment on the American Military in the Matter of the Revolution in Military Affairs," 477. See also the following reports published by RAND for the Office of Net Assessment and the military services: Digby, "Precision-Guided Munitions"; Digby, "The Technology of Precision Guidance"; Digby, "Precision Weapons"; Digby, "Precision-Guided Weapons"; Digby, "Precision-Guided Weapons"; Dudzinsky, Jr. and Digby, "Qualitative Constraints on Conventional Armaments"; "Workshop on Asymmetries in Exploiting Technology as Related to the U.S.-Soviet Competition"; Dudzinsky, Jr., and Digby, "The Strategic and Tactical Implications of new Weapons Technologies."

[30] Digby, "PGMs," 36.

[31] Weigley, *History of the United States Army*, 585; Thompson, *To Hanoi and Back*. Barry Watts argues that World War Two efforts to develop guided munitions are a better reference point for force structure discussions; Watts, *Six Decades of Guided Munitions and Battle Networks*.

Odom, Army Assistant Chief of Staff for Intelligence between 1981 and 1985, and Director of the National Security Agency between 1985 and 1988, also closely read Soviet writing about the RMA.[32] Rosen suggests that Odom may have provided Marshall access to this intelligence beginning in 1981.[33]

Over the course of more than 40 years directing the Office of Net Assessment, Marshall pursued many lines of analysis in parallel – and applied ideas from one research interest to inform another. At RAND, he was interested in organizational behavior, including the work of Herbert A. Simon, James G. March, and Richard M. Cyert.[34] He later applied concepts and ideas from the study of formal organizations to the problem of how to facilitate the management of military innovation.[35] It is not clear when Marshall began to commission comparative analyses of military and organizational innovation to support his thinking about military revolutions and military innovation; one comparative study of military innovation was completed in 1982 and was later cited by a member of Marshall's staff for its diagnosis and comparison of military innovation in different countries.[36]

In 1990, Marshall asked then-Lieutenant Colonel Andrew F. Krepinevich to assess the Soviet argument that technological developments would generate a period of major advances in military capabilities and warfighting. The resulting study, "Soviet Conceptions of Revolutions in Military Affairs," was completed in 1992 and initiated the American RMA debate.[37] The 1991 performance of US military forces against Iraq during the 43-day Persian Gulf War provided empirical data to assess the hypothesis that US military forces' information and sensor

---

[32] See, for example, Odom's unclassified published research which refers to Soviet thinking about the RMA, in "Soviet Force Posture,"; Odom, "Soviet Military Doctrine."
[33] Rosen, "The Impact of the Office of Net Assessment on the American Military in the Matter of the Revolution in Military Affairs," 479.
[34] Krepinevich and Watts, 46.
[35] Augier, "Thinking about War and Peace"; Mandeles, "Interdisciplinary Inquiry and Preparation for Net Assessment."
[36] This chapter's author did not have access to a complete chronology of studies related to RMA concept commissioned by Marshall. An early comparative study of emerging RMAs was an analysis of the introduction of carrier-based aviation in the US Navy, Royal Navy, and Imperial Japanese Navy, by Hone and Mandeles, "*Innovation in Naval Aviation.*" This study was cited by Rosen in *Winning the Next War* – an important early examination of ideas underlying discussion about military innovation and the RMA. According to Krepinevich and Watts, Marshall was especially intrigued by the discussion of the development of carrier-based aviation; Krepinevich and Watts, *The Last Warrior*, 204.
[37] Krepinevich, Jr., "The Military-Technical Revolution,"; Krepinevich and Watts, *The Last Warrior*, 196–213; Rosen, "The Impact of the Office of Net Assessment on the American Military in the Matter of the Revolution in Military Affairs," 480.

technologies had created powerful military possibilities – highly lethal weapons and tactics – although appropriate intellectual frameworks did not yet exist to infer how these new information technologies would perform in future combat. A 1991 Defense Intelligence Agency study of Russian military theorists' analysis of the Persian Gulf War revealed Russians had concluded that the nature of war had changed radically, and that the US had created the first reconnaissance-strike complex[38] by integrating long-range and wide-area sensors, command, control, and communications systems with precision-guided conventional high-explosive munitions. Marshall speculated that RMAs would develop over periods of 20 to 30 years; and would alter (1) organizational structures and processes involved in design, production, and deployment of new equipment; (2) operational concepts and organizational structures to execute new operational concepts; (3) training materials and organizations to conduct training; and (4) selection and promotion criteria appropriate to new organizations, equipment, and missions.

The November 1992 presidential election of William J. Clinton introduced a new set of senior officials to discussions about RMAs, including Secretary of Defense Les Aspin, Deputy Secretary of Defense William J. Perry, and Under Secretary of Defense John M. Deutch. Some Defense Department officials disagreed with inferences Krepinevich made in his 1992 assessment, "Soviet conceptions of Revolutions in Military Affairs." Meetings and discussions commenced to examine implications of the hypothesis that a revolutionary period of military capabilities was underway. In July 1993, Krepinevich completed an update to his 1992 assessment.[39]

In 1993 and 1994, Marshall wrote at least three memoranda for record to capture ideas he discussed with senior Defense Department officials. He argued that some people had misinterpreted the concept "military-technical revolution" to emphasize technology. The concept of "military revolution" also was misinterpreted to imply that the "revolution" had been completed.[40] Marshall reasoned that while technology enabled a revolution, the elements of new operational

---

**38** Lambeth, "Desert Storm and Its Meaning." Israel's 1982 destruction of the Syrian Air Force provided a clue to the lethal role of information and sensor technologies. Soviet analysts misinterpreted operational implications of the Bek'aa Valley air combat. Lambeth, "Moscow's Lessons from the 1982 Lebanon Air War."
**39** Krepinevich and Barry Watts, *The Last Warrior*, 215–220.
**40** Herman Kahn also remarked that "the idea that a 'revolution' in warfare has taken place is imprecise; it gives the impression that the changes caused by the development [of military technology] . . . occurred all at once." There are many "revolutions" in technology and social factors "bringing about changes as great as those that occurred between the Civil War and World War I, or between World War I and World War II"; Kahn, *Thinking About the Unthinkable in the 1980s*, 83.

concepts and new military organizations were essential to realize higher operational capability. He also maintained discussion of future warfare would be clearer if we referred to an "emerging military revolution" or "potential military revolution" to account for the recognition that the "full nature of the changes in that character of warfare have not yet fully emerged."[41]

Marshall noted that Bill Perry and John Deutch were

> interested in the notion that a military revolution may be underway, or may be possible. I want to go beyond what I talked about with them to a fuller description of what might be undertaken if they and other top-level officials become convinced that, in fact we are in the early stages of a major change in the nature of warfare. . . . It might be better to think in terms of our moving into a special period in military affairs, a period during which a major transition between regimes of warfare will take place. . . . [The Russians] point to examples in the past, such as the 20's and 30's, or the period immediately following World War II. If one looks at the whole sweep of military history, one can pick out special periods where what happens is that the available technology leads to major changes in the character of warfare in various areas of warfare. By contrast, in other periods the technology as it changes is used mainly to do what one is already doing, to carry out the kind of operations that one is already undertaking somewhat better. Then you come to periods of revolution where the character of warfare itself changes. The interwar years may be one of the most interesting periods for us to reflect on today; then, aviation technology, the tank, and the exploitation of various other technologies, such as radio and radar, led to really big changes in concepts of operation, and new military organizations were created to fully exploit the new systems. The character of warfare in World War II is very different from that of World War I. The technologies brought changes in almost all areas of warfare . . . some new areas of warfare were created: for example, long-range aerial bombardment.[42]

Marshall recognized that advanced technology, regardless of how exciting policymakers believe such developments might be, does not constitute a military revolution:

> The most important competition is not the technological competition, although one would clearly want to have superior technology if one can have it. The most important goal is to be the first, to be the best in the intellectual task of finding the most important innovations in concepts of operation and making organizational changes to fully exploit the technologies already available and those that will be available in the course of the next decade. . . . Some military establishments do much better in developing the appropriate

---

41 Marshall, "Memorandum for the Record" (July 27, 1993); Marshall, "Memorandum for the Record" (August 23, 1993); Marshall, "Memorandum for the Record" (May 2, 1994). As an aside, one might wonder whether a good deal of the quibbling in the historical community might have been avoided if Michael Roberts had titled his essay "An Emerging Military Revolution, 1550–1660."

42 Marshall, "Memorandum for the Record" (August 23, 1993).

concepts of operation, making the organizational changes, and creating the doctrine and practices that fully exploit the available technologies.[43]

Marshall concluded his August 1993 memo with a discussion of specific actions the Clinton Defense Department might undertake to develop a long-term strategy to exploit future opportunities. In 1994, Bill Perry replaced Aspin as Secretary of Defense and established a group to coordinate a Defense Department effort. Marshall's May 1994 memorandum for the record listed the Office of Net Assessment's actions to support and inform the military services' thinking about innovation, including the following: (1) develop knowledge about organization design able to apply future technology-operational concepts available in 15–20 years, (2) conduct historical analyses of major changes in warfare, (3) investigate how to improve analysis of prospective competitors, and (4) organize roundtable meetings with the Army, Air Force, and Navy and Marine Corps to discuss how each service was thinking about preparing for the future. Marshall also argued that "for the next few years, the central task is to increase the effort devoted to the search for innovative concepts of operation and the organizational changes that will exploit current and likely future technologies and systems."[44]

In 1995, in Senate testimony, Marshall recapitulated earlier discussions and outlined tasks to investigate and develop means to implement the RMA. He noted that the US is at the beginning of a 20 to 30-year period of "significant change." Innovations come about over time within large organizations that have stocks of equipment, internal and external standard operating procedures, doctrines, and relationships with other organizations – all of which constrain the rate and magnitude of change. For the near term, Marshall argued that the search for concepts of operation appropriate to evolving technologies and a changing security environment is an issue of "where do we think we really want to go." Later, as goals become clearer, the Defense Department should focus on experimentation and efforts to align the "acquisition system [better to] support innovation and experimentation." He suggested "gaming and simulation" tools might facilitate field experimentation with new equipment and concepts of operation.

In his Senate testimony (and on other occasions), Marshall reiterated implications of his analysis of the future status of US forces – "the most compelling lesson . . . is that some military establishments were much better than others in developing appropriate concepts of operation, making the organizational changes, and

---

43 Marshall, "Memorandum for the Record" (August 23, 1993).
44 Marshall, "Memorandum for the Record" (May 2, 1994).

creating the doctrine and practices that fully exploit the available changes; and that superiority in these respects had large military consequences."[45]

The preceding discussion of Soviet and American officials' appreciation of the MTR and RMA concepts shows that neither US nor Soviet officials relied on historians' research on military revolutions in early modern Europe to inform weapons development programs. Soviet officials operated in an institutional context that restricted the range of thought and limited empirical examination and tests of hypotheses and theories. Soviet military theorists devoted a great deal of effort to elaborate a Marxist-Leninist framework to assess information and analysis concerning modern weapons system performance.

Office of Net Assessment analysts were aware of military revolutions dating to the European early modern period;[46] they devoted more attention and research to the period between World War One and World War Two.[47] US officials, operating in an institutional context that allowed many roughly independent streams of activity, were incrementally able to produce actionable information and evidence to justify development of (1) new platforms, (2) enabling sensors, (3) operational concepts, and (4) training regimens. The resulting US technology-operational concept-organizational relationships were "works in progress" that informed – but did not determine – US operations in Afghanistan and Iraq. Of course, no set of emerging technology-concept-organization relationships are fault-free – and those developed during military operations in Afghanistan and Iraq provide examples of successes and failures. State and non-state actors have identified at least two shortfalls of an emerging technology-concept-organization relationship comprising a precision-strike military revolution. The first is the failure of militaries conducting precision-strike operations across a theater to develop reliable knowledge relating organizational design to outcomes.[48] The second is the failure to recognize how operational advantages of precision-strike regimes are mitigated by international propaganda campaigns that enable and permit

---

**45** Marshall, "Testimony," 256–258.
**46** Krepinevich noted that as many as ten military revolutions materialized between the fourteenth century and the present; Krepinevich, Jr., "Cavalry to Computer."
**47** Such as Rosen, *Winning the Next War*, and Murray and Millett, ed., *Military Innovation in the Interwar Period*. Knox and Murray's edited volume, *The Dynamics of Military Revolution*, contains chapters on periods preceding the nineteenth century; Knox and Murray, ed., *The Dynamics of Military Revolution*.
**48** Mandeles, *The Future of War*. This book was based on an Office of Net Assessment study entitled "Organizational Structures to Exploit the Revolution in Military Affairs" (Fairfax: The J. de Bloch Group, 2000).

non-state actors to locate military headquarters, and weapons production and storage facilities in midst of civilian housing or other civilian sites.[49]

# Feasibility of Efforts to Foster an RMA through Technological Development

It remains necessary to review the feasibility of efforts to deploy and implement revolutionary military advances. Deliberate assessments of feasibility address whether historical laws of innovation exist to generate revolutionary operational capability, and whether current management strategies, organizational structures, and processes might guide revolutionary changes. We will thus consider three issues concerning deliberate government efforts to advance revolutionary military capability: unpredictability of the growth of knowledge in systems acquisition, over-emphasis on technology development programs, and enduring organizational-bureaucratic conflicts between authority of expertise and political authority.

## Unpredictability of Growth of Knowledge in Systems Acquisition

The unpredictability of the growth of knowledge is a significant limiting condition for the feasibility of all scheduled development programs established to push the frontiers of knowledge and to advance revolutionary new capabilities. Andrew W. Marshall provided a useful example of the unpredictability of growth of knowledge, on more than one occasion offering an analogy between interwar aviation technological and operational opportunities developed and exploited by the US Navy[50] and opportunities for precision-strike made possible by emerging

---

49 Brun, "'While You're Busy Making Other Plans.'" Brun argued that while Western militaries were developing information-based computer and precision-guided missile technologies, other nation-states and non-state actors were developing doctrine, operational concepts, and techniques to counter the West's technological advantages. Brun focused on ideas developed by military theorists in Syria, Iraq, Islamic Republic of Iran, Hizballah, Hamas, and Al-Qa'eda to exploit Western nations' aversion to causing civilian casualties.
50 Hone and Mandeles, "*Innovation in Naval Aviation*," 1982; Hone, Friedman, and Mandeles, *American and British Aircraft Carrier Development*. Originally, Thomas C. Hone, Norman Friedman, and Mark D. Mandeles, *The Introduction of Carrier Aviation into the U.S. Navy and Royal*

computer, communications, and information technologies.[51] The purpose of the analogy was to describe, compare, and make sense of contemporary challenges and opportunities. Marshall did not examine significant differences in the organizational ecology of the two time periods, such as the non-existence in the 1920s and 1930s of the planning, programming, budget system process and associated project management and requirements processes. Nor did Marshall consider the impact of the size difference between the national security community of today with that of the 1920s and 1930s – which translates directly into a larger number of organizational interactions and potential "veto points" that confound contemporary Defense Department development of new weapons and operational concepts.[52] But Marshall expected that information technologies would generate unanticipated impacts, and in the foreword to a 1999 RAND report, he encouraged policy analysts to offer hypotheses about future roles of information technologies.[53]

The logic of Popper's argument against historicism has important implications for government efforts to foster RMAs. It is not surprising that no scholar, policymaker, or policy analyst predicted, say, in 1999 – the date of the RAND report on the changing role of information in warfare – the US Treasury Department's improving tools to track complex transactions and enforce secondary sanctions.[54] Economic sanctions against nuclear weapons proliferators are now more effective at constraining and limiting financial resources used to develop nuclear weapons, and are performing an important role in "shaping the battlespace" and

---

*Navy: Military-Technical Revolutions, Organizations, and the Problem of Decision* (Washington, D.C.: OSD/Net Assessment, 1994).
51 For example, see Marshall, "Memorandum for the Record" (August 23, 1993); Marshall, "Memorandum for the Record" (May 2, 1994); Marshall, "Testimony," 1995.
52 Competing incentives among offices in bureaucracies or between tactical units in wartime is a constant feature of work in bureaucracy. Navy historian Michael J. Crawford noted a conflict that began during John Paul Jones' career: "From the Continental Navy's John Paul Jones on, 'there had always been United States naval officers who were uncomfortable with the association of commerce cruising with privateering. Jones despised privateers, whom he considered 'sordid Adventurers.' Noting that 'Public Virtue is not the Characteristick of the concerned in Privateers,' he observed that privateers failed to bring home many captured enemy seamen who could be used in prisoner exchanges because those captives' consumption of provisions forced them to end their cruises 'before they had glutted their Avarice.' That there were Continental Navy officers 'infected with the same foul Contagion' of the privateers' avarice covered Jones 'with Shame'"; Crawford, "The Abolition of Prize Money in the United States Navy Reconsidered," 119.
53 Marshall, "Foreword."
54 Mead, "Why Russia and China Are Joining Forces."

"imposing costs" against certain authoritarian regimes.[55] Recognizing information technologies capable of tracking complex transactions required knowledge unavailable in 1999.

Elsewhere I proposed design features of an acquisition process that acknowledges and accounts for the growth of knowledge. The core acquisition process problem to be solved – especially how to achieve "revolutionary" advances is: how to employ, exploit, and coordinate the information, knowledge, and products created by public and private sources of discovery, innovation, and analysis. Like it or not, information and knowledge about military capabilities are limited and imperfect. To deal with this situation, a process is needed through which knowledge is acquired, communicated, and applied. Consequently, the broad solution to the problem of organizing the acquisition process is to harness and direct the interactions of people and companies – each of whom possess, more or less, only partial knowledge about the task at hand. This solution overturns the existing acquisition process assumption that the outcome of a process of invention and development may be predicted and scheduled.[56]

Compare the problem definition in the previous paragraph to the traditional definition of the acquisition problem. What problems does the DoD try to solve with its major weapon systems acquisition process? In congressional discussions about authorizations for weapon systems and appropriations for those weapons, the acquisition process seeks to produce new or improved capabilities through discovery, invention, and improvements of components and technologies, at reasonable cost and on a schedule that meets needs of national strategy, and the military commanders and forces that implement the strategy. This conception of the acquisition problem focuses attention on designing organizational structures, processes, and procedures – such as creating an office to support the Director of Cost Assessment and Program Evaluation (CAPE) – that will improve initial program cost estimates and thereby preclude "waste, fraud, and abuse."[57]

---

[55] In the early 1990s, economic sanctions against nuclear weapons proliferators were ineffective; Mandeles, "Between a Rock and a Hard Place," 238.

[56] Mandeles, "System Design and Project Management Principles to Meet the Needs of Operational Forces"; Hayek, "The Use of Knowledge in Society." See also Lofgren, "The DoD Budget Process."

[57] In a memorandum written to US JFCOM, General James N. Mattis urged greater attention to problem definition and less attention to staff processes. Mattis argued that unchallenged acceptance of a procedural template precludes the "critical and creative thinking directed at understanding, visualizing, and describing complex problems"; Mattis, "Memorandum for U.S. Joint Forces Command," 1–2.

Major Defense Acquisition Programs (MDAP) "program of record" entail potentially conflicting incentives and goals specific to different individuals and organizational actors, with each entity working from their own assumptions about aligning means to ends, information and knowledge about the threat, and sometimes incomplete and ambiguous information about a program's status. These various entities comprise an administrative hierarchy for each MDAP that performs at least five functions: (1) creating plans and schedules to develop a new capability; (2) disbursing funds; (3) preparing and distributing information about the program to various audiences; (4) monitoring and managing the activities of many different contractors and sub-contractors; and (5) reporting programmatic activity to DoD and congressional authorities.

Policy makers throughout the legislative and executive branches believe that (1) relevant and unbiased information about costs, schedules, and technological performance is available or can be developed as required; (2) it is possible to prepare an accurate ranked set of near- and future-term threats, and (3) acquisition managers have appropriate analytic tools to compare alternative options to achieve military tasks and missions.[58] These assumptions are at variance with features of everyday bureaucratic practice, such as conflicting assumptions, objectives, and incentives in different offices or organizational levels, and complexity of the task and rigidity of organizational arrangements to execute that task.

Commissions and "blue ribbon" study teams that developed recommendations to overhaul and modify the acquisition process have conceived and justified their work as an effort to make the acquisition process rational – a process in which goals are set, ways and means are identified to achieve the goal, courses of action compared, and the best solution chosen.[59] Recommendations to improve acquisition developed in the "Weapons Systems Acquisition Reform

---

58 Similarly, economists have often believed that with good analysis they could reliably choose *ex ante* the best program or option from among a set of alternatives. Yet experience shows that has seldom been enough *ex ante* information to make a proper decision; Nelson, "Issues and Suggestions for the Study of Industrial Organization in a Regime of Rapid Technical Change," 49.

59 For example, The Commission on Organization of the Executive Branch of the Government, *The National Security Organization*; Commission on Organization of the Executive Branch of the Government, *Business Organization of the Department of Defense*; The Blue Ribbon Defense Panel, *Report to The President and the Secretary of Defense on the Department of Defense*; Commission on Government Procurement, *Report of the Commission on Government Procurement*; The Office of the Secretary of Defense Task Force, *The President's Private Sector Survey on Cost Control*; The President's Blue Ribbon Commission on Defense Management, *A Quest for Excellence*.

Act of 2009" recapitulate the assumptions and logic used by previous commissions about the design of a rational process.

Efforts to achieve a rational process have been a disappointment. Post-World War Two American defense planning processes have not operated as their designers assumed and expected; many programs have suffered budget overruns, schedule delays, and performance shortfalls.[60] As then Secretary of Defense Gates observed,

> [w]hen it comes to procurement . . . the trend has gone towards lower numbers as technology gains made each system more capable. In recent years these platforms have grown ever more baroque, ever more costly, are taking longer to build, and are being fielded in ever dwindling quantities.[61]

Budget overruns, schedule delays, and performance shortfalls occur because acquisition programs have been designed under the incorrect – but widely held – assumption that the future growth of scientific knowledge and technical "know-how" can be planned and scheduled. The assumption ensures that during the decades-long periods to develop new major classes of ships, aircraft, and ground vehicles, these platforms would be eclipsed by the tempo of technological development of command, control, communication, computer, and intelligence ($C^4I$) capabilities. By the time the platforms are delivered, the technological capabilities originally associated with the platforms have become obsolete. The logical impossibility of predicting the growth of scientific knowledge makes it equally impossible to accurately estimate program costs and to predict the schedule and tempo of work to create new capabilities.[62] There have thus been many notorious failed predictions about the feasibility and rate of developing technologies, such as the practicality and operational utility of powered flight, rocketry, machine guns, submarines, and armored vehicles and tanks.[63]

---

60 For example, see Decker and Wagner, *Army Strong*; see also annual US Government Accountability Office assessments of weapons systems, such as *Weapon Systems Annual Assessment*.
61 Gates, "Speech."
62 Popper, *The Poverty of Historicism*. Popper developed a logical proof that shows "*no scientific predictor – whether a human scientist or a calculating machine – can possibly predict, by scientific methods, its own future results*" [emphasis in the original], vii. New techniques to compare expert forecasts of technologies may aid policymakers' deliberations about competing technologies; Farmer and Lafond, "How Predictable is Technological Progress?" Alexander Kott's claim to have developed an accurate method of technological forecasting, based on the Delphi Technique, also may help relieve policymakers' anxiety about the future. Kott's model does not refute Popper's argument; Kott, "The Future of War Technology Whispers to Us From the Past, and We Must Listen Better."
63 Cerf and Navasky, *The Experts Speak*, 203–264.

A brief final example of a failed attempt to plan and promote development of revolutionary military innovation concerns the efforts of the Chief of Naval Operations' (CNO) Strategic Studies Group (SSG). In 1995, CNO Adm. Jeremy M. Boorda asked Adm. James R. Hogg (retired) to lead a re-purposing of the SSG from preparing senior Navy captains to become admirals to developing innovative warfare concepts. The CNO's SSG annual reports produced between 1995 and 2016 (when the SSG was disestablished by CNO Adm. John Richardson) demonstrate conscientious and diligent work. While each SSG paid lip service to the goal of promoting revolutionary military innovation, the methodology employed to search for revolutionary technology could not advance and support transformational warfighting concepts or technology. In practice the SSGs delivered something quite valuable – the SSGs helped the CNOs think about near-term issues and problems they faced during their terms.[64]

## Over-Emphasis on Technology: The Third Offset

In 2014, then-Secretary of Defense Chuck Hagel announced the "Third Offset Strategy" – a set of programs to maintain American military superiority over current and potential foes by developing new operational concepts and technologies. The impetus for the Third Offset was the perceived need to counter rapid Russian and Chinese progress in weapons development and military modernization, and to deter aggression from those countries.[65] During the 1950s, President Dwight D. Eisenhower proposed the First Offset, a program to build US nuclear forces to deter and counter the USSR's conventional force numerical superiority. In the mid- to late-1970s, Secretary of Defense Harold Brown guided the Second Offset – the development of stealth, precision-guided munitions, and intelligence, surveillance, and reconnaissance (ISR) systems to counter the USSR and Warsaw Pact's improving military capabilities and numerical superiority of forces in central Europe.

Third Offset research and development (R&D) funding of about $3.6 billion was included in the FY 2017 defense budget, and a similar amount in the FY 2018 defense budget to continue development of previous Third Offset R&D investments and programs.[66] Third Offset programs emphasize technology development

---

[64] Hanley and Mandeles, *Adapting to a New Era*; Hanley and Mandeles, *Working for Innovation*; Hanley and Mandeles, *Notes from the CNO Strategic Studies Group Archives*.
[65] Work, "The Third U.S. Offset Strategy and its Implications for Partners and Allies"; Carter, "Remarks Previewing the FY 2017 Defense Budget."
[66] Gady, "New US Defense Budget"; Blakeley, *More Money on the Horizon?*

rather than thinking about operational concepts or organizational design,[67] and its technologies include nanotechnology, artificial intelligence, robotics, and genetic engineering. In particular, the Third Offset Strategy includes the following five themes around which development would take place: (1) autonomous "deep learning" machines and systems to improve early warning of events, (2) human-machine collaboration for decision-making, (3) assisted-human technologies to enable people to operate more effectively, (4) network-enabled cyber-hardened weapons, and (5) human-teaming with unmanned systems.[68]

Technologies developed and weapons produced under the Third Offset Strategy are intended to enhance the operation of "precision-strike regimes" composed of precision-guided weapons, advanced sensors, computers, and communications, and organizational structures, processes, and procedures to enable targeting, coordinating forces, and assessing damage.[69] In a review of more than 20 articles and essays about the Third Offset for this chapter,[70] most of the authors assumed that technical goals to be achievable and that higher technical performance is equivalent to higher operational capability. Only one article raised the possibility of glitches in the human-machine collaboration initiative.[71]

---

67 Rochlin, *Trapped in the Net*, 166–168.
68 Clark and Freedberg Jr., "Robot Boats, Smart Guns & Super B-52s"; Mehta, "Work Outlines Key Steps in Third Offset Tech Development."
69 Watts, *The Evolution of Precision Strike*.
70 Brimley, "Offset Strategies & Warfighting Regimes"; Czarnecki, "Against a Tech-Centric Offset"; Freedberg Jr., "Adversaries Will Copy 'Offset Strategy' Quickly"; Hagel, "'Defense Innovation Days' Opening Keynote (Southeastern New England Defense Industry Alliance)"; Hagel, "Subject: The Defense Innovation Initiative,"; Hagel, "A Game-Changing Third Offset Strategy"; Sander, "Exploring a New Offset Strategy"; Scharre, "How to Lose the Robotics Revolution"; Sweetman, "'Third Offset' Addresses Operational and Economic Challenges"; Work and Brimley, *20YY*; Krepinevich, "Testimony [Defense Strategy]"; Weisgerber, "Pentagon Wants to Pair Troops with Machines to Deter Russia, China"; Mehta, "Work Outlines Key Steps in Third Offset Tech Development"; Pellerin, "DoD Seeks Novel Ideas to Shape its Technological Future"; Freedberg Jr., "People, Not Tech"; Gouré, "Directed Energy Weapons Will Be The Key To A Successful Third Offset Strategy"; Anonymous, "U.S. Department of Defense Third Offset, Standard Market Taxonomy"; Hayden, "Video: Transformation and The Third Offset (NSA-CIA)"; Mehta, "Pentagon No. 2"; Simón, "The 'Third' US Offset Strategy and Europe's 'Anti-access' Challenge"; Walton, "Securing The Third Offset Strategy"; Wiitala, "Seizing the Defensive". My thanks and appreciation to CDR (ret.) Paul S. Giarra for identifying and finding the majority of these references.
71 Czarnecki, "Against a Tech-Centric Offset." For an uncritical description of plans to develop a human-machine collaboration capability, see Weisgerber, "Pentagon Wants to Pair Troops with Machines to Deter Russia, China."

The Third Offset assumes a military transformation of some sort is inevitable. For instance, in 2016, General Mark A. Milley, US Army Chief of Staff,[72] told a Center for Strategic and International Studies audience that

> We are on the cusp of a fundamental change in the character of ground warfare. . . . It will be of such significance that it will be like the rifling of a musket or the introduction of a machine gun or it will have such significant impact as the change from horse to mechanized vehicles. . . . Exactly what that's going to look like, I don't know. I just know that we're there. We're on the leading edge of it. . . . I think that by 2025, you're going to see armies – not only the American Army but armies around the world – will be fundamentally and substantively different than they are today.[73]

Advanced technologies promise qualitative improvements in military operational capabilities. The rub is that RMAs do not materialize simply from application of sophisticated scientific and engineering research. Attention limited to emerging technologies and development of new weapons obscures factors that have played a decisive role in invention, adoption, deployment, and employment of qualitative advances in war – the organizational, social, political, cultural, historic, and economic context within which weapons and ways of fighting are developed.[74]

## Expert Authority vs. Political Authority

The third factor that affects feasibility of promoting RMAs concerns an enduring issue in bureaucracies: conflicts between expert and political authority. Sociologist Max Weber examined conflict in bureaucracies between elected officials and technical experts, especially when officials issue decrees "ignored" by bureaucrats charged to implement them. In Weber's words, "the political 'master' always finds himself, vis-à-vis the trained official, in the position of the dilettante facing the ex-

---

[72] General Milley was confirmed by the Senate on 25 July, 2019 to be Chairman, Joint Chiefs of Staff.
[73] Majumdar, "Is the US Army Ready for a Shocking Technological Revolution in Land Warfare?"; Woodford, "So You Still Think You Want A Revolution In Military Affairs?"; Milley, "Priorities for Our Nation's Army with General Mark A. Milley."
[74] Mandeles, *The Future of War*; Mandeles, *Military Transformation Past and Present*; Hone, Friedman, and Mandeles, *American and British Aircraft Carrier Development*; see also Nofi, *Recent Trends in Thinking About Warfare*, 27. Landau and Chisholm, "The Arrogance of Optimism," 69.

pert."⁷⁵ Admiral Hyman Rickover addressed this issue frequently in his interactions with his fellow officers, and in his 1974 speech, "The Role of Engineering in the Navy," to the National Society of Former Special Agents of the Federal Bureau of Investigation.⁷⁶ Admiral Rickover's argument involved three issues. First, the Navy's reliance on technologies of all kinds was increasing. Second, to take advantage of technology, the Navy must raise standards of knowledge and performance for all personnel. Third, the Navy permitted receding standards of technical competence. In doing so, the Navy increased its dependence on industry, and relied on reorganizations and management fads to compensate for lower standards of technical competence.

Admiral Rickover explained shortfalls in Navy leadership by arguing that Navy's leaders have, at potential historical turning points, "misread history." They misunderstood the necessity of applying empirical premises to all manner of problems that derive from the Navy's purpose – to defend our Nation.

Rickover developed this observation about the necessity of applying an empirical attitude and demonstrable knowledge to problems by presenting a conceptual history of Navy Department decision-making. He began with the period following the Civil War when navy leaders retained "faith in [Monitor-type vessels] as major combatant ships long after other leaders of other nations' navies had recognized that they were only a brilliant improvisation addressing a specific problem. The main line of naval progress remained in Europe. Navy leaders had misread the naval results of the Civil War." During the 1880s, when the navy was rebuilding, "the worst errors were caused by the imposition of the opinions of line officers on technical matters." Rickover argued, "the rising tide of technological complexity has engulfed the design engineer ashore as well as the line officer engineer at sea. In both areas, these men now face demands far beyond those which confronted their predecessors." In Rickover's view, young officers must be able to understand the technical details of their equipment; they cannot do this without learning the basics of engineering and science.⁷⁷

Of course, once one learns the basics, one must devote the time and effort to maintain proficiency. In the course of the *Challenger* shuttle investigation, Nobel Laureate Richard P. Feynman, noted that managers, who earlier in their careers had been engineers, estimated the likelihood of a shuttle failure at 1 in 100,000, and working engineers estimated likelihood of failure at 1 in 100.⁷⁸ The three-order of magnitude difference in estimates made by working engineers and

---

75 Weber, "Bureaucracy," 991–993.
76 Rickover, "The Role of Engineering in the Navy."
77 Rickover, "The Role of Engineering in the Navy."
78 Feynman, *"What Do You Care What Other People Think?"*.

managers reflects the type of issue Admiral Rickover highlighted in his history of conflict between line and engineering and engineering duty officers over what premises should guide decisions about development and use of technology in the navy.[79]

A crucial problem faced by commanding officers is that knowledge requirements for command have grown. All services face this problem. General Raymond T. Odierno addressed this issue in an interview conducted in Baghdad in 2009. The increasing complexity of wartime decision-making involves overseeing and managing staff structures and processes to propose lines of operation and calculate and compare impacts, interactions, and tradeoffs of many policies and programs. The complexity of aligning the commander's staff structures, processes, procedures, and lines of operation with the task environment requires developing approaches to operational assessments and analyses that help commanders understand their mission(s); organizational structures, processes, and people; the operational environment; the ways and means to achieve desired ends; and the feasibility and wisdom of mission goals.[80] And commanders still have to defeat the enemy.

Rickover's political battles with much of the navy's leadership are one case of the conflict between authority of knowledge and of rank. As military organizations increasingly employ technologies, organization, and tactics that must be operated "under the rule of expert knowledge," it is inevitable that disagreements and conflicts will erupt between technical and non-technical officials. Practical implications of this conflict are revealed in choices not made about how to foster search processes for technologies (relevant to operational missions), how to build knowledge about interfaces in weapons technology-organization-human relationships, and how to conduct research that incorporates perspectives gained from experiments, exercises, war gaming, and analysis. In this matter, one indicator that the US is not better prepared than its adversaries to identify, develop, and deploy advanced technology-operational concept-organizational innovations is that the only Defense Department research organization specializing in operational-level analysis, the Joint Center for Operational Analysis, was disestablished in 2011.

---

**79** There also is the related issue of the validity of managers' perceptions – a topic about which little has been written. See Starbuck and Mezias, "Opening Pandora's Box."
**80** Mandeles, Memorandum, To Joint Center for Operational Analysis/U.S. Joint Forces Command.

## Conclusion

Military revolutions are most clearly identified with hindsight, despite efforts to anticipate and plan the pace and direction of military technology and doctrinal development.[81] Specific secondary- and higher- order effects of self-generating technological and social change are impossible to predict.[82] Ironically, too often organizational and decision processes capable of flexible responses to dynamic environments are rejected because they look messy and undesirable.

In US national security policy and budget discussions, DoD and military service bureaucracies and their congressional, corporate, and think tank supporters incorrectly assume an identity relationship between higher technical performance and operational capability. They used this simplification of the RMA concept to promise that military dominance would be achieved by expensive advanced weapons acquisition programs. This marketing decision made the RMA concept vulnerable to criticism for a technological focus that excludes considerations of human nature and downplays the complexity of wartime interactions of technology with organizational, operational, and human factors. The RMA thus became associated with paper acquisition and budgeting plans and elaborately expressed goals for new military capabilities that have not translated into higher operational capability – or clear victory. In this process, potential useful insights have been obscured within the RMA research program to guide management and co-evolution of organizational structures with technological and operational advances.

What problems have to be solved when policymakers seek qualitative improvements in military capability? At a minimum, they have to direct or guide the production and growth of knowledge about physical phenomena (i.e., technologies),[83] and how to manage co-evolution of self-correcting and self-evaluating

---

81 Hindsight makes identification of military revolutions seem easier than it is, because, with hindsight (1) order is seen in random behavior, (2) confirmatory evidence is counted more heavily than disconfirmatory evidence, and (3) memory is selective and limited. Misplaced reliance on hindsight was examined in detail in Hone, Friedman, and Mandeles, *American and British Aircraft Carrier Development*.

82 Economists might call this phenomenon an example of "increasing returns on margin." Other synonyms include self-reinforcement, deviation-amplifying mutual causal processes, and cumulative causation; Arthur, "Self-Reinforcing Mechanisms in Economics," 10; Maruyama, "The Second Cybernetics."

83 Since World War Two, most weapons research and development programs have been conducted in government laboratories or funded by government agencies. Mazzucato argued that the most risky and consequential research and development was undertaken or funded by government laboratories or agencies. Small "grassroots" firms have only a small role and

organizational structures[84] with new or recombined technologies,[85] and associated operational concepts. Policymakers also have to (1) set conditions to store, catalog, access, and share relevant knowledge, (2) produce operational knowledge through experimentation, simulation and war games, and operational and historical analysis of combat experience, and (3) enable search of knowledge bases and interaction among various knowledge domains. Impacts of each of these factors are complicated by chance and contingency,[86] such as history of individual nation-states, their geography, natural resources, financial resources, intellectual capital, institutions and associated organizational relationships, alliances, adversaries and adversaries' capabilities. These factors and their interactions render trivial any generalizations posited about deliberate management of military revolutions.

The different organizational roles among the policymaker, policy analyst, and military historian offer little space in which fruitful intellectual exchange may occur concerning how to organize thinking and research to foster an RMA. Andrew W. Marshall noted often enough that the "hard work" of fostering an RMA is in the intellectual task of finding innovations in concepts of operation and making organizational changes to exploit existing and future technologies. Marshall's concern, in contrast to the Third Offset's emphasis on technology, provides additional evidence for the minor role of historical analysis and analogies in devising programs to foster an RMA.

For policymakers, scholars, and policy analysts there may be no satisfactory answer to the question of feasibility of deliberate government guidance of revolutionary military innovation. In renowned historian Geoffrey Parker's 2007 essay, "The Limits to Revolutions in Military Affairs," he warned that "no single mind, no single group can master all of the intellectual complexity" involved in deliberate and conscious efforts to build knowledge relevant to technology and organizations of war. Cultures (and nation-states) "that seek truth in revelation rather than in experiment, or those where the state micromanages all research" cannot keep pace with cultures that seek truth in observation and

---

impact in inventing or developing transformational military technology. See Mandeles, "Technology: The Historiography of Technology Since 1950"; Mazzucato, *The Entrepreneurial State*.

84 No simple managerial guide exists to design or reorganize organizational structures into fault-resisting or self-correcting modes. Wildavsky and Landau outlined the issues involved in such a design task, in Wildavsky, "The Self-Evaluating Organization"; Landau, "On the Concept of a Self-Correcting Organization."

85 Mandeles, "System Design and Project Management Principles to Meet the Needs of Operational Forces."

86 Mokyr, "King Kong and Cold Fusion"; Parker and Tetlock, "Counterfactual History."

experiment. Parker advocated a reasonable outlook for people thinking about the RMA – effective and coherent thought is facilitated by conducting "sustained dialogue" among "politicians, generals, and intellectuals."[87]

# Bibliography

Adamsky, Dima. *The Culture of Military Innovation: The Impact of Cultural Factors on the Revolution in Military Affairs in Russia, the US, and Israel*. Stanford: Stanford University Press, 2010.
Alexander, Arthur J. "R&D in Soviet Aviation." Santa Monica, CA: RAND Corporation, 1970.
Alexander, Arthur J. "Weapons Acquisition in the Soviet Union, United States, and France." Santa Monica, CA: RAND Corporation, 1973.
Alexander, Arthur J. "The Process of Soviet Weapons Design." Santa Monica, CA: RAND Corporation, 1978.
Alexander, Arthur J. "The Linkage between Technology, Doctrine, and Weapons Innovation: Experimentation for Use." Santa Monica, CA: RAND Corporation, 1981.
Alexander, Arthur J. "Soviet Science and Weapons Acquisition." Santa Monica, CA: RAND Corporation, 1982.
Alexander, Arthur J. "Patterns of Organizational Influence in Soviet Military Procurement." Santa Monica, CA: RAND Corporation, 1982.
Alexander, Arthur J. "Knowing about Soviet Weapons Acquisition and Strategic Weapons." Santa Monica, CA: RAND Corporation, 1986.
Anonymous. "U.S. Department of Defense Third Offset, Standard Market Taxonomy." *Govini Analyst Report*, 2016.
Arthur, W. Brian. "Self-Reinforcing Mechanisms in Economics." In *The Economy as an Evolving System*, edited by Philip W. Anderson, Kenneth J. Arrow, and David Pines, 9–31. Redwood City: Addison-Wesley Publishing Co., 1988.
Augier, Mie. "Thinking about War and Peace: Andrew Marshall and the Early Development of the Intellectual Foundations for Net Assessment." *Comparative Strategy* 32 (2013): 1–17.
Avey, Paul C., and Michael C. Desch. "What Do Policymakers Want From Us? Results of a Survey of Current and Former Senior National Security Decision Makers." *International Studies Quarterly* 58 (2014): 227–246.
Berliner, Joseph S. *Factory and Manager in the USSR*. Cambridge: Harvard University Press, 1957.
Blakeley, Katherine. *More Money on the Horizon? Analysis of the FY 2018 Defense Budget Request*. Washington, D.C.: Center for Strategic and Budgetary Assessments, 2017.
Bloch, Jean de. *The Future of War in its Technical, Economic and Political Relations*, 6 volumes. Paris: Imprimerie Paul Dupont, 1898.
Blue Ribbon Defense Panel, *Report to The President and the Secretary of Defense on the Department of Defense* (Fitzhugh Commission). 1 July 1970.

---

87 Parker, "The Limits to Revolutions in Military Affairs."

Brimley, Shawn. "Offset Strategies & Warfighting Regimes." *War on the Rocks* (15 October, 2014). Accessed May27, 2016, http://warontherocks.com/2014/10/offset-strategies-warfighting-regimes/.

Brun, Brig. Gen. Itai. "'While You're Busy Making Other Plans' – The 'Other RMA.'" *The Journal of Strategic Studies* 33 (2010): 535–565.

Buchholz, Arnold, and T. J. Blakeley, "The Role of the Scientific-Technological Revolution in Marxism-Leninism," *Studies in Soviet Thought* 20 (1979): 145–164.

Carter, Ashton. "Remarks Previewing the FY 2017 Defense Budget." As delivered by Secretary of Defense Ash Carter, Washington D.C., 2 February, 2016. Accessed February 24, 2016, http://www.defense.gov/News/Speeches/Speech-View/Article/648466/remarks-previewing-the-fy-2017-defense-budget.

Cerf, Christopher, and Victor Navasky. *The Experts Speak: The Definitive Compendium of Authoritative Misinformation*. New York: Pantheon Books, 1984.

Cherednichenko, Maj. Gen. M. "Conventional Weapons and the Prospects of Their Development." In *Scientific-Technical Progress and The Revolution in Military Affairs*, edited by Col. Gen. N.A. Lomov. Military Publishing House, Ministry of Defense, USSR, Moscow, 1973. Translated under auspices of the USAF. Washington, DC: US Government Printing Office, 1974.

Clark, Colin, and Sydney J. Freedberg, Jr. "Robot Boats, Smart Guns & Super B-52s: Carter's Strategic Capabilities Office." *Breaking Defense* (5 February, 2016). Accessed 24 February, 2016, http://breakingdefense.com/2016/02/carters-strategic-capabilities-office-arsenal-plane-missile-defense-gun/.

Commission on Organization of the Executive Branch of the Government. *The National Security Organization* ("First" Hoover Commission) (15 February, 1949).

Commission on Organization of the Executive Branch of the Government. *Business Organization of the Department of Defense* ("Second" Hoover Commission) (20 June, 1955).

Commission on Government Procurement. *Report of the Commission on Government Procurement* 1 (31 December, 1972).

Crawford, Michael J. "The Abolition of Prize Money in the United States Navy Reconsidered," *The Journal of Military History* 81 (2017): 105–132.

Czarnecki, Jon. "Against a Tech-Centric Offset." *War on the Rocks* (October 29, 2014). Accessed 29 May, 2016, http://warontherocks.com/2014/10/against-a-tech-centric-offset/.

Decker, Gilbert F., and Gen. Louis C. Wagner (USA, ret.). *Army Strong: Equipped, Trained and Ready*. Final Report of the *2010 Army Acquisition Review* (2011).

deLeon, Peter, and James Digby, eds. *Workshop on Asymmetries in Exploiting Technology as Related to the U.S.-Soviet Competition: Unclassified Supporting Papers*. Santa Monica: RAND Corporation, 1976.

Digby, James. *Precision-Guided Munitions: Capabilities and Consequences*. Santa Monica: RAND Corporation, 1974.

Digby, James. *The Technology of Precision Guidance: Changing Weapon Priorities, New Risks, New Opportunities*. Santa Monica: RAND Corporation, 1975.

Digby, James. *Precision Weapons: Lowering the Risks with Aimed Shots and Aimed Tactics*. Santa Monica: RAND Corporation, 1975.

Digby, James. *Precision-Guided Weapons: New Chances to Deal with Old Dangers*. Santa Monica: RAND Corporation, 1975.

Digby, James. *Precision-Guided Weapons*. Santa Monica: RAND Corporation, 1975.

Digby, James. "PGMs: Changing Weapons Priorities, New Risks, New Opportunities." *Astronautics & Aeronautics* 14 (1976): 36–46.
Dudzinsky, Jr., S.J., and James Digby. *Qualitative Constraints on Conventional Armaments: An Emerging Issue*. Santa Monica: RAND Corporation, 1976.
Dudzinsky, Jr., S.J., and James Digby. *The Strategic and Tactical Implications of new Weapons Technologies*. Santa Monica: RAND Corporation, 1976.
Dziak, John J. *Soviet Perceptions of Military Power: The Interaction of Theory and Practice*. New York: Crane, Russak & Company, 1981.
Farmer, J. Doyne, and François Lafond. "How Predictable is Technological Progress?" *Research Policy* 45 (2016): 647–665.
Farrell, Henry. "If policymakers had listened to political scientists, we wouldn't have invaded Iraq." *The Washington Post* (24 September, 2019). Accessed 15 December, 2019, https://www.washingtonpost.com/news/monkey-cage/wp/2014/09/24/if-policymakers-had-listened-to-political-scientists-we-wouldnt-have-invaded-iraq/.
Feynman, Richard P. *"What Do You Care What Other People Think?": Further Adventures of a Curious Character*. New York: Bantam Books, 1988.
Fissel, Mark Charles. "Military Revolutions." In *Oxford Bibliographies in Military History*, edited by Kaushik Roy. New York: Oxford University Press, https://www.oxfordbibliographies.com/view/document/obo-9780199791279/obo-9780199791279-0212.xml.
Freedberg Jr., Sydney J. "Adversaries Will Copy 'Offset Strategy' Quickly: Bob Work." *Breaking Defense* (19 November, 2014). Accessed 29 May, 2016, http://breakingdefense.com/2014/11/adversaries-will-copy-offset-strategy-quickly-bob-work/.
Freedberg Jr., Sydney J. "People, Not Tech: DepSecDef Work On 3$^{rd}$ Offset, JICSPOC." *Breaking Defense* (9 February, 2016). Accessed 29 May, 2016, http://breakingdefense.com/2016/02/its-not-about-technology-bob-work-on-the-3rd-offset-strategy/.
Gady, Franz-Stefan. "New US Defense Budget: $18 Billion for Third Offset Strategy." *The Diplomat* (10 February, 2016). Accessed 25 August, 2019, https://thediplomat.com/2016/02/new-us-defense-budget-18-billion-for-third-offset-strategy/.
Galay, Nikolai, "The Soviet Approach to the Modern Military Revolution." In *The Military-Technical Revolution: Its Impact on Strategy and Foreign Policy*, edited by John Erickson, Edward L. Crowley, and Nikolai Galay, 20–33. New York: Frederick A. Praeger, Publishers, 1966.
Gates, Robert M. "Speech." Presented to National Defense University, Washington, DC (29 September, 2008). Accessed 30 June, 2009, http://www.defenselink.mil/speeches/speech.aspx?speechid=1279.
Gouré, Daniel. "Directed Energy Weapons Will Be The Key To A Successful Third Offset Strategy." *Real Clear Defense* (29 March, 2016). Accessed 20 April, 2016, http://www.realcleardefense.com/articles/2016/03/29/directed_energy_weapons_will_be_the_key_to_a_successful_third_offset_strategy_109199.html.
Hacker, Barton C. "Engineering a New Order: Military Institutions, Technical Education, and the Rise of the Industrial State." *Technology and Culture* 34 (1993): 1–27.
Hacker, Barton C. "Horse, Wheel and Saddle: Recent Works of Two Ancient Military Revolutions." *International Bibliography of Military History* 32 (2012): 175–191.
Hagel, Chuck. "A Game-Changing Third Offset Strategy." *War on the Rocks* (17 November, 2014). Accessed 29 May, 2016, http://warontherocks.com/2014/11/a-game-changing-third-offset-strategy/.
Hagel, Chuck. "'Defense Innovation Days' Opening Keynote." As delivered by Secretary of Defense Chuck Hagel, Southeastern New England Defense Industry Alliance

(3 September, 2014). Accessed 27 May, 2016, http://www.defense.gov/News/Speeches/Speech-View/Article/605602.

Hagel, Chuck. "Subject: The Defense Innovation Initiative." Secretary of Defense Memorandum for Deputy Secretary of Defense (15 November, 2014).

Hanley, John T., and Mark Mandeles. *Adapting to a New Era: The SSG's Transition to an Innovation Cell*. Arlington: Center for Naval Operations, 2018.

Hanley, John T., and Mark Mandeles. *Working for Innovation: SSG Practices and Impacts* Arlington: Center for Naval Operations, 2018.

Hanley, John T., and Mark Mandeles. *Notes from the CNO Strategic Studies Group Archives*, Vol. 3 (*SSGs XXV–XXIX*). Arlington: Center for Naval Operations, 2018.

Hayden, Michael. "Video: Transformation and The Third Offset (NSA-CIA)." *Real Clear Defense* (29 March, 2016). Accessed 3 May, 2016, http://www.realcleardefense.com/video/2016/03/29/michael_hayden_transformation_and_the_third_offset_nsa-cia.html.

Hayek, Friedrich A. *The Road to Serfdom*. Chicago: The University of Chicago Press, 1944.

Hayek, Friedrich A. "The Use of Knowledge in Society." *American Economic Review* 35 (1945): 519–530.

Hayek, Friedrich A. "The Pretence of Knowledge." Nobel Prize lecture in Economic Science (11 December, 1974). Accessed 11 October 2011, http://www.nobelprize.org/nobel_prizes/economics/laureates/1974/hayek-lecture.html.

Hayek, Friedrich A. *The Counter-Revolution of Science: Studies on the Abuse of Reason* Indianapolis: Liberty Press, 1979.

Herspring, Dale R. "Nikolay Ogarkov and the Scientific-Technical Revolution in Soviet Military Affairs." *Comparative Strategy* 6 (1987): 29–59.

Hone, Thomas C., and Mark D. Mandeles. *Innovation in Naval Aviation*. Final Technical Report. Delex Systems, Inc., 1982.

Hone, Thomas C., Norman Friedman, and Mark D. Mandeles. *American and British Aircraft Carrier Development, 1919–1941*. Annapolis: Naval Institute Press, 1999.

Howard, Michael. "Men Against Fire: Expectations of War in 1914," *International Security* 9 (1984): 41–57.

Joravsky, David. "The History of Technology in Soviet Russia and Marxist Doctrine," *Technology and Culture* 2 (1961): 5–10.

Kahn, Herman. *Thinking About the Unthinkable in the 1980s*. New York: Simon and Schuster, 1984.

Knox, MacGregor, and Williamson Murray, eds. *The Dynamics of Military Revolution, 1350–2050*. Cambridge: Cambridge University Press, 2001.

Kott, Alexander. "The Future of War Technology Whispers to Us From the Past, and We Must Listen Better." *War on the Rocks* (3 December, 2019). Accessed 8 December, 2019, https://warontherocks.com/2019/12/the-future-of-war-technology-whispers-to-us-from-the-past-and-we-must-listen-better/.

Krepinevich, Jr.Andrew F., *The Military-Technical Revolution: A Preliminary Assessment*. Washington, DC: Center for Strategic and Budgetary Assessments, 1992.

Krepinevich, Jr., Andrew F. "Cavalry to Computer: The Pattern of Military Revolutions." *The National Interest* (1994): 30–42.

Krepinevich, Jr., Andrew F., and Barry Watts. *The Last Warrior: Andrew Marshall and the Shaping of Modern American Defense Strategy*. New York: Basic Books, 2015.

Krepinevich, Andrew F. "Statement [Defense Strategy]." As delivered to Senate Armed Services Committee (28 October, 2015). Washington, DC: Center for Strategic and Budgetary Assessments, 2015.
Lambert, Andrew. "Review of *Torpedo: Inventing the Military-Industrial Complex in the United States and Great Britain* by Katherine C. Epstein." *The Journal of Military History* 81 (2017): 243–245.
Lambeth, Benjamin S. *Moscow's Lessons from the 1982 Lebanon Air War*. Santa Monica: RAND Corporation, 1984.
Lambeth, Benjamin S. *Desert Storm and Its Meaning: The View from Moscow*. Santa Monica: RAND Corp., 1992.
Landau, Martin. "On the Concept of a Self-Correcting Organization." *Public Administration Review* 33 November–December 1973: 533–542.
Landau, Martin, and Donald Chisholm, "The Arrogance of Optimism: Notes on Failure-Avoidance Management." *Journal of Contingencies and Crisis Management* 3 (1995): 67–80.
Lieberstein, Samuel. "Review of *Chelovek-nauka-tekhnika (opyt marksistskogo analisa nauchno-tekhnicheskoi revoliutsii)*" [Man-science-technology: toward a Marxist analysis of the scientific-technical revolution]. *Technology and Culture* 16 (1975): 691–693.
Lindblom, Charles E., and David K. Cohen. *Usable Knowledge: Social Science and Social Problem Solving*. New Haven: Yale University Press, 1979.
Lofgren, Eric M. "The DoD Budget Process: The Next Frontier of Acquisition Reform." Fairfax: George Mason University, Center for Government Contracting, School of Business, (2020). Accessed 15 August, 2020, https://business.gmu.edu/images/GovCon/White_Papers/The_DoD_Budget_Process.pdf.
Lomov, Col. Gen. N.A., ed. *Scientific-Technical Progress and The Revolution in Military Affairs*. Military Publishing House, Ministry of Defense, USSR, Moscow, 1973. Translated under auspices of the USAF. Washington, DC: US Government Printing Office, 1974.
Majumdar, Dave. "Is the US Army Ready for a Shocking Technological Revolution in Land Warfare?" *The National Interest* (23 June, 2016). Accessed 20 June, 2017, http://nationalinterest.org/blog/the-buzz/the-us-army-ready-shocking-technological-revolution-land-16703.
Mandeles, Mark D. "Between a Rock and a Hard Place: Implications for the U.S. of Third World Nuclear Weapon and Ballistic Missile Proliferation." *Security Studies* 1, (1991): 235–269.
Mandeles, Mark D. "Military Technology, Tactics and Operations, and Social Change: The Continued Relevance of Bloch's Approach." Paper presented at the Peace Historical Society Symposium, Hague Appeal for Peace Conference, The Hague, Netherlands, 11–15 May, 1999.
Mandeles, Mark D. *The Future of War: Organizations as Weapons*. McLean: Potomac Books, 2005.
Mandeles, Mark D. *Military Transformation Past and Present: Historical Lessons for the 21$^{st}$ Century*. Westport: Praeger Publishers, 2007.
Mandeles, Mark D. "Memorandum. To Joint Center for Operational Analysis/U.S. Joint Forces Command. Subject: JCOA and Operational Analysis for US Forces and Their Commanders." (31 March, 2011), https://www.academia.edu/9593159/JCOA_and_Operational_Analysis_for_US_forces_and_Their_Commanders.
Mandeles, Mark D. *System Design and Project Management Principles to Meet the Needs of Operational Forces*. Fairfax, VA: The J. de Bloch Group, 2011, https://www.academia.edu/

9592932/System_Design_and_Project_Management_Principles_to_Meet_the_Needs_of_Operational_Forces.

Mandeles, Mark D. *Interdisciplinary Inquiry and Preparation for Net Assessment*. Fairfax: The J. de Bloch Group, 2014, https://www.academia.edu/50731179/Essay_2_Interdisciplinary_Inquiry_and_Preparation_for_Net_Assessment_by_Mark_D_Mandeles_PhD_President_THE_J_DE_BLOCH_GROUP.

Mandeles, Mark D. "Technology: The Historiography of Technology Since 1950, with a Focus on the Navy." In *Needs and Opportunities in the Modern History of the Navy*, edited by Michael J. Crawford, 309–368. Washington, DC: Naval History and Heritage Command, 2019.

Mantzavinos, C., Douglass C. North, and Syed Shariq. "Learning, Institutions, and Economic Performance." *Perspectives on Politics* 2 (2004): 75–84.

Marshall, Andrew W. "Memorandum for the Record, Subject: Some Thoughts on Military Revolutions." (27 July, 1993).

Marshall, Andrew W. "Memorandum for the Record, Subject: Some Thoughts on Military Revolutions – Second Version." (23 August, 1993).

Marshall, Andrew W. "Memorandum for the Record, Subject: RMA Update." (2 May, 1994).

Marshall, Andrew W. "Testimony," U.S. Congress, Senate, Committee on Armed Services, Subcommittee on Acquisition and Technology, Hearings on S. 1026, 104 Cong. 1 sess., "Revolution in Military Affairs," S. Hrg. 104–387, Pt. 5 (5 May, 1995): 256–258.

Marshall, Andrew W. "Foreword." In *Strategic Appraisal: The Changing Role of Information in Warfare*, edited by Zalmay Khalilzad, John P. White, and Andrew W. Marshall. Santa Monica: RAND Corporation, 1999, https://www.rand.org/pubs/monograph_reports/MR1016.html.

Maruyama, Magorah. "The Second Cybernetics: Deviation-Amplifying Mutual Causal Processes." *American Scientist* 51 (1963): 164–179.

Mattis, James N. "Memorandum for U.S. Joint Forces Command. Subject: Vision for a Joint Approach to Operational Design." (6 October, 2009).

Mazzucato, Mariana. *The Entrepreneurial State: Debunking Public vs. Private Sector Myths*. London/New York: Wimbledon Publishing Co, Anthem Press, 2013.

Mead, Walter Russell. "Why Russia and China Are Joining Forces." *The Wall Street Journal* (29 July, 2019). Accessed 30 July, 2019, https://www.wsj.com/articles/why-russia-and-china-are-joining-forces-11564441072.

Mehta, Aaron. "Work Outlines Key Steps in Third Offset Tech Development." *Defense News* (14 December, 2015). Accessed 24 February, 2016, http://www.defensenews.com/story/defense/innovation/2015/12/14/work-third-offset-tech-development-pentagon-russia/77283732/.

Mehta, Aaron. "Pentagon No. 2: How to Keep Third Offset Going in the Next Administration." *Defense News* (2 May, 2016). Accessed 3 May, 2016, http://www.defensenews.com/story/defense-news/2016/05/02/pentagon-no-2-how-keep-third-offset-going-next-admininistration/83851204/.

Milley, Gen. Mark A. "Priorities for Our Nation's Army with General Mark A. Milley" (23 June 2016). Center for Strategic and International Studies. Accessed 23 June, 2016, https://www.csis.org/events/priorities-our-nations-army-general-mark-milley.

Mokyr, Joel. "King Kong and Cold Fusion." In *Unmaking the West: "What-If?" Scenarios That Rewrite World History*, edited by Philip E. Tetlock, Richard Ned Lebow, and Geoffrey Parker, 277–322. Ann Arbor: The University of Michigan Press, 2006.

Murray, Williamson, and Allan R. Millett, eds. *Military Innovation in the Interwar Period*. Cambridge: Cambridge University Press, 1996.

Nelson, Richard R. "Issues and Suggestions for the Study of Industrial Organization in a Regime of Rapid Technical Change." In *Policy Issues and Research Opportunities in Industrial Organization in a Regime of Rapid Technical Change*, edited by Victor Fuchs, 34–58. New York: Columbia University Press, 1972.

Nofi, Albert A. "Recent Trends in Thinking About Warfare." Alexandria: Center for Naval Analyses, 2006.

Odom, William E. "Soviet Force Posture: Dilemmas and Directions." *Problems of Communism*, 34 (1985): 1–14.

Odom, William E. "Soviet Military Doctrine." *Foreign Affairs* 67 (1988): 114–134.

Office of the Secretary of Defense Task Force. *The President's Private Sector Survey on Cost Control* (Grace Commission). 13 July, 1983.

Ogarkov, N. V. *Military Encyclopedic Dictionary, Vol. 1*. Foreign Broadcast Information Service (8 January, 1985), JPRS-UMA-85-002-L.

Parker, Geoffrey, and Philip E. Tetlock, "Counterfactual History: Its Advocates, Its Critics, and Its Uses." In *Unmaking the West: "What-If?" Scenarios That Rewrite World History*, edited by Philip E. Tetlock, Richard Ned Lebow, and Geoffrey Parker, 363–392. Ann Arbor: The University of Michigan Press, 2006.

Parker, Geoffrey. "The Limits to Revolutions in Military Affairs: Maurice of Nassau, the Battle of Nieuwpoort (1600), and the Legacy." *The Journal of Military History* 71 (2007): 367–372.

Pellerin, Cheryl. "DoD Seeks Novel Ideas to Shape its Technological Future." *DoD News* (24 February, 2015). Accessed 30 May, 2016, http://www.defense.gov/News-Article-View/Article/604159/dod-seeks-novel-ideas-to-shape-its-technological-future.

Popper, Karl R. *The Poverty of Historicism*. New York: Harper Torchbooks, 1964.

Popper, Karl R. "Chapter 10." In *Conjectures and Refutations: The Growth of Scientific Knowledge*, 215–250. New York: Harper Torchbooks, 1968.

Popper, Karl R. "Chapter 16." In *Conjectures and Refutations: The Growth of Scientific Knowledge*, 336–346. New York: Harper Torchbooks, 1968.

Popper, Karl R. *The Open Society and Its Enemies* Vol. 2, *The High Tide of Prophecy: Hegel, Marx, and the Aftermath*. Princeton: Princeton University Press, 1971.

President's Blue Ribbon Commission on Defense Management. *A Quest for Excellence* (Packard Commission). 30 June, 1986.

Rickover, Admiral Hyman G. "The Role of Engineering in the Navy." Presented to the National Society of Former Special Agents of the Federal Bureau of Investigation, Seattle, Washington,(30 August, 1974).

Ricks, Thomas E. "Put down that broad brush, Steve!: I remain unpersuaded by Saideman's valiant defense of political science." *Foreign Policy* (24 September, 2014). Accessed 15 December, 2019, https://foreignpolicy.com/2014/09/24/put-down-that-broad-brush-steve-i-remain-unpersuaded-by-saidemans-valiant-defense-of-political-science/.

Roberts, Michael. "The Military Revolution, 1560–1660." In *Essays in Swedish History*, Michael Roberts. Minneapolis: University of Minnesota Press, 1967.

Rochlin, Gene I. *Trapped in the Net: The Unanticipated Consequences of Computerization*. Princeton: Princeton University Press, 1997.

Rosen, Stephen P. *Winning the Next War: Innovation and the Modern Military*. Ithaca: Cornell University Press, 1991.

Rosen, Stephen P. "The Impact of the Office of Net Assessment on the American Military in the Matter of the Revolution in Military Affairs." *The Journal of Strategic Studies* 33 (2010): 469–482.

Rosengarten, Jr., Adolf G. "John Bloch – A Neglected Prophet," *Military Review* 37 (1957): 27–39.

Sander, Alexandra. "Exploring a New Offset Strategy: What the Experts Say." *War on the Rocks*. (4 December, 2014). Accessed 29 May, 2016, http://warontherocks.com/2014/12/exploring-a-new-offset-strategy-what-the-experts-say/.

Scharre, Paul. "How to Lose the Robotics Revolution." *War on the Rocks* (29 July, 2014). Accessed 27 May, 2016, http://warontherocks.com/2014/07/how-to-lose-the-robotics-revolution/.

Simón, Luis. "The 'Third' US Offset Strategy and Europe's 'Anti-access' Challenge." *The Journal of Strategic Studies* 39 (2016): 417–445.

Starbuck, William A., and John M. Mezias. "Opening Pandora's Box: Studying the Accuracy of Managers' Perceptions." *Journal of Organizational Behavior* 17 (1996): 99–117.

Sokolovskiy, Vasily [Marshal of the Soviet Union] and Maj. Gen. M. Cherednichenko, "The Revolution in Military Affairs, Its Significance and Consequences." *Krasnaya Zvezda* (28 August, 1964), trans. Library of Congress, ATD U-64-104, 24 September, 1964.

*Sovietskaya Voyennaya Entsiklopediya* [Soviet Military Encyclopedia]. 2nd Edition, Vol. I (A–Byulov). Moscow: Military Publishing House, 1990.

Sweetman, Bill. "'Third Offset' Addresses Operational and Economic Challenges." *Aviation Week & Space Technology* (3 November, 2014). Accessed 4 November, 2014, http://aviationweek.com/defense/third-offset-addresses-operational-and-economic-challenges.

Thompson, Wayne. *To Hanoi and Back: The U.S. Air Force and North Vietnam 1966–1973*. Washington, DC: Smithsonian Institution Press, 2000.

U.S. Government Accountability Office. *Weapon Systems Annual Assessment: Limited Use of Knowledge-Based Practices Continues to Undercut DoD's Investments*, GAO-19-336SP, (May 2019). Accessed 15 June, 2019, https://www.gao.gov/products/GAO-19-336SP.

Walton, Timothy A. "Securing The Third Offset Strategy: Priorities For Next US Secretary Of Defense." *Joint Force Quarterly* 82 (2016). Accessed 6 August, 2016, http://ndupress.ndu.edu/JFQ/Joint-Force-Quarterly-82/.

Watts, Barry D. *Six Decades of Guided Munitions and Battle Networks: Progress and Prospects*. Washington, DC: Center for Strategic and Budgetary Assessments, 2007.

Watts, Barry D. *The Evolution of Precision Strike*. Washington, DC: Center for Strategic and Budgetary Assessments, 2013.

Weber, Max. "Bureaucracy." In *Economy and Society*, edited by Guenther Roth, and Claus Wittich, 956–1005. Berkeley: University of California Press, 1978.

Weigley, Russell F. *History of the United States Army*. Enlarged Edition. Bloomington: Indiana University Press, 1984.

Weisgerber, Marcus. "Pentagon Wants to Pair Troops with Machines to Deter Russia, China." *Defense One* (8 November, 2015). Accessed 9 November, 2015, http://www.defenseone.com/technology/2015/11/pentagon-wants-pair-troops-machines-deter-russia-china/123498/?oref=defenseone_today_nl.

Wildavsky, Aaron. "The Self-Evaluating Organization." *Public Administration Review* 32 (1972): 509–520.

Wiitala, Josh. "Seizing the Defensive: A Balanced Approach for the Third Offset." *The Strategy Bridge* (14 June, 2016). Accessed 23 June, 2016, http://www.thestrategybridge.com/the-bridge/2016/6/14/seizing-the-defensive-a-balanced-approach-for-the-third-offset#_edn1.

Woodford, Shawn. "So You Still Think You Want A Revolution In Military Affairs?" (8 August, 2016). Accessed 13 June, 2017, http://www.dupuyinstitute.org/blog/2016/08/08/so-you-still-think-you-want-a-revolution-in-military-affairs/.

Work, Robert O., and Shawn Brimley. *20YY: Preparing for War in the Robotic Age*. Washington, DC: Center for New American Security, 2014.

Work, Bob. "The Third U.S. Offset Strategy and its Implications for Partners and Allies." As delivered by Deputy Secretary of Defense Bob Work, Willard Hotel, Washington, DC (28 January, 2015). Accessed 24 February, 2016, http://www.defense.gov/News/Speeches/Speech-View/Article/606641/the-third-us-offset-strategy-and-its-implications-for-partners-and-allies.

Zvorikine, A. "The History of Technology as a Science and as a Branch of Learning: A Soviet View." *Technology and Culture* 2 (1961): 1–4.

Zvorikine, A. "Technology and the Laws of Development." *Technology and Culture* 3 (1962): 443–458.

João Vicente
# The Dilemma of Human Interference in War: The Coming Revolution of Autonomous Air Warfare

## Introduction

Air power capabilities have experienced revolutionary advances throughout their short history. Air power concerns the military exploitation of air and space by man, not necessarily with man.[1] Such a paradigm has projected unmanned air systems as a transformational competency with disruptive implications, forcing a new debate about the strategic significance of this ongoing revolution. The promises of future warfare conducted from remote locations, without risks, and increasingly performed by machines accentuates the need for debate. The emergence of Autonomous Aircraft Systems (AAS) is one of those revolutionary moments in history that demonstrate that the difference between science fiction and reality is only time. Moreover, insightful application of technological capability is the dilemma, not the technology itself.

Judging by the trends of increasing defense budgets, the drone industry is booming. Platform miniaturization, enhancement of sensory apparatus, increasingly lethal weaponry, improved computing capabilities, and instantaneous network connectivity will enable drone weapons systems to outstrip human capability to properly analyse the flood of data and make combat decisions.

The emergence of AAS will irrevocably redefine human interference in armed conflict and transform the state's strategic culture, with inherent sociopolitical ramifications. The perfection of 'small, many, smart, and lethal' systems render obsolescent many skills fostered by extant military organizations. The full exploitation of autonomous lethal systems constitutes therefore a genuine military revolution. Military and political leadership are compelled to exploit operational capability that further reduces risk to their forces whilst simultaneously attaining a combat superiority. AAS (incorporating miniaturization, weaponization, and greater autonomy) transforms warfare by potentially authorizing machines to make lethal combat decisions. Consequently, commanders must address the aim, context, accountability and implications of greater autonomy.

---

[1] Mason, "Unmanned Aerial Vehicles," 123.

Hence, the development and employment of lethal AAS will be irresistible, inevitable and ultimately irreversible. The fine line between innovation and revolution is gradually being crossed with the development of autonomous, modular, and multifunction systems. However, there are several limiting factors to the ubiquity of AAS and the optimization of its operational product, which will have to be mitigated by reducing uncertainty and improving technological and operational maturity, leading to greater cultural acceptance for increasing levels of autonomy. When this happens, we will be facing a turning point in the future of air warfare, making it less constrained, and with it threatening to change human involvement in future war from an active participant (doer), to a supervisor, and ultimately, to a mere observer.

This debate should take place before the materialization of the expected technological solutions, at the risk of repeating past mistakes, when mankind developed and employed technologies with devastating effects, without proper previous consideration. For example, this applies to nuclear power technology, which gave way to employment of nuclear weapons before the necessary ethical/strategic reflections. Disruptive fields such as genetics, cyberwarfare, and Artificial Intelligence (AI) pose similar moral dilemmas. Consequently, we are at the right time to question the nature of this revolution, addressing its concerns, and consciously choosing the most tolerable future. If our thesis is validated then it will affect ethics, moral and legal domains, besides operational results, as well as political discourse and decision-making.

# From Innovation to Revolutions in Warfare: Setting the Framework

The transdisciplinary approaches on military innovation vary in their classification according to historical, civilizational, sociological, and technological dimensions. To increase the conceptual complexity, they relate this concept with different levels of military science from the strategic to the tactical domains. The vast array of frameworks conditions our analysis. Instead of trying to address all the diverse conceptual approaches, we prefer to point out the different lines of interaction, based on the process definition, the quantification of its occurrence, and the magnitude of its effects.

Stephen Rosen argues that the promotion of a military culture of innovation should not focus exclusively on financial factors.[2] According to "talented military personnel, time, and information have been the key resources for innovation."[3] In addition, Max Boot argues that military superiority often results from clever adaptation and employment of known technologies rather than innovation.[4] Thus, technology could lead to change, but how these technological changes are incorporated in the capability development process, doctrine, training, operations and strategy will determine who achieves victory in the future, as it did in the past.

The original term "military revolution" can be traced back to Michael Roberts' 1955 thesis.[5] From that point onward there has been great contention among the historians regarding the conceptualization, scope and magnitude of military revolutions. As suggested by Jeremy Black, the definitions of military revolutions have varied greatly in terms of duration, content and impact, as well as in tactical, operational and strategic effects, affecting both the military and non-military dimensions.[6]

According to Murray and Knox, the defining feature of a military revolution is that it fundamentally changes the framework of war by recasting society and the state as well as military organizations, altering the ability of states to develop and use force.[7] Throughout these periods of rapid systemic change, some authors argue for disproportionate effects far beyond the military domain, which are manifested in overwhelming, random, and unexpected ways.[8]

Considering the different conceptual approaches, the classification of military revolutions is not a consensual issue in the academic community. Murray and Knox consider five;[9] Hoffman expands the previous authors' list to seven;[10]

---

2 Rosen, *Winning the Next War*, 252.
3 Rosen, *Winning the Next War*, 252.
4 Boot, *War Made New*, 459.
5 The original essay along with other key contributions is reprinted in Rogers, ed., *The Military Revolution Debate*, 13–35.
6 Black, "The Revolution in Military Affairs." Fissel, "Military Revolutions."
7 Murray and Knox, "Thinking About Revolutions in Warfare," 6–7.
8 Morillo and Pavkovic, *What is Military History?*, 73.
9 Seventeenth-century modern state and military institutions; French Revolution; Industrial Revolution; World War One; Nuclear weapons and ballistic missiles. Murray and Knox, "Thinking About Revolutions in Warfare," 13.
10 Expanding Murray and Knox's taxonomy, Hoffman adds Information Revolution and Autonomous Revolution to the list. Hoffman, "Will War's Nature Change in the Seventh Military Revolution?," 20.

Krepinevich points a total of ten;[11] Max Boot identifies four;[12] Alvin and Heidi Toffler[13] argue for three military revolutions associated with three waves: agrarian, industrial and informational, directly related to the transition from the medieval, modern, and information societies. Therefore, we have to agree with Colin Gray assertion that the more historians peer through military experience, the larger the number of revolutions they find.[14] Regardless of the subjectivity of its classification, what is important to retain is that the disruption of values and processes of warfare and its associated organizations, as the result of technological advances, causes changes in the social relations of war.

Bearing in mind the magnitude of the change and its effects, we may envision, alongside and within military revolutions clusters of less all-embracing changes, best conceptualized as Revolutions in Military Affairs (RMA).[15] One way of looking this relationship in a broader historical context is that a military revolution occurs after a series of "anticipatory" RMA's have occurred.[16] Therefore, while RMA concerns primarily to the tactical and operational domains, the broader phenomenon of military revolution translate an all-encompassing change, which has a multidimensional impact in the human endeavor.

Building upon the thesis of the Soviet Marshal Nikolai Ogarkov on military technical revolution, Andrew Marshall, considered the mastermind of the Western RMA concept, argued that such phenomena "occurs when the application of new technologies into military systems combines with innovative operational concepts and organizational adaptation to alter fundamentally the character and conduct of military operations."[17] Under this framework, revolutions comprise four elements: "technological change, military systems evolution, operational innovation, and organizational adaptation."[18]

According to Richard Hundley, the defining characteristic of an RMA can be stated as a "paradigm shift in the nature and conduct of military operations which either renders obsolete or irrelevant one or more core competencies of a dominant player, or creates one or more new core competencies, in some new

---

11 Krepinevich, "Cavalry to Computer: the Pattern of Military Revolutions," 30–42.
12 Max Boot examines four military revolutions over the past 500 years: the Gunpowder Revolution (1500–1700), the Industrial Revolution (1750–1900), the Second Industrial Revolution (1900–1945), and the Information Revolution (1970-present), in Boot, *War Made New*.
13 Toffler and Toffler, *War and Anti-War*.
14 Gray, "RMA's and the Dimensions of Strategy," 51.
15 See the relationship between military revolutions and RMA in Murray and Knox, "Thinking About Revolutions in Warfare," 12–13.
16 Morillo and Pavkovic, *What is Military History?*, 77.
17 Krepinevich, The *Military-Technical Revolution*, 3.
18 Krepinevich, *The Military-Technical Revolution*, 3.

dimension of warfare, or both."[19] Hundley continues by highlighting and explaining the importance of the relationship between paradigm and competency.[20] Operational paradigms are standard models for military operations. The arrangement of the infantry units during Napoleonic Wars or the distinctive positioning of naval units in parallel routes during naval engagements are some of the examples which survived up to the First World War. Also, core competencies are specific features of military capabilities. For example, one of the dominant competences of a modern air force is to identify and engage, with precision, enemy targets using a variety of manned platforms. In this sense, a paradigm shift occurs when such operational skills are changed, creating new operational templates or new ways of warfare. The introduction of the aircraft carrier translates such a shift as it pushed the naval combat beyond visual range. Likewise, the dominant paradigm of the German forces, *Blitzkrieg*, profoundly changed the prevailing ground maneuver operational model, rendering obsolete the static defenses for infantry and artillery units. Similarly, the introduction of intercontinental ballistic missiles has given rise to a new competency, capable of inflicting nuclear destruction from remote locations.

The modern RMA, as presented by Geoffrey Parker, enables the employment of "precision violence," as an essential feature of modern military forces, as a consequence of the synergistic interaction between systems that collect, process, and disseminate information and those who apply lethal force.[21] The introduction of unmanned systems has challenged the fundamental competence of modern air forces and induced systemic changes in the lethal matters of war, which revolutionized two essential air power skills: situational awareness and attack.

However, as Richard Hallion indicates, mere technological superiority does not guarantee, by itself, a combat revolution. It must be accompanied by innovative concepts of employment that catalyze change.[22] Hallion presents several examples in history where nations failed to develop adequate doctrine to leverage the potential of innovative technology. Such instances include the navies' slow adaptation to the potential of the newly introduced center line turrets, insisting in conducting less effective parallel-course fighting tactics. Another example of uninspiring doctrine emerged, during First World War, by the devastating losses suffered by the Allies while insisting on making frontal attacks against artillery and machine-gun fire.

---

19 Hundley, *Past Revolutions*, 9.
20 Hundley, *Past Revolutions*, 9–11.
21 Parker, "The Future of Western Warfare," 419.
22 Hallion, "Doctrine, Technology, and Air Warfare."

In this context, an RMA causes changes at three levels: in the structure of power; in its character; and in its nature or function. Thus, it is associated with three main components: organization, doctrine and technology. This is the relationship that Max Boot describes by portraying the revolutions in the last 500 years as periods in which new technologies, combined with new tactics and organizational structures have fundamentally changed the face of war and with it the overall balance of power.[23]

Finally, this analysis will take into consideration Jeremy Black's advice about the importance of contingent and contextual factors to determine the impact and effectiveness of the transformational capability, that is, taking into consideration its "fitness for purpose" when judging the applicability of new technology.[24]

## The Infancy of the Unmanned Revolution

The technological advances of a society are mirrored in its way of fighting, with the use of increasingly effective and lethal weapon systems. Over the past 500 years, we have witnessed a gradual increase of the technological changes introduced in combat, in contrast with the relative stagnation of the previous thousand years.[25] This exponential progression in the ability to search, identify and destroy has expanded the dimensions and lethality of the battle space. Singer highlights this exponential rise by comparing the half a million times increase of combat power from a Roman legionnaire to a modern bomber aircraft or the increase by a factor of 20, just in the past century, of the range and effectiveness of artillery.[26]

If we read through Dohuet's or Mitchell's writings it is possible to verify that theory sometimes precedes technology. For example, Billy Mitchell, in 1925, argued for the remote control of a fleet of bombers to attack urban targets.[27] Furthermore, for every technological advance there is a matching military application with increased lethality and effectiveness. This relationship leads Van Creveld to point out technology's determining factor in the war's transformation.[28] Nevertheless, the more or less linear progression which has accompanied the history of

---

**23** Boot, *War Made New*.
**24** Black, "Military History and the Whig Interpretation," 26.
**25** Van Creveld, *Technology and War*, 20.
**26** Singer, *Wired for War*, 100.
**27** Mitchell, *Winged Defense*, 165.
**28** Van Creveld, *Technology and War*.

war and technology is bound to disappear, as the result of the exponential acceleration in technological innovations. Such advances may lead to a "singularity" in which the speed and depth of technological change makes it impossible to predict how life will be in the coming decades.[29]

The increase in computing power, advances in genetics, robotics, AI, additive and molecular manufacturing, reveal the different strands of the technological developments. They also reveal that this is a process of interdisciplinary synergies and intra-disciplinary innovation.[30] That is, the interaction between these branches of knowledge will unveil radical new applications, enabling in the short-term performances of the machines similar to the human brain. This technological acceleration acts as a catalyst for new paradigms, which threaten to permanently change the strategic environment, and ultimately, the face of future war.

The information age has brought dramatic changes in the speed of decision and reaction, as well as in reducing the number of systems required to achieve the desired effects. Within this second machine age[31] lays the power to leverage the synergies of AI and the digital connection of all battlespace entities. Thus, effects that used to take months to be achieved and consumed vast human and material resources are now attainable by a single platform in a fraction of the time. The first century of aviation was devoted to developing systems capable of attacking any target, anywhere, in any weather conditions, and accurately. Today, and in the future, finding and locating the desired target, with pinpoint accuracy, becomes the main challenge.

Earth's orography diminishes sensor's effectiveness inflicting additional challenges to surveillance and targeting on surface's environment; hence, simple technology and numerical preponderance cannot dominate the surface war.[32] Despite ongoing discussions that RMA can reduce this opacity, it does not make the land environment transparent, unlike other combat areas (air and sea). Indeed, air warfare (and partly the surface of naval warfare) has different dynamics, simpler than the terrestrial domain, since in the air there is no place to hide (at best only to delay detection using stealth technology).[33] Thus, as Biddle asserts, in the air environment, technology and the preponderance of numbers have the greatest impact.[34]

---

29 Kurzweil, *The Singularity is Near*.
30 Kaku, *Visions*, 5.
31 Brynjolfsson and McAfee, *The Second Machine Age*.
32 Biddle, *Military Power*, 72.
33 Biddle, *Military Power*, 269.
34 Biddle, *Military Power*, 269.

In a comprehensive perspective, we can qualify the operational utility of Unmanned Aircraft Systems, or "drones,"[35] as very relevant in 5D environments, that is "dull, dirty, dangerous, demanding, and different"[36] in which the human dimension becomes the limiting factor. Monotonous activities, with repetitive tasks, such as long flights, which can lead to crew fatigue, are likely to be automated. The possibility of rotating operators while the same platform remains airborne, assures the necessary persistence required for surveillance missions. Similarly, when operating in contaminated environments (with nuclear, biological or chemical agents), the machine does not suffer any kind of limitations, nor does it require the use of protective equipment.

Another situation is the carrying out of dangerous activities that involve high risk to crews, like operations in highly defended airspace, in which aircraft attrition is high. By removing the human element from danger, we are preserving a highly specialized and costly resource, both from a financial perspective and political exploitation. In demanding activities involving speed, accuracy and reliability, the physiological capabilities of the pilot limit the performance of the aircraft. For example, operations at high altitude or under high G load increase the operational risk to the pilot, requiring the use of additional support systems to mitigate it.

Finally, different activities include all the ones which are not possible for manned aircraft, such as the reconnaissance missions carried out by micro drones, in confined spaces and inaccessible to larger aircraft. Just as the evolutionary path of their manned relatives, drones have evolved from observation tasks to attack roles, as instruments of strategic bombing, interdiction and close air support. In this sense, we can distinguish an operational employment template with emphasis on load capacity and persistence and another modality with an interest in autonomy, survival and weapons employment.

The sheer increase in platform numbers we have witnessed in the last few years does not reflect the true nature and magnitude of the phenomena.[37] The exponential and synergistic effects of modularity, autonomy levels and consequent

---

**35** The term Unmanned Aircraft System (UAS) expresses the whole system perspective including the aerial platform (Unmanned Aerial Vehicle – UAV), the payload, the human element that controls the system, the control station and communication links. The reference to the jargon 'drone' applies to the aerial platform.
**36** Alkire et al., *Applications for Navy Unmanned Aircraft Systems*, 25–26.
**37** As of March 2020, there were 181 types of active military UAVs, produced by 108 entities in 43 countries, totaling more than 30,000 platforms, operated by 102 countries (compared with 60 countries in 2010). Since the 1980s, at least 28 countries have deployed UAVs beyond their borders, ten of which are believed to have conducted aerial strikes. Currently, more than 40 countries have in their inventories armed drones. Gettinger, *Drone Databook Update: March 2020*

innovative operational templates transcend the isolated results of technological innovation. Such attributes will allow for better operational effectiveness as the result of the merge of platforms, sensors, weapons and command and control methods. This approach will provide greater agility, flexibility, adaptability, and capacity for future growth, allowing for capability improvement without the need to develop new systems.[38]

The current technophilic images confirm the trend towards a vertical, horizontal, and qualitative proliferation of drones. First, there is a diversity of shapes and sizes, from few centimeters to tens of meters wingspan.[39] Second, we are witnessing an expansion of the spectrum of missions and users, from the military to the civilian dimensions. Finally, machines are gaining increasing levels of autonomy, paving the way for the autonomous use of force. This proliferation trinity, compounded by the growing weaponization of platforms, configures this moment as an "event horizon," or a point of no return, which will have disproportionate consequences that go beyond simple operational effectiveness, threatening to transversely affect all dimensions of human interaction. Such evolution towards the "age of autonomy" has the potential to change both the nature and the character of war.[40] This tipping point is most apparent in the inevitability of the hegemony of lethal AAS, and therefore requires society to understand its full ethical, legal, and political consequences.

## Defining Autonomy

Although its ubiquity, it is difficult to reach a universal consensus over the precise definition and taxonomy of autonomous systems.[41] The semantics' complexity increases when "autonomy" and "automation" are given similar connotations, and

---

(New York: Center for the Study of the Drone at Bard College, 2020). For additional analysis regarding drone proliferation, see also Bergen, Salyk-Virk, and Sterman, "World of Drones."
**38** USAF Flight Plan, *Unmanned Aircraft Systems Flight Plan 2009–2047*, 33. Further developments are refined in USAF, *RPA Vector*; USAF, *Small UAS Flight Plan*.
**39** Within the military domain, the sizes range from the combat proven Black Hornet with only 18g up to the US Navy's MQ-4C Triton with a wingspan of 130ft which is greater than a Boeing 757.
**40** Hoffman, "Will War's Nature Change," 19.
**41** Several academics, military and non-governmental organizations have different conceptual perceptions regarding autonomy. For diverse perspectives, see Scharre and Horowitz, *An Introduction to Autonomy in Weapon Systems*; Ilachinski, *AI, Robots, and Swarms*; UK Ministry of Defence, *Joint Doctrine Publication 0–30.2*; US DoD, *Unmanned Systems Integrated Roadmap FY2017–2042*; US DoD, *Summer Study on Autonomy*.

used interchangeably, thus confusing the nature and effects of autonomous and automatic systems. Whilst the latter systems have the ability to execute specific tasks in a predictable way, they miss the decision-making skills which enables an autonomous machine, free from human interference, to seek the optimal solution for a sequence of actions.

A simple way to define an autonomous weapon system could be as one that "once activated, can select and engage targets without further intervention by a human operator."[42] Yet, such human-centric definition does not translate the full complexity of this concept. Therefore, a more detailed construct of an autonomous system would be one that is able to "independently compose and select among different courses of action to accomplish goals based on its knowledge and understanding of the world, itself, and the situation."[43] However, all the features encapsulated in this possible definition need to be seen within a sliding scale, in which variables as the mission's complexity, the machine's interaction with the environment or the level of human-machine collaboration, will impact systems' performance and intricacy.

Following the US Department of Defense (US DoD) taxonomy as one of the most disseminated examples,[44] we can differentiate four basic levels of autonomy:

| Level 1 – Human Operated |
|---|
| The human operator makes all the decisions. The system has no autonomous control of its environment although it may have information-only responses to sensed data. |
| Level 2 – Human Delegated |
| The vehicle can perform many functions independently of human control when delegated to do so. This level encompasses automatic controls, engine controls, and other low-level automation that must be activated or deactivated by human input and must act in mutual exclusion of human operation. |
| Level 3 – Human Supervised |
| The system can perform a wide variety of activities when given top-level permissions or direction by a human. Both the human and the system can initiate behaviours based on sensed data, but the system can do so only if within the scope of its currently directed tasks. |
| Level 4 – Fully Autonomous |
| The system receives goals from humans and translates them into tasks to be performed without human interaction. A human could still enter the loop in an emergency or change the goals, although in practice there may be significant time delays before human intervention occurs. |

---

42 US DoD, *Directive 3000.09*.
43 US DoD, *Report of the Defense Science Board*, 4.
44 US DoD, *Unmanned Systems Integrated Roadmap FY2011–2036*, 46.

Through the lenses of the human interference in the machine's decision-making process to employ lethal force it is possible to highlight some impacts of the increasing autonomy levels.[45] From the most restrictive control type, "man-in-the-loop," reserves to humans the final decision to engage targets with deadly force. As we move up in the autonomy scale, machines gain the ability to select targets and the "man-on-the-loop," in a supervisory role, will sanction the ones that may be engaged. Reaching the apex of autonomy, machines possess the AI skills which allows them to select and engage targets with deadly force, relegating humans to an "out-of-the-loop" role. Within this framework, one needs to correlate the autonomy levels with the types of operational activities and functions in order to assess the magnitude of its effects and thus, determine the appropriate human intervention in the decision cycle.

## Glimpses of Future Autonomous Air Warfare

Roff and Moyes identified 256 weapon systems using varying degrees of autonomy, ranging from independent movement, employment of weapons, and the ability to autonomously modify or set goals.[46] Some of these systems have been in use for several decades, such as the Tomahawk Land Attack Missile able to strike targets hundreds of miles away without human interference.[47] The Israeli Iron Dome air defense system is capable of identifying, tracking, intercepting and destroying short-range threats such as rockets, artillery fire, and even drones. On the maritime domain, the Aegis weapon system is capable of tracking and engaging enemy air targets such as aircraft and even ballistic missiles providing a protective shield to specific high value assets. Other automatic-capable close protection systems such as the Phalanx are able to engage incoming high-threat targets increasing the survivability of maritime vessels. Other emerging examples include unmanned loitering munitions, such as the Israeli drone Harpy, able to carry beyond the line-of-sight lethal strikes, with growing autonomy levels.[48] In a more rudimentary

---

45 Sharkey, "Towards a Principle for the Human Supervisory Control of Robot Weapons," 305–324.
46 Roff and Moyes, "Meaningful Human Control, Artificial Intelligence and Autonomous Weapons."
47 This weapon system has been the weapon of choice to engage targets in deep enemy territory. In April 2017, 59 Tomahawk missiles were fired at a Syrian air base destroying several targets.
48 Gettinger and Michel, *Loitering Munitions*.

form, also anti-personnel mine or an improvised explosive device represent a basic form of autonomy.

Autonomous systems controlled by humans have proliferated to at least 30 states, mainly under defensive "man-on-the-loop" control types designed to provide protection of high value installations and assets against fast moving time sensitive threats.[49] However, the current level of autonomy is still low because man supervises machine's operation in real time and, if malfunctions or malicious actions occur, is able to override the systems. Nonetheless, a myriad of countries, including China,[50] Russia,[51] France,[52] Germany, Israel, Australia, the UK[53] and the US, are currently pursuing the development of systems with greater autonomy in combat situations.[54]

The US, as the leading developer and user of advanced military technologies, translates its future ambitions in a new iteration of an offset strategy. The *Third Offset Strategy*,[55] building upon the foundations for a disruptive operational template, centers the development efforts on human-machine collaborative-networked

---

[49] Scharre and Horowitz, *An Introduction to Autonomy*, 18; CNAS, *Autonomous Weapons & Human Control*, 4.

[50] Kania, *Battlefield Singularity*.

[51] The Russian chief of General Staff, Gerasimov, was quoted as saying that, "in the near future it is possible a fully robotized unit will be created capable of independently conducting military operations." Freedberg, "Robot Wars." Also, Vladimir Putin acknowledged that "future wars would be fought by countries using drones" and, when referring to AI, "Whoever becomes the leader in this sphere will become the ruler of the world." Vincent, "Putin Says the Nation That Leads in AI 'Will Be the Ruler of the World.'".

[52] France launched in 2003 a project for a technological demonstrator of an Unmanned Combat Air Vehicle (UCAV), designated nEUROn, in cooperation with Italy, Sweden, Spain, Greece and Switzerland.

[53] Current UK policy is clear stating that the "UK does not possess armed autonomous aircraft systems and it has no intention to develop them." However, "the UK's view is that increasing automation, not autonomy, is required to improve capability." UK Ministry of Defence, *Joint Doctrine Publication 0-30.2*, 42–43.

[54] Besides Nations, also International Organizations such as NATO are recognizing the nature and magnitude of this revolution and have already started assessing the potential legal, ethical and strategic impacts of lethal autonomous systems. As an example, see Kuptel and Williams, *Policy Guidance*; JAPCC, *Future Vector & Future Unmanned System Technologies*.

[55] Building upon two previous successful iterations, the Third Offset Strategy intends to replicate the military-technical advantage against a peer competitor. The first offset strategy, through nuclear deterrence in the 1950s, leveraged US nuclear superiority to compensate for the numerical inferiority of ground forces in Europe. As Soviet Union reached nuclear parity, the second offset strategy, precision guided technology in the early 1970s, leveraged US monopoly of advanced technologies to develop long-range precision strike weapons. Jackson, "The Pentagon's Third Offset Strategy."

environments aiming to gain military advantages against likely adversaries.[56] Therefore, it postulates that "advances in artificial intelligence and autonomy – autonomous systems – is going to lead to a new era of human-machine collaboration and combat teaming."[57] This aims to merge the tactical acuity of a computer to enhance human decisions with the employment of manned and unmanned systems.[58] Confronted with Anti-Access and Area-Denial[59] adversary weapons and tactics, the US seeks to develop the means to offset the proliferation of such advanced technologies.

Under the renovated offset strategy umbrella and leveraged by massive injections of funds, technological research proceeds at a fast pace unveiling the ascendancy of "performance" over "execution" in autonomous missions which will endow a machine with decision-making capability.[60] Transcending preprogrammed activity, the system self-decides how to operate itself to accomplish the human-directed mission goals, optimizing its behaviour and performance in unforeseen and changing environments. This enhancement in autonomy levels will ultimately set the stage for revolutionary operational concepts to emerge and induce a fundamental shift in the air warfare paradigm. For the time being, there is still a caveat to avoid delegating authority to machines for the employment of deadly force.[61]

However, advances in AI further diminish human interference in the tactical domain. In the context of air-to-air combat, time delays associated with the use of remote operation of unmanned platforms constrain critical decisions.[62] Aiming to enhance real-time decision-making capabilities, several proof of concept trials have demonstrated the ability to control multiple drones in air combat missions within a realistic simulated environment.[63] In a 2016 demonstration, the first

---

56 Work, "The Third U.S. Offset Strategy and Its Implications for Partners and Allies."
57 Work, "Art, Narrative, and the Third Offset."
58 Work, "Art, Narrative, and the Third Offset."
59 "If anti-access strategies aim to prevent US forces entry into a theater of operations, then area-denial operations aim to prevent their freedom of action in the more narrow confines of the area under an enemy's direct control. Area-denial operations thus include actions by an adversary in the air, on land, and on and under the sea to contest and prevent US joint operations within their defended battlespace." Krepinevich, Watts, and Work, *Meeting the Anti-Access and Area-Denial Challenge*, ii.
60 US DoD, *Unmanned Systems Integrated Roadmap FY2013–2038*, 66–67.
61 "The only time we will delegate a machine authority is in things that go faster than human reaction time, like cyber or electronic warfare." Work, "WATCH.."
62 Byrnes, "Nightfall," 49.
63 Ernest, et. al., "Genetic Fuzzy based Artificial Intelligence for Unmanned Combat Aerial Vehicle Control in Simulated Air Combat Missions."

human vs. AI encounter in the simulator resulted in a machine victory.[64] In a more complex and dynamic environment, DARPA's 2020 competition, "designed to demonstrate advanced algorithms capable of performing simulated, within-visual-range air combat maneuvering",[65] several AI agents fought virtual engagements against an experienced F-16 pilot. The machine won, unequivocally, all five virtual "dogfights" against the human. These promising results have prompted a contracts award, in November, to five companies to further develop AI agents to enable "mixed teams of manned and unmanned combat aircraft to conduct aerial dogfighting autonomously."[66]

Reaping the benefits of these AI breakthroughs, several trends are emerging which will shape the future character of air warfare. Following the human-machine teaming framework, the development of the "loyal wingman" operational template is underway and expected to deliver full operational capabilities early next decade. Besides the US and other western Allies, both China and Russia are keen to unleash and harness the benefits of human-machine teaming technologies.

In 2019, Russia successfully tested the loyal wingman concept during a flight of a Su-57 fifth-generation combat aircraft teamed with an S-70 Okhotnik UAV.[67] More recently, in 2020, unveiled the development of the Grom UAV as a loyal wingman platform that would provide integrated strike capability in advance of a force package of friendly manned aircraft.[68]

In a synergistic process, the USAF is developing the Skyborg project aiming "to deploy a modular, fighter-like aircraft that can be used to quickly update and field iteratively more complex autonomy to support the warfighter."[69] This AI-driven "computer brain" will establish the architecture which will allow the operation of an aerial combat ecosystem, ranging from a loyal wingman platform to fully autonomous unmanned combat air vehicles, and even migrating inside the cockpit of a manned aircraft, acting as a virtual co-pilot.[70] In December 2020,

---

64 Reilly, "Beyond Video Games."
65 DARPA, "AlphaDogfight Trials Foreshadow Future of Human-Machine Symbiosis."
66 Nurkin, "The Importance of Advancing Loyal Wingman Technology."
67 Trevithick, "Watch Russia's S-70 Unmanned Combat Air Vehicle Fly With an SU-57 For the First Time."
68 Butowski, "Russia reveals Loyal Wingman Concept."
69 Trevithick, "USAF Plans for Its 'Skyborg' AI Computer Brain to Be Flying Drones in the Next Two Years."
70 Trevithick, "Glitzy Air Force Video Lays Out 'Skyborg' Artificial Intelligence Combat Drone Program."

a contract was awarded to build loyal wingman-type platforms to start fielding AI-systems developed under the Skyborg program.[71]

These small incremental steps will allow the exploitation of increasing levels of autonomy, enhancing the integration of mixed control type assets within the same mission. In the short term, we may see fifth-generation manned fighters, such as the F-35 and the F-22, having its capabilities augmented by operating together with several robot wingmen, including unmanned versions of legacy jets. This operational template would offer additional benefits, increasing the overall mission effectiveness, while expanding firepower and survivability of the manned platform, which will have the responsibility to orchestrate an extended network of combat assets. The expansion of this concept to other mission types and platforms is also under scrutiny, by either adding to the team unnamed cargo and bomber platforms or providing supplementary sensors that could enhance the awareness of the mother ship.[72] Eventually, ultimate questions about the delegation of lethal authority to machines will continue to emerge.

Hence, the current trends that focus on employing attritable, reusable, low-cost, modular, AI-enabled UAVs will close the capability spectrum gap between expandable platforms and sophisticated systems, offering the possibility to enlarge mission sets, and improving the survivability and lethality in an ever more congested and contested peer competitor battlespace.[73] Simultaneously, a more worrisome operational template starts to emerge, in which drones, with fully autonomous capabilities, collaborate within a "swarm." Such disruptive innovations, that have been emerging in the civilian sector, are quickly, and inevitably, migrating into military applications.

In the civilian arena, China set in 2017 the world record for the largest drone swarm by displaying an autonomous choreography of 1,180 drones.[74] In 2018, this record increased to 1,374, and by the end of 2019, Chinese companies were able to simultaneously fly close to 2,000 drones. At the same time, the US

---

[71] Trevithick, "These Three Companies Will Build Drones to Carry the Air Force's 'Skyborg' AI Computer Brain."
[72] This example can also be applied to a Special Operations Aircraft that will use drones as scouts in order improve its survivability. Swarts, "Air Force Looking at Autonomous Systems to Aid War Fighters."
[73] Gunzinger and Autenried, "Understanding the Promise of Skyborg and Low-Cost Attritable Unmanned Aerial Vehicles."
[74] Lin and Singer, "China is Making 1,000-UAV Drone Swarms Now."

company Intel had flown 2,066 drones,[75] and Russia flew 2,198.[76] Meanwhile, several exhibitions in China set the current record on March 29, 2021, by flying 3,281 drones simultaneously.[77]

On the military front, the US DoD successfully unleashed in 2017 one of the world's largest micro-drone swarms, releasing 103 Perdix drones from three F/A-18 Super Hornets, in a demonstration of "advanced swarm behaviours such as collective decision-making, adaptive formation flying, and self-healing."[78] At a recent demonstration, in 2020, DARPA achieved another milestone of its Project Gremlins, in its quest to launch and recover dozens of cheap, reusable unmanned aircraft, specially tailored for coordinated swarm attacks or intelligence, surveillance, and reconnaissance missions.[79] The UK is also pursuing the benefits of such an operational template, with a futuristic and ambitious vision under the Alvina programme, which has "exercised swarms of over 20 ultra-low cost drones operating together against threat systems to brilliant effect."[80] China provided recently an example of similar innovations, by demonstrating the capability to deploy from a truck, in a few seconds, a swarm of 48 loitering munitions, aiming to blind, confuse, and overwhelm a target area.[81] Furthermore, as research and development progress, new applications towards miniaturization and autonomous precision guided targeting are being explored. Such "slaughterbots," a few centimeters long, are fitted with explosive warheads and offer the ability to engage singular targets based on specific biometrics.[82]

These novel "swarming" operational templates will offer increased effectiveness and resilience throughout a broad mission spectrum, including friendly

---

75 Weaver and Black, "Behind the Scenes as Intel Sets a World Record for Flying Over 2,000 Drones at Once."
76 Inside GNSS, "Two thousand simultaneous drones position themselves with GPS and GLONASS."
77 Zhan, "3,281 Drones Break Dazzling Record for Most Airborne Simultaneously."
78 According to the Strategic Capabilities Office Director, William Roper, "Perdix are not pre-programmed synchronized individuals, they are a collective organism, sharing one distributed brain for decision-making and adapting to each other like swarms in nature. Because every Perdix communicates and collaborates with every other Perdix, the swarm has no leader and can gracefully adapt to drones entering or exiting the team." The project's ultimate goal is to increase the swarm numbers up to 1,000 drones. US DoD, "Department of Defense Announces Successful Micro-Drone Demonstration."
79 Cohen, "DARPA's Gremlins Program Accomplishes First Flight."
80 Wigston, "The Chief of the Air Staff's Speech at the Global Air Chiefs' Conference 2021," Speech delivered at the Global Air Chiefs Conference, London, July 14, 2021.
81 Trevithick, "China Conducts Test of Massive Suicide Drone Swarm Launched From a Box on a Truck."
82 Welsh, "Air Operations at the Level of Boots on the Ground."

forces protection, wide area search and rescue, identification and tracking of multiple targets, suppression of enemy air defences, surface attack, electronic warfare, etc., enhancing both human decision-making as well as machine awareness of the environment. Within this framework, "a human can give general inputs and guidance to the swarm, and be confident that in general, relying on advanced computational capability, the swarm will behave as required."[83] Meanwhile, the well-publicized technology demonstrators continue to be secretly expanded into full-fledged military capabilities by a myriad of covert governmental projects.[84] Currently, such disruptive US developments are being turned over to military branches to support various follow-on programs of massive drone swarms.[85]

Autonomous systems confront humanity with increasingly intricate operational environments and conflicts widened by non-state actors' access to commercially produced technologies.[86] Currently, we can witness a gradual expansion of automated systems beyond traditional Nation state users.[87] ISIS exploits the versatility of drones by using them to capture images for propaganda as well as for transporting explosives.[88] Similarly, Hezbollah employed a surveillance drone to reconnoiter Israeli nuclear facilities, and in 2006 executed a land-to-sea attack against an Israeli naval vessel via a drone fixed with explosives; thus, exemplifying how non-state agencies have widened their operational boundaries through drones.[89] More recently, Yemen's Houthi rebels employed a coordinated attack

---

[83] David Deptula, cited by Colin Clark, "Artificial Intelligence Drone Defeats Fighter Pilot."
[84] Lamothe. "Veil of Secrecy Lifted on Pentagon Office Planning 'Avatar' Fighters and Drone Swarms."
[85] Trevithick, "Pentagon Unveils Details on Effort to Equip Its Services With Massive Swarms of Deadly Drones."
[86] Several drones are easily available and can be built in any home workshop using a 3D printer and some basic skills. These innovative techniques have also captured the interest of modern military organizations. For example, a new US Army project "aims to give soldiers the ability to 3D-print swarms of mini-drones to specific specifications within 24 hours. . . . Future capabilities could include combining 3D printing, drones, and artificial intelligence; . . . even allowing robots to 3D print themselves." Tucker, "US Army Looking to 3D-Print Minidrones in 24 Hours."
[87] Grossman, *Drones and Terrorism*.
[88] On 2 October, 2016, in Irbil, Iraq, a drone flown by ISIS killed two Kurdish soldiers and injured two French paratroopers. The attack is possibly the first where a drone has inflicted casualties on troops from a Western nation. Atherton, "IED Drone Kills Kurdish Soldiers, French Commandos."
[89] Shkolnik, "The Drone Threat to Israeli National Security."

with ten drones against two key Saudi Arabia's oil installations.[90] The prospects of non-state actors and even individuals employing swarms of autonomous drones will disrupt the traditional view of threats, with a catastrophic impact in air warfare and the associated collateral damages. Such operational vignettes have led some authors to classify armed autonomous drone swarms as weapons of mass destruction, based on its potential harm and inability to discriminate between military and civilian targets.[91]

In summary, command and control considerations work against human supervision of drones and in favor of autonomy. First, human frailties such as weariness and error undermine precise collaboration between man and machine. Second, communications links pose a vulnerability, which can be jammed, hacked, or hijacked by hostile parties.[92] Additional technical difficulties and congested electromagnetic spectrum arise from the need to have sufficient bandwidth to sustain increasing needs for real-time video, while ensuring communications to control both the platform and the payload.[93] Further, swarming strategies involve more devices than available operators. Thus, technical factors alone are enough to make the case for fully autonomous systems.

Aerial battlespace dynamics exceed the human capacity to digest information and physically react. Hence, AI and autonomy make tactical air combat deadlier.[94] Even though, for the near future, AAS will not replace manned aviation, they will definitely continue to expand their operational functions; either in collaboration with traditional aircraft or, operating by their own, with various levels of human integration, in specifically tailored scenarios. In order to fully exploit the operational benefits of these emerging concepts, it is necessary to create an adequate "Air Manned-Unmanned Battle" doctrine which orchestrates the appropriate capability vectors within multidomain engagements. Such developments will transform force structures and the ways the military organize, train and procure equipment to implement the human-machine teaming operational templates. Ultimately, such enhancements in military effectiveness, through the employment of swarms of thousands of small, low-cost, expandable autonomous

---

90 Hubbard, Karasz, and Reed, "Two Major Saudi Oil Installations Hit by Drone Strike, and U.S. Blames Iran."
91 Kallenborn, *Are Drone Swarms Weapons of Mass Destruction?*.
92 Several documented reports include the intercept of video feeds from US drones flying over Iraq in 2009 by Iranian-backed militias using $26 off-the-shelf software, or in 2011, reports that a virus had infected some drone control systems at Creech Air Force Base, Nevada. Foust, "Soon, Drones May Be Able to Make Lethal Decisions on Their Own."
93 USAF Flight Plan, *Unmanned Aircraft Systems Flight Plan 2009–2047*. 43.
94 Byrnes, "Nightfall," 49.

systems will provide more time, and willingness, for humans to focus on the ethics of war.[95]

## Impediments and Challenges to the Emergence of Autonomous Air Warfare

The aforementioned discussion demonstrated the need for systems with increasing degrees of autonomy. However, this desire is not free of obstacles and boomerang effects which will influence the coming revolution. Viewing reality through technophilic lenses, we are on the verge of achieving the levels of AI that will allow complete autonomy of drones, and with it, the employment of lethal force without human intervention. However, for most technophobes, despite the technological evolution in this field there are several limitations that make it impossible, for now, to develop fully autonomous systems. Besides the technical issues or the associated cost, organizational and cultural barriers are also strong impediments for the full-fledged adherence to lethal autonomous systems.[96]

Turning to history as a useful and wise guide, we should be reminded that the introduction of novel weapons into battle, without proper experimentation of its effects nor evaluation of its morality, can transform the character of warfare. As in previous cases, the operational employment of lethal autonomous systems will be dependent of the technological evolution as well as the society willingness to accept the delegation of lethal decision to machines. As always, motivated by strategic and operational necessity, humanity has been able to overcome warfare's technological challenges, thus requiring the focus of our attention to the discussion of possible effects of such revolution. This debate is premonitory of the profound impact that the AAS will have in the coming decades, causing a qualitative transformation of human interference in the conduct of war.

---

**95** Brose, "The New Revolution in Military Affairs," 131–132.
**96** The debate about organizational and cultural barriers is framed in Mary Cummings, *Artificial Intelligence and the Future of Warfare*.

## Compliance with Universal Legal and Ethical Standards

Removing humans from the equation quickens the decision cycle, magnifies destructive capability, enhances intelligence gathering along with cooperation with allied autonomous systems, and may preserve civilians' lives and property by reducing or eliminating pilot error. Long-range autonomous systems capable of independently navigating, identifying and attacking mobile targets offer a major conventional deterrence, particularly in scenarios dominated by opposing anti-access and area denial tactics.

The human dilemma facing adoption of AAS is this: does increased precision and speed promise to reduce indiscriminate loss of life, both combatants and non-combatants? Or, is the abdication of the human monopoly on the application of violence an irreversible moral lapse? These ethical concerns are inseparable from the potential economy and efficiency of autonomous systems.[97]

Ronald Arkin, one of the most prominent researchers in the field of AI, argues that robots can be more human in battle than the soldier himself.[98] The quest to humanize war led man to create a set of rules to criminalize those who behave beyond the acceptable international standards. Nevertheless, in the heat of battle, we witness a constant violation of these rules, by either fear, frustration, revenge or the desire to win with no regard to costs. According to him, the development of machines that do not express these emotions, that behave more humanely in action than man himself, and that adhere to the laws of war better than the soldiers themselves, can lead, ultimately, to a reduction of non-combatant casualties which plague hostile conflict. Within this framework, he argues for the implementation of an ethical control system that can regulate the lethal actions of robotic autonomous systems, concluding that it is possible to frame its behavior within the range prescribed by the laws of armed conflict and rules of engagement.

This ability to discern the legitimacy of the targets and apply proportional lethal force implies and is dependent on technological development. In this perspective and in theory, autonomous systems would allow for better compliance with legal and ethical standards of war. However, the technological accomplishment of the essential ethical and legal requirements still seems to lie within the

---

[97] Larkin, "Brave New Warfare: Autonomy in Lethal UAVs," 26–27.
[98] Arkin, *Governing Lethal Behavior in Autonomous Robots*.

field of fiction, as the compliance with the distinction[99] and proportionality[100] principles is particularly complex and ambiguous. Even for the sophisticated human judgment.

Contrary to the view of technophiles, Noel Sharkey highlights the insufficient discrimination between combatants and non-combatants and the lack of proportionality of the response, as main dissociative factors to the emergence of lethal AAS.[101] Compounding the human dilemma is the proliferation of asymmetric warfare and hybrid conflict, for example the kind of combat occurring in Syria where prediction of a party's intent and identification of their affiliation (if any) cannot be discerned.

With regard to proportionality, it will be difficult to calculate objectively and quantitatively what is a proportional response. At this moment, there is no metric that can objectively quantify the superfluous, unnecessary, and disproportionate suffering.[102] This task still requires human judgment. Therefore, we are still far from reaching a technological maturity that allows AAS to successfully pass the "innocent distinction test," and set an intuitive logic that can be programmed to manage the application of lethal force.[103]

By helping to eliminate human fear, frustration or anger from the equation, AAS systems may transform fighting into an act of greater rationality. However, autonomous systems can be programmed to act according to ethical standards, the reverse can also happen, turning machines into ruthless killers. Moreover, the loss of autonomous systems and their capture by the adversary could facilitate the proliferation of this technology. And here lies one of the main concerns: how long will this technology be owned by a limited number of states? And what will happen when non-state actors acquire these capabilities?

Although machines could be programmed with a code of conduct to comply with universal ethical standards, there is still an ultimate dilemma. If the machine acts contrary to the programmed instructions, violating the laws of

---

**99** Rule 1 of customary international humanitarian law: "The parties to the conflict must at all times distinguish between civilians and combatants. Attacks may only be directed against combatants. Attacks must not be directed against civilians"; ICRC, "Customary IHL Database."
**100** Rule 14 of customary international humanitarian law: "Launching an attack which may be expected to cause incidental loss of civilian life, injury to civilians, damage to civilian objects, or a combination thereof, which would be excessive in relation to the concrete and direct military advantage anticipated, is prohibited"; ICRC, "Customary IHL Database."
**101** Sharkey, "Weapons of Indiscriminate Lethality," 26–29.
**102** Rule 70 of customary international humanitarian law: "The use of means and methods of warfare which are of a nature to cause superfluous injury or unnecessary suffering is prohibited"; ICRC, "Customary IHL Database."
**103** Sharkey, "Weapons of Indiscriminate Lethality."

conduct, the answer is obvious: reprogram it, or ultimately, destroy it. However, should this happen to a human, he would be judged since only him could be held liable. Brown summarizes this dilemma by stating that although we teach robots to kill, we do not teach them to commit murder.[104] This author asserts that only humans can commit a war crime, and only they can be blamed. The liability issues in case of error are assumed vitally important, since the fine difference between an accident, an error and a crime can be reduced to intention. Considering that free will means the ability of a drone to search, locate, identify and attack a target without human interference then it is the factor of intentionality that transforms autonomous systems in such a terrifying vision.

Therefore, as we move towards the reality of full autonomous systems, and as a consequence displacing human interference further away from the decision cycle, it becomes even harder to ensure accountability and determine who or what to be held liable for mission failure, breach of rules of engagement or unintended casualties. It is a technological inevitability that malfunctions will occur with catastrophic impact on the performance of the autonomous system. Such risk is aggravated whenever the introduction of weapons into combat occurs without thorough testing, and as the result of urgent operational requirements.[105]

Hence, accountability, controllability, and moral responsibility demand that lethal autonomous systems are subject always to "meaningful human control."[106] Scharre and Horowitz highlight three essential components, or "minimum necessary standards," that ensure better informed decisions and actions, while reducing the potential for mistakes:

> (1) Human operators are making informed, conscious decisions about the use of weapons; (2) Human operators have sufficient information to ensure the lawfulness of the action they are taking, given what they know about the target, the weapon, and the context for action; (3) The weapon is designed and tested, and human operators are properly trained, to ensure effective control over the use of the weapon.[107]

---

104 Brown, "War Crimes and Killer Robots."
105 Peter Singer reminds us of some of these tragic but educational moments, usual to the initial processes of technological innovation, and exposing the potential dangers of excessive confidence in the autonomy of weapon systems; Singer, *Wired for War*, 196–199.
106 Topic first addressed by the non-governmental organization 'Article 36.' Article 36, "Killer Robots." For a comprehensive discussion about the "meaningful human control" concept, its strengths and weaknesses, as well as other conceptual and policy-oriented approaches that address these concerns, see UNIDIR, *The Weaponization of Increasingly Autonomous Technologies*; Scharre. *Army of None*.
107 Scharre and Horowitz, *Meaningful Human Control in Weapon Systems*, 4. However, there are other interpretations regarding the essential principles for meaningful human control which

Thus, assessing lethal autonomous weapon systems' compliance against such cumulative principles could ensure legitimate target selection and proportionate response, while guaranteeing a sufficient framework of human accountability throughout the use of lethal force.[108] However, achieving a high degree of "meaningful human control" over autonomous machines will require a tailored amount of information that is scenario driven and dependent on multiple variables and indicators. Hence, contextual awareness about weapons' employment will be required in order to determine the most suitable human intervention in the decision cycle. As an example of such contextual diversity, it will be almost impossible to compare the meaningful human control requirements between an air-to-air engagement, where most elements of the kill-chain (search, identify, track, target, engage and assess) are heavily reliant on a networked information environment, with an urban scenario where an infantry soldier is engaging an enemy combatant in close proximity.[109]

Implicit among countries competent in the utilization of autonomous systems is a concurrence that humans ultimately must retain control of machines that kill.[110] This consensus is also reflected in studies and resolutions produced within the United Nations.[111] More explicitly, the US DoD Directive on Autonomy in

---

may increase the complexity of its implementation. The International Committee for Robot Arms Control establishes that, "[f]irst, a human commander (or operator) must have full contextual and situational awareness of the target area and be able to perceive and react to any change or unanticipated situations that may have arisen since planning the attack. Second, there must be active cognitive participation in the attack and sufficient time for deliberation on the nature of the target, its significance in terms of the necessity and appropriateness of attack, and likely incidental and possible accidental effects of the attack. Third, there must be a means for the rapid suspension or abortion of the attack." Garcia, "Technical Statement by the International Committee for Robot Arms Control Convention on Conventional Weapons Meeting of Experts on Lethal Autonomous Weapons Systems." Roff and Moyes also offer the cumulative principles for meaningful human control, such as: "[S]ufficient predictability, reliability, and transparency in the technology, sufficient confidence in the information that is guiding the human judgments being made, sufficient clarity of human action and potential for timely intervention and a sufficient framework of accountability"; Roff and Moyes, "Meaningful Human Control," 6.
108 Roff and Moyes, "Meaningful Human Control," 6.
109 Scharre and Horowitz, "Meaningful Human Control," 12.
110 Scharre, "Statement to the UN Convention on Certain Conventional Weapons on Way Ahead."
111 The UN debate on Lethal Autonomous Weapon Systems, at the Convention on Certain Conventional Weapons, has focused on analyzing the risk of employment of lethal autonomous systems. Some authors suggest going beyond the discussion on procedural aspects, such as human control, and focus instead on the mitigation of inadvertent engagements; Lewis, *Redefining Human Control*.

Weapon Systems[112] embodies the first publicly disclosed policy regarding the use of lethal autonomous systems. It establishes guidelines, assigns responsibilities for the development and employment of these weapons, in order to minimize the probability and consequences of failures in autonomous and semi-autonomous weapon systems that could provoke unintended engagements and gratuitous collateral damage contrary to the laws of armed conflict, rules of engagement, and commander's intent.[113] Such policies encourage the use of autonomous systems for non-lethal missions, with the caveat that all options for AAS are on the table in crisis situations.[114] Emergencies, however, result from unforeseen scenarios, and one cannot experimentalize for something not yet conceptualized. While this US Directive provides international guidance for the development and employment of autonomous systems, it cannot presume to solve the human dilemma posed by this military revolution in terms of ethics and, frankly, the abdication of the human monopoly on the application of deadly force. The laws of war in general, and most relevantly international humanitarian law endeavour to temper destructiveness according to humanistic criteria such as compassion, common sense, and judgment. Although rather effective when performing quantitative assessments, they have as yet limited qualitative abilities to fathom the nuances of human moral values and civilized sensibilities.[115]

Urban environments pose complications in identifying enemies for machines as they do for human combatants, especially autonomous systems employing swarming tactics. Can AAS recognize and respond to situations which international humanitarian law might classify as inflicting "superfluous injury" or "unnecessary suffering"? Given these perplexities, including the "unintended" consequences of the recent employment of remote operated drones under the operational template of "targeted killings," ethical and operational problems compound as systems are liberated from human supervision. No matter how resilient the AI may be, it is impossible to predict how swarms of systems will behave when confronted with each other. Autonomous systems guided by AI hasten the tempo of combat and may, perhaps exponentially increase the frequency, scale, proportionality and sheer destructiveness of warfare. Given contemporary forms of political and military organization, including asymmetric war by non-state actors (some of whom may be isolated individuals) we enter a new era of affixing responsibility for atrocity and predacious actions. Therefore,

---

112 US DoD, *Directive 3000.09*.
113 US DoD, *Directive 3000.09*, 15.
114 US DoD, *Directive 3000.09*, 3.
115 Heyns, *Report of the Special Rapporteur on Extrajudicial, Summary or Arbitrary Executions*, 10.

currently, and in the near future, it is reasonable to say that the subjective nature of morality seems difficult to codify in software.[116]

Such dilemmas and paradoxes are spurring a worldwide debate and awareness about the proliferation of lethal autonomous weapons, thus leading a growing number of citizens, researchers and policy makers to pledge for a ban of fully autonomous weapons, trying to prevent a future vision which would dangerously destabilize every country and individual.[117]

Alongside this, the debate within the United Nations has been slowly progressing, from 2014 to 2019, although without consensus, exposing nations' concerns about the requirement to continue to retain meaningful human control over the use of force.[118] Meanwhile, a recent UN report on Libya, which implied the wartime debut of lethal autonomous weapons systems, a Turkish Kargu-2 drone, further fueled the debate about the rise of autonomous warfare, thus reinforcing the requirement to regulate this unstoppable proliferation.[119]

## Political Ambivalence (Seduction/Resistance) to Lethal Autonomous Systems

The earliest definitions of military revolution ascribed its profundity to the degree it reconstructed the relationship between the military and its political framework, ultimately transmogrifying the state.[120] Politically, AAS makes accessible national aspirations by reducing economic costs, and distances the horrors of war from politicians and the public, as well as soldiers. Targeted killings have proven less offensive to democratic societies by insulating the population waging war from the participation of its citizens and the public spectacle of men in battle

---

116  Asaro, "Ethical and Societal Perspectives on the Development of Autonomous Weapons," 219.
117  Currently, 247 organizations and 3,253 individuals have signed a pledge requiring strong international norms, regulations and laws against lethal autonomous weapons, on the moral argument that the decision to take a human life should never be delegated to a machine; Future of Life Institute, "Lethal Autonomous Weapons Pledge."
118  The results of the discussion held in the United Nations' annual meetings is encapsulated in: United Nations, *Draft Report of the 2019 Session of the Group of Governmental Experts on Emerging Technologies in the Area of Lethal Autonomous Weapons Systems.*
119  Choudhury, *Final Report of the Panel of Experts on Libya Established Pursuant to Security Council Resolution 1973.*
120  Palmer, *Autonomous UAS*, 12.

communicated through mass media.[121] Autonomy extends the user's base, transforming lethal autonomous systems into an affordable solution to those willing to engage in targeted killings anywhere in the globe. State sovereignty becomes increasingly vulnerable to incursions and, correspondingly, governments more susceptible to paranoia. Thus, transferring the use of force to machines increases the risk of "dehumanizing armed conflict even further and precludes a moment of deliberation in those cases where it may be feasible."[122]

The West's growing reliance upon unmanned aircraft systems, and the development of AAS colors society's percipience of war. War can be sanitized and rendered largely invisible in a brave new world free of conscription, press releases divulging casualties, and mounting defense budgets. The decampment of living soldiers from the field of battle, along with diminished public awareness of the carnage of war through mass media, coupled with the cost-effectiveness of future war anesthetizes the human conscience. Humankind is less capable of the compassion and restraint that the World Court, the United Red Cross and the United Nations have advocated in restraining gratuitous bloodshed. Not only would humans abdicate their primordial monopoly on the infliction of violence, but potentially they would be immune to the suffering of others.

One could argue these developments accentuate what Edward Luttwak dubbed as "post-heroic war."[123] Despite modern armed forces being structured for large-scale war, the demographic base of the advanced post-industrial societies, with small families, reduces tolerance for casualties.[124] This stagnation (and even decline) of the population in western countries, as well as the gradual distancing of society from the military values, constitute compelling arguments for the increasingly selective use of human resources in war in the face of the risk posed by the future operational environment. However, acceptance of casualties in combat depends on the interests at stake, the perception of the importance of war and even the ability of political leaders to shape public perceptions about the rationales for conflict.

---

**121** Regarding the topic of 'Targeted Killings' we can find a wide range of references that address the legal arguments (pro/con). Besides the governmental references, for an in-depth discussion, see Anderson, *Targeted Killing in U.S. Counterterrorism Strategy and Law*; Fisher, "Targeted Killing, Norms, and International Law," 711–758; Melzer, *Targeted Killing in International Law*; Raemdonck, "Vested Interest or Moral Indecisiveness?" *IAI Working Papers 1205*; Solis, "Targeted Killing and the Law of Armed Conflict," 127–146. For an appraisal involving the use of long-range armed drones in targeted killing, see Davis, McNerney, and Greenberg, *Clarifying the Rules for Targeted Killing*.
**122** Heyns, *Report of the Special Rapporteur*, 17.
**123** Luttwak, "Towards Post-Heroic Warfare," 109–122.
**124** Luttwak, "Dead End," *Harper's Magazine*, 33–42.

Therefore, autonomous warfare template aggravates the ongoing distancing from what Luttwak has called as the eternal realities of combat – mortal sacrifice, bravery, fear and cohesion, morale and leadership.[125] Even the elementary fact that war is about killing and being killed suffered an irremediable unilateral change. Checking the historical evolution, one can see that war has become increasingly remote to Western societies. Since it became a professional activity, war has been privatized, using private military companies and mercenaries, transforming itself, figuratively and literally, in a remote war.[126] Additionally, the indifference that drones cause in society, transforming the citizen-soldier in a mere spectator, and war in another televised spectacle that competes for share with other recreational activities, lead to think that the public scrutiny of this vital activity may have diminished, and with it, to favor the frequency of war.

Public acceptance of autonomy depends mainly on safety concerns and the perceived peaceful nature of this technology. As the latter matures and more civilian and military AI applications are introduced, so will popular trust increase and with it, acceptance of increasing degrees of autonomy. Public acceptance began with the civilian domain technologies and it will gradually extend to military applications. Urgent operational requirements, like those of the Afghanistan and Iraq conflicts, will catapult armed drones, consequently accelerating the development of lethal autonomous systems to a point of no return.

In summary, the emerging tri-dimensional proliferation (vertical, horizontal, and qualitative) of autonomous systems will develop horizontally between states and vertically, from states to non-state actors, leading to greater "democratization" of war, both increasing its frequency an adding a new portfolio of effects to warfare. Thomas Hammes argues that the advances of commercial off the shelf technologies have set the stage for the proliferation of "small, smart, cheap, and long-range drones capable of carrying significant payloads."[127] The author further advocates that due to increased standards for target discrimination, West will be inhibited from employing cheap lethal AAS; thus, providing an incentive for less technological advanced actors, with lower ethical constrains, to field lethal AAS, allowing them to have an initial advantage and affect a broader spectrum of targets. This will be of particular concern for rogue nations, non-state actors or even single individuals who are likely to seize the

---

**125** Luttwak, "Post-Heroic Warfare and its Implications," 132.
**126** Engelhardt, "How Drone War Became the American Way of Life."
**127** Hammes, "The Future of Warfare: Small, Many, Smart vs. Few & Exquisite?" Hammes highlights that the "dramatic improvements in the fields of robotics, artificial intelligence, additive manufacturing, biology, and nano-materials are changing the cost/effectiveness calculation in favor of the 'many and simple' against the 'few and complex.'".

initiative, employing autonomous systems in novel operational templates, including terrorist attacks.[128] Consequently, the use of such systems could become a future technophilic alternative, free of sacrifice, to the suicide bomber. Again, the reversal of polarity in affordability of autonomous destructive power most likely spawns greater chaos, not a peace or a monolithic stability enforced by sovereign states.

Therefore, the West will be confronted with an emergent paradigm shift, "from the exquisite and very few to cheap and very many,"[129] and as a result, creating additional incentives for the most advanced states to preventively develop lethal autonomous systems while they can still maintain an asymmetric advantage. Within this revolutionary warfare paradigm, quantity will, once again, have a quality all of its own.

The emerging robotics' revolution in warfare threatens the current world order with an uncertain and perhaps nihilistic future. The painful reality is that insisting a human being remain in decision-making position regarding the use of force limits the development of potentially useful warfighting technologies.[130] The maturation of autonomous systems is a process, not a moment in time. This unique military revolution must be reconciled with its inevitable transformation of the society which spawned it.

As the path towards autonomy and complexity progresses, so does the risk of technical failures, jeopardizing machines' ability to comply with the discrimination requirements. Such failures could be acceptable when discussing defensive autonomous engagements. However, this legal and ethical framework barely stands against the complexity and fluidity of offensive engagements, especially in surface environments. Offensive missions in urban environments are likely incapable of effective AAS engagement whilst obeying absolutely and flawlessly to current discrimination requirements.[131] Thus, prior evaluation of weapons before fielding them would reject offensive lethal autonomous systems, given rigid contemporary standards of legal and ethical application of force.[132] As military revolutions and RMAs go, the human dilemma of AAS inhibits a steady ascendancy of adoption. The pragmatic reciprocation between technological advance and

---

128 Krishnan, *Killer Robots*, 155.
129 Hammes, "The Future of Warfare."
130 Under this framework, "US leaders should support development of operationally effective lethal autonomous weapons systems now with the dual objectives of maintaining strategic capability overmatch today and participating in eventual arms control negotiations about these systems from a position of strength." Hill and Thompson, "Five Giant Leaps for Robotkind."
131 Boothby, "Automated and Autonomous Weapons:," 207.
132 Boothby, "Automated and Autonomous Weapons," 207.

legal/ethical policy imposes developmental stages. What is militarily possible must reconcile with what is ethically permissible, though non-state actors will not be bound by sovereign state conventions. This synthesis has a determinist outcome, making the proliferation of lethal autonomous systems unavoidable. When this happens, we will be faced with a true military revolution.

## Conclusion

> There is a real possibility that, after many false starts and broken promises, a technological tipping point is approaching that may well deliver a genuine revolution in military affairs.
> Joint Doctrine Note 2/11: The UK approach to Unmanned Aircraft Systems. London: United Kingdom, Ministry of Defence, 2011, iii

The changing nature of man's relationship with the machine pervades air power's short history. From the very first moment when man helped propel airborne a heavier than air flying apparatus, rapidly took control of the machine and begun exploring the combat potential of such disruptive capability. Spurred by technological innovations, strategic utility and operational effectiveness, machines have been gradually acquiring higher levels of performance threatening to push man further away from the cockpit into supervisory roles. In hindsight, the changing human-machine relationship promises to gradually redefine man's role in war from the physical domain to the cognitive and ethical realms.

The eclipse of millennia of human-fought wars by autonomous systems guided by humans and/or AI proceeded out of the RMA of the 1990s. Like the latter series of developments, the warfighting transformation that is the subject of this chapter resembles the original military revolution as defined by Roberts and Parker. In terms of transforming strategy, affecting the state (in this case via fiscal and policy mutations), and ultimately upon society by simultaneously making warfare more destructive to those who suffer its wrath, yet palatable to the populations waging war. Thus, the emergence of autonomous air warfare has spurred a fundamental transformation of war. Such innovative operational templates leveraged by human-machine teaming will augment operational effectiveness and lethality, shaping the future fighter prototype, changing human preeminence in war, and as a consequence, reducing human interference both at tactical and political levels. Ultimately, this revolution will bring with it a redefinition of the human role in air war: from a practitioner, to a supervisory role, and ultimately to a mere observer, hopefully reserving for himself the final authorization to use lethal force.

The military revolution posed by autonomous systems (in their many facets) embodies a point of no return unveiled by a vertical, horizontal and qualitative proliferation of drones. The diversity of forms and sizes, extending the spectrum of missions and the user base, and increasing levels of autonomy on par with the prospect of weaponization and autonomous use of force, result in disproportionate consequences. The entire world faces the dawn of an RMA with epic implications across the nature of conflict and in the very essence of human existence. Dehumanization of air power results from new capabilities hitherto unthinkable before man inhabited a cockpit. Deciding what will be feasible, appropriate and acceptable in a future air strategy in which the preeminence of AAS will be even broader than today, raises unprecedented dilemmas for those who decide when and how to wage war.

The operational reasons for the migration towards autonomous systems are obvious. The need to perform increasingly complex and risky military missions with lower costs (human and economic) determines the interest in autonomous weapons systems. This is because the differential between recruiting, training and sustaining the human element is higher when compared with the alternative. The human factor thus becomes the main obstacle, since his direct intervention in the operation and exploitation of drones prevents the maximization of the system capability.

The inexorable proliferation of swarming strategies and technologies threatens with obsolescence human controllers who must contend with (1) cognitive exhaustion, (2) an exponential increase in information, (3) escalation of the tempo and frequency of operations, (4) the narrowing window of time allowed for humans to make life-and-death decisions, and (5) swarming tactics that require all-inclusive and instantaneous management of many platforms.

Slowly, man will abandon the execution function (in-the-loop) to oversee the behavior of the machine and authorize the use of lethal force (on-the-loop). With the increasing speed of the decision cycle, one can anticipate a future in which the human reaction time is not appropriate to the conduct of air warfare. Humans will monitor the cycle execution, performed at the machine's speed. Largely, machines, programmed to follow the commander's intent, will gradually take responsibility for combat decisions, while man supervises operations. Inexorably, humans become the lowest common denominator within an autonomous system. If (or when) that happens, human interference in the conduct of war will be of mere observation (out-of-loop). Under this framework, it is at least desirable that man may remain part of a wider loop, ensuring the definition of governing laws of the autonomous systems which could bound machines' behavior and keeping to himself the ultimate prerogative on whether to allow machines to use force. So, the emergence of lethal autonomous systems constitutes a singular military revolution that blurs the line between weapons and warriors.

Viewing war as the continuation of politics, consideration must be given to the emergence of AAS and their role as catalysts for warfare proliferation, where the operational, social and political costs of wielding the military instrument in pursuit of national objectives are severely reduced. Such framework would imply a reduction of political accountability, as a result of a growing disconnection of society, which in turn could transform preventive military responses as a seductive political option.

Additionally, by employing an autonomous system in geopolitical environments of great sensitivity, we will exacerbate the political and military risks, to the extent that a technical failure may cause an inadvertent escalation of the adversary's response. AAS are not fail-safe, a troubling characteristic that has been explored in the deployment of nuclear weapons. Like nuclear weapons, autonomous systems possess a psychological advantage in that they extend destructive force to almost any corner of the globe at almost any time.

However, it is the irresistibility of its surgical, non-apocalyptic character, at reduced costs, which turns the proliferation of autonomous systems irreversible, threatening to increase hostility and danger of the future environment, already complex and adverse. It is precisely the prospect of a "risk-free warfare," which may increase the frequency of war. Instead of seeking ways to eradicate the problems that lead to war, we are witnessing a trend which takes man further away from the battlefield, and ultimately, from the consequences of his actions. By providing their proponents the opportunity to engage in perpetual vigilance and selective attacks on any part of the globe, discreetly and without risk to human life of the offender, allows states to favor preventive military postures in conflict situations.

Conversely, considering the human behavior variables in war some may argue that autonomous systems can contribute to an increase of ethics on the battlefield, where human limitations on combat effectiveness, including susceptibility to error, fear, and the desire for survival, do not limit the performance of the machines. However, it is precisely this subjective nature of morality that seems difficult to codify in software.

At one end of the spectrum, the operation of lethal autonomous systems seems entangled in legal, moral and technological constraints, hardly soluble in the medium term. However, as noted in the evolution of manned air power and slowly replicated in unmanned capabilities, the natural tendency in the development of autonomous systems will occur primarily in the areas of reconnaissance and surveillance, naturally progressing to more complex and dangerous activities, as the technology matures and confidence increases.

Considering the operational needs of such systems within an incremental development framework, it is possible to foresee a hypothetical evolution. Public and political acceptance can be improved by restricting the employment of AAS in areas with low probability of collateral damage or using to the maximum extent possible non-lethal ordnance. Concomitantly, adequate "meaningful human control" protocols must be implemented, allowing commanders to enforce the necessary span of command over AAS and retain the ability to select the appropriate level of autonomous operation.

However, the emergence of lethal autonomous systems will be primarily driven by political endorsement rather than by purely technological achievements. For now, there is still a consensus that humans should continue to maintain control over the use of deadly force, ensuring accountability over the use of autonomous machines. Additionally, it is generically agreed that international humanitarian law should govern such systems and that commanders would have command responsibility for their robots, as they do for their soldiers. Hence, at this stage, human-machine teaming appears to provide the desired solution. Nevertheless, as the technological equalization emerges between adversaries, so will increase the drivers to develop fully autonomous systems. Eventually, as AI progresses and autonomous machines proliferate, social and political constrains will diminish, opening the way for the rise of autonomous warfare, where humans could become an operational liability when engaging technological advanced adversaries.

Considering the analysis, we now have sufficient facts that allow us to qualify with greater certainty the nature of change. We have adopted the concept of RMA as a paradigm shift that leads to the obsolescence of traditional skills of military organizations. Therefore, it reflects a fundamental change in the structure and operation of military organizations as the result of drastic changes in the variables of war. Hence, the increase in military effectiveness is normally associated with a triad of new technologies employed using innovative concepts and operations, and supported by organizational change.

RMAs, like military revolutions, rarely render accustomed weapons systems entirely obsolescent, or inoperable, or even redundant. The development of ordnance in the thirteenth, fourteenth, fifteenth, and sixteenth centuries (initially for siege but over time also for battle) did not lead to the extinction of armored horsemen. Nor did the evolution of handguns displace even within a century the operational efficacy of archery. The introduction of the aircraft carrier did not imply the withdrawal of the battleships from service but relegated them to auxiliary functions of bombardment of the coast and escort. Imbricating weapon systems do frequently force accustomed and even prevailing military technologies out of some roles and into others. This is abundantly true with AAS, which reduces the need for manned aircraft in surveillance and reconnaissance functions, precision strikes, and (in the immediate future) also in air-to-air combat. Though the autonomous operational template did not eliminate the fundamental competence of manned aviation, it promises to alter profoundly the character, the lethality, the political utility, and transversely, the experience and role of man in air war.

This transformational change in the application of air power transcends the mere operational domain and affects, in a multidimensional manner, war itself, as an expression of human interaction. In this perspective, the term revolution does not qualify a rapid change but the magnitude of its effects. Thus, what has started as an embryonic RMA of "precision violence" has the potential to transform itself into a genuine military revolution translating an all-encompassing change that has a multidimensional impact on human endeavor.

RMAs and the military evolutions that preceded them are more profound than simply adapting more efficient gadgetry. An authentic military revolution transforms organizational culture. The spread of AAS technology seems inevitable and predestined. However, the proliferation of autonomous systems collides with the organizational culture of an independent Air Force focused on pilots as fighters and leaders. In an age of austerity in which the changes are more easily justified, it is not unreasonable to question the need for military pilot skills. Ultimately, a future full of autonomous systems will lead to total irrelevance of the fighter pilot. Still, this trend is part of a broader revolution in the actors and the means of future war, in which the effects of cyber-warriors overlap the traditional physical contact amongst modern fighters.

History has proven that ethical and moral barriers to weapons' employment can be surpassed by urgent operational requirements. Such conundrum will drive the efforts to field systems which further expand the distance of engagements, while reducing the physical risk to combatants. Technological innovation and trust on the machine's decision-making ability will be the catalysts for the coming age of autonomous air warfare. Meanwhile, the prevailing operational templates are likely to combine manned and autonomous platforms, while still ensuring the continuous and desirable human interference and interaction in the lethal matters of war.

A future filled with lethal autonomous systems represents a paradigm shift in terms of the use of force. Its employment may fundamentally change the nature of war dynamics, irrevocably transforming the strategic cultures of states. When that happens, we will be in the presence of an RMA of epic proportions. This will be part of an emerging technophilic futuristic vision, poised to materialize within the next decades, with more comprehensive and profound effects, in which the robot war will elevate the hostile phenomenon to a post-human level.

# Bibliography

Alkire, Brien, et al. *Applications for Navy Unmanned Aircraft Systems*. Santa Monica: RAND, 2010, https://www.rand.org/pubs/monographs/MG957.html.

Anderson, Kenneth. *Targeted Killing in U.S. Counterterrorism Strategy and Law*. Washington DC: Brookings Institution, the Georgetown University Law Center, and the Hoover Institution, 2009, https://www.brookings.edu/research/targeted-killing-in-u-s-counterterrorism-strategy-and-law/.

Arkin, Ronald. *Governing Lethal Behavior in Autonomous Robots*. Boca Raton: Taylor and Francis Group, 2009.

"Article 36. Killer Robots: UK Government Policy on Fully Autonomous Weapons. 2013," https://article36.org/wp-content/uploads/2013/04/Policy_Paper1.pdf.

Asaro, Peter. "Ethical and Societal Perspectives on the Development of Autonomous Weapons." In *Conduct of Hostilities: The Practice, the Law and the Future*, edited by Edoardo Greppi, 218–222. San Remo: International Institute of Humanitarian Law, 2015.

Atherton, Kelsey. "IED Drone Kills Kurdish Soldiers, French Commandos." *Popular Science* (October 11, 2016), http://www.popsci.com/booby-trapped-isis-drone-kills-kurdish-soldiers-french-commandos.

Bergen, Peter, Salyk-Virk, Melissa, and David Sterman. "World of Drones". *New America*. Last updated on July 30, 2020, https://www.newamerica.org/international-security/reports/world-drones/who-has-what-countries-with-armed-drones/.

Biddle, Stephen. *Military Power*. Princeton: Princeton University Press, 2004.

Black, Jeremy. "Military History and the Whig Interpretation." *Nuova Antologia Militare* 1.0 (2020): 3–26, https://www.nam-sism.org/Articoli/NAM%20521%20Articolo%20n.%200%20-%20BLACK.pdf.

Black, Jeremy. "The Revolution in Military Affairs: The Historian's Perspective." *Journal of Military and Strategic Studies* 9.2 (Winter 2006/07), https://jmss.org/article/view/57696/43367.

Boot, Max. *War Made New: Technology, Warfare, and the Course of History, 1500 to Today*. London: Gotham Books, 2006.

Boothby, William. "Automated and Autonomous Weapons: What Military Utility in Which Operational Scenario?" In *Conduct of Hostilities: The Practice, the Law and the Future*, edited by Edoardo Greppi, 205–210. San Remo: International Institute of Humanitarian Law, 2015.

Brose, Christian. "The New Revolution in Military Affairs." Foreign Affairs, May/Jun (2019). https://www.foreignaffairs.com/articles/2019-04-16/new-revolution-military-affairs.

Brown, Andrew. "War Crimes and Killer Robots." *The Guardian* (18 March, 2009). http://www.guardian.co.uk/commentisfree/andrewbrown/2009/mar/18/religion-robots.

Brynjolfsson, Erik, and Andrew McAfee. *The Second Machine Age: Work, Progress, and Prosperity in a Time of Brilliant Technologies*. New York: W. W. Norton & Company, 2014.

Butowski, Piotr. "Russia reveals Loyal Wingman Concept." *Aviation Week* (4 September, 2020), https://aviationweek.com/special-topics/air-dominance/russia-reveals-loyal-wingman-concept.

Byrnes, Michael. "Nightfall: Machine Autonomy in Air-to-Air Combat." *Air & Space Power Journal* (May/June 2014): 48–75, https://apps.dtic.mil/sti/pdfs/ADA602090.pdf.

Choudhury, Lipika. *Final Report of the Panel of Experts on Libya Established Pursuant to Security Council Resolution 1973 (2011)*. New York: United Nations Security Council, 2021, https://digitallibrary.un.org/record/3905159.

Clifford J. Rogers, ed. *The Military Revolution Debate: Readings on the Military Transformation of Early Modern Europe*. Boulder: Westview Press, 1995.

CNAS. *Autonomous Weapons & Human Control*. Washington DC: Center for a New American Security, 2016, https://www.files.ethz.ch/isn/196780/CNAS_Autonomous_Weapons_poster_FINAL%20(1).pdf.

Cohen, Rachel. "DARPA's Gremlins Program Accomplishes First Flight." *Air Force Magazine* (21 January, 2020), https://www.airforcemag.com/darpas-gremlins-program-accomplishes-first-flight/.

Cummings, Mary. *Artificial Intelligence and the Future of Warfare*. London: Chatham House, 2017.

DARPA. "AlphaDogfight Trials Foreshadow Future of Human-Machine Symbiosis" (26 August, 2020), https://www.darpa.mil/news-events/2020-08-26.

Davis, Lynn E., McNerney, Michael J., and Michael D. Greenberg. *Clarifying the Rules for Targeted Killing*. Santa Monica: RAND, 2016, https://www.rand.org/pubs/research_reports/RR1610.html.

Deptula, David. "Artificial Intelligence Drone Defeats Fighter Pilot: The Future?" Breaking Defense.com (8 August, 2016), http://breakingdefense.com/2016/08/artificial-intelligence-drone-defeats-fighter-pilot-the-future/.

Engelhardt, Tom. "How Drone War Became the American Way of Life". *Al Jazeera* (1 March, 2012), https://www.aljazeera.com/opinions/2012/3/1/how-drone-war-became-the-american-way-of-life/.

Ernest, Nicholas, et al. "Genetic Fuzzy based Artificial Intelligence for Unmanned Combat Aerial Vehicle Control in Simulated Air Combat Missions." *Journal of Defence Management* 6.1 (2016), doi:10.4172/2167-0374.1000144.

Fisher, Jason. "Targeted Killing, Norms, and International Law." *Columbia Journal of Transnational Law* 45.3 (2007): 711–758.

Fissel, Mark Charles. "Military Revolutions." In *Oxford Bibliographies in Military History*, edited by Kaushik Roy. New York: Oxford University Press, https://www.oxfordbibliographies.com/view/document/obo-9780199791279/obo-9780199791279-0212.xml.

Foust, Joshua. "Soon, Drones May Be Able to Make Lethal Decisions on Their Own." *National Journal* (8 October, 2013), http://www.nationaljournal.com/national-security/soon-drones-may-be-able-to-make-lethal-decisions-on-their-own-20131008.

Freedberg, Sydney. "Robot Wars: Centaurs, Skynet, & Swarms." *Breaking Defence* (31 December, 2015), http://breakingdefense.com/2015/12/robot-wars-centaurs-skynet-swarms/.

Future of Life Institute. "Lethal Autonomous Weapons Pledge". Accessed 8 September, 2021. https://futureoflife.org/lethal-autonomous-weapons-pledge/.

Garcia, Denise. "Technical Statement by the International Committee for Robot Arms Control Convention on Conventional Weapons Meeting of Experts on Lethal Autonomous Weapons Systems." Statement on 2014 UN CCW Expert Meeting, Geneva, 13–16 May, 2014, https://www.icrac.net/icrac-statement-on-technical-issues-to-the-2014-un-ccw-expert-meeting/.

Gettinger, Dan, and Arthur H. Michel. *Loitering Munitions*. New York: Center for the Study of the Drone at Bard College, 2017, https://dronecenter.bard.edu/files/2017/02/CSD-Loitering-Munitions.pdf.

Gettinger, Dan. *Drone Databook Update: March 2020*. New York: Center for the Study of the Drone at Bard College, 2020, https://dronecenter.bard.edu/files/2020/03/CSD-Databook-Update-March-2020.pdf.

Gray, Colin. "RMA's and the Dimensions of Strategy." *Joint Forces Quarterly* 17 (Autumn/Winter 1998): 50–54, https://apps.dtic.mil/sti/pdfs/ADA356343.pdf.

Grossman, Nicholas. *Drones and Terrorism*. New York: I.B.Tauris & Co. Ltd, 2018.

Gunzinger, Mark, and Lukas Autenried. "Understanding the Promise of Skyborg and Low-Cost Attritable Unmanned Aerial Vehicles." *Mitchell Institute Policy Paper* 24 (2020), https://mitchellaerospacepower.org/wp-content/uploads/2021/01/a2dd91_2a1da65374434775b321619daf50a0a3.pdf.

Hallion, Richard. "Doctrine, Technology, and Air Warfare: A Late Twentieth – Century Perspective." *Air Power Journal* I.2 (1987): 16–27, https://www.airuniversity.af.edu/Portals/10/ASPJ/journals/Volume-01_Issue-1-4/1987_Vol1_No2.pdf.

Hammes, Thomas. "The Future of Warfare: Small, Many, Smart vs. Few & Exquisite?" *War on the Rocks* (16 July, 2014), http://warontherocks.com/2014/07/the-future-of-warfare-small-many-smart-vs-few-exquisite/.

Heyns, Christof. *Report of the Special Rapporteur on Extrajudicial, Summary or Arbitrary Executions. A/HRC/23/47*. New York: United Nations, 2013, https://documents-dds-ny.un.org/doc/UNDOC/GEN/G13/127/76/PDF/G1312776.pdf?OpenElement.

Hill, Andrew, and Gregg Thompson. "Five Giant Leaps for Robotkind: Expanding the Possible in Autonomous Weapons." *War on the Rocks* (28 December, 2016), https://warontherocks.com/2016/12/five-giant-leaps-for-robotkind-expanding-the-possible-in-autonomous-weapons/.

Hoffman, Frank. "Will War's Nature Change in the Seventh Military Revolution?" *Parameters* 47.4 (2017): 19–31, https://publications.armywarcollege.edu/pubs/3554.pdf.

Hubbard, Ben, Palko Karasz, and Stanley Reed. "Two Major Saudi Oil Installations Hit by Drone Strike, and U.S. Blames Iran." *The New York Times* (14 September, 2019), https://www.nytimes.com/2019/09/14/world/middleeast/saudi-arabia-refineries-drone-attack.html.

Hundley, Richard. *Past Revolutions, Future Transformations*. Santa Monica: RAND, 1999, https://www.rand.org/pubs/monograph_reports/MR1029.html.

ICRC. "Customary IHL Database," https://ihl-databases.icrc.org/customary-ihl/eng/docs/home.

Ilachinski, Andrew. *AI, Robots, and Swarms: Issues, Questions, and Recommended Studies*. Arlington: Center for Naval Analyses, 2017, https://www.cna.org/cna_files/pdf/DRM-2017-U-014796-Final.pdf.

Inside GNSS. "Two thousand simultaneous drones position themselves with GPS and GLONASS." *Inside GNSS* (23 November, 2020), https://insidegnss.com/two-thousand-simultaneous-drones-position-themselves-with-gps-and-glonass-rtk/.

Jackson, Van. "The Pentagon's Third Offset Strategy: What US Allies and Partners Need to Know." *The Diplomat.com* (28 April, 2015), http://thediplomat.com/2015/04/the-pentagons-third-offset-strategy-what-us-allies-and-partners-need-to-know/.

JAPCC. *Future Vector & Future Unmanned System Technologies: Legal and Ethical Implications of Increasing Automation*. Kalkar: Joint Air Power Competence Center, 2016.

Kaku, Michio. *Visions: How Science Will Revolutionize the 21st Century*. New York: Anchor Books, 1997.

Kallenborn, Zachary. *Are Drone Swarms Weapons of Mass Destruction?* Maxwell Air Force Base: United States Air Force Center for Strategic Deterrence Studies, 2020, https://media.defense.gov/2020/Jun/29/2002331131/-1/-1/0/60DRONESWARMS-MONOGRAPH.PDF.

Kania, Elsa. *Battlefield Singularity: Artificial Intelligence, Military Revolution, and China's Future Military Power*. Washington DC: Center for a New American Security, 2017, https://s3.us-east-1.amazonaws.com/files.cnas.org/documents/Battlefield-Singularity-November-2017.pdf?mtime=20171129235805&focal=none.

Krepinevich, Andrew, Barry Watts, and Robert Work. *Meeting the Anti-Access and Area-Denial Challenge*. Washington DC: Center for Strategic and Budgetary Assessments, 2003, https://csbaonline.org/uploads/documents/2003.05.20-Anti-Access-Area-Denial-A2-AD.pdf.

Krepinevich, Andrew. "Cavalry to Computer: The Pattern of Military Revolutions." *The National Interest* 37 (1994): 30–42, https://nationalinterest.org/article/cavalry-to-computer-the-pattern-of-military-revolutions-848.

Krepinevich, Andrew. *The Military-Technical Revolution: A Preliminary Assessment*. Washington DC: Center for Strategic and Budgetary Assessments, 2002, https://csbaonline.org/uploads/documents/2002.10.02-Military-Technical-Revolution.pdf.

Krishnan, Armin. *Killer Robots: Legality and Ethicality of Autonomous Weapons*. Surrey: Ashgate, 2009.

Kuptel, Artur, and Andy Williams. *Policy Guidance: Autonomy in Defence Systems*. Norfolk: Supreme Allied Commander Transformation, 2014.

Kurzweil, Ray. *The Singularity is Near: When Humans Transcend Biology*. New York: The Viking Press, 2005.

Lamothe, Dan. "Veil of Secrecy Lifted on Pentagon Office Planning 'Avatar' Fighters and Drone Swarms." The Washington Post.com (8 March, 2016), https://www.washingtonpost.com/news/checkpoint/wp/2016/03/08/inside-the-secretive-pentagon-office-planning-skyborg-fighters-and-drone-swarms/?utm_content=link&utm_medium=website&utm_source=fark.

Larkin, Matthew. "Brave New Warfare: Autonomy in Lethal UAVs." Master's dissertation. Naval Postgraduate School, 2011, https://calhoun.nps.edu/bitstream/handle/10945/5781/11Mar_Larkin.pdf?sequence=1&isAllowed=y.

Lewis, Larry. *Redefining Human Control: Lessons from the Battlefield for Autonomous Weapons*. Arlington: Center for Autonomy and AI, 2018, https://www.cna.org/cna_files/pdf/DOP-2018-U-017258-Final.pdf.

Lin, Jeffrey, and Peter Singer. "China is Making 1,000-UAV Drone Swarms Now." *Popular Science* (9 January, 2018), https://www.popsci.com/china-drone-swarms.

Luttwak, Edward. "Dead End: Counterinsurgency Warfare as Military Malpractice." *Harper's Magazine* (February 2007): 33–42.

Luttwak, Edward. "Post-Heroic Warfare and its Implications." In *Proceedings of International Symposium on Security Affairs – War and Peace in the 21st Century: Reflections upon the Century of War*, 127–139. Tokyo: National Institute for Defense Studies, 2000, http://www.nids.mod.go.jp/english/event/symposium/pdf/1999/sympo_e1999_5.pdf.

Luttwak, Edward. "Towards Post-Heroic Warfare." *Foreign Affairs* 74.3 (1995): 109–122, https://www.foreignaffairs.com/articles/yugoslavia/1995-05-01/toward-post-heroic-warfare.

Mason, Richard. "Unmanned Aerial Vehicles: Progress and Challenge." In *Air Power: UAVs – The Wider Context*, edited by Owen Barnes, 116–123. London: Royal Air Force, 2009.

Melzer, Niels. *Targeted Killing in International Law*. New York: Oxford University Press, 2008.

Mitchell, William. *Winged Defense: The Development and Possibilities of Modern Air Power-Economic and Military*. Mineola: Dover Publications, 1988.
Morillo, Stephen, and Michael Pavkovic. *What is Military History?* Malden: Polity Press, 2006.
Murray, Williamson, and McGregor Knox. "Thinking about Revolutions in Warfare." In *The Dynamics of Military Revolution 1300–2050*, edited by McGregor Knox, and Williamson Murray, 1–14. Cambridge: Cambridge University Press, 2001.
Nurkin, Tate. "The Importance of Advancing Loyal Wingman Technology." *Defence News* (21 December, 2020), https://www.defensenews.com/opinion/commentary/2020/12/21/the-importance-of-advancing-loyal-wingman-technology/?fbclid=IwAR3PYnB81rA%E2%80%A6.
Palmer, Adam. *Autonomous UAS: a Partial Solution to America's Future Airpower Needs*. Montgomery: Air Command and Staff College, 2010, https://apps.dtic.mil/sti/pdfs/AD1018416.pdf.
Parker, Geoffrey. "The Future of Western Warfare." In *The Cambridge History of Warfare*, edited by Geoffrey Parker, 423–442. New York: Cambridge University Press, 2005.
Raemdonck, Nathalie. "Vested Interest or Moral Indecisiveness? Explaining the EU's Silence on the US Targeted Killing Policy in Pakistan." In *IAI Working Papers 1205*. Roma: Istituto Affari Internazionali, 2012, https://www.files.ethz.ch/isn/141630/iaiwp1205.pdf.
Reilly, M. "Beyond Video Games: New Artificial Intelligence Beats Tactical Experts in Combat Simulation." *University of Cincinnati Magazine* (27 June, 2016), http://magazine.uc.edu/editors_picks/recent_features/alpha.html.
Roff, Heather, and Richard Moyes. "Meaningful Human Control, Artificial Intelligence and Autonomous Weapons." *Briefing Paper prepared for the Informal Meeting of Experts on LAWS, UN Convention on Certain Conventional Weapons, Geneva, April 11–15*, 2016, https://article36.org/wp-content/uploads/2016/04/MHC-AI-and-AWS-FINAL.pdf.
Rosen, Stephen. *Winning the Next War*. Ithaca: Cornell University Press, 1991.
Scharre, Paul, and Michael Horowitz. *An Introduction to Autonomy in Weapon Systems*. Washington DC: Center for a New American Security, 2015, https://www.files.ethz.ch/isn/188865/Ethical%20Autonomy%20Working%20Paper_021015_v02.pdf.
Scharre, Paul, and Michael Horowitz. *Meaningful Human Control in Weapon systems: A Primer*. Washington DC: Center for a New American Security, 2015, https://www.files.ethz.ch/isn/189786/Ethical_Autonomy_Working_Paper_031315.pdf.
Scharre, Paul. "Statement to the UN Convention on Certain Conventional Weapons on Way Ahead." *CNAS* (17 April, 2015), https://www.cnas.org/publications/blog/statement-to-the-un-convention-on-certain-conventional-weapons-on-way-ahead.
Scharre, Paul. *Army of None: Autonomous Weapons and the Future of War*. New York: W.W. Norton & Company, 2018.
Sharkey, Noel. "Towards a Principle for the Human Supervisory Control of Robot Weapons". *Politica & Società* (May-August 2014): 305–324.
Sharkey, Noel. "Weapons of Indiscriminate Lethality." *FIfF Kommunikation* 1/09 (2009): 26–29, https://www.fiff.de/publikationen/fiff-kommunikation/fk-2009/fiff-ko-1-2009/fiko_1_2009_sharkey.pdf.
Shkolnik, Michael. "The Drone Threat to Israeli National Security." *War on the Rocks* (10 January, 2017), https://warontherocks.com/2017/01/the-drone-threat-to-israeli-national-security/.
Singer, Peter. *Wired for War*. New York: Penguin Press, 2009.

Solis, Gary. "Targeted Killing and the Law of Armed Conflict." *Naval War College Review* 60.2 (2007): 127–146, https://digital-commons.usnwc.edu/cgi/viewcontent.cgi?article= 2007&context=nwc-review.

Swarts, P. "Air Force Looking at Autonomous Systems to Aid War Fighters." *Air Force Times* (17 May, 2016), https://www.airforcetimes.com/story/military/2016/05/17/air-force-looking-autonomous-systems-aid-warfighters/84502234/.

Toffler, Alvin, and Heidi Toffler. *War and Anti-war: Survival at the Dawn of the 21st Century.* New York: Little, Brown and Company, 1993.

Trevithick, Joseph. "China Conducts Test of Massive Suicide Drone Swarm Launched From a Box on a Truck." *The Drive* (14 October, 2020), https://www.thedrive.com/the-war-zone /37062/china-conducts-test-of-massive-suicide-drone-swarm-launched-from-a-box-on-a-truck?fbclid=IwA%E2%80%A6.

Trevithick, Joseph. "Glitzy Air Force Video Lays Out 'Skyborg' Artificial Intelligence Combat Drone Program." *The Drive* (24 June, 2020), https://www.thedrive.com/the-war-zone /34351/glitzy-air-force-video-lays-out-skyborg-artificial-intelligence-combat-drone-program.

Trevithick, Joseph. "Pentagon Unveils Details on Effort to Equip Its Services With Massive Swarms of Deadly Drones." *The Drive* (16 March, 2021), https://www.thedrive.com/the-war-zone/39814/pentagon-unveils-details-on-effort-to-equip-its-services-with-massive-swarms-of-deadly-drones.

Trevithick, Joseph. "These Three Companies Will Build Drones to Carry the Air Force's 'Skyborg' AI Computer Brain." *The Drive* (7 December, 2020), https://www.thedrive.com/ the-war-zone/38015/these-three-companies-will-build-drones-to-carry-the-air-forces-skyborg-ai-computer-brain?fbclid=%E2%80%A6.

Trevithick, Joseph. "USAF Plans for Its 'Skyborg' AI Computer Brain to Be Flying Drones in the Next Two Years." *The Drive* (20 March, 2019), https://www.thedrive.com/the-war zone/27067/ usaf-plans-for-its-skyborg-ai-computer-brain-to-be-flying-drones-in-the-next-two-years.

Trevithick, Joseph. "Watch Russia's S-70 Unmanned Combat Air Vehicle Fly With an SU-57 For the First Time." *The Drive* (27 September, 2019), https://www.thedrive.com/the-war-zone /30053/watch-russias-s-70-unmanned-combat-air-vehicle-fly-with-an-su-57-for-the-first-time.

Tucker, Patrick. "US Army Looking to 3D-Print Minidrones in 24 Hours." Defenceone.com (10 January, 2017), http://www.defenseone.com/technology/2017/01/us-army-looking -3d-print-minidrones-24-hours/134494/?oref=d-topstory.

UK MoD. *Joint Doctrine Publication 0-30.2: Unmanned Aircraft Systems*. London: United Kingdom Ministry of Defence, 2017, https://assets.publishing.service.gov.uk/govern ment/uploads/system/uploads/attachment_data/file/673940/doctrine_uk_uas_jdp_0_ 30_2.pdf.

UNIDIR. *The Weaponization of Increasingly Autonomous Technologies: Considering How Meaningful Human Control Might Move the Discussion Forward.* Geneva: United Nations Institute for Disarmament Research, 2014, https://unidir.org/publication/weaponization-increasingly-autonomous-technologies-considering-how-meaningful-human.

United Nations. *Draft Report of the 2019 Session of the Group of Governmental Experts on Emerging Technologies in the Area of Lethal Autonomous Weapons Systems*. Geneva: United nations, 2019, https://undocs.org/pdf?symbol=en/CCW/GGE.1/2019/CRP.1/REV.2.

US DoD. "Department of Defense Announces Successful Micro-Drone Demonstration." *US Department of Defense* (9 January, 2017), https://www.defense.gov/News/News-

Releases/News-Release-View/Article/1044811/department-of-defense-announces-successful-micro-drone-demonstration.
US DoD. *Directive 3000.09: Autonomy in Weapon Systems*. Washington DC: United States Department of Defense, 2012, https://www.esd.whs.mil/portals/54/documents/dd/issuances/dodd/300009p.pdf.
US DoD. *Report of the Defense Science Board Summer Study on Autonomy*. Washington DC: Department of Defense, 2016, https://www.hsdl.org/?view&did=794641.
US DoD. *Unmanned Systems Integrated Roadmap FY2011-2036*. Washington DC: United States Department of Defense, 2011, https://irp.fas.org/program/collect/usroadmap2011.pdf.
US DoD. *Unmanned Systems Integrated Roadmap FY2013-2038*. Washington DC: United States Department of Defense, 2013, https://www.hsdl.org/?view&did=747559.
US DoD. *Unmanned Systems Integrated Roadmap FY2017-2042*. Washington DC: United States Department of Defense, 2017, https://s3.documentcloud.org/documents/4801652/UAS-2018-Roadmap-1.pdf.
USAF Flight Plan. *Unmanned Aircraft Systems Flight Plan 2009–2047*. Washington DC: Headquarters United States Air Force, 2009, https://irp.fas.org/program/collect/uas_2009.pdf.
USAF. *RPA Vector: Vision and Enabling Concepts 2013-2038*. Washington DC: Headquarters United States Air Force, 2014, https://www.af.mil/Portals/1/documents/news/USAFRPAVectorVisionandEnablingConcepts2013-2038.pdf.
USAF. *Small UAS Flight Plan: 2016-2036*. Washington DC: Headquarters United States Air Force, 2016, https://apps.dtic.mil/sti/pdfs/AD1013675.pdf.
Van Creveld, Martin. *Technology and War: from 2000 B.C. to the Present*. New York: The Free Press, 1991.
Vincent, James. "Putin Says the Nation That Leads in AI 'Will Be the Ruler of the World.'" *The Verge* (4 September, 2017), https://www.theverge.com/2017/9/4/16251226/russia-ai-putin-rule-the-world.
Weaver, Darren, and Erin Black. "Behind the Scenes as Intel Sets a World Record for Flying Over 2,000 Drones at Once." *CNBC* (18 July, 2018), https://www.cnbc.com/2018/07/17/intel-breaks-world-record-2018-drones.html.
Welsh, Sean. "Air Operations at the Level of Boots on the Ground." *The Central Blue* (23 March, 2018), http://centralblue.williamsfoundation.org.au/air-operations-at-the-level-of-boots-on-the-ground-sean-welsh/.
Wigston, Mike. "The Chief of the Air Staff's Speech at the Global Air Chiefs' Conference 2021". Speech delivered at the Global Air Chiefs Conference, London, 14 July, 2021, https://www.gov.uk/government/speeches/the-chief-of-the-air-staffs-speech-at-the-global-air-chiefs-conference-2021.
Work, Robert. "Art, Narrative, and the Third Offset." Speech delivered at the 2016 Global Strategy Forum, Atlantic Council. Webcast, 2 May, 2016, https://www.atlanticcouncil.org/unused/webcasts/2016-global-strategy-forum/.
Work, Robert. "The Third U.S. Offset Strategy and Its Implications for Partners and Allies." *US Department of Defense* (28 January, 2015), http://www.defense.gov/News/Speeches/Speech-View/Article/606641/the-third-us-offset-strategy-and-itsimplications-for-partners-and-allies.
Work, Robert. "WATCH: David Ignatius and Pentagon's Robert Work on the Latest Tools in Defense." Washington Post Live (30 March, 2016), https://www.washingtonpost.com/

blogs/post-live/wp/2016/02/29/securing-tomorrow-with-david-ignatius-whats-at-stake-for-the-world-in-2016-and-beyond/?utm_term=.1bace9de7bab.

Zhan, Echo. "3,281 Drones Break Dazzling Record for Most Airborne Simultaneously." *Guiness Book of Records* (17 May, 2021), https://www.guinnessworldrecords.com/news/commercial/2021/5/3281-drones-break-dazzling-record-for-most-airborne-simultaneously-655062.

# Appendix: On the Cover Illustration

**Figure 1:** Holy Roman Emperor Maximilian I (1459–1519) as the fictional character Theuerdank makes evident his mastery of the tools of the gunpowder revolution. The poem proclaims that in the skirmish Theuerdank "shot many a man". *Theuerdank* (Nuremberg 1517), Woodcut engraving from the editor's collection.

# Appendix: On the Cover Illustration

The cover of the present volume simultaneously intimates the setting of the sun on medieval warfare and the dawning of the gunpowder revolution. Or does it? The veracity of what are conjectured to be military revolutions depends upon an identification of authentic qualitative shifts in reality that can be distinguished within the duration of historical time. The ensuing challenge is to confirm the degree to which that change can be demarcated, and thus sequenced, resulting in periodization.

Our cover illustration is from *Theuerdank* (Nuremberg 1517), composed by Holy Roman Emperor Maximilian I and his court secretaries, Melchior Pfinzing and Marx Treitz-Saurwein. The epic poem details the hero's odyssey as he undertakes some 80 trials which "The Last Knight" has to overcome to win the hand of his lady. The woodcut engravings enlivening the verses convey courtly values and nubilous reflections of events from the sixteenth century. The Theuerdank character personifies Maximilian I. Throughout the poem one finds subsumed allusions to Maximilian's courtship of Mary of Burgundy in 1477, his territorial perambulations, and the defense of Christendom via crusade.[1] The Emperor's secretary Melchior Pfinzing fashioned these resonances between the literary tale and Maximilian's actual experiences, in effect serving as intermediary and editorial commentator. For example, illustration 26, where the image is clearly Maximilian (e.g., the nose, the costume, and what resembles the Order of the Golden Fleece) is referenced by Pfinzing in relation to a near-mishap on a Swabian spiral staircase in a "dilapidated high tower." Likewise, in illustration 39, in Austria ("under the Enns") Maximilian escapes injury when close to the muzzle of a loaded cannon and a burning torch. Illustration 50, wherein an armored Theuerdank touches off one of a trio of cannon is, according to Pfinzing, an "incident [that] stands for all the various dangers with heavy artillery that Theuerdank was exposed to in Picardy."[2] Illustrations 58 and 60 depict near fatal accidents with stored gunpowder in Geldern and Upper Tyrol respectively. A town unleashes cannonades and flaming pitch against Maximilian, explicitly attributed to a nasty reception he received in Hungary, represented by illustration 91.

Our dust cover features illustration 79, "Theuredank in a gun battle in front of a city wall."[3] Theuerdank/Maximilian skirmishes with a similarly armed enemy contingent that has emerged from the town's gate.[4] The verses relate that

---

[1] Füssel, *Theuerdank*, 9–10, 49–50, 54–55, 57, 59, and 62. Grateful thanks to Ian Copestake for editing this appendix. All mistakes, however, are the sole responsibility of the author.
[2] Füssel, *Theuerdank*, 221.
[3] Füssel, *Theuerdank*, 287.
[4] James O'Neill's essay herein (188–189) where he points out that the art of skirmishing has been underappreciated by historians.

Theuerdank fires his weapon expertly, based on the casualties he inflicts, but withdraws (again dodging bullets) when his band becomes outnumbered and outgunned.[5] Secretary Pfinzing's gloss asserts that "this episode stands for all the skirmishes in which Theuerdank was involved with 'small arms.'"[6]

To the modern eye, a heavily armored knight taking close aim with a firearm seems incongruous. Splendidly accoutered men-at-arms belong on horseback, hefting lances. Guns are for soiled but nimble infantrymen, humble but expertly wielding the weapon of the future. However, historical evidence contradicts that tacit and binary assumption. What has been purported to be the earliest documentation (February 11, 1326) of Western guns, the Florentine *pilas seu palloctas ferreas et canones de mettallo*, plants gunpowder-based military revolution squarely in the medieval world. Visual evidence from the same year such as the illustrations associated with the Millimete MSS, the firing of an arrow-like projectile ignited by mailed warriors, makes the image of a knight discharging a firearm in the second decade of the sixteenth century both appropriate and emblematic in interpreting the amalgamations of the art of war described in the opening chapters of this anthology.

In a book on the military revolution, we profit from contemplating diverse depictions of the tools of the gunpowder revolution. These are marvelously illustrated in the chivalric context of *Theuerdank*, a classic of German Renaissance literature. *Der Weisskunig*, similarly composed by Maximilian and his team of secretaries c. 1505 to 1516, also exhibits representations of the gunpowder revolution. The prominent luminaries involved in creating illustrations for both above-mentioned works were Leonhard Beck and Hans Burgkmaier, artists known to military historians for woodcuts of soldiers and battles.[7] The depictions of warfare handcrafted by Maximilian's stable of artists, although inevitably idealized, nevertheless captured realistic detail. *Theuerdank*, like the *Iliad*, is thus an epic poem grounded in the realities of war even if it is neither history nor chronicle as defined in a modern historicist context. There is no intention to achieve an authentic and precisely accurate factual narrative. Maximilian's figure on an armed quest is the focal point as it should be. Pfinzing's interweaving of poetic fiction with vignettes from historical events embellished Maximilian's reputation as

---

[5] Füssel, 5, 286–287. Thanks to Steve Walton for translating and theorizing about the nature of the images.
[6] Füssel, *Theuerdank*, 287. This woodcut's iconography must be approached thoughtfully because its artist is neither Leonhard Beck nor Hans Burgkmaier but most likely Wolf Traut, according to Füssel, *Theuerdank*, 75.
[7] For an exposition of examples, see von Hefner, ed. *Hans Burgkmairs Turnier-Buch* and the second volume of H.T. Musper, ed. *Kaiser Maximilians I.*

much as it celebrated the values of the age in which he lived.[8] Maximilian's *Ruhmeswerken* ("works of fame" or self-glorifications) coupled with his self-memorialization (associated with the concept of *Gedächtnis*) were the Emperor's guiding principles. Like Caesar's *Commentaries*, *Theuerdank* and *Der Weisskunig* are primarily exaltations of leadership.[9] The Emperor and his court secretaries justifiably exercised customary poetic license and embellishment for literary purposes. Maximilian was a master craftsman and manipulator of the infant printed medium, and those skills were largely honed to leave a memorial that would (and has) persisted to the present day. In the process, a rich visual legacy was left for military historians. *Theuerdank* and *Der Weisskunig* complement historical evidence because these literary works reveal how practitioners of the art military conceived of warfare. Glimpses into the consciousness of agents of the gunpowder revolution emerge in Maximilian's woodcuts.

The *Theuerdank* selection is especially sonorous with Hyeok Hweon Kang's essay on technology and global military history. Asian archaeological evidence establishes the existence of "hand-cannons" in 1287 to 1288 in Heilongjiang province and at Xanadu in 1298 (though Tonio Andrade and others caution that these may be better classified as fire-lances than as "guns"). The origins of the gunpowder revolution that created the modern world is unquestionably medieval, a fact that our frequently anachronistic perspective obscures. Scores of medieval representations, some of contemporary events and others depicting antiquity, feature armored foot soldiers manning primitive siege guns and aiming arquebuses. Our dust cover serves as a reminder that from the vantage point of the era of revolutions in military affairs one sees decisive change and discontinuity. Those living through the dawn of the gunpowder age perceived no jarring historical shift (in contrast, say, to those witnessing events in August 1945). Theuerdank firing a gun 230 years after the development of firearms should not strike us as a curiosity, and archaeological evidence sustains Maximilian's sense of continuity and integration.

While burnishing his image Maximilian remained a steely-eyed soldier who understood and articulated the contemporary art of war within its social (chivalric) context. He quite deservedly earned the title "father of the *landsknechts*."[10] In 1486 the Emperor applied his military organizational skills to establishing an infantry force among his subjects capable of protecting themselves from incursions by invaders such as the Swiss. Such actions demonstrate that Maximilian comprehended contemporary military science and was as much an actor in the

---

8 Tennant, "Productive Reception," 295–348.
9 See Müller, "Le jeune Maximilien et l'internationalisation de la guerre," 342; Tennant, "Productive Reception: *Theuerdank* in the Sixteenth Ceentury," 295–348.
10 See Hof, *Die Ritterideologie Kaiser Maximilians I. im "Theuerdank."*

military revolution as Machiavelli was in the latter's experiments in mobilizing a militia for defensive purposes.[11] With few exceptions, the military revolutions that historians have proposed were infantry-based, and Maximilian's armies featuring pikemen were in themselves a "glorification of imperial military strength," illustrated graphically by Erhard Schön's *A Column of Mercenaries* (c. 1532) and, earlier, the marching troops in the famed massive woodcut of *The Triumph of Maximilian*.[12] Theuerdank/Maximilian shouldering a firearm amid a group of foot soldiers (who also discharge guns) demonstrates that the Emperor's commitment to the promotion of an infantry revolution co-existed with his advocacy of chivalric equine warfare. Although soldiering during what many consider the apex of the military revolutions, the Emperor comes to terms rather comfortably with a weapons system that was undergoing profound change throughout his lifetime. If any man was inclined to make note of how gunpowder weaponry had transformed warfare quantitatively (in terms of scale) and qualitatively (in terms of gunpowder-based tactics), lending credibility toward the existence of a military revolution, it was Maximilian I. The Holy Roman Emperor "controlled a territory that extended from Eastern Europe to the center of the Low Countries and from Switzerland to Northern Italy, and embarked on constant military enterprises becoming proficient in the seven languages spoken in these vast areas so that he could lead armies of soldiers from a variety of different linguistic backgrounds."[13] This *uomo universale* also developed a lexicon of the militaria of the age and studied (and thus compared) regional forms of war.

Maximilian's *Gedächtnis* presumed to adhere to the purest and highest standards of the chivalric ideal. Simultaneously he was quick to dirty his hands to grasp the technical and mechanical intricacies of artillery, as depicted in illustration 39. Maximilian investigates innovations in artillery design in illustration 50, in this case a "snake box."[14] In illustration 57 Maximilian/Theuerdank "tinkers" with a mounted "hatchet gun" that has been overcharged with powder.[15] The Catherine's Wheel emblazoned on the cloak of his servant Ehrenhold, invariably turned in the "Last Knight's" favor. He survives near misses by bullets and cannonballs (e.g., illustrations 78, 80, 84, 91), braves ordnance aimed squarely at him when traversing water (illustration 76), thwarts potential explosions in handling powder, etc. He is, surely, bulletproof. The ubiquitous wheel of fortune (it appears at least 102 times in *Theuerdank*, and in illustrations 11 and 25

---

11 Oman, *A History of the Art of War*, 75.
12 Moxey, *Peasants, Warriors, and Wives*, 71.
13 See Müller, "Le jeune Maximilien et l'internationalisation de la guerre," 342.
14 Füssel *Theuerdank*, 221.
15 Füssel, *Theuerdank*, 235.

the image is incised on the hero's cuirass) also served as a warning against hubris concerning one's invincibility. In the latter case, we are reminded of the triumphal procession of a victorious Roman general, supremely confident but constantly aware that hubris will bring a fall from grace.[16] The only determinism in *Theuerdank* is the steadfastness of God's grace for the faithful in enduring the dangerous vicissitudes of life. Gunpowder weaponry is no more than an impotent accoutrement.

How does the poem *Theuerdank* characterize gunpowder? Perilous. An unpredictable commodity.[17] The accident-prone gadgetry and noxious fumes of the gunpowder revolution are part and parcel of the multitudinous dangers of waging war. Gunpowder weaponry, despite its comparatively complex evolution decade by decade, embellishes (and somewhat complicates) the contemporary art of war rather than rendering obsolete centuries-old implements of destruction. Modern definitions of military revolutions and RMAs are frequently contingent upon new weapons making traditional tools of war obsolescent. Theuerdank's guns are just an additional category of armament in the knightly panoply (for example, see illustration 87, where each of the four sentries inspected by Maximilian/Theuerdank carries a representative specimen of contemporary weaponry, without qualitative distinction: sword [apparently], pike, arquebus, and crossbow). Continuity reigns. In the *Der Weisskunig* a consultation with the armorers dovetails with a close inspection of the gunmakers in their foundry.[18] Everything is in its proper place under the paternal watch of the Emperor.

Maximilian I's universalist views would hardly understand, let alone embrace, the argument that a fundamental disjunction in the art of war occurred in the second decade of the 1500s. That is not to suggest that the Emperor suffered from myopia and that historians enjoy perfect vision. The posthumous *nom de guerre* of the "Last Knight" does a disservice to Maximilian.[19] Characterizations of Theuerdank/Maximilian as the "Last Knight" are brazenly anachronistic, inflicting periodization with its temporal boundaries and characterizations of eras

---

[16] Steven Walton shared thoughts on such matters. Any misinterpretations or errors are my own, however.
[17] For example, illustrations 39, 50, 57, 58, 60, etc. in Füssel, *Theuerdank*.
[18] Compare the Emperor's treasured collections of armor and ordnance: Terjanian, "The Currency of Power," 17–37, 45 (the latter page depicting a scene from the *Inventory of Maximilian I's Armaments in Tyrol*); Füssel, *Theuerdank*, 7.
[19] On the evolution of the name bestowed upon the Emperor, see Krause's discussion in "The Making of the 'Last Knight,'" 53–54.

upon historical reality. The unkindest cut of all is to approximate the Theuerdank figure with that of Cervantes' Don Quixote.[20] Certainly contemporaries had a notion of societal decline and decay. Francesco Petrarca (1304–1374) springs to mind. Regarding the eclipse of chivalry, Cervantes' vantage point (as an individual with both military experience and familiarity with bureaucracy) overlooked the Europe of the wars of religion. It was not the harmonious worldview from an imperial court at the zenith of the High Renaissance. Maximilian shared the Renaissance perspective that reconciled pagan mythology, Roman imperial imagery, with Christianity. He was simultaneously Hercules Germanicus, Julius Caesar, and the Lord's chosen tool to restore a broken Christendom.[21] Where historians see division and categorization, the Emperor saw integrality rather than a staccato of discontinuities.

Like other High Renaissance men, such as Michelangelo Buonarotti and Leonardo da Vinci, Maximilian ranged from appreciating the highest levels of aesthetics to expertise in the brutal business of poliorcetics.[22] Firearms and cannon were modified over decades and their improvements seemed less discordant to the soldiery than to historians. In an early proof of illustration 50, a Theuerdank figure that is unquestionably Maximilian I (based on physiognomy and garb) sets a burning match to the touchhole of a cannon.[23] Theuerdank personally fires gunpowder weapons (illustrations 50, 57, 79, 88). Thus, Theuerdank skirmishing with a matchlock as depicted on our dustcover is consistent with the chivalric ethos. Since Theuerdank is a paragon of virtue, one must conclude that there is nothing immoral or devilish about gunpowder weaponry. Despite the horrors of the early gunpowder age, the spiritual purity associated with the medieval military ideal persisted. God's grace watched over Theuerdank (and the Emperor himself). *Fortuna* smiles upon Maximilian, though he would attribute good fortune to his faith in God (which seems dangerously Lutheran from our perspective).

This exemplary contemporary saw continuity where we historians have theorized radical cleavages in the art of war. Maximilian and his Court expressed no belief that the battlefield had changed. There was nothing new under the sun. In the midst of the "epicenter" of the military revolution and at the cusp of the Protestant revolution, no alarms were sounded in the pages of *Theuerdank*

---

**20** See Hof, *Die Ritterideologie Kaiser Maximilians I. im "Theuerdank."*
**21** Julius Caesar in the sense that the latter's *Commentaries* were akin to Maximilian's *Ruhmeswerken* and *gedechtnus/Gedächtnis*.
**22** Füssel, *Theuerdank*, 7.
**23** For a comparison of the early proof with the final product, see Dodgson. "Some Undescribed States of *Theuerdank* illustrations," 131–138, 131.

other than the clarion call to defend Christendom from Islam. On the other hand, the transformation of the world by the printing press seems to have dawned on *Der Weisskunig* given the energy and funds he plowed into *Theuerdank* and related projects devoted to *Ruhmeswerk*. Maximilian's devotion to memory and chronicle reveals his reverence toward history. The unfolding years were neither subdivided nor in conflict. Analysis from a perspective centuries later reveals macrocosmic trends that contemporaries, as they freeze-framed episodes in their lifetime, were more subtle in perceiving. Maximilian's perception of continuum was not an optical illusion; rather it is a factor to consider when striving to see the historical episodes as they might have appeared to the actors. That is why Maximilian's efforts to characterize himself, and his age, must qualify our notions of military revolution.

The *imperator litteratus* was more familiar with late fifteenth- to early sixteenth-century warfare to a degree than we can ever achieve. If salient commentary on warfare, an honored literary tradition in antiquity as well as through the Middle Ages, might have garnered for the Emperor yet more accolades then Maximilian would have critiqued early-modern warfare.[24] Undoubtedly *Theuerdank* was composed and revised in conscious awareness of the significance of the battle of Ravenna (1512), where victory was decided by artillery. Maximilian was keenly aware of the events that culminated on April 11, 1512. Indeed, the commander of Nemour's *landsknechts*, Jacob Empser, received a personal dispatch from the "father of the *landsknechts*," Maximilian recalling him from French service because the Emperor planned to use these infantrymen himself in a war against France. Maximilian's recall threatened to dissolve the allied army just as it was poised to fight and thus precipitated the immediate hazarding of battle.[25] The French victory at Ravenna arguably saw the military revolution at its height, in terms of infantry disposition and the tactical incorporation of gunpowder weaponry. The defeated looked to traditional strategies and associations to remove the invader. A veteran of that bitter contest, Marcellus Polonius, appealed to Maximilian via an ornate manuscript *Oratio* to act as Roman emperor and impose peace upon the Italian peninsula by ejecting the French. In late 1515 and the early months of 1516, Maximilian forged an anti-French alliance. His Sforza in-laws had lost Milan. Maximilian's venture to seize the city (March 1516) from the forces of Francis I stumbled and failed. This setback did not prevent other humanists (Ulrich von Hutten and Riccardus Bartholinus) from seconding Polonius' plea

---

24 See Müller: *Gedechtnus. Literatur und Hofgesellschaft um Maximilian I*.
25 Oman, *A History of the Art of War*, 133–134.

for imperial deliverance.[26] The point is that Maximilian was enmeshed in the Northern Italian wars that have famously exhibited the hallmarks of the military revolution; he was a firsthand witness and participant. Principal actors Maximilian and Machiavelli both looked to tradition for inspiration in order to find solutions for the French incursions (though Machiavelli, like Maximilian, understood that artillery had an appreciable impact). The Florentine, too, did not believe that gunpowder had changed warfare forever, and four years after *Theuerdank* had been completed, he circulated those views in *The Art of War* (1521).[27]

In the second decade of the sixteenth century, a time frame which occupies much of the content of this anthology, contemporaries looked back to antiquity and found history to be harmonious with their own existence and cognizant that all epochs were punctured by the ravages of war. Ironically, in the West, 1517 would forever change the course of human affairs, in a fashion similar to that which would occur after 1789 and subsequently to European events circa 1917. Given the omnipresence of historical memory in studies of military revolutions, we ought to consider the silent, somewhat distracted figure in the lower left-hand corner of our cover illustration. If professional historians cannot claim heroic Theuerdank as our man, we embrace Ehrenhold. Historians should feel kinship with Theuerdank's companion, the scholarly Ehrenhold. More than the "Last Knight's" attendant or "good genius," Ehrenhold is also a servant of Clio. He records Theuerdank's exploits so that an accurate account of Maximilian's heroics is eye witnessed and recorded for posterity. Ehrenhold's wheel of fortune livery would suit a historian. The essays of Aliaksandr Kazakou and Vladimir Shirogorov document twists of fate on the battlefield. We are Ehrenhold's heirs in that historians are tasked with identifying and explaining what contemporaries did not see; military revolution studies do indeed draw the big picture whilst acknowledging history's wheel of fortune. Ehrenhold's presence is a sympathetic tribute to the historian's eye. Accurate and objective chronicling of military campaigns (in this case Maximilian's odyssey to prove himself worthy of his bride-to-be) is a noble calling. Reliable narrative, of course, is the first step toward making sense of history. *Theuerdank* implicitly subscribes to a worldview in which combat and history-writing are seamlessly connected.

In a book documenting multiple military revolutions, it is imperative that we acknowledge that the major actors in our foremost military revolution's events would likely be skeptical of our enterprise. *Theuerdank* reflects, through

---

**26** Füssel, *Theuerdank*, 39–43.
**27** Heuser, "Denial of Change," 3–27; Neill, "Ancestral Voices," 487–520; Cassidy, "Machiavelli and the Ideology of the Offensive," 381–404.

the circulation of ideas via the printed book, how those living through the prodigious changes wrought by the early sixteenth-century military revolution reconciled chivalric ideals with the gritty reality of gunpowder revolution described in this anthology. Our cover image accentuates continuity in the practice of arms at the height of the the gunpowder age's military revolution. Simultaneously its scene piques the historian's sense of the inherent and apparent incongruity of the end of the ancient world we have lost and the commencement of centuries of devastating gunpowder warfare that followed.

# Bibliography

Cassidy, Ben. "Machiavelli and the Ideology of the Offensive: Gunpowder Weapons in The Art of War." *The Journal of Military History* 67.2 (April 2003): 381–404.

Dodgson, Campbell. "Some Undescribed States of *Theuerdank* illustrations." *The Dome* 6.16 (April 1900): 131–138.

Füssel, Stephan. *Theuerdank. The Epic of the Last Knight* Cologne: Taschen, 2018.

Heuser, Beatrice. "Denial of Change: The Military Revolution as seen by Contemporaries." *International Bibliography of Military History* 32 (2012): 3–27 [pagination varies with format and mode].

Hof, Anna-Franziska. *Die Ritterideologie Kaiser Maximilians I. im "Theuerdank"* Marburg: GRIN Publishing, 2011.

Krause, Stefan. "The Making of the 'Last Knight'. Maximilian I's Commemorative Projects and their Impact." In *The Last Knight: The Art, Armor, and Ambition of Maximilian I*, edited by Pierre Terjanian, 53–61. New Haven: Yale University Press, 2019.

Moxey, Keith. *Peasants, Warriors, and Wives. Popular Imagery in the Reformation* Chicago: University of Chicago Press, 1989.

Müller, Jan-Dirk *Gedechtnus. Literatur und Hofgesellschaft um Maximilian I*. Forschungen zur Geschichte der Älteren Deutschen Literatur 2. Munich: Fink 1982.

Müller, Jan-Dirk "Le jeune Maximilien et l'internationalisation de la guerre. A propos du Weiskunig de Maximilien 1er." In *Les mots de la guerre dans l'Europe de la Renaissance*, edited by Marie Madeleine Fontaine and Jean-Louis Fournel, 21–27, 341–342. Geneva: Librairie Droz, 2015.

Musper, H.T., ed. *Kaiser Maximilians I. "Weisskunig"*, 2 volumes. Stuttgart: W. Kohlhammer Verlag 1956.

Neill, Donald A. "Ancestral Voices: The Influence of the Ancients on the Military Thought of the Seventeenth and Eighteenth Centuries." *Journal of Military History* 62.3 (July 1998): 487–520.

Oman, Sir Charles. *A History of the Art of War in the Sixteenth Century* New York: E.P. Dutton, 1979.

Tennant, Elaine C. "Productive Reception: *Theuerdank* in the Sixteenth Century." In *Maximilians Ruhmeswerk: Künste und Wissenschaften im Umkreis Kaiser Maximilians I.*, edited by Jan-Dirk Müller, and Hans-Joachim Ziegeler, 295–348. Berlin, München, Boston: De Gruyter, 2015.

Terjanian, Pierre, ed. *The Last Knight: The Art, Armor, and Ambition of Maximilian I*. New Haven: Yale University Press, 2019.

von Hefner, J.H., ed. *Hans Burgkmairs Turnier-Buch* Sigmund Schirmer or Schmerver [2 different spellings]. 1853. Dortmund: Taschen, 1978.

# List of Contributors

**Mark Charles Fissel** (PhD, California, Berkeley 1983) is Professor Emeritus of History at Augusta University (GA). He has authored *The Bishops' Wars: Charles I's Campaigns against Scotland 1638–1640* (Cambridge 1994) and *English Warfare 1511–1642* (Routledge 2001). He edited *War and Government in Britain, 1598–1650* (Manchester 1991). In 2006, he co-edited with D.J.B. Trim *Amphibious Warfare 1000–1700. Commerce, State Formation and European Expansion* (Brill); and in 2007, with Buchanan Sharp, *Law and Authority in Early Modern England* (Delaware). Forthcoming works include "Asia's ancient Mediterranean littoral as a military contact zone 2500-498 BCE" in *Handbook of Asian Military History*, edited by Kaushik Roy and Dennis Showalter, Oxford University Press – India. Fissel is a Fellow of the Royal Historical Society.

**Hyeok Hweon Kang** is Assistant Professor of East Asian Languages and Cultures at Washington University in St. Louis, Missouri. His current book project, *The Artisanal Heart: Craft and Vernacular Science in Early Modern Korea*, recasts the history of early modern science from the perspective of artisans and practitioners in Chosŏn Korea (1392–1910). His previous work on the subject includes "Crafting Knowledge: Artisan, Officer, and the Culture of Making in Chosŏn Korea, 1392–1910" (PhD Dissertation, Harvard University), for which he was the laurate of the 2021 Turriano Prize and the 2021 International Convention of Asia Scholars Book Prize (Best Dissertation in the Humanities). His research on the military revolution has appeared in the *Journal of World History* and *The Journal of Chinese Military History*.

**Aliaksandr Kazakou** received his Candidate of Sciences degree from Belarusian State University in 2011 for the dissertation on migration from Muscovy to the Grand Duchy of Lithuania in early modern times. His field of interest also embraces military history of Eastern Europe in the fifteenth and sixteenth centuries with an emphasis on the wars between Lithuania and Muscovy. He is currently an independent scholar.

**Wayne E. Lee**'s most recent book is the co-authored *The Other Face of Battle: The Experience of Combat in America's Forgotten Wars* (Oxford 2021). Other books include *Waging War: Conflict, Culture, and Innovation in World History* (Oxford 2016), and *Barbarians and Brothers: Anglo-American Warfare, 1500–1865* (Oxford 2011). He has edited two volumes on world military history and has published extensively on subjects ranging from Native American warfare to Mongol conquest methods. He has an additional career as an archaeologist, with field work in Greece, Albania, Hungary, Croatia, and Virginia, and is a principal author and a co-editor of *Light and Shadow: Isolation and Interaction in the Shala Valley of Northern Albania*, winner of the 2014 Society for American Archaeology's book award.

**Dr. Mark D. Mandeles** is author of *The Development of the B-52 and Jet Propulsion* (Air University Press, 1998), *The Future of War: Organizations as Weapons* (Potomac Books, 2005), and *Military Transformation Past and Present: Historical Lessons for the $21^{st}$ Century* (Praeger, 2007), and co-author of *American and British Aircraft Carrier Development, 1919–1941* (Naval Institute Press, 1999), *Innovation in Carrier Aviation* (Naval War College, 2011), and *Managing "Command and Control" in the Persian Gulf War* (Praeger, 1999). He was an independent consultant to the Department of Defense, including the Office of Net Assessment, and an adjunct professor at George Washington University's Elliott School of International Affairs.

**James O'Neill**, a native of Belfast, worked as a contract archaeologist for 16 years prior to pursuing his doctorate in history. He received a First in earning his baccalaureate in modern history at Queens University Belfast (QUB) in 2007. His MA and PhD degrees followed in 2009 and 2013. Subsequently, while at University College Cork, O'Neill published *The Nine Years War, 1593–1603. O'Neill, Mountjoy and the Military Revolution*, Four Courts Press (Dublin 2017). He has also published a score of articles about Irish military history in the sixteenth and seventeenth centuries. O'Neill was a co-director of archaeological contractors which provided specialist archaeological advice and skills to the Northern Ireland Environment Agency: Built Heritage (NIEA:BH). He developed the North Ireland Battlefield Database as well as the Defence Heritage Database for NIEA which are used to inform archaeological mitigation for planning and development issues across Northern Ireland.

**Vladimir Shirogorov** authored the three-volume *Ukrainian War. The Armed Conflict over Eastern Europe in the 16$^{th}$-17th Centuries*, published in Russian by Molodaya Gvardiya (Moscow) in 2017, 2018, and 2019, respectively. The series includes *The Melee of Rus* (volume one), *Turkish Onslaught: Balkans – Black Sea – Caucasus* (volume two), and *Head-to-head Offensive: Baltics – Lithuania – Steppes* (volume three). In 2021, his monograph *War on the Eve of Nations. Conflicts and Militaries in Eastern Europe, 1450–1500*, was published in English by Lexington Books (New York and London). Shirogorov is a member of the Società Italiana di Storia Militare.

**Colonel João Vicente** has authored *Guerra Aérea Remota: A revolução do poder aéreo e as oportunidades para Portugal* (Fronteira do Caos, 2013); *"Beyond-the-box" Thinking on Future War: The Art and Science of Unrestricted Warfare* (VDM, 2009) and *Guerra em Rede: Portugal e a Transformação da NATO* (Prefácio, 2007). He has co-edited, authored and co-authored several articles in areas related with Air Power and Unmanned Aircraft Systems. He is a faculty member of the Military Sciences PhD course at the Military University Institute, in Lisbon.

# Index

How to use this anthology's indexes

Given the geographical and chronological breadth of this volume, a traditional index was not an option. The editor and the Press believe it most utilitarian (though ahistorical) to provide in the printed edition an index that focuses almost exclusively on topics and themes. These have been cross-referenced for the most part. Some topics (e.g., strategy, tactics, warfare, infantry, musketeers, soldiers, state formation, weapons/weaponry, peripheries, militia) are so ubiquitous in this work that they are difficult to manage in a succinct index. In many instances, however, such terms do appear in the sub-listings within a comparatively narrower topic. We have been sparing in explicit citations of "systems" as well, conceptual constructs that permeate much of the anthology. The qualitatively different pair of indexes referenced below should somewhat ameliorate unavoidable minimalizations and omissions.

Historical personages and geographical locations (the latter including battles) have been relegated to an electronic data base as separate indexes. These may be found on the publisher's webpage

https://www.degruyter.com/document/isbn/9783110661415/html?lang=en

The publisher, editor, and authors consent to allow (indeed, encourage) downloading of these indexes as pdf files.

*ablegate* 93
acquisition process
– European peripheries, differences on 354
– growth of knowledge and 385–86, 393–94
– naval military revolutions and 316–17
– Revolution in Military Affairs (RMA) and 346
– predictability of outcomes 354
– state's role in 327
– United States, in 316, 381
air conditioning 329
ammunition 78–79, 85, 106, 185, 197
– muskets, and 32, 35, 40, 47, 52, 57
– naval military revolution, in 226, 274, 33
amphibious operations
– aside build-up 228–30, 243–45, 246, 273–77, 280, 285–88, 290, 291, 293
– assault ship 208–10, 228, 239, 240, 242, 259, 281, 286, 293 (See also battleship, caravel, carrack, galley)
– attack 168, 170, 208–10, 212, 224, 239, 273–84
– craft 180, 211, 225, 227, 236, 237, 239–40, 258, 278, 279–82, 319, 328 (See also battleship; *boyer*; buss; caravel; carrack; flyboat; galley; *krafell*, *strugi*)
– direct assault 228–31, 234, 244, 259–69 passim, 273–76, 281, 285–88, 291, 293
– doctrine 210, 274, 276, 277, 286
– escort design 214–21, 245
– expedition 210, 221, 223, 234, 235, 239, 243, 257, 277, 286
– force 208–21, 224, 228, 236 (See also deck-to-shore gunfire; artillery)
– landing 210
– operation 208–10, 214–24
– troops 208, 209, 210, 221, 224, 228, 234, 235, 239–45 passim, 258, 259, 273–80 passim, 284–85 (See also marines)
– venture 214, 235, 279, 287

– warfare 22, 207–312, 339
anachronism 337
antiquity, classical, inspiration and authority 339, 344, 348
– military revolutions, role in 338
arquebus (See also harquebus)
– accuracy of 49
– *archibugio* 75
– *arquebuse à croc* 75
– gunpowder revolution, part of 16
arquebusiers 97
– infantry, part of 292
– significance of 194
– tactical utility of in warfare 101, 194
art of war 2, 6–8, 11, 21, 314, 323, 334, 336, 448–56
– amphibious 285–86, 291
– Sarnicki, Stanislaw and 98
Artificial Intelligence (AI) 19, 433
– autonomy of drones, allows 423
– battlespace, and 411
– defensive value of 416
– ethical issues raised by 424
– human control, meaningful 415, 417, 426, 427, 429, 436
– human-machine teaming, and 389, 414, 416–17, 422, 435
– loyal wingman 418
– moral dilemmas posed by 406
– role in shaping future war 411, 418
– Skyborg Project, and 418–19
– swarming, tactical 420, 422, 428
– (See also Autonomous Aircraft Systems [AAS])
artillery
– bastion design, part of 118
– battle of Orsha, in 67–112
– coastal 120
– defensive value 78
– English, use by 180, 195, 197
– field 67, 89, 96, 180
– fortification, role in 9, 78–81
– fortress 126–43
– French, use by 66, 102, 106, 117–18
– gunpowder revolution, part of 4
– increase in destructive power 66
– increasing importance of 67
– Ireland, limited use in 180, 197, 199 201, 209
– kolomborna 152
– Lithuania, use by 78, 84, 96
– Lviv, use by 74
– manufacture of 9
– Moldavia, use by 86
– Muscovy, use by 81, 87–88
– Ottomans, use by 103–5, 113–76 passim
– Parker, Geoffrey, ideas about 177, 178
– Poland, use by 85, 96
– Russians, use by 153
– siege 66, 87–89, 91, 102, 116, 118,
– Spanish, use by 102
– strategic culture 15
– technology, changes in 9
– *trace italienne*, and 116–23
– (See also artillery revolution; artillery tower; deck-to-shore gunfire)
artillery revolution 340, 353
artillery tower 120
– Ottoman fortress, use of 131, 140, 143
Asianists 32
– Geoffrey Parker and Michael Roberts, responding to 32
autonomy 313
– navies, of 329
– of state in Marxist theory 296
– regional in Ottoman empire 166
– (See also Autonomous Aircraft Systems [AAS]; Artificial Intelligence [AI])
autonomous
– autonomous military technology and autocracy 405, 406, 408, 412
– Science and Technology Studies sees technology as 360
– (See also Autonomous Aircraft Systems [AAS])
Autonomous Aircraft Systems (AAS) 406–46 passim
– autonomous military technology and autocratic governments 405, 406, 408, 412
– challenges to 423–33
– definition of 413–15
– ethical dilemmas in 426–29
– inevitability of 389, 406
– Iraq 431

- irreversibility of 406
- lethality of 389, 405, 406, 409
- MRA, part of 409,
- Third Offset, and 389
- unmanned aircraft, role in 360, 389, 409, 410–13, 417, 430, 433
- (See also Artificial Intelligence [AI])

bastion
- angular 118
- fortress see bastion fortress
- mixed elements in 150
- offensive power of 118, 144–46
- platform 137
- tower 140,
- *trace italienne* forts, use in 118, 137, 143, 150, 157
- triangular 122, 123, 133
- water battery, part of 151
- (See also bastion fortress)
bastion fortress 66, 155–62, 170, 171, 196, 198, 199
- incorporate mixed elements 140, 150, 154
- offensive power of 133, 144–46
- Ottoman and European styles, difference between 133, 143
- *trace italienne*, design 118
battlefleet revolution 332, 345
battlespace 331, 351, 384, 411, 419, 422
- See also Autonomous Aircraft Systems (AAS)
battleships
- "Geoffrey Roberts" paradigm 325
- British, innovators of 324
- evolutionary character of 335
- punctuated equilibrium, and 331–32, 335
black box 327, 347
bomb ketch 324
bombardment
- Battle of Orsha, at 96
- from sea 208, 225, 227, 235, 240, 273, 280, 332
- military revolution, and 380, 436
- Ottoman fortress, protected against 127
- punctuated equilibrium, and 332
- (See also amphibious operations; deck-to-shore gunfire)

borders
- creation of *obrona potoczna*, led to 72
- diffusion of technology, shaped by 312
breech-loader
- amphibious actions, importance to 227, 237
- naval military revolution, part of 341
bureaucracy
- crucial for military change 296
- cultural context, seen in 8
- conduct of war, shaped by 24
- limits military change 6, 9–10, 171
- military infrastructure, shaped by 171
- Revolution in Military Affairs (RMA), and 370
bureaucratization 8
buss 235–36, 236

cannon (See artillery)
capitalism 289, 290, 319, 327, 356
caravel
- amphibious war, value in 228, 235
- artillery, usage of 227
- Portuguese use of 226
carriage 66, 78, 128, 242, 323, 329
- amphibious actions, importance to 227, 237
- four-wheel 240
- guns, breech-loader, use of 227, 237, 241,
- guns, muzzle-loader, use on 227, 237
- muzzle-loader 227, 237
- two-wheel 127, 227, 235, 237, 238, 239, 240
carriers, aircraft
- commercial ships remade into 332
- future war, part of 334
- naval warfare, and 409
- Revolution in Military Affairs (RMA), part of 333
carrier revolution 333–34
cast iron ordnance 66
Catherine's Wheel 451
cavalry 74, 90–106 passim, 292
- amphibious attacks, presence in 212–14, 292
- Battle of Orsha, at 65, 91, 200
- battle, utility of in 55–57
- East Asia, distinctive usage of 21
- firearms, vulnerability to 67
- infantry, relative size to 76, 83, 90, 90

- Japanese target practice, role in 38
- Korean musket training, role in 46
- Lithuania, use of by 90
- mercenaries, presence in 72
- modernization, and 103–5
- Muscovy, use of by 81
- Poland, use of by 76
- state sponsored 73
- strategic culture, and 15
- vulnerable to harquebusiers 194
- *zemskaya sluzhba*, consist of 72
- (See also horsemen, light horse)

challenge-response mechanism 20, 286, 331, 334, 340

*Challenger* shuttle investigation 391

*chorągiew* 93–4, 96

civil war
- English 19, 20, 54
- United States 328, 335, 391

command, control, communications, computer, intelligence capabilities (C4I) 379

commercial revolution 324–28 passim, 339

comparative history 59

concept of operations/operational concepts 24, 372, 379, 381, 382, 383, 384, 388, 394
- experimentation, and 408, 417
- role of in Revolutions in Military Affairs (RMA) 394

*Consilium rationis bellicae* (1558) 77

continuous reinvention 322, 330

convergence
- convergent evolution 329
- competing military systems, among 114
- history of Ottomans and Europeans 114
- gigantism 114
- naval evolution 331
- Revolution in Military Affairs, causes 356
- strategic culture, and 314
- technology, capitalism, nationalism 342

Corps of Royal Architects, Ottoman 149, 152

correlation of forces 315, 375–76

corvettes 351

cost-benefit calculation 115

crossbow
- displaced 96

- Poland, use of 76–7, 82
- power of 116, 208

cruisers 328, 332

cultural exchange 67
- military revolutions, and 301, 320, 338–39
- technological developments, and 201, 292–93, 317

cyberwarfare 3, 22, 333, 351, 356, 406

cyclical patterns in history 297, 313–14, 333, 335

deck-to-shore-gunfire 240, 258, 259
- amphibious operations, crucial to 247, 257
- caravel, used by 209, 210, 217, 223, 224–28
- escort design, change 257
- galley, used by 281
- gunpowder revolution, and 286
- innovations in 291, 293
- naval military revolution, part of 234
- success of 287
- (See also amphibious operations)

*delo* 78

determinism 18
- technological determinism, and 10, 18, 23
- technological determinism and cultural determinism 18

determinism, technological 68
- cultural factors, and 18
- limited explanatory power of 10
- Revolutions in Military Affairs (RMA), and 22, 313–15
- parity, and 58

*deti boyarskie* 81

divergence 7
- culture, military, in 51
- fortification design, in 115
- Japan and Europe, between 31, 57
- Musketry, training in 51–2, 55
- Technological 32
- training, military, in 51–2

dockyards 331, 333, 335

doctrine 210, 274, 276, 277, 286, 375

dragoons 83

dreadnoughts 332, 335

drill 169, 242

- difference between European and East Asian types of 32
- East Asian emphasis on 30
- marksmanship, emphasis on 31, 41–51
- military revolution, and 29, 113–14
- parity, and 51–5
- speed, emphasis on 33, 37, 3

Eighty Years' War 237
electricity 329
empires
- British 338
- Holy Roman 246, 257, 264, 447–56
- Mughal 313, 316, 326
- inner Asian 317
- of the weak 17, 19
- Ottoman 75, 114, 122–26 passim, 131, 144–50 passim, 154–59 passim, 164–69 passim, 289, 296, 299, 351
- Portuguese 228, 230, 277, 284
- Russian 153, 292
- Spanish 230
- Swedish 277
- predatory 356
expansionary states 3, 11, 15, 21, 239, 317
expedient design 147, 168
experimentation 10, 35, 49
- evolution change, as 258
- failures, lessons of 23
- firearms, with 35, 48–49
- gunpowder, with 242
- improvisation, as 122, 135, 143
- iron making, in 320
- naval 221–23, 331–33.
- operational concepts, an 394
- peripheries, on 339
- physical terrain, effect on 340,
- Soviet doctrine, discouraged by 375
- United States, growing interest in 38

fighting
- "mutant practices" of 297
- amphibious actions, value in 208, 210, 221, 224, 279, 286,
- autonomous Aircraft Systems (AAS), transformed by 425, 432, 437

- convergent evolution, causes innovation in 329
- firearms, handheld, usefulness in 106, 213,
- Greece, naval techniques in 328, 332, 339,
- Gunpowder revolution, ushers in new types of 244, 301, 322
- Lithuania, capacity of 106
- muskets, English techniques with 39–40
- muskets, Irish success with 14, 186
- muskets, Korean techniques with 37
- pike-and-shot, example of innovation in 301
- technological innovations, and 323, 378, 409
- ways of war, determined by social context 410
- (See also wagon camp)
fighting wagon. See *oboz, tabur*, wagon camp
financiers 325
firearms
- adaptation of 317, 351
- amphibious warfare, changes in use of 242–44, 274, 290
- amphibious warfare, defense against 226
- amphibious, warfare, effectiveness in 221, 225, 240, 242–43, 291
- Battle of Orsha, at 91–101
- defensive value of 66, 77
- East Asia, use in 31–32, 41–51
- gigantism, and 114
- infantry and 222, 234
- Ireland, use in 21, 180–83, 189, 191, 194, 199
- Lithuania, usage of 78, 82
- military revolution, and 3, 66, 201
- Moscow, usage of 81, 83, 87–90
- Ottoman, usage of 169, 301
- Parker, Geoffrey, central to theories of 113–14, 55, 201
- Poland, usage of 75, 76, 77, 80, 83–87, 301
- strategic culture, and 5
- tactical innovations, made possible by 101–7

- technology, regional variation 18, 32–51, 58–59
- transformations in 209, 355, 453
- usage, regional variation in 7, 9, 10–12, 32–51, 58–59
- (See also arquebus; *Hackenbüchse*; *hakownica*; harquebus; matchlocks, hatchet gun; muskets; pike and shot; pistol gun; *rusznica*; *tyufyaki*; *piszczels*)

fiscal-military state
- bastion fortress, impact on 207
- military revolutions, and 29, 207, 297, 299
- naval warfare, and 207
- Parker, Geoffrey, ideas about 202, 207
- ruling class, and 355
- state formation, and 207, 299, 433

fiscalism 327, 342
flintlocks 33
flyboat 237
force projection 11, 209, 213, 221, 273, 286, 319, 322, 326, 329, 332, 342–45, 354, 360

fortification
- amphibious warfare, and 208–9, 212, 223, 234, 236, 239–40, 258–59, 273–75, 280–84, 298–99
- architecture, European 30
- artillery, role in 9, 78–91
- Bashtovë (Albania), at 135, 137–38, 160
- Battyán, at 146
- Buzim (Bosnia-Herzegovina), at 133, 137
- Bender, at 125, 165-6
- Cow Tower at Norwich 120–21
- Eastern European 85–91
- Fetislam (Serbia), at 135, 137–38
- Germany, in 67
- Henry VIII of England, of 120–21
- Iberia, in 67
- Ireland, in 16, 177–201
- Italy, in 67
- Kinsale, at 199
- Kızılhisar, at 147
- Korean 55–7
- Multi-Gun Tower 120
- Mytilini, at 140, 142
- Native American 115
- Ottoman 16, 113–71

- palankas; 147–8
- Rio (Messenia), at 133–34, 139, 141, 151, 167
- Salses, at 119
- Smederovo (Serbia), at 140, 160
- Travnik (Bosnia Herzegovina), at 133–34, 136
- Vál, at 146
- zone at Buda (Budun) 146
- Zsámbek, at 146

fortification vocabulary (Ottoman) 154, 169

fortress(es)
- amphibious attacks, and 208, 228, 234–37, 273, 276, 278, 278, 282, 283
- Anadolu Hisari 127–28
- Avlonya (Vlorës), at 125, 137, 139, 165–66
- artillery, Ottoman use of 131, 140, 143
- artillery, used against 66, 89, 106
- Brăila (Romania), at 157
- construction, innovation in 113–17
- Danube frontier (Eger, Esztergom, Kaniszsa, and Székesfehérvár) 153
- Dniester, at 125
- English forts in Ireland (Newry, Carrickfergus, Corkbeg, Charlemont, Mountnorris, Philipstown, Maryborough, and Blackwater Fort) 198
- firearms, in 80, 85, 114
- fortress, bastion. See bastion fortress
- gigantism and 114
- Greek (Kelefa, Passavas, and Zarnata) 160, 162–64, 168–69
- Hungarian (Eger, Kassa, Komárom, Csókakő, Gesztes, Győr, and Gyula) 146–47
- Kale-i Sultaniye 129–31, 135, 140
- Kilidülbahir 129–31, 133, 140–1, 151, 158, 164, 167
- Kumkale 157–58, 167
- Özi, at 125, 153–57, 160, 164–66, 168
- Parker, Geoffrey, ideas about 19, 165, 177
- Rumeli Hisari 127–8, 151, 158
- revolution 330, 353
- Seddülbahir 157–9, 167–69
- Segedin (Szeged), at 125, 165–66
- *trace italienne* (See also *trace italienne* fortress)
- Yedikule 129, 1313–2, 135, 164

fortress design 113–76 passim
- defensive elements of 119, 127, 157, 170–71
- offensive elements of 118, 120, 127, 129, 133, 146, 150, 152, 157, 160, 171
- New Navarino (Pylos, or Anavarin-i Cedid) 150–52, 169
- transitional period of 118, 122
frontiers 5, 6, 14, 74, 88, 145, 147, 154, 167, 342, 383

galleons 225, 227
galleys
- amphibious craft, as poor 279–81
- French use of 211
- Ireland, use in 180
- La Goletta, attack on 227
- naval transformation, part of 321
- Ottomans, use of 273
- Spanish, use by 276
gigantism 114–15
global military history 29, 30, 31–32, 59, 287, 313, 327, 341, 343, 351, 352, 355, 371
- (See also parity)
*gulyaj-gorod* 293, 301
- (See also pike and shot; *oboz*; *tabur*; wagon camp)
gunfire
- amphibious, battles, in see deck-to-shore gunfire
- pike-and-shot, from 200
- sharpshooting, as 30, 34, 39–59, 102
- (See also marksmanship; arquebus; harquebus; gunnery; firearms; matchlocks; pistol guns)
gunlocks 7
gunnery 240
- amphibious assault, and 284
- design, changing 331
- Hospitallers, use of 282
- naval 240
- Ottomans, success with during naval attacks 259
- success of, relative, in naval attacks 274
- Spain, use of 283
gunpowder revolution 16–21

- amphibious warfare, impact on 11, 208, 224–41
- infantry, changes in usage of 8
- Ireland, in 19–20, 181
- Japan, in 7
- Korea, in 7
- Muscovy, in 10,
- Ottoman empire, in 113
- Poland, in
- Lithuania, in 78
- military revolution, and 10–11, 313–25 passim
- nation building, and 15
- naval military revolution, and 11
- Parker, Geoffrey, emphasis on 66, 177
- peripheries, different on 13
- regional variation 5–6
- Revolution in Military Affairs (RMA), and 22–23
- Roberts thesis, and 8–9
- state, transformations in 8, 18
- strategic cultures, occurs differently in 5, 17, 18
- *trace italienne*, and 66
gunpowder
- "gunpowder empire," Ottoman 75
- amphibious warfare, impact on 2–7, 312
- Battle of Orsha, at 91–101
- Eastern Europe, impact of in 68
- Hundred Years' War, in 66
- innovations in 242
- Ireland, use in 180, 186, 191, 197, 200, 201
- Lithuania, usage in 78, 89–90, 91–101
- Muscovy, use in 82–84, 87–89, 91–101
- naval military revolutions, part of 316–25
- Ottoman empire, use in 103–5
- Ottoman fortress, and 113–76 passim
- Poland, use in 78–80, 84–87, 90–91, 91–101
- *Reconquista*, in 66
- state, role in distributing 79
- *trace italienne*, and 66
- (See also arquebus; artillery; firearms; harquebus; hatchet gun; musket; pike-and- shot, pistol gun)

guns
- Battle of Orsha, at 94, 91–101 passim
- fortification, used during 87
- increasing success with 105
- Ireland, usage in 183, 193, 197, 201
- Ottoman empire, in 115
- Scotland, usage in 181
- siege of Smolensk, at 88, 91
- (See also: arquebus; artillery; breech-loader, firearms, *Hackenbüchse*, *hakownica*, harquebus, hatchet guns, muskets, muzzle-loader, ordnance, pike and shot, pistols, *rusznica*, *tyufyaki*, *piszczels*)

*Hackenbüchse* 67
*hakownica* 67, 75–78, 81–82
handgunners 56
- amphibious warfare, in 239, 242, 291–92, 299
harquebus
- amphibious assault, role in 242–43
- Eastern Europe, first used in 67, 75
- military tactics, encourages new 101–5
- (See also arquebus)
harquebusiers 67
- cavalry, value against 194
- infantry, part of 103, 292
- significance of 194
- tactical utility of in warfare 101–5, 243, 292–93
- (See also arquebusiers)
hatchet gun 451
hetman 65, 80, 86, 93, 98, 101, 292
*Hetman Books* 98
High Renaissance 336, 453
*histoire événementielle* 14, 18
historicism 314, 352, 384
historiography
- firearms, transformation of 209
- fortification, on 143
- Ireland, early modern, on 179, 200
- military revolution 2, 66, 312, 314, 322
- Ottoman, amphibious power of 279
hoplite revolution 338–39
horsemen (troops) 34, 57, 81, 103, 212, 213, 291, 436

- amphibious warfare, in 211–12
- Battle of Orsha, at 91–101 passim
- Ireland, utility in warfare 189
- Japanese, use of 56
- Poland, use of 73
- (See also cavalry)
*hospodar* 70, 74, 80, 86, 94
Hundred Years' War 66–67, 370
hybridization/hybridity 317, 201, 292–93
hydraulics 332, 335

imperialism 6, 12, 13, 15, 19–23 passim, 54, 315, 318, 326, 327, 333, 338, 341–44, 373
industrial revolution 34, 323–29 passim., 331, 333, 342, 344, 355
industrialization 333, 366
- imperialism, importance to 327
- military revolutions, crucial to 315, 325, 328, 329, 342, 350
- nationalism, and 341–48 passim
- naval military revolution, crucial to 319, 324, 331, 332
- peripheries, rendered extinct by 24
- Revolutions in Military Affairs (RMAs), foster 315
- state formation, necessitated by 355
- strategic culture, and 341–44
innovation
- aircraft, in 333
- amphibious warfare, in 226, 228, 245, 286, 290
- Automated Aircraft Systems (AAS), in 406
- cannon/artillery, in 66
- cultural exchange spurs 3, 24, 67, 317
- European role in 29–30
- fortress construction, in 113–17
- gunpowder usage, in 4, 29, 67
- hybridization, cause of 3, 317, 201
- identifying historical laws of 369, 383
- imperialism and nationalism, encouraged by 315, 342
- improvisation, through 14
- military evolution, and 330, 358
- military tactics, in 4, 29, 67
- military technology and organization, in 369, 378, 380

- musketry, in 34–58, 29–64 *passim*.
- nationally/regionally specific 29–30, 33–34, 58, 66, 106, 166–67, 178, 228
- new technology, driven by 24
- operational 408, 312–13
- peripheries and semi-peripheries, on 3, 14,
- Revolution in Military Affairs (RMA), as 334, 381, 394
- social change, requires 297
- state formation, facilitated by 327
- technology, not determinative of 293, 347, 357–58, 407

ironclads 335, 36
Islam/Islamic 224, 234, 383
Italian Wars (1494–1559) 63, 67, 91, 455

janissaries-harquebusiers 104
jousts 336

*kapukulu* 74
*kazionnye pishchalniki* 82
khorasan/horasani mortar 144, 169–70
knowledge, authority of
- Revolution in Military Affairs (RMA), in 390, 392

knowledge, growth of
- shaped by acquisition process 385, 386, 393–94
- slowed by Soviet ideology 373, 375
- spurred by Office of Net Assessment 381
- through interaction at port cities 321
- through nation state 327, 386, 387, 394–94
- unpredictability limits Revolution in Military Affairs (RMA) 383

knowledge, requirements for command
- higher after recent Revolution in Military Affairs (RMA) 392

*Krasnaya Zvezda (Red Star)* 373

light horse 56 (See also horsemen, cavalry)
linear patterns in history 3, 9, 15, 277, 313, 314, 330, 336, 337, 340, 344, 348, 358, 410
*listy przypowiedne* 86
Long War (1593) 146–47

*ludwisarnia* 74
littoral 282–86
- African, vulnerable to Portuguese amphibious attacks 228
- amphibious war, site for 210
- carrack, utility in attacks on 225
- firearms, revolutionary effect in attacking 242
- framed military revolutions 3–6
- naval attacks on, difficulty of 242
- naval evolution, increases attacks on 332

logistics
- amphibious assault, importance in 211, 212, 214, 221, 273, 287, 342
- English, success with 178
- firearms, problems with 85
- helps resist European powers 21, 30, 179
- naval war, importance in 201
- Ottoman military, problems with 166

Major Defense Acquisition Programs (MDAP) 386
marines 208, 211, 213
- amphibious assault, important during 240, 286
- amphibious assault, coordinated role in 244
- amphibious assault, changing roles in 291
- gunpowder era, increasing importance in 241, 274, 284, 290

marksmanship 4, 7, 30, 34, 39–59, 84
- accuracy, emphasis on in drill 31, 34, 44, 46–51, 54
- speed, emphasis on in drill 41, 57
- (See also sharpshooters)

mercantile/-ism/-ist 11
- carrack 318
- class 354
- expansion 324, 325
- governmental policy and institutions, and 326, 327–28
- ideology 319
- states, and 342

Marxism-Leninism
- theory as a barrier to military innovation 371–73

Marxism
- dialectic 356
- gunpowder, as revolutionary "productive force" in 353
- interpretive framework, of 371

Marxist class struggle
- military revolutions, to understand context of 314
- state strategic concerns, role in shaping 355

matchlocks 36, 54, 242
mercenaries 431
- amphibious combat, in 213, 243
- Battle of Orsha, at 93–101 passim
- infantry, dominated by 90
- Ireland, army in 194
- Lithuania: in army 73, 90, 92, 106
- Poland, in army 72, 292
- Teutonic Order, in army 85

metallurgy 115, 320, 335
microhistory 314, 358
military revolution
- "Ideal Type" paradigm, limited utility of 288–93, 302
- aircraft carriers and 33
- amphibious war, importance of 207, 237, 242, 243, 273, 277, 285–93
- artillery, role of 74, 180, 330
- becomes Revolution in Military Affairs (RMA) 333, 342
- bureaucracy and 326
- commercial expansion and 324, 326
- comparative history, importance of 113
- comparative study of 68, 355
- core-periphery model, inadequacy of 290
- cultural exchange and 301, 320, 338, 339
- determinism and 36–37, 310, 314, 356–58
- East Asia, experience of 41–57
- fighting, innovations in 297
- firearms, role in 201
- fiscal-military state, role of 207, 299
- fortress, bastion 207, 288
- gunpowder, role of 17–18, 242–43, 288, 341
- Hundred Years' War, contributes to 370
- imperialism/colonialism and 324, 338
- industrialization, and 315, 325, 328, 342
- Ireland, experience in 179, 180, 192, 201, 207
- love of the ancients, and 336
- military evolution, and 354, 356–57
- multiple types of 3
- Muscovy, in 74
- mutant practices 293, 296, 297, 299, 313
- nationalism, role of 1, 5, 22, 315, 333, 342–45
- Ottoman empire, distinctiveness of 143, 165, 113–71 passim, 273
- parity among regions 51–59
- Parker, Geoffrey, ideas about 18, 66, 207,
- pike and shot innovations, example of 179
- regional differences, significance 4, 6, 7, 9, 12. 16. 21, 24, 29–30, 33–34, 47, 50, 55, 58, 66, 106, 166–67, 178, 228, 301
- Revolution in Military Affairs (RMA) and 1, 408, 429–33
- Roberts thesis 29, 243
- social change and 20, 350
- social context of 297, 314, 317, 348
- Soviet ideas of 371–76
- spurred by specific conflicts 228
- state sponsorship, role of 319, 327
- strategic culture and 20, 313
- Sweden, example of 277
- technological determinism 36–37
- technology transfer, role of 74
- *trace italienne* fortresses, significance of 18, 66
- variants of 3
- variations outside of western Europe 30, 143, 165, 341
- Western Europe as paradigm of 30

Military-Technical Revolution (MTR) 371–73, 376, 379
mortars 324
muzzle-loader 227, 237, 328
muskets
- divergence, in training with 51–52, 55
- English success with 39–40
- innovations in 34–58, 29–64 passim
- Irish success with 14, 186
- Japanese, musketry training at 48, 53, 55
- Korean training, in 46, 37

Napoleonic Wars 324
nationalism 5, 22
- industrialization and imperialism, interacts within warfare 333, 341–48
- naval military revolutions, and 315
nation-building
- Eastern Europe, preceded military revolution 10
- military revolution, relation to 341, 343
- Roberts thesis, role in 5, 29, 202
- strategic culture, shaped by 341
- warfare, effect on 18
- (See also fiscal-military state; military revolution, Revolutions in Military Affairs)
naval military revolution
- acquisition process, and 316–17
- breech-loader, part of 341
- carrier revolution, part of 334
- commercial, aspects of 324–25
- evolutionary, aspects of 323, 330–32, 334–35
- experimental, character of 331
- gunpowder, impact on 322, 323–24, 326
- industrial revolution, role of 315, 325
- land-based military revolutions, distinct from 319
- nationalism and industrialism, interaction with 1, 315
- specificity, geographic 326
- state, involvement of 319, 327
- strategic culture, and 336, 339, 341–45, 348–51, 354, 357, 359–61
- tactics and weaponry, aspects of 339
nuclear weapons 359, 373–74, 384, 406, 435

*oboz* 213, 292, 293, 301
- (See also *gulyaj-gorod*)
*obrona potoczna* 72–73
Office of Net Assessment 370, 376
- comparative study of military innovation, by 378, 381
- formation of 376–77
- study of history, by 382
organizational ecology 384
ordnance 80, 83, 85, 101–2
- artillery, power of 66, 80, 82, 84

- Battle of Orsha, at 94–101 passim
- cast-iron, utility of in warfare 66
- development over time 66, 78
- French, use of 66
- Lithuania, use of 78, 81
- manufacture of 9, 79, 106
- mobility of 66
- Muscovy, use of 9, 87, 89
- state, role of in distributing 79
- (See also artillery)
Ordnance Office (English) 350
*ostrog* 309
Ottoman War of 1499–1503 321

paradigmatic armies 8, 114, 289
parity 51–59
*pavises* 76
*pavisiers* 77, 97, 99
Persian Gulf War 378, 380
philhellenism 338
pike and shot 179, 290–93
pikes
- Irish, usage by 171, 189–90
*pishchalniki* 82, 83, 88
pistol gun 67, 105, 189
*piszczels* 83, 89, 76, 78, 82
planning, programming, budget system 384
*pochet* 72, 80, 93
poliorcetics 278, 330, 453
*pomeshchik* 81
*pomestie* 81
*pomestnaya* cavalry 81
*pospolite ruszenie* 72
precision-guided munitions (PGMs) 388, 389, 420, 424, 437
precision-strike military revolution 389
- Autonomous Aircraft Systems (AAS), aided by 436–37
printing
- Europe, in 318
- Ottoman empire, in 116, 124, 149, 169
propaganda campaign 382, 421
punctuated equilibrium 330–35
*pushechnaya izba* 81

rangefinders 329
*Reconquista* 107

refrigeration 329
regionality 313
Revolutions in Military Affairs (RMA) 384, 369–405 passim
– authority of knowledge in 390, 392
– conflict between military experts and politicians, shaped by 390
– emerge within historical and social contexts 390
– emphasis on technology rather than history 382, 393
– Soviet ideas of 371–76
– Third Offset Strategy, and 388–90
– unpredictable knowledge of growth, limited by 383
revolution in siege warfare 330
*ruchnaya piszczel* 83
*rusznica* 75, 83, 87, 94–99 passim

sailing gunship 207
screw propulsion 329
seapower 322–29 passim, 332–34
searchlights 329
*Sejm* 72–73
sharpshooters 7, 51, 54, 58 (See also marksmanship)
*shlange* guns 80
siegecraft 16, 65, 84, 88–89, 124, 144, 330
snake box 451
social construction of technology (SCOT) 350
social constructivism 58, 349, 356
social shaping of technology (SST) 349, 350, 356–57
springalds 116
spears (English royal) 74, 292
steel 98, 185, 191, 329, 332, 354
strategic calculation 5, 10, 15, 167, 355
strategic culture
– amphibious 214
– antiquity, in 336
– artillery, and 15
– evolution, naval military and 329–30
– industrialization, and 331–44
– military revolution, and 20, 313, 316, 361
– naval military revolution, and 316, 336, 339, 341–45, 348, 341–45, 348–51, 354, 357, 359–61

– Revolutions in Military Affairs (RMA) 341–51 passim, 354, 369, 361
– Russian 239, 293
– social context, relationship to 317
– state, influence of on 327, 361
– technology transfer, and 317
– wealth, and 316
– weapons and tactics, relationship with 315
strategic predicament 4, 7, 15, 331, 337, 339
*streltsy* 83, 243, 292
*strug* 237, 239
submarines 334, 374, 387
sword-and-buckler 336
synchronization 17, 101, 192
*szlachta* 72, 93, 96, 106, 107

*tabur/tabor* 213, 301, 303
– (See also wagon camp; *gylyaj-gorod*)
taxonomy 11, 222, 413, 414
technological advances, interwar aviation 383–84, 390, 393
technological choice 36, 323, 349
technological determinism 10, 18
– different in pre-industrial and industrial societies 348
future war, in 23
– social constructivism, relationship to 58, 349, 356
– military revolution, and 68, 105, 313, 345–46, 356, 360
– Revolution in Military Affairs, and 22, 313, 315, 337, 343, 345
– Western Europe, paradigmatic example of 29
technologies, information 379
technology transfer
– artillery use, in 82
– contact zones, in 21
– global 327
– gunpowder empire, relationship to 75
– maritime spaces, in 320
– port cities, at 5, 321
– regional differences, importance of 104
– Revolution in Military Affairs (RMA), and 321, 352, 357
– strategic culture, as a component of 20, 317

technology-operational concept-organization relationship  381, 382, 392
*Theuerdank*  447–456
Third Offset strategy (U.S.)  388, 389, 390, 394
tipping point  329–30, 413, 443
torpedoes  239, 334
*trace italienne*
– amphibious warfare, role of  239
– banquette in  129, 131, 133, 139, 140
– concentric design of  119, 129, 113, 135
– crenellations in  118, 119, 129, 131, 135, 158, 161, 163, 169, 170
– defensive role of  170
– effectiveness of  198
– embrasures in  118, 120, 127, 133, 137, 139, 158, 164, 170
– expense of  168
– machicolations  118–19, 180
– military revolution, role in  18, 66
– orillons  122, 150
– scarping  119, 143
– water battery  127–33
Treaty of Karlowitz (1699)  148

trebuchets  116
*tyufyaki*  81

*Voyennaya Strategiya (Military Strategy)*  373

wagenburg. See *tabur/tabor.*
wagon camp
– defensive value of  301
– regions, relative usage in  213, 242
– pike and shot technique, as variant of  291–92
– *Gulyaj-gorod*, hybrid version of  301
– (See also *tabur/tabor*, *gulyaj-gorod*, pike and shot)
war and society  99, 348
War of the Holy League (1683–1699)  154, 156
Whigs  325

Yom Kippur War (1973)  377

*zemskaya sluzhba*  72, 90, 92, 94
*zholners*  84

www.ingramcontent.com/pod-product-compliance
Lightning Source LLC
Chambersburg PA
CBHW061923220426
43662CB00012B/1792